P9-EAJ-838

Handbook of Experimental Pharmacology

Volume 109

Principles and Treatment of Lipoprotein Disorders

Contributors

D.H. Blankenhorn, H.B. Brewer, G.A. Coetzee, S.L. Connor
W.E. Connor, J. Davignon, A. Gaw, A.J.R. Habenicht
E. von Hodenberg, H.N. Hodis, D.R. Illingworth
U. Janßen-Timmen, D. Kritchevsky, K.J. Lackner, H.J. Menzel
G. Olivecrona, T. Olivecrona, A.G. Olsson, C.J. Packard
J.R. Patsch, W.O. Richter, P.B. Salbach, E.B. Schmidt
P. Schwandt, D. Seidel, J. Shepherd, G. Utermann
W.J.S. de Villiers, D.R. van der Westhuyzen

Editors

Gotthard Schettler and Andreas J.R. Habenicht

Springer-Verlag
Berlin Heidelberg New York London Paris
Tokyo Hong Kong Barcelona Budapest

Professor Dr. Dr. Dres.h.c. GOTTHARD SCHETTLER
University of Heidelberg
Klinisches Institut für Herzinfarktforschung
an der Medizinischen Universitätsklinik
Bergheimer Straße 58
D-69115 Heidelberg
Germany

Professor Dr. ANDREAS J.R. HABENICHT
University of Heidelberg
School of Medicine, Department of Medicine
Division of Endocrinology and Metabolism
Bergheimer Straße 58
D-69115 Heidelberg
Germany

With 75 Figures and 68 Tables

ISBN 3-540-57121-3 Springer-Verlag Berlin Heidelberg New York
ISBN 0-387-57121-3 Springer-Verlag New York Berlin Heidelberg

CIP data applied for

Typesetting: Best-set Typesetter Ltd., Hong Kong
27/3130/SPS – 5 4 3 2 1 0 – Printed on acid-free paper

List of Contributors

BLANKENHORN, D.H., Atherosclerosis Research Institute, University of Southern California, CSC-132, 2250 Alcazar St., Los Angeles, CA 90033, USA

BREWER, H.B., National Institutes of Health, National Heart, Lung, and Blood Institute, Molecular Disease Branch, Building 10, Room 7N117, 9000 Rockville Pike, Bethesda, MD 20982, USA

COETZEE, G.A., MCR/UCT Unit for the Cell Biology of Atherosclerosis, Department of Medical Biochemistry, University of Cape Town Medical School, Observatory, Cape Town 7925, South Africa

CONNOR, S.L., The Division of Endocrinology, Diabetes and Clinical Nutrition, Oregon Health Sciences University, 3181 SW S. Jackson Park Road, L465, Portland, OR 97201-3098, USA

CONNOR, W.E., The Division of Endocrinology, Diabetes and Clinical Nutrition, Oregon Health Sciences University, 3181 SW S. Jackson Park Road, L465, Portland, OR 97201-3098, USA

DAVIGNON, J., Hyperlipidemia and Atherosclerosis Research Group, Clinical Research Institute of Montreal, 110 Pine Avenue, West, Montreal, Quebec, Canada H2W 1R7

GAW, A., Department of Molecular Genetics, University of Texas, Southwestern Medical Center, 5323 Harry Hines Blvd, Dallas, TX 75235, USA

HABENICHT, A.J.R., University of Heidelberg, School of Medicine, Department of Medicine, Division of Endocrinology and Metabolism, Bergheimerstr. 58, D-69115 Heidelberg, Germany

HODENBERG, E. VON, University of Heidelberg, School of Medicine, Department of Medicine Division of Cardiology, Bergheimerstr. 58, D-69115 Heidelberg, Germany

HODIS, H.N., Atherosclerosis Research Institute, University of Southern California, CSC-132, 2250 Alcazar St., Los Angeles, CA 90033, USA

ILLINGWORTH, D.R., Division of Endocrinology, Diabetes and Clinical Nutrition, Department of Medicine, L465, Oregon Health Sciences University, 3181 SW Sam Jackson Park Rd., Portland, OR 97201-3098, USA

JANSSEN-TIMMEN, U., University of Heidelberg, School of Medicine, Department of Medicine, Division of Endocrinology and Metabolism, Bergheimerstr. 58, D-69115 Heidelberg, Germany

KRITCHEVSKY, D., The Wistar Institute of Anatomy and Biology, 3601 Spruce Street, Philadelphia, PA 19104-4268, USA

LACKNER, K.J., Institute of Clinical Chemistry, University of Regensburg, Franz-Josef-Strauß Allee 11, D-93053 Regensburg, Germany

MENZEL, H.J., Institute for Medical Biology and Human Genetics, University of Innsbruck, Schöpfstr. 41, A-6020 Innsbruck, Austria

OLIVECRONA, G., Department of Medical Biochemistry and Biophysics, University of Umea, S-90187 Umea, Sweden

OLIVECRONA, T., Department of Medical Biochemistry and Biophysics, University of Umea, S-90187 Umea, Sweden

OLSSON, A.G., Department of Internal Medicine, University of Linköping, Faculty of Health Science, S-581 85 Linköping, Sweden

PACKARD, C.J., Institute of Biochemistry, Royal Infirmary, Glasgow G4 OSF, Great Britain

PATSCH, J.R., Medical Clinic, Anichstraße, A-6020 Innsbruck, Austria

RICHTER, W.O., Medical Clinic III, Ludwig Maximilians-Universität, Marchioninistr. 15, D-81377 München, Germany

SALBACH, P.B., University of Heidelberg, School of Medicine, Department of Medicine, Division of Cardiology, Bergheimerstr. 58, D-69115 Heidelberg, Germany

SCHMIDT, E.B., Division of Endocrinology, Diabetes and Clinical Nutrition, Department of Medicine, L465, Oregon Health Sciences University, 3181 SW Sam Jackson Park Rd., Portland, OR 97201-3098, USA

SCHWANDT, P., Medical Clinic III, Ludwig Maximilians-Universität, Marchioninistr. 15, D-81377 München, Germany

SEIDEL, D., Institute of Clinical Chemistry, Ludwig Maximilians-Universität, Marchioninistr. 15, D-81377 München, Germany

SHEPHERD, J., Institute of Biochemistry, Royal Infirmary, Glasgow G4 OSF, Great Britain

UTERMANN, G., Institute for Medical Biology and Human Genetics, University of Innsbruck, Schöpfstr. 41, A-6020 Innsbruck, Austria

Villiers, W.J.S. de, MRC/UCT Unit for the Cell Biology of Atherosclerosis, Deparatment of Medical Biochemistry, University of Cape Town Medical School, Observatory, Cape Town 7925, South Africa

Westhuyzen, D.R. van der, MRC/UCT Unit for the Cell Biology of Atherosclerosis, Department of Medical Biochemistry, University of Cape Town Medical School, Observatory, Cape Town 7925, South Africa

Preface

Recent advances in research on the molecular mechanisms underlying primary and secondary hyperlipidemias, the development of effective dietary treatment regimens, and the discovery of new classes of lipid-lowering drugs have allowed both investigators and physicians to gain new insights into the pathogenesis and clinical management of cardiovascular disease. This volume provides a selection of chapters dealing with three central areas: (1) the physiology and pathophysiology of lipid metabolism; (2) lipid-lowering therapy with emphasis on the rationale to treat and dietary strategies of treatment; and (3) major lipid-lowering drugs that are presently being used clinically.

Section A of the book is introduced by two chapters: in the first, J. Patsch provides the reader with an update on the principles of the biochemistry and biology of blood lipids and plasma lipoproteins, and in the second, H.B. Brewer covers the ongoing research on lipoprotein metabolism with a focus on potential roles of plasma lipoprotein subpopulations in carrying out specific tasks related to lipid transport and metabolism. Since most plasma lipoproteins can interact with and/or are catabolized through specific cell surface receptors, W.J.S. de Villiers et al. review the advances in this important field with a focus on the classical low-density lipoprotein receptor. The progress in molecular mechanisms of lipoprotein disorders, which has the potential to relate molecular events in cells to epidemiologically defined risks for large populations, is discussed by G. Utermann and H.J. Menzel, followed by A.J.R. Habenicht et al., who review the evidence that lipoproteins can directly interact with arterial wall cells to trigger the disease. T. Olivecrona and G. Olivecrona then portray recent aspects of lipoprotein and hepatic lipases, some of which are apparently not related to their enzymatic activities. Section A concludes with a description by D. Kritchevsky of how animal models can be used to search for molecular mechanisms of lipoprotein disorders and what can be learned from them to study the human disease.

Section B begins with two chapters dealing with the important issue of "to treat or not to treat" hyperlipidemia. The first of these, by D.H. Blankenhorn and H.N. Hodis, discusses the rationale to treat hyperlipidemias while the second, by W.E. Connor and S.L. Connor, examines the important role that dietary intervention plays in lowering high lipid levels in patients at risk of developing cardiovascular disease. D. Seidel then describes the recently developed technique of "lipid apheresis," which directly removes plasma lipoproteins from the circulation, and discusses what is presently known about the clinical feasability, safety, and metabolic sequelae of this technique.

Section C begins with a chapter by E. Illingworth and E. Berg Schmidt, who report the recent progress in the field of the mechanisms of action and clinical safety of 3-hydroxy-3-methylglutaryl coenzyme A (HMG-CoA) reductase inhibitors, which have increasingly been used alone or in combination with ion exchange resins in patients with various types of hypercholesterolemia. The latter group of substances is dealt with in a separate chapter by P. Schwandt and W.O. Richter. A. Gaw et al. follow by covering one of the widely used lipid-lowering drugs in hyperlipidemia, the family of fibrates. A.G. Olsson describes recent aspects of the use of nicotinic acid derivatives, and J. Davignon summarizes the pharmacology of probucol, which appears to be effective in animal models of hypercholesterolemia without displaying major lipid-lowering effects, but appears to owe its antiatherogenic effects in animals to its antioxidant properties. Finally, K.J. Lackner and E. von Hodenberg review what is known about lipid-lowering drugs whose use is limited to small numbers of patients or whose side effects prevent their use in larger populations.

We wish to point out that the basis of any clinical intervention and management of hyperlipidemic patients should consist in reducing high lipid levels through dietary means or, in the case of secondary hyperlipidemias, in treating the underlying disease. Only if these attempts fail should drug treatment be considered. While there are several million patients around the world presently being treated with various types or combinations of lipid-lowering drugs, our understanding of whom and when to treat and which drug for which type of hyperlipidemia and at which dose is still rather limited. It is, therefore, another objective of this volume to increase the reader's awareness of such unsolved questions in order to avoid false ideas and the erroneous interpretation of epidemiological data. Only carefully controlled prospective trials of large groups of patients and the analysis of separate patient subgroups that have been treated for prolonged periods of time will provide the basis for answers to these questions.

It is the environment of uncertainty that needs to be recognized and in which this book could serve as one source of information to stimulate discussions on the principles and management of hyperlipidemia.

Heidelberg,
January 1994

Gotthard Schettler
Andreas J.R. Habenicht

Contents

Section A: Physiology and Pathophysiology of Lipid Metabolism

CHAPTER 5

Interactions Between Lipoproteins and the Arterial Wall
A.J.R. Habenicht, P.B. Salbach, and U. Janssen-Timmen.
With 2 Figures 139

CHAPTER 6

Lipoprotein Lipase and Hepatic Lipase

CHAPTER 7

Animal Models of Lipoprotein Metabolism

Section B: Lipid Lowering Therapy

CHAPTER 8

Rationale to Treat

CHAPTER 16

Miscellaneous Lipid-Lowering Drugs

Section A
Physiology and Pathophysiology of Lipid Metabolism

CHAPTER 1

An Introduction to the Biochemistry and Biology of Blood Lipids and Lipoproteins

J.R. PATSCH

A. Introduction

Atherosclerotic coronary heart disease (CHD) is a leading cause of death and disability in industrialized countries. Epidemiologic evidence has linked changes in blood lipids such as elevated cholesterol, elevated low-density lipoprotein (LDL) cholesterol, or decreased high-density lipoprotein (HDL) cholesterol to the development of CHD. This evidence has been supported by animal studies showing that the progression and regression of atherosclerotic lesions correlate with the rise and fall of cholesterol, respectively.

In primary prevention trials such as the Lipid Research Clinics Coronary Primary Prevention Trial (LRC-CPPT) and the Helsinki Heart Study (HHS), a drug-induced reduction in cholesterol, LDL cholesterol, and triglycerides (TG) and an increase in HDL cholesterol were accompanied by a subsequent decrease in CHD risk (LIPID RESEARCH CLINICS PROGRAM 1984a,b; FRICK et al. 1987). LRC-CPPT and HHS did not show a reduction in total mortality in the treatment groups, possibly because the study periods were too short. In a group of patients from the Coronary Drug Project (a secondary prevention trial) who were given nicotinic acid for the initial 5 years, the overall mortality rate was significantly reduced, as a 15-year follow-up showed (CANNER et al. 1986). More recent studies, such as the Cholesterol-Lowering Atherosclerosis Study (CLAS; BLANKENHORN et al. 1987; CASHIN-HEMPHILL et al. 1990), the Familial Atherosclerosis Treatment Study (FATS; BROWN et al. 1990), the U.S. National Heart, Lung, and Blood institute (NHLBI) Type II Coronary Intervention Study (BRENSIKE et al. 1984), the Report of the Program on the Surgical Control of the Hyperlipidemias (POSCH; BUCHWALD et al. 1990), and the Lifestyle Heart Trial (ORNISH et al. 1990), demonstrated with no exception that the progression of angiographically documented coronary atherosclerotic lesions in native coronary arteries and in bypass grafts is slowed, and in some cases halted, by reductions in plasma cholesterol induced by drugs, by partial ileal bypass surgery, or by changes in lifestyle.

It is generally accepted that hypercholesterolemia and dyslipidemia should be detected and treated (LAROSA et al. 1990; CONSENSUS CONFERENCE 1985). The Adult Treatment Panel of the National Cholesterol Education Program (NCEP) has provided guidelines regarding acceptable, borderline, and high-risk cholesterol levels for adults and given specific recommendations on how and when to use dietary and drug interventions (Expert Panel 1988). Similar guidelines and recommendations have been issued by the study groups of the European Atherosclerosis Society, which differs from the NCEP in that it pays considerably more attention to hypertriglyceridemia (Study Group, European Atherosclerosis Society 1987, 1988), which is increasingly coming into the focus of interest as a risk factor for CHD (Assmann et al. 1991).

Recent biochemical evidence strongly suggests that the two major blood lipids, cholesterol and triglycerides, should not be viewed separately. The identification of lipid transfer proteins has provided a plausible explanation for the well-known finding that triglycerides affect the distribution of cholesterol in the blood; elevated triglycerides cause a depression of the powerful risk factor HDL cholesterol (Patsch 1991). This chapter provides a brief introduction into the biochemistry and biology of blood lipids and lipoproteins.

B. Blood Lipid Transport: Historical Aspects

The five major classes of lipids in blood plasma are cholesterol, cholesteryl esters, phospholipids, triglycerides, and unesterified fatty acids. Unesterified fatty acids are transported with plasma albumin, whereas all other lipids are integrated in micellar structures called lipoproteins. The first to identify a lipid protein complex from plasma was Macheboeuf (1929) at the Pasteur Institute in the 1920s. Treating serum with ammonium sulfate under specific conditions led to the precipitation of a fraction with a relatively constant lipid and protein content. In the 1940s, Oncley and colleagues applied the technique of Cohn fractionation to separate lipoproteins (Oncley 1963). Gofman and coworkers separated lipoproteins on the basis of sedimentation velocity in the analytical ultracentrifuge (Gofman et al. 1949) and found that the levels of LDL were correlated with CHD. They also noted that the levels of intermediate-density lipoproteins (IDL) are associated with CHD and that those of HDL showed an inverse association (Gofman et al. 1949). Because the lipid content reduces the density of the lipoproteins, they can be brought to floatation in the ultracentrifuge under conditions where the lipid-free proteins sediment. In 1955, Havel et al. described a procedure for sequentially isolating lipoproteins at increasing background densities of salt in the ultracentrifuge, which permitted the detailed compositional analysis of defined lipoprotein density classes.

Ultracentrifugation is one of two classic methods used to separate lipoproteins. Chylomicrons collect as a top layer after overnight refrigeration of plasma. In order to be separated from plasma, all other lipoproteins need gravity forces which can only be provided by ultracentrifugation. The density of normal protein-free plasma or serum is approximately 1.006 g/ml. At this density, the very low density lipoproteins (VLDL) float to the top of the ultracentrifuge. By raising the density to 1.019 *g*/ml and 1.063 *g*/ml, IDL and LDL, respectively, float in an ultracentrifugal field. Finally, at a density of 1.21 *g*/ml, HDL float to the top of the tube. HDL can be divided into two major subclasses, i.e., HDL_2 (density, 1.063–1.125 *g*/ml) and HDL_3 (density, 1.125–1.210 *g*/ml). There are various techniques of ultracentrifugation which can be specifically employed to isolate and characterize lipoproteins: (a) sequential preparative ultracentrifugation at predetermined densities (Havel et al. 1955), (b) isolation of lipoproteins through salt gradients (Kelley and Kruski 1986), and (c) zonal ultracentrifugation (Patsch et al. 1974; Patsch and Patsch 1986), which can both analyze and prepare lipoproteins. Properties of ultracentrifugally isolated lipoproteins are illustrated in Table 1.

Other methods of lipoprotein isolation include precipitation, gel filtration chromatography, and affinity column chromatography. None of these methods are perfect, and the choice of technique depends on the purpose for which lipoproteins are to be isolated. Precipitation techniques using various combinations of reagents are very fast and economical, as well as being very valuable for the subsequent isolation of apolipoproteins. However, they may alter certain structural and biological properties of some lipoproteins. Gel

Table 1. Classification and properties of plasma lipoproteins

Lipoprotein class	Major lipids	Apolipoproteins	Density (*g*/ml)	Diameter (Å)	Electrophoretic mobility
Chylomicrons	Dietary triglycerides, cholesteryl esters	A-I, A-II, A-IV, B-48, C-I, C-II C-III, E	<0.095	800–5000	Origin
Remnants	Dietary cholesteryl esters	B-48, E	<1.006	>300	Origin
VLDL	Endogenous triglycerides	B-100, C-I, C-II, C-III, E	<1.006	300–800	Pre-β
IDL	Cholesteryl esters, triglycerides	B-100, E	1.006–1.019	250–350	Pre-β/β
LDL	Cholesteryl esters	B-100	1.019–1.063	180–280	β
HDL_2	Cholesteryl esters, phospholipids	A-I, A-II, C-I, C-II, C-III, E	1.063–1.125	90–120	α
HDL_3	Cholesteryl esters, phospholipids	A-I, A-II, C-I, C-II, C-III, E	1.125–1.210	50–90	α

VLDL, very low density lipoprotein; IDL, intermediate-density lipoprotein; LDL, low-density lipoprotein; HDL, high density lipoprotein.

filtration appears to cause less protein loss and is certainly valuable when the composition and structural integrity of the intact lipoprotein particles are critical. Antibody affinity columns are useful for smallscale analytical procedures in which lipoproteins containing specific apolipoproteins can be isolated.

The second classic method used to separate and identify lipoproteins is electrophoresis. LEES and HATCH (1963) improved the technique of lipoprotein electrophoresis by introducing albumin into the electrophoretic buffer. With this technique, chylomicrons remain at the origin, VLDL migrate to the front of the β-globulins and are therefore called pre-β-lipoproteins. The LDL comigrate with β-globulins and are therefore also referred to as β-lipoproteins. IDL also have β-mobility. HDL migrate with the α-globulins, hence α-lipoproteins. All lipoproteins which contain apolipoprotein B (apoB-48 and/or apoB-100, see below) can be precipitated under specified conditions, leaving HDL, which is devoid of apoB, in solution. This method makes it possible to determine HDL cholesterol content in plasma after the removal of all the other lipoproteins. The method usually employs either heparin and manganese or phosphotungstate (BURSTEIN et al. 1970). FREDRICKSON et al. (1967) used paper electrophoresis in conjunction with heparin/manganese precipitation and/or preparative ultracentrifugation to identify the elevation of one or more lipoprotein classes in order to establish a system for classifying plasma lipoprotein disorders or dyslipoproteinemias. This approach is called β-quantification. The hyperlipoproteinemia phenotypes and their associations with genetic and other disorders are summarized in Table 2.

C. Blood Lipids, Apolipoproteins, and Lipoproteins

I. Lipids

The major blood lipids, triglycerides and cholesterol, are water insoluble and need special transport vehicles – called plasma lipoproteins – for their transport through the plasma compartment. The general structure of a lipoprotein particle is illustrated in Fig. 1. Each lipoprotein particle has a nonpolar core containing various amounts of triglycerides and cholesteryl esters that form an oily droplet. The core is surrounded by a monolayer of phospholipids. The polar-head groups of phospholipids are oriented towards the aqueous phase of the plasma, which stabilizes the lipoprotein particle so that it can remain in solution in plasma. Unesterified cholesterol and proteins are also present in the surface monolayer. The proteins are amphipathic in character: they are partly embedded in the lipid domain of the lipoprotein and partly protruding from the surface of the lipoprotein into the aqueous phase of the plasma. The latter property allows the lipoprotein particle to be soluble and to interact with enzymes and cell-membrane receptors

Table 2. Hyperlipoproteinemia phenotype: definitions and associations with genetic and other disorders

Phenotype	Lipid increased	Laboratory definition	Associated with genetic disorders	Conditions associated with secondary hyperlipoproteinemia
Type I	Triglycerides greatly	Hyperchylomicronemia and absolute deficiency of LPL or PHLA	Familial LPL deficiency ApoC-II deficiency	Dysglobulinemia, pancreatitis, poorly controlled diabetes mellitus
Type IIa	Cholesterol	LDL increased	Familial hypercholesterolemia (LDL receptor deficiency) Familial defective apoB-100 Familial combined hyperlipidemia Polygenic hypercholesterolemia	Hypothyroidism, acute intermittent porphyria, nephrosis, idiopathic hypercalcemia, dysglobulinemia, anorexia nervosa
Type IIb	Cholesterol, Triglycerides	LDL increased VLDL increased	Familial hypercholesterolemia Familial combined hyperlipidemia	
Type III	Cholesterol, Triglycerides	β-VLDL, VLDL cholesterol/VLDL triglyceride >0.35 ApoE-II homozygote on isoelectric focusing	Familial dys-β-lipoproteinemia	Diabetes mellitus, hypothyroidism, dysglobulinemia (monoclonal gammopathy)
Type IV	Triglycerides	VLDL increased	Familial hypertriglyceridemia Familial combined hyperlipidemia	Glycogen storage disease, hypothyroidism, disseminated lupus erythematosus, diabetes mellitus, nephrotic syndrome, renal failure, ethanol abuse
Type V	Triglycerides greatly, Cholesterol	Chylomicrons and VLDL increased LPL present	Familial hypertriglyceridemia Familial combined hyperlipidemia	Poorly controlled diabetes mellitus, glycogen storage disease, hypothyroidism, nephrotic syndrome, dysglobulinemia, pregnancy, estrogen administration in women with familial hypertriglyceridemia

LPL, Lipoprotein lipase; LDL, low-density lipoproteins; VLDL, very low density lipoproteins.
Adapted from PATSCH and GOTTO 1987, 1989; GOTTO et al. 1986; PATSCH et al. 1989.

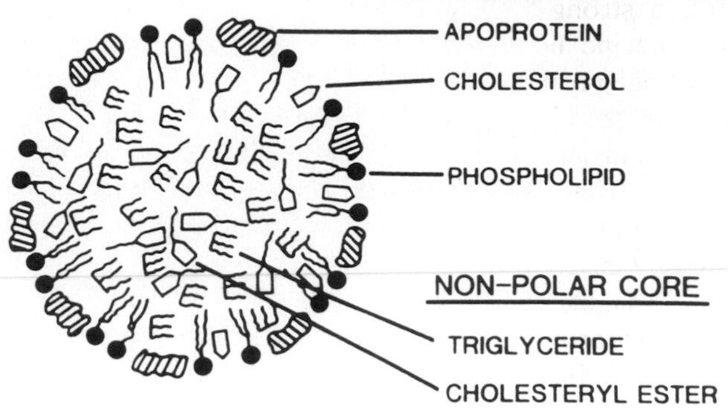

Fig. 1. Model of a lipoprotein particle

(ATKINSON and SMALL 1986; BRADLEY and GOTTO 1978). Thus, the apolipoproteins hold a key function for the metabolic fate of the lipoproteins.

The chemical composition of lipoproteins according to core lipids and surface components is illustrated in Table 3. Triglycerides dominate the core of chylomicrons and VLDL, and cholesteryl esters are the major core constituent of LDL and HDL. Since the thickness of the monomolecular surface coat is the same in all lipoproteins, i.e., about 20 Å, the variability in particle size is due to variability in core size. With the exception of the extremely small HDL particles, the hydrophobic core accounts for most of the mass of a lipoprotein particle. The larger a lipoprotein, the larger is its core of neutral lipids and, as a consequence, the lower is its density (Table 1).

Table 3. Chemical composition of normal human plasma lipoproteins[a]

	Surface components			Core lipid	
	Cholesterol[b] (mol%)	Phospholipids (mol%)	Apolipoprotein (mol%)	Triglycerides (mol%)	Cholesteryl esters (mol%)
Chylomicrons	35	63	2	95	5
VLDL	43	55	2	76	24
IDL	38	60	2	78	22
LDL	42	58	0.2	19	81
HDL_2	22	75	2	18	82
HDL_3	23	72	5	16	84

VLDL, very low density lipoproteins; IDL, intermediate-density lipoproteins; LDL, low-density lipoprotein; HDL, high-density lipoproteins.
[a] adapted from SMITH et al. (1983)
[b] may be distributed between the surface and core and in the case of large chylomicrons, more cholesterol may be in the core than on the surface.

The fatty acid compositions of lipids vary in different human lipoproteins. However, a strong similarity usually exists between the fatty acid composition of specific lipid classes. This may be due to the effects of several lipid exchange proteins, which have been identified in plasma (BARTER et al. 1982; MORTON and ZILVERSMIT 1983; ALBERS et al. 1984; TALL 1986). Oleic and linoleic acids are the major fatty acids of cholesteryl esters. These are derived from the action of lecithin cholesterol acyltransferase (LCAT) on phosphatidylcholine, which almost invariably contains saturated and unsaturated fatty acids in the sn-1 and -2 positions, respectively (GLOMSET and NORUM 1973; GLOMSET 1968, 1972). The degree of unsaturation of phospholipids in lipoproteins is sufficiently high for them to always be in a fluid state at 37°. In contrast, many cholesteryl-ester-rich lipoproteins undergo thermal transitions around 37°C. The exact temperature of these transitions is a function of triglyceride content and the fatty acid composition of the cholesteryl esters.

II. Apolipoproteins

In the mid-1960s the role of the protein components of lipoproteins, termed apolipoproteins (apo), became the focus of much interest and methods for their isolation and chemical and physical characterization were increasingly explored. Nomenclature, distribution, and tissue source of major apolipoproteins are summarized in Table 4.

Analyses of apolipoprotein sequences at the DNA and amino acid level by various algorithms show internal repeats within the amino acid sequence of individual apolipoproteins as well as extensive homologies among different apolipoproteins (BOGUSKI et al. 1986; LUO et al. 1986). Such similarities indicate that apolipoproteins comprise a multigene family dispersed into the genome (BARKER and DAYHOFF 1977). The ancestral apolipoprotein gene was probably half the size of apoC-I (LUO et al. 1986), and as a result of gene duplication, deletion, and exon shuffling, other apolipoproteins evolved and acquired highly specialized functions. Table 5 lists chromosomal localization and some functions of apolipoproteins.

The apolipoproteins have the ability to solubilize lipids for transport in an aqueous surrounding such as the blood plasma. Apolipoproteins appear to contain regions which have a high affinity for a lipid–water interface. The hydrophobic effect drives the association of the apolipoprotein with phospholipids, which depends on the exclusion of both the apolar amino acid side chains of the apolipoprotein and the fatty acid groups of the phospholipids from the aqueous phase. The polar-head groups of phospholipids appear not to be involved in binding to the apolipoproteins. The primary structure of the apolipoproteins contains specific regions which have the potential of forming so-called amphipathic helices (SEGREST et al. 1974). The regions contain both polar and nonpolar amino acid residues, which – after formation of an α-helix – are a distributed on opposite sides of the

Table 4. Characteristics of human plasma apolipoproteins in normal fasting plasma

	Plasma concentration		Distribution in lipoproteins				Major tissue source	Molecular mass (Da)
	(mg/dl)	(mol%)[a]	HDL (nmol%)[b]	LDL (nmol%)[b]	IDL (nmol%)[b]	VLDL (nmol%)[b]		
ApoA-I	130	43	100				Liver and intestine	28016
ApoA-II	40	22	100				Liver and intestine	17414
AopA-IV							Liver and intestine	44465
ApoB-48	80	5		90	8	2	Intestine	264000
ApoB-100							Liver	550000
ApoC-I	6	9	97		1	2	Liver	6630
ApoC-II	3	3	60		10	30	Liver	8900
ApoC-III	12	13	60	10	10	20	Liver	8800
ApoD	10	5	100					22000
ApoE-II								
ApoE-III	5	2	50	10	20	20	Liver	34145
ApoE-IV								

HDL, high-density lipoprotein; LDL, low-density lipoprotein; IDL, intermediate-density lipoprotein; VLDL, very low density lipoprotein.
[a] Based on total plasma concentration.
[b] For each apolipoprotein.

Table 5. Properties and functions of apolipoproteins

Name	Chromosomal localization	Tissue expression	Length (amino acids) of mature protein	Function
ApoA-I	11	Liver, intestine	243	Structural; activator of LCAT
ApoA-II	1	Liver, intestine	77	Structural; activator of hepatic lipase (?)
ApoA-IV	11	Liver, intestine	377	Unknown
ApoB-100	2	Liver	4536	Structural, secretion of VLDL, ligand for LDL receptor
ApoB-48	2	Intestine	2152	Structural, secretion of chylomicrons
ApoC-I	19	Liver, intestine	57	Activator of LCAT
ApoC-II	19	Liver, intestine	79	Activator of lipoprotein lipase
ApoC-III	11	Liver, intestine	79	Inhibits premature removal of TG-rich lipoproteins
ApoD	3	Liver, intestine, pancreas, kidney	169	Cholesteryl ester exchange
ApoE	19	Liver, macrophages	299	Ligand for LDL receptor and putative chylomicron remnant receptor

LCAT, lecithin cholesterol acyltransferase; VLDL, very low density lipoprotein; TG, triglyceride; LDL, low-density lipoprotein.

helix. The nonpolar amino acids are thought to penetrate into the lipid surface of the lipoproteins. The magnitude of the affinity of an apolipoprotein for the phospholipid surface is a function of the hydrophobicity of the nonpolar phase of the amphipathic helix of the apolipoprotein. The polar residues on the opposite side of the helix are in contact with the aqueous phase in which the lipoprotein is suspended. The ability to form the amphipathic helix with polar and nonpolar amino acid residues on opposite sides affords the apolipoprotein detergent-like properties. The polar amino acid residues keep the apolipoprotein at the surface of the lipoprotein particle, thus facilitating the transfer of the apolipoproteins between lipoprotein particles and, by keeping them on the surface, allowing the apolipoproteins to exert their specific functions on the lipoprotein surface such as activation of enzymes and exposure of binding domains for cellular surface receptors. Table 6 illustrates some functions of apolipoproteins.

III. Lipoproteins

With the exception of albumin-bound fatty acids and a few other specialized carriers such as retinol-binding protein, the plasma lipoproteins are the only known transport vehicles of lipid in the circulation. The plasma lipoproteins consist of five major classes, some of which are divided in subclasses because of readily recognizable heterogeneity. However, each lipoprotein class when separated by ultracentrifugation displays some features of heterogeneity with respect to size and composition.

Normal lipoproteins are spherical and can be visualized by electron microscopy using negative staining or shadowing of fixed material. An exception to this general spherical shape are nascent HDL newly synthesized from the liver, the intestine, or cholesterol-laden macrophages. These nascent HDL are devoid of neutral lipids such as triglycerides or cholesteryl

Table 6. Functions of plasma apolipoproteins in lipid transport

Function	Apolipoprotein
Lipoprotein biosynthesis/secretion	B-48 (intestine) B-100 (liver)
Enzyme activation	
Lipoprotein lipase	C-II
Lecithin-cholesterol acyltransferase	A-I C-I A-IV
Interaction of lipoproteins with cellular receptors	
LDL receptor recognition (B/E receptor)	B-100
Chylomicron remnant receptor recognition (E receptor)	E
Inhibition of interaction with hepatic receptors	C-I C-II C-III

LDL, low-density lipoprotein.

esters and consist only of surface components; in electron microscopy, in negatively stained preparations, they appear as discs tending to form rouleaux (HAMILTON et al. 1976). These nascent HDL assume a spherical shape when a critical amount of their phosphatidylcholine and cholesterol is esterified to cholesteryl esters through the action of LCAT.

The largest lipoprotein particles are the chylomicrons (Table 1), which are synthesized by the intestine and which transport mainly dietary triglycerides constituting most of their lipids. VLDL are synthesized in the liver and transport endogenously synthesized triglycerides and cholesterol. The IDL represent an intermediate with respect to size and density between VLDL and LDL. Compared with VLDL, they are relatively depleted in triglycerides. The LDL are mainly degradation products of VLDL and IDL and carry about two-thirds of the plasma cholesterol in plasma of normolipidemic individuals. The HDL are the smallest and densest lipoprotein particles (Table 1) and contain about 50% protein and 50% lipid by weight. The major lipids of HDL are cholesteryl esters and phosphatidylcholine. HDL are traditionally divided into two major subclasses, HDL_2 and HDL_3 (PATSCH et al. 1980). HDL_2 have a strong correlation with HDL cholesterol and, therefore, are elevated in individuals with high HDL cholesterol (PATSCH et al. 1983) and show a stronger inverse statistical relationship with CHD than HDL_3 do (BALLANTYNE et al. 1982; JOHANSSON et al. 1991).

Lipoproteins can be viewed as micellar structures that are in a state of dynamic equilibrium with respect to each other as well as with membranes of various tissues in the body. Virtually all components of plasma lipoproteins (with the exception of apoB) are transferable from one lipoprotein particle to another or to membranes. However, the individual components of plasma lipoproteins are not transferred with equal readiness: phospholipids and unesterified cholesterol, for instance, equilibrate rapidly between various lipoprotein particles and cell membranes, whereas cholesteryl esters equilibrate comparably slowly and require a transfer factor. The same is true for triglycerides. The general structure of a lipoprotein particle is shown in Fig. 1. According to this model (EDELSTEIN et al. 1979), a core of neutral lipids, i.e., cholesteryl esters and/or triglycerides is surrounded and in this way separated from the external aqueous environment of blood plasma by a surface monolayer made up of apolipoproteins and polar lipids, mostly phospholipids. However, appreciable amounts of the cholesterol may be distributed into the core, and small amounts of cholesteryl esters and triglycerides are found in the surface monolayer (SMALL and SHIPLEY 1974).

D. Enzymes and Transfer Factors Involved in the Biochemistry of Blood Lipid Regulation

The enzyme responsible for the hydrolysis of triglycerides of triglyceride-rich lipoproteins is lipoprotein lipase (LPL, EC 3.1.1.34). LPL hydrolyzes lipoprotein triglycerides to monoglyceride and fatty acids, and in this way unloads 70–150 g of triglycerides per day from chylomicrons and VLDL (NILSSON-EHLE et al. 1980; JACKSON 1983; OLIVECRONA and BENGTSSON-OLIVECRONA 1989; ECKEL 1989). The human enzyme is encoded by a gene spanning about 30 kb of chromosome 8, with its coding sequence split into ten exons, and is produced mainly in fat and muscle cells (ECKEL 1989; SPARKES et al. 1987; WION et al. 1987; DEEB and PENG 1989). To allow interactions between the enzyme and circulating triglyceride-rich lipoproteins, LPL is secreted from its sites of synthesis and attaches to heparan sulfate anchors of the vascular endothelium (OLIVECRONA and BENGTSSON-OLIVECRONA 1989), the site of its action. The enzyme requires the activation of apoC-II (HAVEL et al. 1970; LAROSA et al. 1970), one of the small exchangeable apolipoproteins. ApoC-II is associated mainly with HDL and transferred to chylomicrons and/or VLDL as they are secreted, rendering them active substrates for LPL. Familial LPL deficiency is a major cause of the phenotypic hyperlipoproteinemia type I (Table 2), and numerous mutations at the LPL gene locus have been defined (for review, see HAYDEN et al. 1991; LALOUEL et al. 1992), disrupting the function of the enzyme at the levels of synthesis, endothelial attachment, or catalytic competence (AUWERX et al. 1989). ApoC-II deficiency has also been described in some subjects with severe hypertriglyceridemia (COX et al. 1978; BRECKENRIDGE et al. 1978), which clinically presents also as type I hyperlipoproteinemia phenotype.

The second lipolytic enzyme is hepatic lipase, which – like LPL – is released into the blood stream by heparin and which also hydrolyzes phospholipids and triglycerides (NILSSON-EHLE et al. 1980). However, hepatic lipase differs from LPL in that it does not require apoC-II for activation and that it is not inhibited by salt. In familial hepatic lipase deficiency, IDL and HDL_2 accumulate in the plasma (BRECKENRIDGE et al. 1982), suggesting that these two lipoprotein classes are preferred physiologic substrates for the enzyme. Indeed, the enzyme shows a high affinity for HDL_2 (SHIRAI et al. 1981) and can convert triglyceride-enriched HDL_2 into the smaller HDL_3 (PATSCH et al. 1984).

The third major enzyme involved in the biochemistry of blood lipid regulation is LCAT, which is responsible for the formation of nearly all plasma cholesteryl esters in normal humans (GLOMSET 1968). Substrates for LCAT are phosphatidylcholine and cholesterol, giving rise to the products cholesteryl esters and lysolecithin. Both steps of the catalytic activity, hydrolysis and transesterification, are activated by apoA-I, the major protein component of HDL (Table 6). HDL (or rather a subclass of HDL) are the putative physiologic substrate for the enzyme (ARON et al. 1978). Other lipoproteins such as apoC-I and apoA-IV have also been suggested as activators for LCAT. Fatty acids located at the sn-2 position of phosphatidylcholine are the preferred, though not the exclusive, site of cleavage. The enzyme prefers unsaturated fatty acyl groups.

Acyl coenzyme A-O-acyltransferase (ACAT) is, as opposed to LCAT, not a plasma enzyme, but rather an intracellular enzyme located in the endoplasmic reticulum (GOODMAN et al. 1964) mostly of enterocytes, and can contribute in this way to the pool of cholesteryl esters in plasma. For instance, as a result of a cholesterol-rich diet, ACAT forms cholesteryl esters in the intestinal wall which later appear in the plasma as part of cholesteryl-ester-rich chylomicrons of intestinal origin and, in some mammalian species, of VLDL of hepatic origin.

Lipid transfer proteins are a group of proteins which mobilize lipids from one compartment to another. The transfer of lipids among plasma lipoproteins and between plasma lipoproteins and cell membranes is a fundamental component for the proper function of the lipid transport by plasma lipoproteins. Three major mechanisms have been proposed for lipid exchange and transfer. One of these involves the formation of a collision complex between surfaces of lipoproteins and/or membranes, thus allowing the exchange or transfer of lipids. Although this mechanism cannot be excluded, there is not much evidence to support it. A second mechanism requires the rate-limiting desorption of a single lipid molecule from a membrane or a lipoprotein surface into the surrounding aqueous phase, from which it rapidly diffuses to another accessible lipid surface. This mechanism appears to operate with sparingly soluble substances such as cholesterol (McLEAN and PHILIPS 1981) and short-chained phosphatidylcholines (MASSEY et al. 1982), while it becomes less important with increasing chain length or hydrophobicity of the lipid molecules (MORTON and ZILVERSMIT 1983). The relative importance of the desorption process is usually difficult to evaluate in vivo because of competing processes, such as hydrolysis of lipids and receptor-mediated cellular uptake of lipoproteins, etc. A third mechanism, the importance of which has become increasingly recognized in recent years, involves the transfer of very insoluble lipids between lipoproteins and, possibly, membrane surfaces. This transfer of lipid molecules requires the action of specific transfer proteins in plasma (BARTER et al. 1982; MORTON and ZILVERSMIT 1983; ALBERS et al. 1984; TALL 1986).

E. Lipoprotein Metabolism

I. Transport of Exogenous Lipids

In the intestine, dietary cholesteryl esters are mixed with a similar amount of endogenous cholesterol from saliva, gastric secretions, bile, and sloughed intestinal epithelial cells (for review see PATSCH 1987). Cholesterol can also be synthesized de novo by enterocytes (SVIRIDOV et al. 1986), Within the enterocyte, dietary triglycerides and cholesterol are packaged into a core and wrapped into a surface film of phospholipids, cholesterol, and apoB-48 to form a chylomicron with apoB-48 as the principal apolipoprotein. The nascent chylomicron particles are secreted from the Golgi vacuoles into the intercellular spaces, where they reach the intestinal lymph, pass the thoracic duct, and enter the general circulation through the subclavian vein. The appearance of chylomicrons in plasma begins within 1 h of fat ingestion and is called postprandial lipemia. Chylomicrons transport about 100 g triglyceride as dietary fat and 0.5–1.0 g cholesterol per day.

The catabolism of chylomicrons is basically a two-step process. First, LPL hydrolyzes the triglycerides from the chylomicron particles to form the triglyceride-depleted chylomicron remnants. The $t_{1/2}$ of chylomicron triglycerides in the circulation is about 5 min. Second, chylomicron remnants are taken up by remnant receptors in the liver. After entering the circulation, a nascent chylomicron undergoes rapid modification, with some components released from the particle and others joining it. ApoB-48 remains an integral component, but the soluble apolipoproteins A-I, A-II, and A-IV quickly leave the chylomicron to associate with HDL. Newly secreted chylomicrons gain apolipoproteins C-I, C-II, C-III, and E from VLDL and HDL.

In the capillaries, the chylomicrons adhere to binding sites on the vessel walls, where they are exposed to the enzyme lipoprotein lipase (LPL; Jackson 1983). For LPL to act on the chylomicron particle, apoC-II is required. The enzyme causes lipolysis of triglycerides resulting in the formation of free fatty acids and monoglycerides, which are quickly removed from the blood. The fatty acids pass through the endothelial cell layer and enter the underlying cells such as adipocytes or muscle cells. The major sites of LPL activity are adipose tissue, skeletal muscle, the myocardium, and the mammary gland. These tissues are known to take up triglycerides for storage (adipose tissue), oxidation (muscle work), or secretion (milk).

The chylomicron remnant is the end result of the first step of chylomicron catabolism (i.e., lipolysis) and contains cholesteryl esters as the predominant core component. As lipolysis proceeds, additional modifications occur (Patsch 1987), and as the size of the chylomicron is reduced, surface phospholipids become redundant and are transferred to HDL. The activator protein of LPL, apoC-II, is also shuttled back to HDL. The same is true for apoC-III, which functions to prevent premature interaction of the chylomicron with the hepatic receptor (Table 6). Thus, most chylomicron constituents not cleared with the remnant are transferred to HDL. With most of the triglycerides hydrolyzed and many of the surface components dissociated, the chylomicron particle will have lost about 95% of its mass. The remaining chylomicron remnant, enriched in cholesteryl esters and containing apoB-48 and apoE, dissociates from the capillary endothelium to reenter the circulation.

The remnant particle is removed from the circulation by the liver. This uptake appears to be mediated by apoE. The remnant's other major apolipoprotein, B-48, is necessary for secretion of the chylomicron by the enterocyte, but is not required for chylomicron removal. Although B-48 is a product of the same gene as B-100, which binds to the LDL receptor, it lacks the COOH-terminal receptor binding domain of B-100 (CHEN et al. 1987). Chylomicron remnants are internalized into the hepatocytes by receptor-mediated endocytosis and subsequently degraded within lysosomes. The cholesterol directed through chylomicron remnants to the liver is converted to bile acids or excreted into the bile without conversion to bile acids. An additional fraction of the chylomicron cholesterol can be incorporated in VLDL particles and secreted by the liver into the bloodstream.

The body synthesizes 60%–80% of its cholesterol, primarily in the liver and intestine, and derives the remainder from the diet. The uptake of cholesterol by the liver from chylomicron remnants reduces both the hepatic synthesis of cholesterol and LDL receptor activity (ANGELIN et al. 1983). Chylomicron remnants are more effective than LDL or HDL in suppressing cholesterol biosynthesis by the liver (ANDERSEN et al. 1979). Excess dietary cholesterol not only suppresses the liver's endogenous synthesis of cholesterol, but also decreases the activity of the LDL receptors, which bind apoB-100 and apoE. Chronic suppression of hepatic LDL receptor activity by dietary cholesterol or saturated fat may lead to elevated levels of LDL in the blood and thus predispose to atherosclerosis.

Chylomicron remnants can also interact with receptors other than those on hepatocytes. The uptake of these remnants – by macrophages in particular – can lead to the overloading of cells with both triglyceride and cholesterol (BATES et al. 1984), raising the possibility that these lipoproteins contribute to atherosclerosis (ZILVERSMIT 1979).

II. Transport of Endogenous Lipids

Triglycerides synthesized in the liver are incorporated into VLDL and secreted into the bloodstream. This synthesis is enhanced in the postprandial state or when the diet contains excess carbohydrates. In these situations, excess fatty acids are generated and incorporated by the liver into triglycerides. In addition to triglycerides, cholesteryl esters are also packed into the VLDL core and contribute about one-tenth to the mass of the core.

The principal apolipoprotein of VLDL is B-100. With 4536 amino acid residues, B-100 is a very large protein molecule (YANG et al. 1986; KNOTT et al. 1986; HIGUCHI et al. 1987). It contains many lipid-binding domains and also a domain on its COOH-terminal end that binds to the LDL receptor (YANG et al. 1986). When the VLDL particles reach the capillaries, they are hydrolyzed by LPL. Analogous to chylomicron metabolism, the VLDL particle loses most of its apolipoprotein components, but not B-100, during lipolysis. Redundant surface phospholipids, generated during the reduction of VLDL size through lipolysis, are transferred to HDL. The same is true for apoC-II and apoC-III. Removal of much of the triglyceride from the VLDL particle core results in the formation of the VLDL remnant particle called IDL. The IDL particle is similar to the chylomicron remnant, but has a different metabolic fate. Only some of the IDL particles are catabolized by the liver after interacting with LDL receptors on hepatocyte membranes. The ligand for this interaction is apoB-100 or apoE on the surface of the IDL particle. From the IDL particles remaining in the plasma compartment, triglyceride is removed until LDL particles are formed whose core contains almost only cholesteryl esters and whose only protein component is apoB-100. One function of LDL is to supply cholesterol to cells for the synthesis of cell membranes and some special functions like the synthesis of steroid hormones and bile acids. For this purpose, LDL bind to LDL receptors to be internalized into the cells by receptor-mediated endocytosis (BROWN and GOLDSTEIN 1986). Subsequently, LDL receptors are recycled to the cell surface. In the lysosomes, the cholesteryl esters of LDL are hydrolyzed by a lysosomal cholesteryl esterase (acid lipase) into unesterified cholesterol, thus converting cholesterol from its transport form to its metabolically active form. The number of LDL receptors is thought to be a major determinant of the concentration of LDL in the blood. Number and activity of LDL receptors are high early in life and tend to diminish with age. As discussed above, a lifelong diet rich in saturated fat and cholesterol may lead to chronic suppression of LDL receptor activity. Factors that diminish intra-hepatic cellular concentration of cholesterol will lead to an increase in receptor activity, whereas an increase in intrahepatic cellular cholesterol will downregulate the receptors.

F. Cholesterol and Atherosclerosis

Hypercholesterolemia due to elevated blood levels of LDL is well established as a major cause of CHD. Lowering elevated levels of cholesterol has been demonstrated to reduce the risk of CHD (LIPID RESEARCH CLINICS PROGRAM 1984a,b; FRICK et al. 1987; CANNER et al. 1986; BLANKENHORN et al. 1987; CASHIN-HEMPHILL et al. 1990; BROWN et al. 1990; BRENSIKE et al. 1984; BUCHWALD et al. 1990; ORNISH et al. 1990). In healthy humans, more than 70% of the LDL particles circulating in plasma are removed through LDL receptors. The remainder are removed by alternative pathways, some receptor mediated and others nonspecific. This alternative removal of LDL from the circulation, chiefly by cells of the reticuloendothelial system, has been termed collectively "the scavenger-cell pathway" (BROWN and GOLDSTEIN 1986); it is thought to function as a removal mechanism for excess LDL and apparently is not needed to satisfy the physiological requirements of cells for cholesterol.

The "scavenger receptor" of the receptor-mediated alternative route is also called the "acetyl-LDL receptor" (KODAMA et al. 1990) because it recognizes chemically modified LDL. Exposure of LDL to endothelial cells results in its peroxidation (STEINBERG et al. 1989). The peroxidation process includes a number of structural changes of LDL, including peroxidation of polyunsaturated fatty acids of its lipid moieties, a conversion of lecithin to lysolecithin, oxidation of cholesterol, and fragmentation of apoB-100, with a decrease in the content of histidine, lysine, and proline. Fragmentation of fatty acids can lead to formation of shorter-chain aldehydes.

Some of these lipid peroxidation products can attach covalently to B-100 (FONG et al. 1987), thus masking lysine e-amino groups of the apolipoprotein (STEINBRECHER et al. 1987; STEINBRECHER 1987). These sructural changes cause an increase in negative charge and in density of the peroxidized LDL particles and hence major changes in biologic activity – including chemotactic activity for circulating blood monocytes, the ability to become substrates for the scavenger receptor, and cytotoxicity (STEINBERG et al. 1989). Oxidation of LDL – or other modifications of LDL, including glycation of B-100 (STEINBRECHER and WITZTUM 1984) – could thus lead to the accumulation of cholesteryl esters in macrophages and smooth muscle cells and the development of atherosclerotic plaque. An alternative mechanism for foam cell formation could be the assembly of immune complexes between oxidized LDL and autoantibodies against oxidized LDL (PALINSKI et al. 1989; MITCHINSON et al. 1988) and the uptake of the complexes by macrophages through the Fc receptor (PARUMS and MITCHINSON 1981; GORDON et al. 1977). It appears that normal, unmodified LDL particles do not go through the scavenger pathway or lead to the formation of foam cells (BROWN and GOLDSTEIN 1986).

G. High-Density Lipoprotein Cholesterol and Reverse Cholesterol Transport

A strong negative correlation exists between plasma level of HDL cholesterol and CHD (GORDON 1977). Most of the variability in HDL cholesterol level reflects the levels of HDL_2 (PATSCH et al. 1983, 1987); plasma levels of the other major HDL subclass, HDL_3, are fairly constant both intra- and interindividually (PATSCH and GOTTO 1987).

HDL particles are not secreted into the circulation as mature, spherical lipoproteins but as discoid, precursor particles devoid of cholesteryl esters (HAMILTON et al. 1976; EISENBERG 1984). These nascent HDL particles – secreted by intestine, liver, and cholesteryl-ester-enriched macrophages – are complexes of apolipoproteins and phospholipids and acquire unesterified cholesterol originating in cell membranes during cell renewal or death. Their major phospholipid is phosphatidylcholine or lecithin.

Both lecithin and unesterified cholesterol serve as substrates for the cholesterol-esterifying enzyme LCAT (GLOMSET 1968, 1972; GLOMSET and NORUM 1973), which circulates with HDL in the plasma. LCAT acts on the nascent HDL particles to generate a core of cholesteryl esters and to effect the structural transition to mature, spherical HDL particles. LCAT also acts on mature HDL_3 particles, i.e., after they have acquired cholesterol and lecithin not only from cell membranes but also from chylomicrons and VLDL during lipolysis of these particles (PATSCH et al. 1978). Generation of cholesteryl esters by the LCAT reaction leads to enlargement of the small HDL_3 particles, such that they are converted into the larger HDL_2 particles. Formation of HDL_2 increases the cholesterol-carrying capacity of HDL, and HDL cholesterol levels rise; this drives the process termed reverse cholesterol transport, in which HDL returns cholesterol from peripheral tissues to the liver for excretion into the bile.

The cholesteryl esters of HDL, particularly of HDL_2 formed de novo by the LCAT action (GLOMSET 1968), need not remain within the HDL core. They can be transferred from HDL to the triglyceride-rich lipoproteins, i.e., chylomicrons and VLDL, in exchange for triglyceride molecules. This heteroexchange of insoluble cholesteryl esters and triglycerides between HDL and triglyceride-rich lipoproteins is catalyzed by the action of cholesteryl ester transfer protein (CETP; MORTON and ZILVERSMIT 1983; ALBERS et al. 1984; TALL 1986).

The transferred triglycerides are hydrolyzed from the HDL core by hepatic lipase (PATSCH et al. 1984), located on the endothelial cells of the liver. Only cholesteryl esters that were not exchanged for triglycerides remain in the core. Hence, HDL_2 particles are converted back into the smaller HDL_3 particles (PATSCH et al. 1987). This mechanism is the basis for the well-established clinical observation that individuals with permanent or temporary hypertriglyceridemia (due to increased levels of VLDL or due to accumulation of chylomicrons in the course of postprandial lipemia) have low HDL_2 and low HDL cholesterol levels (PATSCH et al. 1983).

The cholesteryl esters transferred from HDL to the triglyceride-rich lipoproteins remain with the latter particles along their lipolytic cascade and the endocytotic pathways of their remnants (i.e., the LDL receptor and the scavenger pathways). Thus, transfer of cholesteryl esters from HDL to triglyceride-rich lipoproteins may contribute to the atherogenic potential of chylomicrons and VLDL, in that "good" cholesterol is exchanged for "bad" cholesterol (MIESENBÖCK and PATSCH 1991).

A splicing defect of the CETP gene that leads to CETP deficiency appears to be a frequent cause of increased HDL levels in Japan (BROWN et al. 1989; INAZU et al. 1990). The heritable deficiency prevents triglyceride-rich lipoproteins from directing their triglycerides into HDL_2 (thus preventing their catabolism) and from robbing HDL of its cholesteryl esters, i.e., from switching antiatherogenic into potentially atherogenic cholesterol. Indeed, the CETP-deficient Japanese patients exhibit high HDL cholesterol levels and an overwhelming preponderance of HDL_2 and even larger HDL particles, and they present no evidence of premature atherosclerosis.

H. Triglycerides, High-Density Lipoprotein Cholesterol, and Coronary Artery Disease

As stated above, the risk factor role of triglycerides is increasingly becoming a focus of interest (ASSMAN et al. 1991). It has been demonstrated that low HDL cholesterol levels in normotriglyceridemic individuals signal an individual's poor ability to clear triglyceride after a fatty meal (PATSCH et al. 1983). Hence, it was proposed (PATSCH 1991) that this observation could help to explain the strong negative risk factor role of HDL cholesterol. Two case control studies recently demonstrated that the magnitude of postprandial triglyceride elevation is associated with angiographically verified CHD (GROOT et al. 1991; SIMPSON et al. 1990). A third study demonstrated in a multivariate analysis that postprandial triglyceride levels are independent risk factors for CHD, even when other parameters such as HDL cholesterol are included (PATSCH et al. 1992). This is the first demonstration of an independent risk factor role of plasma triglyceride concentration and was made possible by the fact that triglyceride levels after challenge were used instead of fasting triglyceride.

A substantial flow of cholesterol is continually passing through HDL, although in the steady state only about one-fourth of the total plasma cholesterol pool is actually transported in this lipoprotein (MONSALVE et al. 1990). In an adult male, LCAT esterifies about 6 g cholesterol per day (KOO et al. 1988), a quantity approximately equalling the total plasma cholesterol pool size. Part of the cholesteryl esters formed in HDL, with their linoleic acyl chain serving as the distinctive mark of LCAT origin (GLOMSET and NORUM 1973), are distributed to other plasma lipoproteins. HDL, therefore constitutes a pivot of plasma cholesterol traffic. This pivot is at least in part controlled by triglyceride metabolism made possible by the activity of CETP: when triglyceride metabolism is efficient and triglyceride concentration in plasma remains low, there is no or little triglyceride/cholesteryl ester exchange between triglyceride-rich particles and HDL, and cholesterol will remain a component of HDL. However, when triglyceride concentration inappropriately rises, either continually or even only in the postprandial phase, cholesterol will be translocated to apoB- and apoE-containing lipoproteins. Keeping cholesterol with HDL secures efficient targeting of cholesterol to cells with the capability of excreting cholesterol or of using it for bile acid or steroid hormone synthesis. However, when HDL cholesterol is transferred into accumulated triglyceride-rich lipoprotein, the risk of CAD will rise because the former "good" cholesterol can now be diverted into macrophages and smooth muscle cells. Triglycerides withdraw cholesterol from HDL; as a result, the steady-state HDL cholesterol level decreases. Low HDL cholesterol levels, therefore, appear to indicate two closely linked aberrations of lipid transport: diversion of cholesteryl esters from HDL into triglyceride-rich lipoproteins (i) boosts the atherogenic potential of these cholesterol acceptor particles and (ii) reduces the benefit accomplished by HDL-mediated cellular cholesterol efflux, because the cell-derived cholesterol is shunted from a safe into a hazardous branch of reverse cholesterol transport. Fundamental to this derangement is impaired triglyceride metabolism; hence, the accepted risk constellation "low HDL cholesterol – CHD" is in reality more likely to be "low triglyceride metabolic capacity – low HDL cholesterol – CHD" (MIESENBÖCK and PATSCH 1992).

I. Conclusion

The above introductory chapter makes it clear that each lipoprotein family has a specific lipid transport function, but also that each lipoprotein family affects the other. The plasma levels of lipoproteins are under complex regulation by the dietary intake of cholesterol, monounsaturated and polyunsaturated fats, by lipoprotein synthesis and the pathways of lipoprotein clearance involving specific hepatic receptors, tissue receptors, and low-affinity pathways. A knowledge of the biochemistry of blood lipids and lipoproteins is necessary for a rational understanding of the mechanisms of hyperlipoproteinemia and hyperlipidemia. An excess of circulating chylomicron remnants, IDL or LDL, predisposes to atherosclerosis, whether the mechanism is an oversynthesis of the lipoprotein, a decrease in catabolism, or a combination of the two defects.

Regarding the relation between atherosclerosis and blood lipids, important progress has recently been made in two areas: the recognition of the potential atherogenicity of oxidized or otherwise modified LDL, and the recognition of CETP as the means of equilibration of cholesteryl esters and triglycerides between lipoproteins. The CETP-regulated equilibration process points to the potential atherogenicity of the triglyceride-rich lipoproteins since their metabolism appears to be a major determinant of cholesterol metabolic routing. Our advancing knowledge of lipid transport and lipoprotein metabolism should enable us to formulate more clearly how these phenomena relate to atherosclerosis and to devise better methods of treatment and prevention.

References

Albers JJ, Tollefson JH, Chen CH, Steinmetz A (1984) Isolation and characterization of human plasma lipid transfer proteins. Arteriosclerosis 4:49–58

Andersen JM, Turley SD, Dietschy JM (1979) Low and high density lipoproteins and chylomicrons as regulators of rate of cholesterol synthesis in rat liver in vivo. Proc Natl Acad Sci USA 76:165–169

Angelin B, Raviola CA, Innerarity TL, Mahley RW (1983) Regulation of hepatic lipoprotein receptors in the dog. Rapid regulation of apolipoprotein B,E receptors, but not of apolipoprotein E receptors, by intestinal lipoproteins and bile acids. J Clin Invest 71:816–831

Aron L, Jones S, Fielding CJ (1978) Human plasma lecithin-cholesterol acyltransferase. J Biol Chem 253:7220–7226

Assmann G, Gotto AM Jr, Paoletti R (1991) The hypertriglyceridemias: risk and management. Am J Cardiol 68:1A–42A

Atkinson D, Small DM (1986) Recombinant lipoproteins: implications for structure and assembly of native lipoproteins. Annu Rev Biophys Biophys Chem 15: 403–456

Auwerx JH, Babirak SP, Fujimoto WY, Iverius P-H, Brunzell JD (1989) Defective enzyme protein in lipoprotein lipase deficiency. Eur J Clin Invest 19:433–437

Ballantyne FC, Clark RS, Simpson HS, Ballantyne D (1982) High density and low density lipoprotein subfractions in survivors of myocardial infarction and in control subjects. Metabolism 31:433–437

Barker WC, Dayhoff MO (1977) Evolution of lipoproteins deduced from protein sequencing data. Comp Biochem Physiol 576:309–315
Barter PJ, Hopkins GJ, Calvert GD (1982) Transfers and exchanges of esterified cholesterol between plasma lipoproteins. Biochem J 208:1–7
Bates SR, Murphy PL, Feng ZC, Kanazawa T, Getz GS (1984) Very low density lipoproteins promote triglyceride accumulation in macrophages. Arteriosclerosis 4:103–114
Blankenhorn DH, Nessim SA, Johnson RL, Sanmarco ME, Azen SP, Cashin-Hemphill L (1987) Beneficial effects of combined colestipol-niacin therapy on coronary atherosclerosis and coronary venous bypass grafts. JAMA 257: 3233–3240
Boguski MS, Freeman M, Elshourbagy NA, Taylor JM, Gordon JI (1986) On computer-assisted analysis of biological sequences: proline punctation, consensus sequences, and apolipoprotein repeats. J Lipid Res 27:1011–1034
Bradley WA, Gotto AM Jr (1978) Structure of intact human plasma lipoproteins. In: Dietschy JM, Gotto AM Jr, Ontko JA (eds) Disturbances in lipid and lipoprotein metabolism. American Physiology Society, Bethesda, pp 111–137
Breckenridge WC, Little JA, Steiner G, Chow A, Poapst M (1978) Hypertriglyceridemia associated with deficiency of apolipoprotein C-II. N Engl J Med 298:1265–1273
Breckenridge WC, Little JA, Alaupovic P, Wang CS, Kuksis A, Kakis G, Lindgren F, Gardiner G (1982) Lipoprotein abnormalities associated with a familial deficiency of hepatic lipase. Atherosclerosis 45:161–179
Brensike JF, Levy RI, Kelsey SF, Passamani ER, Richardson JM, Loh IK et al. (1984) Effects of therapy with cholestyramine on progression of coronary arteriosclerosis: results of the NHLBI Type II Coronary Intervention Study. Circulation 69:313–324
Brown G, Albers JJ, Fisher LD, Schaefer SM, Lin JT, Kaplan C et al. (1990) Regression of coronary artery disease as a result of intensive lipid-lowering therapy in men with high levels of apolipoprotein B. N Engl J Med 323: 1289–1298
Brown ML, Inazu A, Hesler CB, Agellon LB, Mann C, Whitlock ME et al. (1989) Molecular basis of lipid transfer protein deficiency in a family with increased high-density lipoproteins. Nature 342:448–451
Brown MS, Goldstein JL (1986) A receptor-mediated pathway for cholesterol homeostasis. Science 232:34–47
Buchwald H, Varco RL, Mattis JP, Long JM, Fitch LL, Campbell GS, Pearce MB, Yellin AE, Edmiston WA, Smink RD, Sawin HS, Campos CT, Hansen BJ, Tuna N, Karnegis JN, Sanmarco ME, Amplatz KA, Castaneda-Zuniga WR, Hunter DW, Bissett JK, Weber JF, Stevenson JW, Leon AS, Chalmers TC, Posch Group (1990) Effect of partial ileal bypass surgery on mortality and morbidity from coronary heart disease in patients with hypercholesterolemia. N Engl J Med 323:946–955
Burstein M, Scholnick HR, Morfin R (1970) Rapid method for the isolation of lipoproteins from human serum by precipitation with polyanions. J Lipid Res 11:583–595
Canner PL, Berge KG, Wenger NK, Stamler J, Friedman L, Prineas RJ et al. (1986) Fifteen year mortality in Coronary Drug Project patients: long-term benefit with niacin. J Am Coll Cardiol 8:1245–1255
Cashin-Hemphill L, Mack WJ, Pogoda JM, Sanmarco ME, Azen SP, Blankenhorn DH (1990) Beneficial effects of colestipol-niacin on coronary atherosclerosis. A 4-year follow-up. JAMA 264:3013–3017
Consensus Conference (1985) Lowering blood cholesterol to prevent heart disease. JAMA 253:2080–2086
Cox DW, Breckenridge WC, Little JA (1978) Inheritance of apolipoprotein C-II deficiency with hypertriglyceridemia and pancreatitis. N Engl J Med 299: 1421–1424

Chen SH, Habib G, Yang CY, Gu ZW, Lee BR, Weng SA et al. (1987) Apolipoprotein B-48 is the product of a messenger RNA with an organ-specific in-frame stop codon. Science 238:363–366
Deeb S, Peng R (1989) Structure of the human lipoprotein lipase gene. Biochemistry 28:4131–4135
Eckel RH (1989) Lipoprotein lipase: a multifunctional enzyme relevant to common metabolic diseases. N Engl J Med 320:1060–1067
Edelstein C, Kezdy F, Scanu AM, Shen BW (1979) Apolipoproteins and the structural organization of plasma lipoproteins: human plasma high density lipoprotein-3. J Lipid Res 20:143–153
Eisenberg S (1984) High density lipoprotein metabolism. J Lipid Res 25:1017–1058
Expert Panel (1988) Report of the National Cholesterol Education Program Expert Panel on detection, evaluation, and treatment of high blood cholesterol in adults. Arch Intern Med 148:36–69
Fong LG, Parthasarathy S, Witztum JL, Steinberg D (1987) Nonenzymatic oxidative cleavage of peptide bonds in apoprotein B-100. J Lipid Res 28:1466–1477
Fredrickson DS, Levy RI, Lees RS (1967) Fat transport in lipoproteins: an integrated approach to mechanisms and disorders. N Engl J Med 276:148–56 contd.
Frick MH, Elo O, Haapa K, Heinonen OP, Heinsalmi P, Helo P et al. (1987) Helsinki Heart Study: primary-prevention trial with gemfibrozil in middle-aged men with dyslipidemia. Safety of treatment, changes in risk factors, and incidence of coronary heart disease. N Engl J Med 317:1237–1245
Gofman JW, Lindgren FT, Elliott H (1949) Ultracentrifugal studies of lipoproteins. J Biol Chem 179:973
Glomset JA (1968) The plasma lecithin: cholesterol acyltransferase reaction. J Lipid Res 9:155–167
Glomset JA (1972) Plasma lecithin: cholesterol acyltransferase. In: Nelson G (ed) Blood lipids and lipoproteins: quantitation, composition, and metabolism. Wiley-Interscience, New York, pp 745–787
Glomset JA, Norum KR (1973) The role of lecithin: cholesterol acyltransferase: perspectives from pathology. Adv Lipid Res 11:1–65
Goodman DS, Deyykin D, Shiratori T (1964) The formation of cholesterol esters with rat liver enzymes. Characterization of cofactor-dependent phospholipase activity. J Biol Chem 239:1335–1345
Gordon T, Castelli WP, Hjortland MC, Kannel WB, Dawber TR (1977) High density lipoprotein as a protective factor against coronary heart disease. The Framingham Study. Am J Med 62:707–714
Gotto AM Jr, Pownall HJ, Havel RJ (1986) Introduction to the plasma lipoproteins. Methods Enzymol 128:3–41
Groot PHE, Van Stiphout WAHJ, Krauss XH, Jansen H, Van Tol A, Van Ramshorst E, Chin-On S, Hofman A, Cresswell SR, Havekers L (1991) Postprandial lipoprotein metabolism in normolipidemic men with and without coronary artery disease. Arterioscler Thromb 11:653–662
Hamilton RL, Williams MC, Fielding CJ, Havel RJ (1976) Discoidal bilayer structure of nascent high density lipoproteins from perfused rat liver. J Clin Invest 58:667–680
Havel RJ, Eder HA, Bragdon JH (1955) Distribution and chemical composition of ultracentrifugally separated lipoproteins in human serum. J Clin Invest 34: 1345
Havel RJ, Shore VG, Shore B, Bier DM (1970) Role of specific glycopeptides of human serum lipoproteins in the activation of lipoprotein lipase. Circ Res 27:595
Hayden MR, Ma Y, Brunzell J, Henderson HE (1991) Genetic variants affecting human lipoprotein and hepatic lipases. Curr Opin Lipidol 2:104–109
Higuchi K, Monge JC, Lee N, Law SW, Brewer HB Jr, Sakaguchi AY et al. (1987) The human apoB-100 gene: apoB-100 is encoded by a single copy gene in the human genome. Biochem Biophys Res Commun 144:1332–1339

Inazu A, Brown ML, Hesler CB, Agellon LB, Koizumi J, Takata K et al. (1990) Increased high-density lipoprotein levels caused by a common cholesteryl-ester transfer protein gen mutation. N Engl J Med 323:1234–1238

Jackson RL (1983) Lipoprotein lipase and hepatic lipase. In: Boyer PD (ed) The enzymes, vol 16. Academic, New York, pp 141–181

Johansson J, Carlson LA, Landou C, Hamsten A (1991) High density lipoproteins and coronary atherosclerosis. A strong inverse relation with the largest particles is confined to normotriglyceridemic patients. Arterioscler Thromb 11:174–182

Kelley JL, Kruski AW (1986) Density gradient ultracentrifugation of serum lipoproteins in a swinging bucket rotor. Methods Enzymol 128:170–181

Knott TJ, Pease RJ, Powell LM, Wallis SC, Rall SC Jr, Innerarity TL et al. (1986) Complete protein sequence and identification of structural domains of human apolipoprotein B. Nature 323:734–738

Kodama T, Freeman M, Rohrer L, Zabrecky J, Matsudaira P, Krieger M (1990) Type I macrophage scavenger receptor contains alpha-helical and collagen-like coiled coils. Nature 343:531–535

Koo C, Wernette-Hammond ME, Garcia Z et al. (1988) Uptake of cholesterol-rich remnant lipoproteins by human monocyte-derived macrophages is mediated by low density lipoprotein receptors. J Clin Invest 81:1332–1340

Lalouel J-M, Wilson DE, Iverius P-H (1992) Lipoprotein lipase and hepatic triglyceride lipase: molecular and genetic aspects. Curr Opin Lipidol 3:86–95

LaRosa JC, Levy RI, Herbert P, Lux SE, Fredrickson DS (1970) A specific apoprotein activator for lipoprotein lipase. Biochem Biophys Res Commun 41: 57–62

LaRosa JC, Hunninghake D, Bush D, Criqui MH, Getz GS, Gotto AM Jr et al. (1990) The cholesterol facts. A summary of the evidence relating dietary fats, serum cholesterol, and coronary heart disease. A joint statement by the American Heart Association and the National Heart, Lung, and Blood Institute. The Task Force on Cholesterol Issues, American Heart Association. Circulation 81:1721–1733

Lees RS, Hatch FT (1963) Sharper separation of lipoprotein species by paper electrophoresis in albumin-containing buffer. J Lab Clin Med 61:518

Lipid Research Clinics Program (1984a) The Lipid Research Clinics Coronary Primary Prevention Trial results: I. Reduction in incidence of coronary heart disease. JAMA 251:351–364

Lipid Research Clinics Program (1984b) The Lipid Research Clinics Coronary Primary Prevention Trial results: II. The relationship of reduction in incidence of coronary heart disease to cholesterol lowering. JAMA 251:365–374

Luo CC, Li WH, Moore MN, Chan L (1986) Structure and evolution of the apolipoprotein multigene family. J Mol Biol 187:325–340

Macheboeuf MA (1929) Recherches sur les phosphoaminolipides et les sterides du serum et du plasma sanguins: entrainement des phospholipides, des sterols et des sterides par les diverses fractions au cours du fractionnement des proteides du serum. Bull Soc Chim Biol 11:268

Massey JB, Gotto AM Jr, Pownall HJ (1982) Kinetics and mechanism of the spontaneous transfer of fluorescent phosphatidylcholines between apolipoprotein-phospholipid recombinants. Biochemistry 21:3630–3636

McLean LR, Philips MC (1981) Mechanism of cholesterol and phosphatidylcholine exchange or transfer between unilamellar vesicles. Biochemistry 20:2893–2900

Miesenböck G, Patsch JR (1991) Coronary artery disease: synergy of triglyceride-rich lipoproteins and HDL. Cardiovasc Risk Fact 1:293

Miesenböck G, Patsch JR (1992) Postprandial hyperlipidemia: the search for the atherogenic lipoprotein. Curr Opin Lipid 3:196–201

Mitchinson MJ, Ball RY, Carpenter KLH, Parums DV (1988) Macrophages and ceroid in atherosclerosis. In: Suckling KE, Groot PHE (eds) Hyperlipidaemia and atherosclerosis. Academic, London, p 117

Monsalve MV, Henderson HE, Roederer G et al. (1990) A missense mutation at codon 188 of the human lipoprotein lipase gene is a frequent cause of lipoprotein lipase deficiency in persons of different ancestries. J Clin Invest 86: 728–734

Morton RE, Zilversmit DB (1983) Inter-relationship of lipids transferred by the lipid-transfer protein isolated from human lipoprotein-deficient plasma. J Biol Chem 258:11751–11757

Nilsson-Ehle P, Garfinkel AS, Schotz MC (1980) Lipolytic enzymes and plasma lipoprotein metabolism. Annu Rev Biochem 49:667–693

Olivecrona T, Bengtsson-Olivecrona G (1989) Heparin and lipases. In: Lane DA, Lindahl U (eds) Heparin. Arnold, London, pp 335–361

Oncley JL (1963) Brain lipids and lipoproteins and leukodystrophies. Elsevier, Amsterdam

Ornish D, Brown SE, Scherwitz LW, Billings JH, Armstrong WT, Ports TA, McLanahan SM, Kirkeeide RL, Brand RJ, Gould KL (1990) Can lifestyle changes reverse coronary heart disease? The Lifestyle Heart Trial. Lancet 336: 129–133

Palinski W, Rosenfeld ME, Yla Herttuala S, Gurtner GC, Socher SS, Butler SW et al. (1989) Low density lipoprotein undergoes oxidative modification in vivo. Proc Natl Acad Sci USA 86:1372–1376

Parums D, Mitchinson MJ (1981) Demonstration of immunoglobulin in the neighbourhood of advanced atherosclerotic plaques. Atherosclerosis 38:211–216

Patsch JR (1987) Postprandial lipaemia. Baillieres Clin Endocrinol Metab 1:551–580

Patsch JR (1991) Postprandial dyslipidemia and coronary artery disease. In: Gotto AM Jr, Paoletti R (eds) Atherosclerosis reviews, vol 22. Raven, New York, pp 47–49

Patsch JR, Gotto AM Jr (1987) Metabolism of high density lipoproteins. In: Gotto AM Jr (ed) Plasma lipoproteins, new comprehensive biochemistry, vol 14. Elsevier, Amsterdam, pp 221–259

Patsch JR, Gotto AM Jr (1989) Biochemistry of lipid regulation. In: Hurst JW (ed) The heart. McGraw Hill, New York, pp 106–111

Patsch JR, Patsch W (1986) Zonal ultracentrifugation. Methods Enzymol 129:3–26

Patsch JR, Sailer S, Kostner G, Sandhofer F, Holasek A, Braunsteiner H (1974) Separation of the main lipoprotein density classes from human plasma by rate-zonal ultracentrifugation. J Lipid Res 15:356–366

Patsch JR, Gotto AM Jr, Olivercrona T, Eisenberg S (1978) Formation of high density lipoprotein2-like particles during lipolysis of very low density lipoproteins in vitro. Proc Natl Acad Sci USA 75:4519–4523

Patsch JR, Karlin JB, Scott LW, Smith LC, Gotto AM Jr (1983) Inverse relationship between blood levels of high density lipoprotein subfraction 2 and magnitude of postprandial lipemia. Proc Natl Acad Sci USA 80:1449–1453

Patsch JR, Prasad S, Gotto AM Jr, Bengtsson-Olivecrona G (1984) Postprandial lipemia. A key for the conversion of high density $lipoprotein_2$ into high density $lipoprotein_3$ by hepatic lipase. J Clin Invest 74:2017–2023

Patsch JR, Prasad S, Gotto AM Jr, Patsch W (1987) High density $lipoprotein_2$. Relationship of the plasma levels of this lipoprotein species to its composition, to the magnitude of postprandial lipemia, and to the activities of lipoprotein lipase and hepatic lipase. J Clin Invest 80:341–347

Patsch JR, Miesenböck G, Hopferwieser T, Mühlberger V, Knapp E, Dunn JK, Gotto AM Jr, Patsch W (1992) The relationship of triglyceride metabolism and coronary artery disease: studies in the postprandial state. Arterioscler Thromb 12:1336–1345

Patsch W, Schonfeld G, Gotto AM Jr, Patsch JR (1980) Characterization of human high density lipoproteins by zonal ultrazentrifugation. J Biol Chem 255: 3178–3185

Patsch W, Patsch JR, Gotto AM Jr (1989) The hyperlipoproteinemias. Med Clin N Am 73(4):859–893

Segrest JP, Morrisett JD, Jackson RL, Gotto AM Jr (1974) A molecular theory of lipid–protein interactions in the plasma lipoproteins. FEBS Lett 38:247–253

Shirai K, Barnhart RL, Jackson RL (1981) Hydrolysis of human plasma high density lipoproteins$_2$ phospholipids and triglycerides by hepatic lipase. Biochem Biophys Res Commun 100:591–599

Simpson HS, Williamson CM, Olivecrona T, Pringle S, Maclean J, Lorimer AR, Bonnefous F, Bogaievsky Y, Packard CJ, Shepherd J (1990) Postprandial lipemia fenofibrate and coronary artery disease. Atherosclerosis 85:193–202

Small DM, Shipley GG (1974) Physical-chemical basis of lipid deposition in atherosclerosis. The physical state of the lipids helps to explain lipid deposition and lesion reversal in atherosclerosis. Science 185:222–229

Smith LC, Massey JB, Sparrow JT, Gotto AM Jr, Pownall HJ (1983) Structure and dynamics of lipoproteins. In: Pifet G, Herak JN (eds) Supramolecular structure and function. Plenum Publishing, New York, pp 205–244

Sparkes RS, Zollman S, Klisak I, Kirchgessner TG, Komaromy MC, Mohandas T, Schotz MC, Lusis AJ (1987) Human genes involved in lipolysis of plasma lipoproteins: mapping of loci for lipoprotein lipase to 8p22 and hepatic lipase to 15q21. Genomics 1:138–144

Steinberg D, Parthasarathy S, Carew TE, Khoo JC, Witztum JL (1989) Beyond cholesterol. Modifications of low-density lipoprotein that increase its atherogenicity. N Engl J Med 320:915–924

Steinbrecher UP (1987) Oxidation of human low density lipoprotein results in derivatization of lysine residues of apolipoprotein B by lipid peroxide decomposition products. J Biol Chem 262:3603–3608

Steinbrecher UP, Witztum JL (1984) Glucosylation of low-density lipoproteins to an extent comparable to that seen in diabetes slows their catabolism. Diabetes 33:130–134

Steinbrecher UP, Witztum JL, Pathasarathy S, Steinberg D (1987) Decrease in reactive amino groups during oxidation or endothelial cell modification of LDL. Correlation with changes in receptor-mediated catabolism. Arteriosclerosis 7:135–143

Study Group, European Atherosclerosis Society (1987) Strategies for prevention of coronary heart disease: a policy statement of the Eurpean Atherosclerosis Society. Eur Heart J 8:77–88

Study Group, European Atherosclerosis Society (1988) The recognition and management of hyperlipidaemia in adults: a policy statement of the European Atherosclerosis Society. Eur Heart J 9:571–600

Sviridov DD, Safonova IG, Talalaev AG, Repin VS, Smirnov VN (1986) Regulation of cholesterol synthesis in isolated epithelial cells of human small intestine. Lipids 21:759–763

Tall AR (1986) Plasma lipid transfer proteins, J Lipid Res 27:361–367

Wion KL, Kirchgessner TG, Lusis AJ, Schotz MC, Lawn RM (1987) Human lipoprotein lipase complementary DNA sequence. Science 235:1638–1641

Yang CY, Chen SH, Gianturco SH, Bradley WA, Sparrow JT, Tanimura M et al. (1986) Sequence, structure, receptor-binding domains and internal repeats of human apolipoprotein B-100. Nature 323:738–742

Zilversmit DB (1979) Atherogenesis: a postprandial phenomenon. Circulation 60: 473–485

CHAPTER 2

Lipoprotein Metabolism

H.B. Brewer

A. Introduction

Plasma lipids are transported by lipoproteins composed of several classes of lipids (including cholesterol, triglycerides, and phospholipids) and proteins designated apolipoproteins. Our understanding of the role of lipoproteins and apolipoproteins in lipid transport has markedly increased over the last two decades. The roles of lipoprotein receptors, enzymes, and apolipoproteins in lipoprotein metabolism have been elucidated, and this new information provides a conceptual framework for the understanding of lipid transport in normal subjects and dyslipoproteinemic patients.

Six major classes of plasma lipoproteins are separated by hydrated density, including chylomicrons, very low density lipoproteins (VLDL), intermediate-density lipoproteins (IDL), low-density lipoproteins (LDL), Lp(a), and high-density lipoproteins (HDL) (Gofman et al. 1954; Berg et al. 1974). HDL are further separated into HDL_2 and HDL_3. The lipoproteins which remain at the origin in electrophoresis are equivalent to chylomicrons, the pre-β-lipoproteins to VLDL, the β-lipoproteins to LDL, and the α-lipoproteins to HDL (Lee and Hatch 1963). Lp(a) migrates in the pre-β position. Triglycerides are transported primarily by chylomicrons and VLDL. Plasma LDL and HDL transport approximately 70% and 20% of plasma cholesterol, respectively. A schematic model of a plasma lipoprotein is illustrated in Fig. 1 (Obsorne and Brewer 1977; Brewer 1981). The surface of the spherical lipoprotein particle contains phospholipids, free cholesterol, and apolipoproteins interdigitated between the polar-head groups of the phospholipids. The core of the lipoprotein particle is similar to a lipid droplet and contains the hydrophobic cholesteryl esters and triglycerides.

Clinically, the concentrations of lipoproteins are most frequently assessed by quantifying the cholesterol moiety of the lipoprotein particle. Apolipoproteins may also be utilized to quantitate lipoproteins particles. Thus, either the cholesterol or the apolipoprotein component of the lipoprotein particle may be employed in the determination of the plasma levels of lipoproteins. The plasma levels of Lp(a) can be ascertained only by the determination of the unique apolipoprotein associated with the Lp(a) particle.

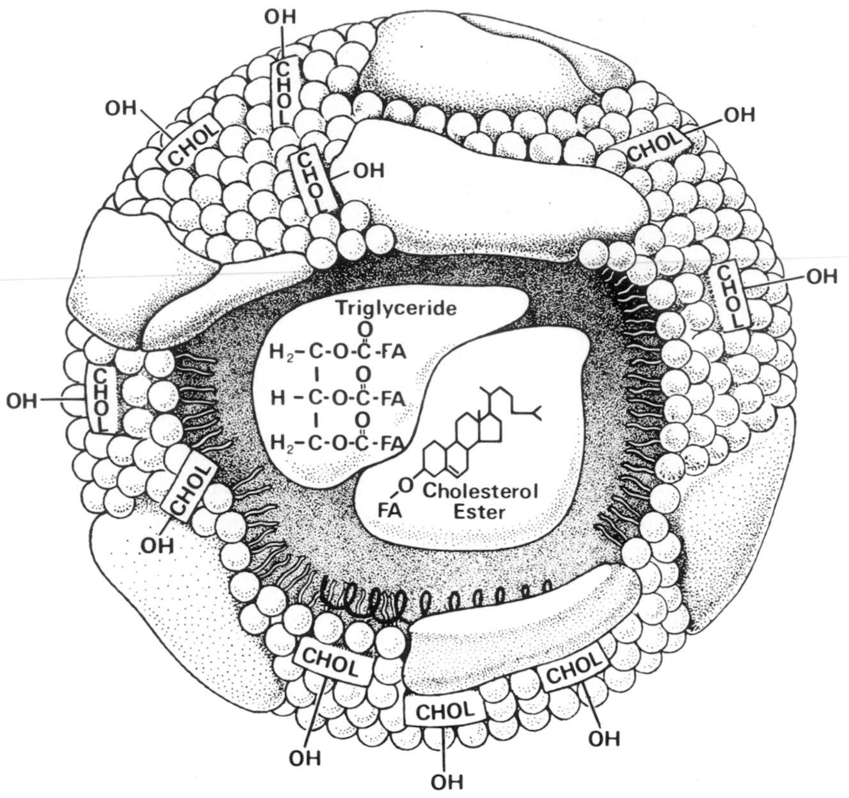

Fig. 1. Schematic model of a plasma lipoprotein. The surface of the lipoprotein particle contains the polar-head groups of the phospholipids. The apolipoproteins as well as cholesterol (*CHOL*) are intercalated between the polar-head groups of the phospholipids. The neutral lipids, cholesteryl esters and triglycerides, are localized in the core of the lipoprotein particle. *FA*, fatty acid

B. Plasma Apolipoproteins

Ten major human plasma apolipoproteins have been identified (Table 1) and their physiological role in lipoprotein metabolism ascertained (Osborne and Brewer 1977; Scanu and Landsberger 1980; Brewer et al. 1988; Breslow 1988; Li et al. 1988). The six most clinically relevant apolipoproteins are A-I, B-100, B-48, C-II, E and apo(a). The two principle apolipoproteins on HDL are apoA-I and apoA-II. In human plasma, apoB exists as two isoproteins, designated apoB-100 and apoB-48, with molecular masses of 512 and 250 kDa, respectively (Kane 1983). ApoB-100 and apoB-48 are synthesized from a single apoB gene (Blackhart et al. 1986). ApoB-100 has 4536 amino acids and is translated from the full-length mRNA. ApoB-48 has 2152 amino acids and is synthesized from an apoB mRNA

Table 1. Major human plasma apolipoproteins

Apolipoproteins	Approx. mol. mass (kDa)	Major site of synthesis	Major density class
ApoA-I	28	Liver, intestine	HDL
ApoA-II	18	Liver	HDL
ApoA-IV	45	Intestine	Chylomicrons
ApoB-48	250	Intestine	Chylomicrons-VLDL-IDL
ApoB-100	500	Liver	Chylomicrons-VLDL-IDL-LDL
ApoC-I	7	Liver	Chylomicrons-VLDL-HDL
ApoC-II	10	Liver	Chylomicrons-VLDL-HDL
ApoC-III	10	Liver	Chylomicrons-VLDL-HDL
ApoE	34	Liver	VLDL-IDL-HDL
Apo(a)	500	Liver	LDL-HDL

HDL, high-density lipoprotein; VLDL, very low density lipoprotein; IDL, intermediate-density lipoprotein, LDL, low-density lipoprotein.

containing a premature in-frame translational stop codon introduced by mRNA editing (Hospattankar et al. 1987; Powell et al. 1987; Chen et al. 1987; Higuchi et al. 1988). Thus two isoproteins, apoB-100 and apoB-48, are synthesized from a single gene by a unique mRNA editing mechanism. The apoB isoproteins are the major structural apolipoproteins on chylomicrons, VLDL, IDL, and LDL. ApoB-100 is the major apolipoprotein on LDL.

Apo(a) is the unique apolipoprotein on Lp(a), and elevated levels of Lp(a) are a risk factor for the development of premature cardiovascular disease (Berg et al. 1974; Kostner et al. 1981; Armstrong et al. 1986; Utermann 1989). ApoE is a 299 amino acid apolipoprotein that is primarily associated with VLDL and HDL (Rall et al. 1982). ApoE is a genetically determined polymorphic apolipoprotein with three common codominantly inherited alleles, designated ε-2, ε-3, and ε-4 (Utermann et al. 1979; Zannis et al. 1981; Mahley et al. 1984; Davignon et al. 1988). The relative frequencies of the three common alleles for the apoE gene locus are 0.073 (E-2), 0.783 (E-3), and 0.143 (E-4). The E apolipoproteins coded for by these alleles are apoE-2, apoE-3, and apoE-4. In the population there are three major classes of homozygous and three heterozygous genotypes, resulting in a total of six phenotypes. The stuctural differences between the common E-2, E-3, and E-4 isoproteins are due to the substitution of amino acids at residues 112 and 158. ApoE-2 contains two cysteines at these two positions, apoE-3 contains a cysteine and arginine, and apoE-4 has two arginines. The charge differences in the apoE isoproteins readily permits the determination of the six common apoE phenotypes by isoelectrofocusing gel electrophoresis (Mahley et al. 1984; Davignon et al. 1988; Gregg and Brewer 1988).

Over the last few years three major physiological functions for the plasma apolipoproteins have been identified (Table 2):

Table 2. Functions of the plasma apolipoproteins in lipoprotein metabolism

Function	Apolipoprotein
I. Structural protein on lipoprotein particles	
Intestinal chylomicrons	ApoB-48, ApoB-100
Hepatic VLDL	ApoB-100
HDL	ApoA-I
II. Ligand on lipoprotein particles for interaction with receptor sites on cells	
Remnant receptor	ApoE
LDL receptor	ApoB-100, ApoE
HDL receptor	ApoA-I, ApoA-II
III. Cofactor for enzyme	
Lipoprotein lipase	ApoC-II
Lecithin cholesterol acyltransferase	ApoA-I

VLDL, very low density lipoproteins; HDL, high-density lipoproteins; LDL, low density lipoproteins.

1. Apolipoproteins function as structural proteins for the biosynthesis and secretion of plasma lipoproteins. ApoA-I has been proposed to be a critical structural protein for the biosynthesis and remodeling of HDL. Patients with a structural defect in the apoA-I gene are unable to synthesize and secrete apoA-I and have a virtual absence of plasma HDL and premature cardiovascular disease (NORUM et al. 1982; SCHAEFER et al. 1985; SCHMITZ and LACKNER 1989; DEEB et al. 1991; BEKAERT et al. 1991). ApoB-100 and apoB-48 are required for the secretion of triglyceride-rich lipoproteins from the liver and intestine. Defects in the biosynthesis or secretion of apoB result in abetalipoproteinemia and homozygous hypo-β-lipoproteinemia, diseases characterized by the absence of plasma chylomicrons, VLDL, IDL, and LDL as well as malabsorption and neurological defects (HERBERT et al. 1978; LACKNER et al. 1986; LEPPERT et al. 1988; ROSS et al. 1988; WETTERAU et al. 1992).
2. Apolipoproteins function as cofactors or activators of enzymes involved in lipid–lipoprotein metabolism. Of particular importance is apoC-II, the activator of lipoprotein lipase, which is responsible for the perivascular hydrolysis of triglycerides contained in lipoproteins to monoglycerides and free fatty acids (LAROSA et al. 1970; HAVEL et al. 1970; JACKSON et al. 1977; HOSPATTANKAR et al. 1984). Lipoprotein lipase is attached to the capillary endothelium by heparin-like proteoglycans, which permits the direct interaction of the lipoprotein lipase enzyme with the circulating triglyceride-rich lipoproteins. ApoA-I modulates the enzymic activity of lecithin cholesterol acyltransferase (LCAT), which catalyzes the esterification of lipoprotein cholesterol to cholesteryl esters (FIELDING et al. 1972).
3. Apolipoproteins also play a critical role in lipoprotein metabolism as ligands on lipoprotein particles for interaction with cellular receptors.

Specific apolipoproteins on lipoprotein particles interact with defined cellular receptors, resulting in either absorptive endocytosis or the initiation of protein kinase C and cholesterol translocation to the cell membrane.

C. Cellular Receptors

I. Low-Density Lipoprotein Receptor

The two major apolipoproteins that serve as ligands for the LDL receptor are apolipoproteins B-100 and E (GOLDSTEIN and BROWN 1979; GOLDSTEIN et al. 1985; DAVIGNON et al. 1988; GREGG and BREWER 1988). Both apolipoproteins B-100 and E are ligands involved in the uptake of apoB containing lipoproteins in VLDL and IDL, whereas apoB-100 is the principle ligand on LDL for interaction with the LDL receptor.

The LDL receptor is a 839 amino acid glycosylated protein containing five separate domains (YAMAMOTO et al. 1984; SUDHOF et al. 1985). These include a 322 residue amino terminal cysteine-rich ligand-binding domain, a 350 amino acid domain which displays a high degree of homology with the epidermal growth factor (EGF) precursor, a 48 amino acid domain containing the O-linked carbohydrates, a 22 amino acid membrane-spanning domain, and a 50 residue cytoplasmic domain (Fig. 2). The expression of the LDL receptor gene is under transcriptional regulation with *cis*-acting DNA sequences required for both basal and sterol-regulated control. A sterol-regulatory element (SRE-1) has been identified in the 5′ flanking region of the LDL receptor gene, which modulates the expression of the LDL

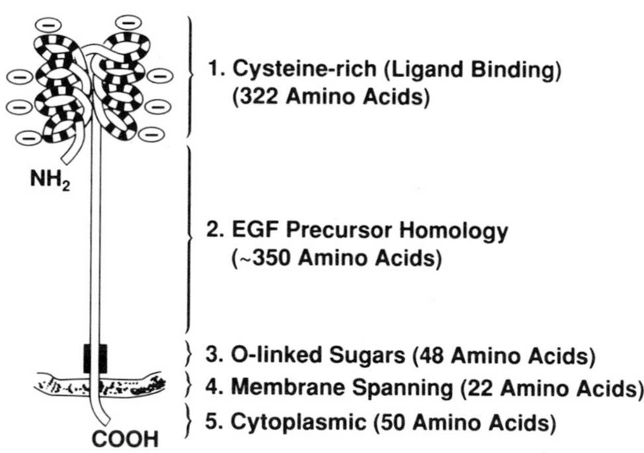

Fig. 2. Schematic model of the low-density lipoprotein (LDL) receptor. The 839 amino acid receptor contains five separate functional domains (see text for further details) (YAMAMOTO et al. 1984)

receptor in response to sterols (Dawson et al. 1988). A similar SRE-1 element has also been identified in 3-hydroxy-3-methylyglutaryl coenzyme A (HMG-CoA) reductase and HMG-CoA synthase.

A reduction in the concentration of intracellular cholesterol initiates a coordinate upregulation of HMG-CoA reductase with increased synthesis of intracellular cholesterol as well as an increased synthesis of the LDL receptor. The upregulation of the LDL receptor results in both an increased number of LDL receptors on the cell surface as well as enhanced binding and degradation of plasma LDL. This compensatory mechanism returns the intracellular cholesterol concentration toward normal. Drugs (e.g., lovastatin) which inhibit HMG-CoA reductase and reduce intracellular cholesterol levels result in an increased expression of the LDL receptor and a subsequent decrease in the plasma levels of LDL.

Familial hypercholesterolemia is due to mutations in the LDL receptor gene. A number of different mutations have been identified that result in a variable loss of LDL receptor function (Goldstein et al. 1985; Russell et al. 1989; Van der Westhuyzen et al. 1990). The residual LDL receptor activity in homozygous patients is the major factor which is responsible for the range of plasma LDL levels between 600 and 1000 mg/dl.

II. Scavenger Receptor(s)

Chemical modification of LDL leads to cellular uptake and degradation of modified LDL via the scavenger receptor pathway which is a separate receptor system from the LDL receptor (Kodama et al. 1990). The scavenger receptor is a trimer of 77-kDa subunits containing six distinct domains (Kodama et al. 1990): the amino terminal 110-residue cysteine-rich domain, a 72-residue collagen-like domain, a 163 amino acid α-helical coiled-coil domain, a 32-residue spacer, a 26 amino acid transmembrane-spanning domain, and a 50 amino acid cytoplasmic domain. It has been proposed that the functional role of the scavenger receptor in lipoprotein metabolism is the clearance of chemically modified lipoproteins (Fig. 3). The scavenger receptor has been proposed to play a central role in the macrophage uptake of modified lipoproteins and the development of the foam cells in the atherosclerotic lesion.

III. Chylomicron Remnant Receptor

The identification of the receptors responsible for the clearance of remnants of triglyceride-rich lipoproteins has been a challenge for the lipoprotein field. The clue to the presence of a receptor pathway separate from the LDL receptor pathway came from clinical observations that patients with homozygous familial hypercholesterolemia cholesterolemia with defects in the LDL receptor had no difficulty in clearing remnants of triglyceride-rich lipoproteins. Initial studies revealed that in addition to the LDL receptor,

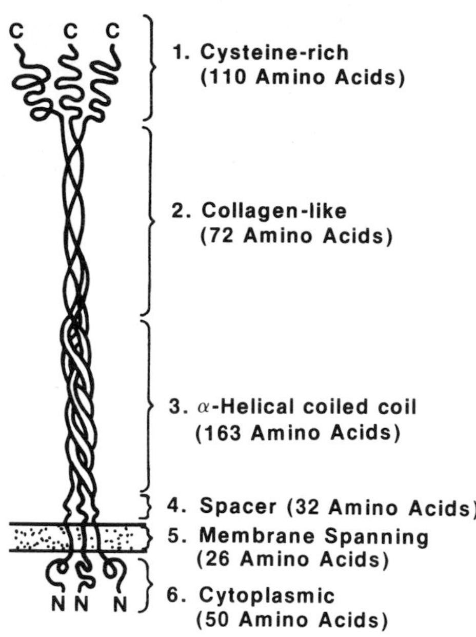

Fig. 3. Schematic structural model of the type I macrophage scavenger receptor. The receptor is a triple helix structure with each subunit containing seven separate domains. The cysteine-rich domains are similar in structure to the low-density lipoprotein (LDL) binding domain of the LDL receptor (KODAMA et al. 1990). *C*, C terminal; *N*, N terminal

hepatic membranes bound apoE and chylomicron remnants (MAHLEY et al. 1981; HOEG et al. 1986). Hepatocytes isolated from normolipidemic subjects as well as patients homozygous for familial hypercholesterolemia were able to bind and internalize lipoproteins independent of the LDL receptor (EDGE et al. 1986). The best candidate for the putative remnant receptor is a glysolyated 600-kDa protein that is designated the LDL receptor related protein (LRP; HERZ et al. 1988). LRP has a high degree of homology with portions of the LDL receptor, as well as the EGF precursor and has been proposed to belong to a broad receptor family. The major ligand for the putative remnant receptor is apoE, and the clearance of triglyceride-rich chylomicron remnants has been proposed to involve an apoE remnant lipoprotein particle pathway.

Recent studies have indicated that the importance of LRP may not be limited to lipoprotein metabolism since LRP is identical to the α_2-macroglobulin receptor which has been shown to bind several ligands (STRICKLAND et al. 1990). Thus, LRP may play an important role in the clearance of several different plasma proteins. Additional studies will be required to definitively understand the mechanisms that are involved in the metabolism and cellular uptake of the remnants of triglyceride-rich lipoproteins.

IV. High-Density Lipoprotein Receptor

The presence of a specific HDL receptor involved in the transport of cholesterol from peripheral cells back to the liver, where it can be excreted from the body, continues to be controversial. An HDL receptor has been postulated based on the presence of saturable, specific binding of HDL to hepatic, endothelial, and adipose cells as well as fibroblasts and steroidogenic tissues (Oram et al. 1983; Suzuki et al. 1983; Schmitz et al. 1985; Barbaras et al. 1987b). Ligand blot analyses have suggested that the HDL receptor has an apparent molecular mass of 80–110 kDa. Several different apolipoprotein ligands have been reported to bind to the putative 110-kDa HDL receptor in different tissues including A-I, A-II, and A-IV. A model cell culture system which has been useful in the analysis of HDL binding is the OB 1771 adipocytes, where detailed studies have suggested a specific saturable HDL-binding site which facilitated cholesterol efflux (Barbaras et al. 1987b).

Recent studies have suggested that the interaction of HDL with a specific cellular receptor in several cells initiates a second messenger pathway with protein kinase C activation (Theret et al. 1990; Mendez et al. 1991). The activation of the protein kinase C pathway is associated with the translocation of intracellular cholesterol to the cell membrane, where it can be removed from the membrane by a variety of different plasma lipoproteins.

The structure of a 110-kDa HDL-binding protein has recently been reported (McKnight et al. 1992). This binding protein has many of the characteristics of the HDL receptor, however, the binding protein does not have a transmembrane domain and does not appear to be present on the extracellular surface of the cell membrane. Additional studies will be required to definitively establish the role this binding protein plays in cholesterol and lipoprotein metabolism and its relationship to the putative HDL receptor.

D. Lipoprotein Metabolism

I. ApoB Metabolic Cascade

A conceptual overview of the pathways for lipoprotein biosynthesis, processing, and catabolism of the apoB-containing lipoproteins including chylomicrons, VLDL, IDL, and LDL is shown schematically in Fig. 4. The metabolism of the major classes of lipoproteins containing apoB-48 and apoB-100 may be considered to consist of two major "apoB metabolic cascades" (for general reviews, see Eisenberg 1984; Tall 1986; Brewer et al. 1988, 1989; Vega et al. 1991). The first apoB cascade involves the stepwise delipidation of triglyceride-rich chylomicron particles which transport dietary cholesterol and triglycerides from the intestine to peripheral

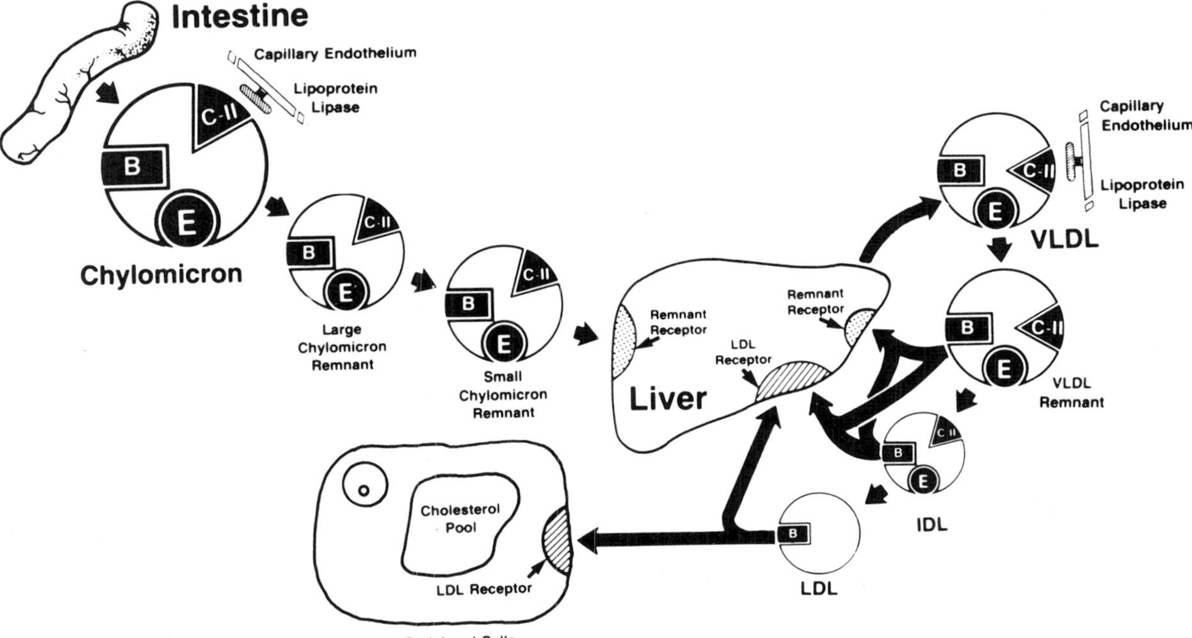

Fig. 4. Overview of the metabolism of the major apoB containing lipoprotein particles. The "intestinal apoB cascade" involves the secretion of triglyceride rich chylomicrons from the intestine. Chylomicron triglycerides undergo hydrolysis by lipoprotein lipase and the chylomicron remnant particles are formed which have a hydrated density of very low-density lipoproteins (*VLDL*) and finally intermediatedensity lipoproteins (IDL). The small chylomicron remnants are taken up by the liver by the putative remnant receptor. The "hepatic apoB cascade" involves the secretion of triglyceride rich VLDL secreted by the liver. The VLDL triglycerides are hydrolyzed by lipoprotein lipase and remnants undergo stepwide delipidation with the formation of particles with a hydrated density of IDL and finally low-density lipoproteins (*LDL*). VLDL remnants are cleared from the plasma by interacting with the remnant and LDL receptors. LDL interacts with the LDL receptor, which initiates receptor-mediated endocytosis and degradation of LDL (see test for further details). *B*, apoB; *E*, apoE; *C-II*, apoC-II

tissues and the liver. Following secretion, chylomicrons acquire apoE and apoC-II primarily from HDL. As reviewed above, apoC-II activates lipoprotein lipase, which results in triglyceride hydrolysis and remodeling of the triglyceride-rich lipoprotein particles. Concomitant with triglyceride hydrolysis of the lipoprotein particles, both apolipoproteins as well as lipid constituents are transferred to HDL. With remodeling of the chylomicron particles, the hydrated density of the chylomicron increases and chylomicron remnants are generated with a hydrated density of VLDL and then IDL. Chylomicron remnants have been proposed to be removed primarily by a putative hepatic remnant receptor (MAHLEY et al. 1984; DAVIGNON et al. 1988; GREGG and BREWER 1988). The 600-kDa LRP has been proposed as a potential candidate for the remnant receptor (HERZ et al. 1988; STRICKLAND et al. 1990).

The second apoB cascade involves triglyceride-rich VLDL containing apoB-100 secreted by the liver. ApoC-II and apoE dissociate from HDL and reassociate with the hepatogenous triglyceride-rich VLDL following secretion from the liver. ApoC-II activates LPL as outlined above and VLDL are serially converted to VLDL remnants, IDL, and finally LDL. During the metabolic conversion of VLDL to LDL, approximately 50% of VLDL remnants and IDL are removed from the plasma through interaction of apolipoproteins E and B with the remnant and LDL receptors.

Hepatic lipase, a second lipolytic enzyme, and apoE have been proposed to be necessary for the conversion of IDL to LDL. Hepatic lipase functions in lipoprotein particle metabolism as both a phospholipase and triglyceryl hydrolase. LDL, the final product of the VLDL cascade, contains virtually only apoB-100, which interacts with the LDL receptor on the plasma membranes of liver, adrenal, and peripheral cells including fibroblasts and smooth muscle cells. The interaction of LDL with the LDL receptor initiates receptor-mediated endocytosis and transport of LDL to lysosomes, where the protein moiety is degraded and cholesteryl esters are hydrolyzed to free cholesterol and transferred to the intracellular cholesterol pool. Approximately 50% of plasma LDL is catabolized by the liver and peripheral cells, respectively.

II. High-Density Lipoprotein Metabolism

Lipoproteins within HDL are synthesized by four major pathways (Fig. 5; for review, see EISENBERG 1984; BREWER et al. 1988, 1989). Nascent HDL, composed primarily of apoA-I phospholipid discs, are secreted from both the human intestine and liver. Nascent HDL acquire excess cholesterol from tissues, and the enzyme LCAT catalyzes the esterification of lipoprotein cholesterol to cholesteryl esters. With the formation of cholesteryl esters, the nascent HDL are converted to spherical lipoproteins with a hydrated density of HDL_3. HDL_3 are converted to the larger HDL_2 by the acquisition of apolipoproteins and lipids released during the stepwise delipidation and

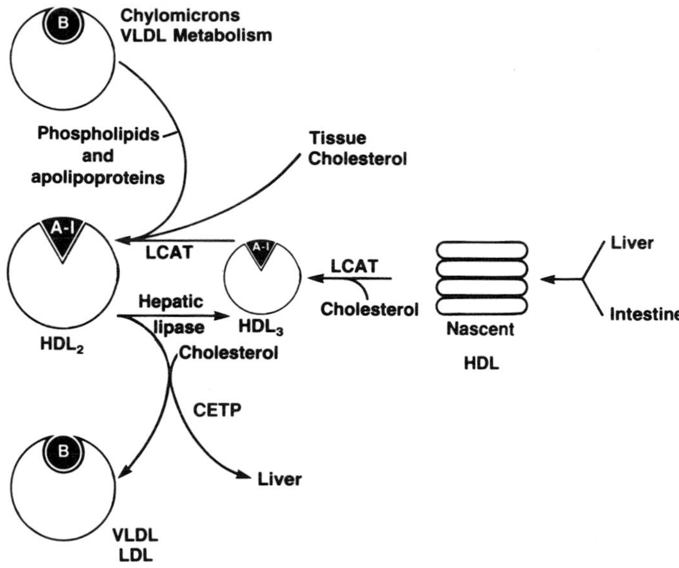

Fig. 5. Pathways for the biosynthesis of high-density lipoproteins (*HDL*). Disk-shaped nascent HDL are synthesized by the intestine and liver. Nascent HDL take up cholesterol and are converted to spherical lipoprotein particles with a hydrated density of HDL_3. The further addition of cholesterol, phospholipids, and apolipoproteins from both the metabolism of triglyceride-rich lipoproteins and the uptake from peripheral tissues result in the conversion of lipoproteins in HDL_3 to particles with a hydrated density of HDL_2. Cholesterol in the lipoprotein particles is converted to cholesteryl esters by the enzyme lecithin cholesterol acyltransferase (*LCAT*). Cholesteryl esters are transferred to very low density–intermediate–density–low-density lipoproteins (*VLDL–IDL–LDL*) by the cholesteryl ester transfer protein (*CETP*). Lipoproteins in HDL_2 are converted to HDL_3 by the enzyme hepatic lipase (see text for additional information). *B*, apoB; *A-I*, apoA-I

remodeling of the triglyceride-rich chylomicrons and VLDL as well as by the esterification of the cholesterol removed from peripheral tissues. HDL_2 is converted back to HDL_3 by the removal of phospholipids and triglycerides by hepatic lipase as well as by the transfer of cholesteryl esters into VLDL and LDL by the cholesterol ester transfer protein (CETP) and by the transfer of cholesterol to the liver and other tissues (Tall 1986).

In this overall process, lipoproteins within HDL are interconverted from HDL_3 to HDL_2 and back to HDL_3 as cholesterol is picked up and transferred from peripheral tissues to the liver or the apoB-containing lipoproteins. This still hypothetical process, termed reverse cholesterol transport, is summarized schematically in Fig. 6 (Glomset et al. 1966; Glomset 1968). VLDL are secreted from the liver and undergo stepwise delipidation to LDL as outlined above. LDL, the major cholesterol-transporting lipoprotein, binds to the LDL receptor in the liver and in peripheral cells, where it supplies cholesterol to the intracellular cholesterol pool. Excess cholesterol

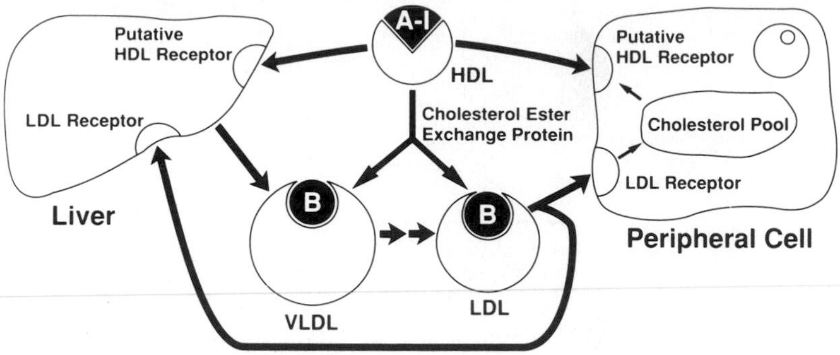

Fig. 6. General overview of high-density lipoprotein (*HDL*) and reverse cholesterol transport. Very low density lipoproteins (*VLDL*) are secreted by the liver and undergo intravascular remodeling with the formation of low density lipoproteins (*LDL*). LDL are taken up by the liver and peripheral cell by interaction with the LDL receptor. Excess intracellular cholesterol is removed by HDL following interaction with the putative HDL receptor. Cholesterol is converted to cholesteryl esters by lecithin cholesterol acyltransferase (LCAT), and the cholesteryl esters are transferred to either the liver or VLDL and LDL by the cholesteryl ester transfer protein (CETP) (see text for additional details). *A-I*, apoA-I; *B*, apoB

is removed from the peripheral cells by HDL. In this proposed model, HDL interacts with a putative HDL receptor that facilitates the transfer of cellular cholesterol to HDL. HDL transports this cholesterol in plasma and delivers it to the liver via the HDL receptor for removal from the body by direct secretion into bile or following conversion to bile acids. A major pathway for the transport of cholesterol from peripheral cells to the liver is the transfer of cholesteryl esters from HDL to VLDL-IDL-LDL by CETP. Thus, cholesterol may be transported back to the liver directly by HDL or following exchange to VLDL and LDL. A variable portion of tissue cholesterol has also been proposed to be transported to the liver by HDL particles containing apoE, which may interact with both the hepatic remnant and LDL receptors.

Over the last few years, several lines of evidence have indicated that HDL are heterogeneous. HDL contains several separate lipoprotein particles which may have different functions in lipoprotein metabolism and reverse cholesterol transport (Alaupovic 1972; Kostner and Alaupovic 1972; Osborne and Brewer 1977). Several different methods including electrophoresis, hydrated density, gradient gel electrophoresis, and affinity chromatography have been employed to separate and characterize the lipoprotein particles within HDL (Kostner and Alaupovic 1972; Nestruck et al. 1983; Cheung and Albers 1984). The most effective current method available to classify lipoprotein particles in HDL is based on apolipoprotein composition (Alaupovic 1972). The two major apoA-I-containing lipoprotein particles within HDL classified by apolipoprotein composition are LpA-

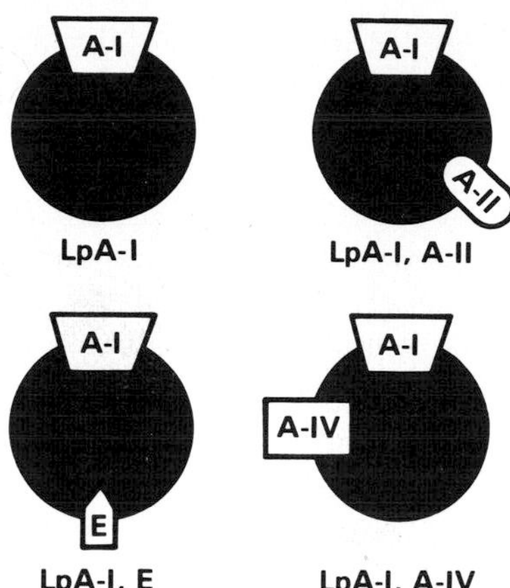

Fig. 7. Schematic model of the apoA-I (*A-I*) containing lipoprotein particles (LpA-I,A-II,LpA-I,E,LpA-I,A-IV) within high-density lipoproteins (HDL)

I and LpA-I,A-II (Fig. 7). Two minor apoA-I containing apolipoproteins include LpA-I,A-IV and LpA-I,E (Fig. 7). The C apolipoproteins and other minor apolipoproteins are also present on the major lipoprotein particles within HDL.

The relative roles of LpA-I and LpA-I,A-II in the protective effect of HDL on the development of premature cardiovascular disease have recently been addressed. LpA-I, but not LpA-I,A-II, has been reported to be inversely correlated with angiographically established coronary atherosclerosis (Puchois et al. 1987). Studies in cell culture have indicated that LpA-I, but not LpA-I,A-II, increased cholesterol efflux from OB1771 adipocytes (Barbaras et al. 1987a). Competitive studies with LpA-I and LpA-I,A-II revealed that LpA-I,A-II not only was not effective in effluxing cholesterol from adipocytes, but in fact inhibited the cholesterol efflux mediated by LpA-I (Barbaras et al. 1987a). Thus, in cell culture studies only LpA-I was effective in facilitating the efflux of cellular cholesterol.

A separate approach to the evaluation of the antiatherogenic properties of LpA-I and LpA-I,A-II has been reported. A comparison of the ability of LpA-I and LpA-I,A-II to protect against the development of diet-induced atherosclerosis has been performed in transgenic mice on a high-fat and high-cholesterol diet (Schultz et al. 1992). Transgenic mice utilizing human apoA-I and apoA-I + apoA-II were developed and the degree of atherosclerosis induced by a diet enriched in cholesterol and fat was quantitated. LpA-I offered better protection than LpA-I,A-II against diet-induced atherosclerosis in the transgenic mouse model.

Additional studies have also revealed different metabolic pathways for LpA-I and LpA-I,A-II. In a comparison of the apoA-I-containing lipoprotein particles, LpA-I,A-II was shown to be a better substrate for hepatic lipase than LpA-I (Mowri et al. 1990). These results are consistent with our previous report that apoA-II increased the enzymic activity of hepatic lipase 1.5-fold in vitro (Jahn et al. 1983). Thus, apoA-II may function as an important modulator of the enzymic activity of hepatic lipase in lipoprotein metabolism. Based on these results, it is proposed that the LpA-I,A-II particles in HDL_2 are preferentially converted to particles within HDL_3. The combined results from all of these studies have been interpreted as indicating that LpA-I is the major antiatherogenic particle within HDL.

Because of the emerging evidence that different lipoprotein particles in HDL may have a variable effect on the protection against atherosclerosis, a new series of metabolic studies were initiated to elucidate the kinetics of LpA-I and LpA-I,A-II (Rader et al. 1991). The kinetics of LpA-I and LpA-I,A-II have been analyzed in normolipidemic controls to gain insight into the potential metabolic differences between these two major apoA-I-containing lipoprotein particles in man. LpA-I and LpA-I,A-II were separated from plasma or HDL by affinity chromatography utilizing antibodies to apoA-I and apoA-II. In these studies apoA-I and apoA-II were radiolabeled, incubated with plasma, and LpA-I and LpA-I,A-II isolated. In a separate approach, isolated LpA-I and LpA-I,A-II were directly radiolabeled as lipoproteins. In the kinetic studies ^{125}I-LpA-I and ^{131}I-LpA-I,A-II were injected simultaneously in normal subjects, and the in vivo catabolism of LpA-I and LpA-I,A-II analyzed over 14 days. LpA-I particles were catabolized at a faster rate than LpA-I,A-II particles (Rader et al. 1991). Based on these results, we have proposed that there are two cascades involved in the metabolism of HDL particles in HDL_2 and HDL_3 (Fig. 8). One cascade involves the metabolism of LpA-I and the other cascade contains the LpA-I,A-II particles. The results of the kinetic data support the view that LpA-I and LpA-I,A-II may have different metabolism and physiological functions in HDL metabolism.

III. Lp(a)

Lp(a) is a cholesterol-rich lipoprotein that closely resembles LDL in lipid composition and has a hydrated density intermediate between LDL and HDL. The protein moiety primarily consists of apoB-100 and a specific apolipoprotein, apo(a) (Utermann and Weber 1983; Gaubatz et al. 1983; Fless et al. 1984). Apo(a) is linked by a single disulfide bridge to apoB-100 on LDL to form Lp(a). Apo(a) is a large glycoprotein ranging in size from 400 to 700 kDa. The amino acid sequence of apo(a) is similar to the sequence of plasminogen and contains cysteine-rich domains of 80–114 amino acids in length called "kringles," due to their structural similarity to a Danish pastry (Eaton et al. 1987; McLean et al. 1987). Apo(a) contains a variable number of copies of kringle 4, a single copy of kringle 5 followed by the protease domain of plasminogen. In contrast to plasminogen, apo(a) has no serine protease enzymic activity. Apo(a) cannot be converted to an active plasmin-like enzyme by tissue plasminogen activator, streptokinase, or urokinase.

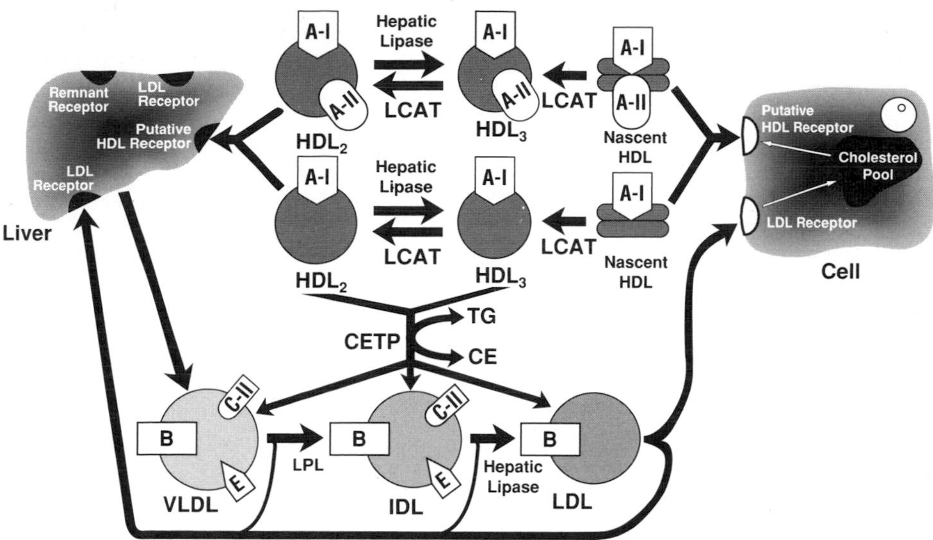

Fig. 8. Overview of the metabolism of lipoproteins LpA-I and LpA-I,A-II particles in high-density lipoproteins (*HDL*). The metabolism of LpA-I and LpA-I, A-II in HDL is shown as two independent cascades. One cascade contains LpA-I particles while the second cascade involves the LpA-I,A-II particles. Both particles may contribute to the transport of HDL cholesterol in the plasma. *B*, *LCAT*, lecithin cholesterol acyltransferase; *CE*, cholesteryl ester; *CETP*, cholesteryl ester transfer protein; *LPL*, lipoprotein lipase; *VLDL*, very low density lipoproteins; *LDL*, low-density lipoproteins; *IDL*, intermediate-density lipoproteins

Of clinical importance is the observation that there is a highly significant association of the individual apo(a) phenotypes with the plasma Lp(a) levels (UTERMANN et al. 1987). The Lp(a) isoproteins of higher and lower molecular weight are associated with lower and higher plasma concentrations, respectively. The different molecular sizes of Lp(a) in human plasma are due to the presence of a variable number of copies of kringle 4 in the amino acid sequence of apo(a) (KOSCHINSKY et al. 1990; AZROLAN et al. 1990). The physiological function(s) of Lp(a) in lipoprotein metabolism is as yet unknown.

E. Major Plasma Atherogenic and Antiatherogenic Lipoproteins

Increased plasma levels of three classes of plasma lipoproteins, LDL, β-VLDL and Lp(a), and decreased levels of HDL have been associated with the development of premature vascular disease. A schematic overview of the interactions of the three major atherogenic lipoproteins and the antiatherogenic HDL is shown in Fig. 9. Increased levels of the atherogenic lipo-

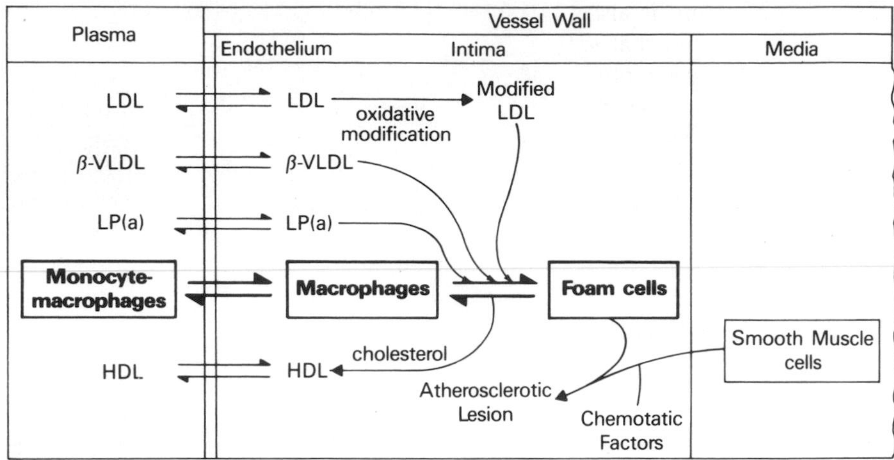

Fig. 9. Schematic model of the interaction of lipoproteins (*LP*) with macrophages and the formation of foam cells, the lesion characteristic of early atherosclerosis. Elevated levels of three major classes of plasma lipoproteins, low-density lipoproteins (*LDL*), β very high-density lipoproteins (*βVLDL*), and Lp(a) have been associated with an increased risk of premature cardiovascular disease. LDL undergoes oxidative modification with the formation of modified LDL, which is rapidly taken up by the macrophage with the formation of foam cells. Elevated intimal levels of β-VLDL and Lp(a) are also associated with foam cell formation. Foam cell formation, macrophage activation, lipid oxidation, and endothelial cell injury all lead to the release of chemotactic agents which contribute to the development of the atherosclerotic lesion (see text for further details). *HDL*, high-density lipoproteins

proteins in the vessel wall have been proposed to increase the intimal macrophage uptake of cholesterol-rich lipoproteins by specific macrophage cellular receptors with the generation of arterial foam cells, which characterize the early atherosclerotic lesion. Smooth muscle cells may also take up atherogenic lipoproteins and undergo conversion to foam cells.

I. Low-Density Lipoproteins

The pathophysiological mechanisms involved in the atherosclerosis associated with LDL have recently been elucidated (for reviews, see HABERLAND and FOGELMAN 1987; VAN LENTEN and FOGELMAN 1990; STEINBERG 1983, 1991). Native LDL is not readily taken up by the macrophage in vitro, and incubation with LDL does not result in the formation of foam cells. It is now known that oxidative modification of LDL in vitro results in markedly enhanced LDL uptake by the scavenger receptor on macrophages with foam cell formation. Oxidative modifications of LDL were observed following in vitro incubation with endothelial cells, smooth muscle cells, and macrophages or following modification with malondialdehyde. Malondialdehyde is a metabolic byproduct of arachidonic acid metabolism in the biosynthesis of prostaglandins and is also formed during lipid peroxidation. Recent studies have also indicated that the oxidized lipids within LDL may play an important role in the pathophysiology of the atherosclerotic lesion by stimulating the secretion of cytokinins and other factors which modulate endothelial cell function as well as facilitating the recruitment of plasma monocytes in the vessel wall. Based on an increasing body of data, it has been proposed that oxidative modification of LDL may be a prerequisite for the macrophage uptake of LDL and foam cell formation.

II. β Very Low Density Lipoproteins

β-VLDL, metabolic remnants of triglyceride-rich chylomicrons and VLDL, accumulated in type III hyperlipoproteinemia and in experimental animals fed diets high in cholesterol and saturated fat (for reviews, see Mahley 1979; Gregg et al. 1981; Havel 1982; Brewer et al. 1983). Efficient clearance of chylomicrons and hepatic VLDL remnants requires apoE. ApoE functions to facilitate hepatic lipoprotein remnant clearance by the liver as the binding ligand for both the putative remnant and LDL receptors. In addition, apoE is required for the effective conversion of IDL to LDL. An absence or structural defects in apoE result in defective removal of remnants of triglyceride-rich lipoproteins and the accumulation of plasma β-VLDL characteristic of type III hyperlipoproteinemia (Mahley 1979; Gregg et al. 1981; Havel 1982; Brewer et al. 1983; Mahley et al. 1984; Davignon et al. 1988; Gregg and Brewer 1988). β-VLDL remnant apolipoproteins have been proposed to be taken up by macrophages to form foam cells by either the LDL receptor or a specific macrophage β-VLDL receptor.

III. Lp(a)

A considerable body of data has accumulated to establish that an elevated plasma level of Lp(a) is an important independent risk factor for the development of premature cardiovascular disease (Berg et al. 1974; Kostner et al. 1981; Armstrong et al. 1986; Utermann et al. 1989). Lp(a) levels in plasma range from less than 1 to more than 100 mg/dl. Approximately 20% of the population have levels above 30 mg/dl, which is associated with a twofold increase in the relative risk of premature vascular disease. The mechanism by which elevated plasma levels of Lp(a) increase the risk of premature cardiovascular disease remains to be established. Lp(a) may be taken up by macrophages resulting in cholesterol deposition and foam cell formation, or the atherogenic properties may be more related to its role in increasing thrombosis. Lp(a) has been reported to interact with fibrin peptides, inhibit thrombolysis, and be a competitive inhibitor of plasminogen for the endothelial plasminogen receptor (Loscalzo 1990; Miles and Plow 1990).

IV. High-Density Lipoproteins

As reviewed above, HDL have been regarded as the principle antiatherogenic lipoprotein in plasma (MILLER and MILLER 1975; GORDON et al. 1977; MILLER et al. 1977; GORDON and RIFKIND 1989). The major mechanism for the protective effect of HDL has ben proposed to be due to the role of HDL in transporting excess cholesterol from peripheral cells back to the liver by reverse cholesterol transport (GLOMSET 1968; ALAUPOVIC 1972). The higher the plasma levels of HDL, the more efficient would be the transport of excess cholesterol from peripheral cells back to the liver. As previously indicated, the importance of the different lipoprotein particles within HDL for the protective effect of HDL on premature cardiovascular remains to be definitively established.

F. Summary

During the last two decades there have been several major advances in our understanding of the pathways of lipoprotein metabolism in normal subjects and patients with dyslipoproteinemias. The covalent structures of ten plasma apolipoproteins have been elucidated. The specific functions of the individual plasma apolipoproteins in lipoprotein metabolism have been determined, and the molecular defects in the individual apolipoproteins which result in characteristic dyslipoproteinemias have been established. The structure and function of the LDL, scavenger, and LRP receptors have been determined and a putative receptor for HDL identified. The metabolic role of lipoprotein and hepatic lipases as well as LCAT have been defined, and the key role that CETP plays in the transfer of cholesteryl esters and triglycerides has been ascertained. The combined results from these studies have provided new insights into lipid transport in normal subjects and have permited a more definitive approach to the diagnosis and treatment of patient with disorders of lipoprotein metabolism.

References

Alaupovic P (1972) Conceptual development of the classification systems of plasma lipoproteins. Protides of the biological fluids. Proceedings of the 19th colloquium 9–19

Armstrong VW, Cremer P, Eberle E, Manke A, Schulze F, Wieland H, Kreuzer H, Seidel D (1986) The association between serum Lp(a) concentrations and angiographically assessed coronary atherosclerosis. Dependence on serum LDL levels. Atherosclerosis 62:249–257

Azrolan N, Gavish D, Breslow JL (1990) Lp(a) levels correlate inversely with apo(a) size and KIV copy number but not with apo(a) mRNA levels in a cynomolgus monkey model. Circulation 82:III–90

Barbaras R, Puchois P, Fruchart JC, Ailhaud G (1987a) Cholesterol efflux from cultured adipose cells is mediated by LpAI particles but not by LpAI:AII particles. Biochem Biophys Res Commun 142:63–69

Barbaras R, Puchois P, Grimaldi P, Barkia A, Fruchart JC, Ailhaud G (1987b) Relationship in adipose cells between the presence of receptor sites for high density lipoproteins and the promotion of reverse cholesterol transport. Biochem Biophys Res Commun 149:545–554

Bekaert ED, Alaupovic P, Knight-Gibson CS, Laux MJ, Pelachyk JM, Norum RA (1991) Characterization of apoA- and apoB-containing lipoprotein particles in a variant of familial apoA-I deficiency with planar xanthoma: the metabolic significance of LP-A-II particles. JLR 32:1587–1599

Berg K, Dahlen G, Frick MH (1974) Lp(a) lipoprotein and pre-beta1-lipoprotein in patients with coronary heart disease. Clin Genet 6:230–235

Blackhart BD, Ludwig EM, Pierotti VR, Caiati L, Onasch MA, Wallis SC, Powell L, Pease R, Knott TJ, Chu ML, Scott MJ, McCarthy BJ, Levy-Wilson B (1986) Structure of the human apolipoprotein B gene. J Biol Chem 261:15364–15367

Brewer HB Jr (1981) Current concepts of the molecular structure and metabolism of human apolipoproteins and lipoproteins. Klin Wochenschr 59:1023–1035

Breslow JL (1988) Apolipoprotein genetic variation and human disease. Physiol Rev 68:85–132

Brewer HB Jr, Zech LA, Gregg RE, Schwartz D, Schaefer EJ (1983) Type III hyperlipoproteinemia: diagnosis, molecular defects, pathology, and treatment. Ann Intern Med 98:623–640

Brewer HB Jr, Gregg Re, Hoeg JM, Fojo SS (1988) Apolipoproteins and lipoproteins in human plasma: an overview. Clin Chem 34:4–8

Brewer HB Jr, Gregg RE, Hoeg JM (1989) Apolipoproteins, lipoproteins, and atherosclerosis. In: Braunwald E (ed) Heart disease: a textbook of cardiovascular medicine, 3rd edn. Saunders, New York, pp 121–144

Chen SH, Habib G, Yang CY, Gu ZW, Lee BR, Weng SA, Silberman SR, Cai SJ, Deslypere JP, Rosseneu M, Gotto AM Jr, Li WH, Chan L (1987) Apolipoprotein B-48 is the product of a messenger RNA with an organ-specific in-frame stop codon. Science 238:363–366

Cheung MC, Albers JJ (1984) Characterization of lipoprotein particles isolated by immunoaffinity chromatography. Particles containing A-I and A-II and particles containing A-I but no A-II. J Biol Chem 259:12201–12209

Davignon J, Gregg RE, Sing CF (1988) Apolipoprotein E polymorphism and atherosclerosis. Arteriosclerosis 8:1–21

Dawson PA, Hofman SL, Van der Westhuyzen DR, Sudhof TC, Brown MS, Goldstein JL (1988) Sterol-dependent repression of low density lipoprotein receptor promoter mediated by 16-base pair sequence adjacent to binding site for transcription factor Sp1. J Biol Chem 263:3372–3379

Deeb SS, Cheung MC, Peng R, Wolf AC, Stern R, Albers JJ, Knopp RH (1991) A mutation in the human apolipoprotein A-I gene. Dominant effect on the level and characteristics of plasma high density lipoproteins. J Biol Chem 266:13654–13660

Eaton DL, Fless GM, Kohr WJ, McLean JW, Xu QT, Miller CG, Lawn RM, Scanu AM (1987) Partial amino acid sequence of apolipoprotein(a) shows that it is homologous to plasminogen. Proc Natl Acad Sci USA 84:3224–3228

Edge SB, Hoeg JM, Triche T, Schneider PD, Brewer HB Jr (1986) Cultured human hepatocytes. Evidence for metabolism of low density lipoproteins by a pathway independent of the classical low density lipoprotein receptor. J Biol Chem 261:3800–3806

Eisenberg S (1984) High density lipoprotein metabolism. J Lipid Res 25:1017–1058

Fielding CJ, Shore VG, Fielding PE (1972) A protein cofactor of lecithin: cholesterol acyltransferase. Biochem Biophys Res Commun 46:1493–1498

Fless GM, Rolih CA, Scanu AM (1984) Heterogeneity of human plasma lipoprotein (a). Isolation and characterization of the lipoprotein subspecies and their apoproteins. J Biol Chem 259:11470–11478

Gaubatz JW, Heideman C, Gotto AM Jr, Morrisett JD, Dahlen GH (1983) Human plasma lipoprotein [a]. Structural properties. J Biol Chem 258:4582–4589

Glomset JA (1968) The plasma lecithin: cholesterol acyltransferase reaction. J Lipid Res 9:155–167

Glomset JA, Janssen ET, Kennedy R, Dobbins J (1966) Role of plasma lecithin: cholesterol acyltransferase in the metabolism of high density lipoproteins. J Lipid Res 7:638–648

Gofman JW, deLalla O, Glazier F et al. (1954) The serum lipid transport system in health, metabolic disorders, atherosclerosis, and coronary artery disease. Plasma 2:413–484

Goldstein JL, Brown MS (1979) The LDL receptor locus and the genetics of familial hypercholesterolemia. Annu Rev Genet 13:259–289

Goldstein JL, Brown MS, Anderson RG, Russell DW, Schneider WJ (1985) Receptor-mediated endocytosis: concepts emerging from the LDL receptor system. Annu Rev Cell Biol 1:1–39

Gordon DJ, Rifkind BM (1989) High-density lipoprotein – the clinical limplications of recent studies. N Engl J Med 321:1311–1316

Gordon T, Castelli WP, Hjortland MC, Kannel WB, Dawber TR (1977) High density lipoprotein as a protective factor against coronary heart disease. The Framingham study. Am J Med 63:707–714

Gregg RE, Brewer HB Jr (1988) The role of apolipoprotein E and lipoprotein receptors in modulating the in vivo metabolism of apolipoprotein B-containing lipoproteins in humans. Clin Chem 34:28–32

Gregg RE, Zech LA, Schaefer EJ, Brewer HB Jr (1981) Type III hyperlipoproteinemia: defective metabolism of an abnormal apolipoprotein E. Science 211: 584–586

Haberland ME, Fogelman AM (1987) The role of altered lipoproteins in the pathogenesis of atherosclerosis. Am Heart J 113:573–57

Havel RJ (1982) Familial dysbetalipoproteinemia. New aspects of pathogenesis and diagnosis. Med Clin North Am 66:441–454

Havel RJ, Shore VG, Shore B, Bier DM (1970) Role of specific glycopeptides of human serum lipoproteins in the activation of lipoprotein lipase. Circ Res 27:595–600

Herbert PN, Gotto AM Jr, Frederickson DS (1978) Familial lipoprotein deficiency (abetalipoproteinemia, hypobetalipoproteinemia, and Tangier disease). In: Stanberg JB, Wyngaarden JB, Frederickson DS (eds) Metabolic basis of inherited disease. McGraw-Hill, New York, pp 544–588

Herz J, Hamann U, Rogne S, Myklebos O, Gausepohl H, Stanley KK (1988) Surface location and high affinity for calcium of a 500 kDa liver membrane protein closely related to the LDL receptor suggest a physiological role as a lipoprotein receptor. EMBO J 7:4119–4127

Higuchi K, Hospattankar AV, Law SW, Meglin N, Cortright J, Brewer HB Jr (1988) Human apolipoprotein B (apoB) mRNA: identification of two distinct apoB mRNAs, an mRNA with the apoB-100 sequence and an apoB mRNA containing a premature in-frame translational stop codon, in both liver and intestine. Proc Natl Acad Sci USA 85:1772–1776

Hoeg JM, Edge SB, Demosky SJ Jr, Starzl TE, Triche T, Gregg RE, Brewer HB Jr (1986) Metabolism of low-density lipoproteins by cultured hepatocytes from normal and homozygous familial hypercholesterolemic subjects. Biochim Biophys Acta 876:646–657

Hospattankar AV, Fairwell T, Ronan R, Brewer HB Jr (1984) Amino acid sequence of human plasma apolipoprotein C-II from normal and hyperlipoproteinemic subjects. J Biol Chem 259:318–322

Hospattankar AV, Higuchi K, Law SW, Meglin N, Brewer HB Jr (1987) Identification of a novel in-frame translational stop codon in human intestine apoB mRNA. Biochem Biophys Res Commun 148:279–285

Jackson RL, Baker HN, Gilliam EB, Gotto AM Jr (1977) Primary structure of very low density apolipoprotein C-II of human plasma. Proc Natl Acad Sci USA 74:1942–1945

Jahn CE, Osborne JC Jr, Schaefer EJ, Brewer HB Jr (1983) Activation of the enzymic activity of hepatic lipase by apolipoprotein A-II. Characterization of a major component of high density lipoprotein as the activating plasma component in vitro. Eur J Biochem 131:25–29

Kane JP (1983) Apolipoprotein B: structural and metabolic heterogeneity. Annu Rev Physiol 45:637–650

Kodama T, Freeman M, Rohrer L, Zabrecky J, Matsudaira P, Krieger M (1990) Type I macrophage scavenger receptor contains alpha-helical and collagen-like coiled coils. Nature 343:531–535

Koschinsky ML, Beisiegel U, Henne-Bruns D, Eaton DL, Lawn RM (1990) Apolipoprotein(a) size heterogeneity is related to variable number of repeat sequences in its mRNA. Biochemistry 29:640–644

Kostner G, Alaupovic P (1972) Studies of the composition and structure of plasma lipoproteins. Separation and quantification of the lipoprotein families occurring in the high density lipoproteins of human plasma. Biochemistry 11:3419–3428

Kostner GM, Avagaro P, Zazzolato G, Marth E, Bittolo-Bon G, Quinci GB (1981) Lipoprotein Lp(a) and the risk for myocardial infarction. Arteriosclerosis 38:51–61

Lackner KJ, Monge JC, Gregg RE, Hoeg JM, Triche TJ, Law SW, Brewer HB Jr (1986) Analysis of the apolipoprotein B gene and messenger ribonucleic acid in abetalipoproteinemia. J Clin Invest 78:1707–1712

LaRosa JC, Levy RI, Herbert P, Lux SE, Fredrickson DS (1970) A specific apoprotein activator for lipoprotein lipase. Biochem Biophys Res Commun 41:57–62

Lee RS, Hatch RT (1963) Sharper separation of lipoprotein species by paper electrophoresis in albumin containing buffer. J Lab Clin Med 61:518–528

Leppert M, Breslow JL, Wu L, Hasstedt S, O'Connell P, Lathrop M, Williams RR, White R, Lalouel JM (1988) Inference of a molecular defect of apolipoprotein B in hypobetalipoproteinemia by linkage analysis in a large kindred. J Clin Invest 82:847–851

Li WH, Tanimura M, Luo CC, Datta S, Chan L (1988) The apolipoprotein multigene family: biosynthesis, structure, structure–function relationships, and evolution. J Lipid Res 29:245–271

Loscalzo J (1990) Lipoprotein(a). A unique risk factor for atherothrombotic disease. Arteriosclerosis 10:672–679

Mahley RW (1979) Dietary, fat, cholesterol, and accelerated atherosclerosis. Atherosclerosis Rev 5:1–34

Mahley RW, Hui DY, Innerarity TL, Weisgraber KH (1981) Two independent lipoprotein receptors on hepatic membranes of dog, swine, and man. Apo-B,E and apo-E receptors. J Clin Invest 68:1197–1206

Mahley RW, Innerarity TL, Rall SC Jr, Weisgraber KH (1984) Plasma lipoproteins: apolipoprotein structure and function. J Lipid Res 25:1277–1294

McKnight GL, Reasoner J, Gilbert T, Sundquist KO, Hokland B, McKernan PA, Champagne J, Johnson CJ, Bailey MC, Holly R, O'Hara PJ, Oram JF (1992) Cloning and expression of a cellular high density lipoprotein-binding protein that is up-regulated by cholesterol loading of cells. J Biol Chem 267:12131–12141

McLean JW, Tomlinson JE, Kuang WJ, Eaton DL, Chen EY, Fless GM, Scanu AM, Lawn RM (1987) cDNA sequence of human apolipoprotein(a) is homologous to plasminogen. Nature 330:132–137

Mendez AJ, Oram JF, Bierman EL (1991) Protein kinase C as a mediator of high density lipoprotein receptor-dependent efflux of intracellular cholesterol. J Biol Chem 266:10104–10111

Miles LA, Plow EF (1990) Lp(a): an interloper in the fibrinolytic system. Thromb Haemost 63:331–335

Miller GJ, Miller NE (1975) Plasma-high-density-lipoprotein concentration and development of ischaemic heart disease. Lancet 1:16–19

Miller NE, Thelle DS, Forde OH, Mjos OD (1977) The Tromso heart-study: high-density lipoproteins and coronary heart-disease: a prospective case-control study. Lancet 1:965–968

Mowri H-O, Patsch W, Smith LC, Gotto AM Jr, Patsch JR (1990) Different reactivities of HDL2 subfractions with hepatic lipase. Circulation 82:558

Nestruck AC, Niedmann PD, Wieland H, Seidel D (1983) Chromatofocusing of human high density lipoproteins and isolation of lipoproteins A and A-I. Biochim Biophys Acta 753:65–73

Norum RA, Lakier JB, Goldstein S, Angel A, Goldberg RB, Block WD, Noffze DK, Dolphin PJ, Edelglass J, Bogorad DD, Alaupovic P (1982) Familial deficiency of apolipoproteins A-I and C-III and precocious coronary-artery disease. N Engl J Med 306:1513–1519

Oram JF, Brinton EA, Bierman EL (1983) Regulation of high density lipoprotein receptor activity in cultured human skin fibroblasts and human arterial smooth muscle cells. J Clin Invest 72:1611–1621

Osborne JC Jr, Brewer HB Jr (1977) The plasma lipoproteins. Adv Protein Chem 31:253–337

Powell LM, Wallis SC, Pease RJ, Edwards YH, Knott TJ, Scott J (1987) A novel form of tissue-specific RNA processing produces apolipoprotein-B48 in intestine. Cell 50:831–840

Puchois P, Kandoussi A, Fievet P, Fourrier JL, Bertrand M, Koren E, Fruchart JC (1987) Apolipoprotein A-I containing lipoproteins in coronary artery disease. Atherosclerosis 68:35–40

Rader DJ, Castro G, Zech LA, Fruchart JC, Brewer HB Jr (1991) in vivo metabolism of apolipoprotein A-I on high density lipoprotein particles LpA-I and LpA-I, A-II. J Lipid Res 32:1849–1859

Rall SC Jr, Weisgraber KH, Mahley RW (1982) Human apolipoprotein E. The complete amino acid sequence. J Biol Chem 257:4171–4178

Ross RS, Gregg RE, Law SW, Monge JC, Grant SM, Higuchi K, Triche TJ, Jefferson J, Brewer HB Jr (1988) Homozygous hypobetalipoproteinemia: a disease distinct from abetalipoproteinemia at the molecular level. J Clin Invest 81:590–595

Russell DW, Esser V, Hobbs HH (1989) Molecular basis of familial hypercholesterolemia. Arteriosclerosis [Suppl I] 9:1–8

Scanu AM, Landsberger FR (1980) Lipoprotein structure. Ann NY Acad Sci 384: 1–436

Schaefer EJ, Ordovas JM, Law SW, Ghiselli G, Kashyap ML, Srivastava LS, Heaton WH, Albers JJ, Connor WE, Lindgren FT, Lemeshev Y, Segrest JP, Brewer HB Jr (1985) Familial apolipoprotein A-I and C-III deficiency, variant II. J Lipid Res 26:1089–1101

Schmitz G, Lackner K (1989) High density lipoprotein deficiency with xanthomas: a defect in apoA-I synthesis. In: Crepaldi G, Baggio G (eds) Atherosclerosis VIII. Tekno, Rome, pp 399–403

Schmitz G, Niemann R, Brennhausen B, Krause R, Assmann G (1985) Regulation of high density lipoprotein receptors in cultured macrophages: role of acyl-CoA: cholesterol acyltransferase. EMBO J 4:2773–2779

Schultz JR, Verstuyft JG, Gong EL, Nichols AV, Rubin EM (1992) ApoAI and apoAI + apoAII trangenic mice: a comparison of atherosclerotic susceptibility. Circulation 86:I-472

Steinberg D (1983) Lipoproteins and atherosclerosis. A look back and a look ahead. Arteriosclerosis 3:283–301

Steinberg D (1991) Antioxidants and atherosclerosis: a current assessment. Circulation 84:1420–1425

Strickland DK, Ashcom JD, Williams S, Burgess WH, Migliorini M, Argraves WS (1990) Sequence identity between alpha2-macroglobulin receptor and low density lipoprotein receptor-related protein suggests that this molecule is a multifunctional receptor. J Biol Chem 265:17401–17404

Sudhof TC, Goldstein JL, Brown MS, Russell DW (1985) The LDL receptor gene: a mosaic of exons shared with different proteins. Science 228:815
Suzuki N, Fidge N, Nestel P, Yin J (1983) Interaction of serum lipoproteins with the intestine. Evidence for specific high density lipoprotein-binding sites on isolated rat intestinal mucosal cells. J Lipid Res 24:253–264
Tall AR (1986) Plasma lipid transfer proteins. J Lipid Res 27:361–367
Theret N, Delbart C, Aguie G, Fruchart JC, Vassaux G, Ailhaud G (1990) Cholesterol efflux from adipose cells is coupled to diacylglycerol production and protein kinase C activation. Biochem Biophys Res Commun 173:1361–1368
Utermann G (1989) The mysteries of lipoprotein(a). Science 246:904–910
Utermann G, Weber W (1983) Protein composition of Lp(a) lipoprotein from human plasma. FEBS Lett 154:357–361
Utermann G, Vogelberg KH, Steinmetz A, Schoenborn W, Pruin N, Jaeschke M, Hees M, Canzler H (1979) Polymorphism of apolipoprotein E: II. Genetics of hyperlipoproteinemia type III. Clin Genet 15:37–62
Utermann G, Menzel HJ, Kraft HG, Duba HC, Kemmler HG, Seitz C (1987) Lp(a) glycoprotein phenotypes. Inheritance and relation to Lp(a)-lipoprotein concentrations in plasma. J Clin Invest 80:458–465
Van der Westhuyzen DR, Fourie AM, Coetzee GA, Gevers W (1990) The LDL receptor. Curr Opin Lipidol 1:128–135
Van Lenten BJ, Fogelman AM (1990) Processing of lipoproteins in human monocyte-macrophages. J Lipid Res 31:1455–1466
Vega GL, Denke MA, Grundy SM (1991) Metabolic basis of primary hypercholesterolemia. Circulation 84:118–128
Wetterau JR, Aggerbeck LP, Bouma ME, Eisenberg C, Munck A, Hermier M, Schmitz J, Gay G, Rader DJ, Gregg RE (1992) Absence of microsomal triglyceride transfer protein in individuals with abetalipoproteinemia. Science 258: 999–1001
Yamamoto T, Davis CG, Brown MS, Schneider WJ, Casey ML, Goldstein JL, Russell DW (1984) The human LDL receptor: a cysteine-rich protein with multiple Alu sequences in its mRNA. Cell 39:27–38
Zannis VI, Just PW, Breslow JL (1981) Human apolipoprotein E isoprotein subclasses are genetically determined. Am J Hum Genet 33:11–24

CHAPTER 3

Lipoprotein Receptors

W.J.S. de Villiers, G.A. Coetzee, and D.R. van der Westhuyzen

A. Introduction

Polypeptide-binding receptors in mammalian plasma membranes may be divided into two categories, class I and class II receptors (Kaplan 1981). Class II receptors mediate the cellular uptake and delivery of ligands to intracellular sites, generally to lysosomes for degradation, and then recycle to the cell surface. In contrast, class I receptors are characterized by their ability to mediate signal transduction. Lipoprotein receptors fall mainly into the class II category. Since the low-density lipoprotein (LDL) receptor was cloned and sequenced in 1984, the structures of two other receptors have also been determined.

Lipoprotein receptors are cell surface proteins that bind lipoproteins with high affinity and specificity via interactions with certain apolipoproteins, resulting in a physiological response that affects both the ligand and the cell. These receptors span the cell membrane and are strategically placed to influence both intracellular lipid and extracellular lipoprotein levels. This dual-purpose function of lipoprotein receptors has somewhat contradictory consequences; for example, excess influx of cholesterol (via synthesis and diet) into the body downregulates LDL receptors by intracellular sterol control mechanisms. This, however, leads to overaccumulation of extracellular LDL and increased atherosclerosis. A better understanding of the molecular and cell biology of lipoprotein receptors will lead to improved therapeutic strategies for the management of hyperlipidaemic patients.

In this chapter, we describe the existing knowledge about four types of lipoprotein receptors. These are the receptors for LDL, chylomicron remnants, and high-density lipoproteins (HDL) as well as the so-called scavenger receptor. Recent reviews include those for the LDL receptor (Hobbs et al. 1990; Soutar and Knight 1990), the chylomicron remnant receptor (Brown et al. 1991; Mahley and Hussain 1991), the HDL receptor (Oram 1990), and the scavenger receptor (Kurihara et al. 1991).

B. The Low-Density Lipoprotein Receptor

The LDL receptor is an oligomeric cell surface glycoprotein which mediates the clearance of plasma LDL through receptor-mediated endocytosis. It was discovered in the early 1970s and then characterized by M.S. Brown and J.L. Goldstein, who received the 1985 Nobel Prize for Physiology or Medicine for their studies (Brown and Goldstein 1986). The LDL receptor can be regarded as the single most important element in the regulation of plasma cholesterol levels, since it mediates the endocytosis of both LDL, the major cholesterol-carrying lipoprotein in human plasma, as well as certain lipoprotein precursors of LDL. LDL is taken up by receptors mainly in the liver, but also by all peripheral cells. Numerous mutations in the LDL receptor gene (both those occurring naturally and those artificially constructed) disrupt normal receptor expression and have contributed in a major way to our understanding of the role of LDL receptors in vivo and of the structure–function relationships of this molecule. LDL receptor expression is controlled at the transcriptional level by intracellular cholesterol levels, and progress has been made in understanding the mechanisms of this regulation.

I. Functions of the Low-Density Lipoprotein Receptor in the Body

The main role of LDL receptors is to mediate LDL clearance from the circulation. This is best illustrated by the consequences of LDL receptor mutations (see Sect. B.IV), which cause familial hypercholesterolaemia (FH). In this disease, hypercholesterolaemia reaches up to two to three times and four to six times the normal levels in FH heterozygotes and homozygotes, respectively (Goldstein and Brown 1989). Numerous studies (reviewed by Brown and Goldstein 1986) have indicated that normally about two-thirds of LDL clearance is mediated through the LDL receptor. Hepatic LDL receptors specifically are responsible for more than half of LDL clearance. Furthermore, in addition to clearing LDL more slowly than normal, FH patients also appear to overproduce LDL (Simons et al. 1975; Bilheimer et al. 1975). The reason for this became apparent when Brown and Goldstein (1986) showed that LDL receptors also normally clear lipoprotein precursors of LDL (intermediate density lipoprotein, IDL) from the circulation. This clearance depends on multiple copies of apoE, another apolipoprotein ligand of the LDL receptor, being present on the IDL particle. Defects in LDL receptors therefore result in the accumulation of IDL particles, which are then available for conversion into LDL.

Recently, it has become clear that LDL receptors also contribute directly to the conversion of IDL to LDL. The conversion of some very low density lipoprotein (VLDL) particles to IDL apparently also involves LDL receptors (Packard et al. 1990), but the mechanism(s) remains unknown. LDL receptors are thought to play a role in VLDL and IDL lipolysis by binding and orienting lipoproteins on the cell surface, allowing efficient access to lipases (Packard et al. 1990).

II. Structure of the Low-Density Lipoprotein Receptor

The LDL receptor is encoded by a single-copy gene present on the distal short arm of chromosome 19 (SÜDHOF et al. 1985a). The gene spans 45 kb of DNA and consists of 18 exons. There is a close correlation between the arrangement of the different exons and the various domains of the receptor protein, indicating that the LDL receptor was assembled in evolution by exon shuffling. The LDL receptor belongs to a family of LDL-receptor-like proteins, of which the LDL receptor related protein (LRP) (see Sect. C) and the Heymann nephritis autoantigen GP330 (RAYCHOWDHURY et al. 1989) are two other members. The 860 amino acid LDL receptor is translated from a 5.3-kb mRNA, about half of which corresponds to an unusually long 3′- untranslated region (YAMAMOTO et al. 1984).

1. Domains

The LDL receptor contains at least five independent domains which have been defined by structural and functional studies (BROWN and GOLDSTEIN 1986; HOBBS et al. 1990). The LDL receptor has not yet been crystallized, however, and its three-dimensional structure remains unknown. A number of regions within these domains have structural homology to sequences found in other proteins, making the LDL receptor a mosaic of different structural protein motifs. The five domains are: (1) the ligand-binding domain, (2) the epidermal growth factor (EGF) precursor-like domain, (3) the O-linked sugar domain, (4) the membrane-spanning domain and (5) the cytoplasmic "tail" domain (Fig. 1).

The ligand-binding domain at the amino end of the receptor is encoded by exons 2–6 (exon 1 encodes a typical cleavable signal peptide). The domain comprises seven repeat sequences of approximately 40 amino acids each. These repeats are homologous to a sequence found in several complement proteins (MARAZZITI et al. 1988; SÜDHOF et al. 1985a). All the repeats are rich in cysteine, with the six highly conserved cysteines of each repeat forming three intrarepeat disulphide bonds. The carboxyl end of each repeat contains a cluster of negatively charged amino acids thought to be involved in the binding of lipoproteins via positively charged amino acids on apoB-100 or apoE. Ligand binding requires the presence of a divalent cation, preferably Ca^{2+}, which probably binds to each one of the repeats (VAN DRIEL et al. 1987a).

The binding domain has been investigated using artificial receptor mutants constructed by site-directed mutagenesis (ESSER et al. 1988; RUSSELL et al. 1989). The repeats appear to form a modular structure in which different repeat combinations are utilized for the binding of apoB-100 and apoE. Each LDL receptor binds a single LDL particle via apoB-100. Repeats 3–7 are critical for this binding, as is repeat A of the EGF homology domain. The binding site on apoB-100 is not yet accurately defined. In

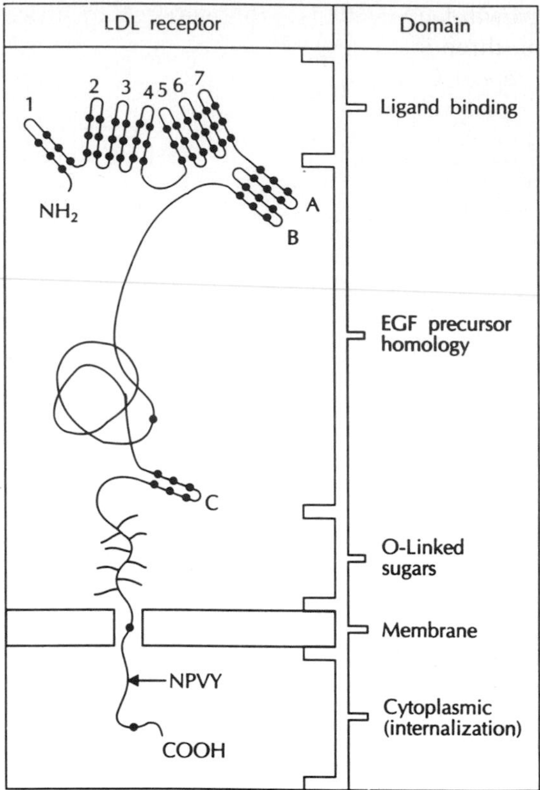

Fig. 1. A schematic illustration of the domain structure of the low-density lipoprotein (*LDL*) receptor. *A*, *B*, *C*, cysteine-rich repeats; *NPVY*, a short sequence necessary for internalization; NH_2, N terminal; *COOH*, C terminal; *EGF*, epidermal growth factor (Innerarity 1991), (from Hobbs et al. 1990, with permission)

contrast, only repeat 5 appears to be directly involved in the binding of apoE. In the known tertiary structure of apoE, the proposed binding sequence (amino acids 130–150) lies appropriately on an exposed face of the apoE molecule (Wilson et al. 1991). Certain lipoprotein particles, such as HDL_c and β-VLDL, are enriched with multiple copies of apoE. These particles apparently interact with a cluster of LDL receptors through multiple apoE-receptor contacts, accounting for a higher affinity (but lower capacity) of receptor binding to these particles than to LDL (Pitas et al. 1979). It is known that LDL receptors can exist as dimers, trimers or larger multimers in the cell membrane (Van Driel et al. 1987b), and they probably also function as multimers during endocytosis and recycling (Basu et al. 1981; Grant et al. 1990).

The EGF-precursor-like domain is a 400 amino acid domain which resembles a portion of the precursor of EGF (Südhof et al. 1985a,b). It includes three growth-factor-like repeats (distinct from the binding repeats) that each contain three intrarepeat disulphide bonds. Repeats A and B are separated from repeat C by a sequence of 280 amino acids which contains five copies of a conserved Tyr-Trp-Thr-Asp motif. In addition to its role in LDL binding, this domain also appears to be responsible for reversible structural changes which allow the acid-dependent dissociation of receptor-ligand complexes, a property necessary for the separation of receptor and ligand in endosomes and receptor recyling (DAVIS et al. 1987a).

The O-linked sugar domain, encoded by exon 15, is 58 amino acids in length and enriched in O-linked glycosylated serine and threonine residues. The role of this region remains obscure. Possibly it functions as a "stalk" to correctly position the other extracellular domains on the cell surface. Exons 16 and part of 17 encode a hydrophobic membrane-spanning sequence. The C-terminal cytoplasmic domain of 50 amino acids is encoded by part of exons 17 and 18. This cytoplasmic "tail" interacts with components of clathrin-coated pits on the cell membrane (PEARSE 1988), thereby participating in surface clustering. The interaction is through a structural motif which is necessary for receptor internalization and which is similar to sequences found in other cell surface receptors (DAVIS et al. 1987b; CHEN et al. 1990).

The LDL receptor contains both N- and O-linked oligosaccharides. The exact role of these sugar chains in LDL receptor function is unclear (reviewed by KUWANO et al. 1991). About one-third of the O-linked sugars are found outside the so-called O-linked sugar domain, and SEGUCHI et al. (1991) have demonstrated that these are crucial for high-affinity binding of LDL. These sugars probably play a role in the correct folding of the different binding repeats of the receptor.

2. Evolution

The primary structure of the LDL receptor is remarkably well-conserved between species, with a particularly high degree of homology in the ligand binding, EGF precursor homology and cytoplasmic domains (reviewed by INNERARITY 1991). For example, the cytoplasmic domain of the frog is 86% identical to its human counterpart (MEHTA et al. 1991a). Also, the three repeats found in the 5′-flanking promoter region (see Sect. B.V.1) of the frog LDL receptor gene are virtually identical to the repeats that mediate sterol-regulated transcription of the human gene (MEHTA et al. 1991b). The LDL receptor and its gene, therefore, probably functioned in a manner similar to present-day LDL receptors at least since the time of amphibian evolution more than 350 million years ago. On the other hand, receptor-mediated clearance of LDL in pigeons is not downregulated by a high cholesterol diet, indicating that the pigeon LDL receptor may have lost sterol control, possibly through a lack of sterol-regulatory elements or through mutation of these elements (REAGAN et al. 1990).

Chickens, interestingly, have evolved two forms of the LDL receptor. Chickens have a typical 130-kDa mammalian-type LDL receptor which is present in somatic cells. However, another distinct receptor mediates the massive uptake of hepatically synthesized vitellogenin and triglyceride-rich VLDL into the oocyte during yolk formation (BARBER et al. 1991). This 95-kDa receptor is immunologically related, and has some sequence similarity, to the LDL receptor.

3. Low-Density Lipoprotein Receptor Polymorphisms

Alleles at the LDL receptor locus can be identified by haplotype analysis using known restriction fragment length polymorphisms (RFLPs). At least 22 RFLPs and 123 haplotypes have so far been identified at this locus (HOBBS et al. 1990). Because of linkage disequilibrium between many RFLP sites, the analysis of only five selected sites is almost as informative as a more detailed anaysis of ten sites (heterozygosity index of 84% vs. 86%; LEITERSDORF et al. 1989a). Such haplotype analysis allows mutant alleles to be traced within families suffering from FH. Efforts to identify LDL receptor alleles that affect plasma cholesterol levels in the general population have not thus far been successful.

III. Low-Density Lipoprotein Receptor Endocytosis, Recycling and Turnover

LDL receptor synthesis takes place in the rough endoplasmic reticulum where mannose-rich N-linked and core O-linked carbohydrate chains are added, either co-translationally or soon after synthesis (CUMMINGS et al. 1983). Processing of newly synthesized receptors continues in the Golgi apparatus, leading to the formation of complex-type N-linked and terminally modified O-linked sugar chains. Maturation of receptors also involves disulphide bond formation, correct protein folding and probably receptor oligomerization.

Receptors on the cell surface bind and internalize available ligands by the process of receptor-mediated endocytosis (GOLDSTEIN et al. 1985). Receptors cluster into clathrin-coated pits on the cell surface from a dispersed population (Fig. 2) and are internalized together with bound ligand in endosomal vesicles. No evidence exists for ligand-induced clustering or endocytosis, and LDL receptors are thought to recycle constitutively. Receptors and ligands dissociate in endosomes as a result of endosomal acidification (DAVIS et al. 1987a). Ligands are delivered to lysosomes for degradation, which includes the hydrolysis of cholesterol esters of LDL. Cholesterol delivered in this way exerts a number of feedback controls which regulate the free cholesterol content of cells (GOLDSTEIN et al. 1985); increased levels of cellular cholesterol are associated with an inhibition of LDL receptor synthesis, an inhibition of 3-hydroxy-3-methylglutaryl

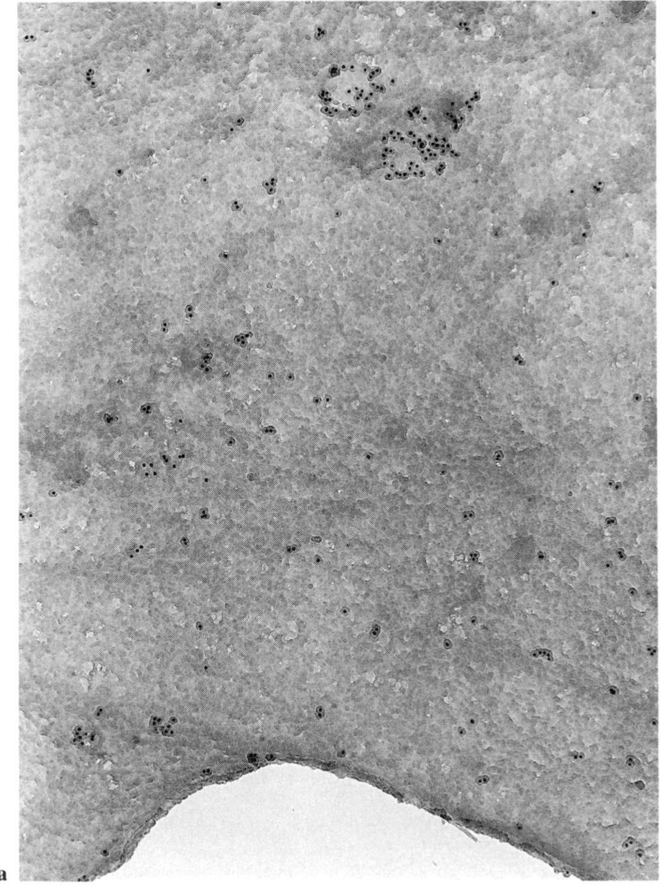

Fig. 2a,b. Electron micrographs showing clustered and dispersed low-density lipoprotein (LDL) receptors on the surfaces of cultured cells as probed with gold–LDL complexes. **a** LDL receptor clusters are generally found "inland" away from the "coastal" region near the edges of cells where dispersed receptors predominate; magnification, ×25,400. **b** (p. 60) Coated pits on the cell surface containing bound gold–LDL probes; magnification, ×65,000. The coated pit (*C*) is almost completely invaginated and communicates with the surface via a neck too narrow to admit gold–LDL probes (SANAN et al. 1987)

coenzyme A (HMG-CoA) reductase expression and an activation of a cholesterol-esterifying enzyme, acyl CoA cholesterol acyltransferase. In contrast to the ligands, LDL receptors recycle back from endosomes to the cell surface and mediate further rounds of endocytosis. The endocytic round trip of the receptor takes about 12 min (GOLDSTEIN et al. 1979a). The LDL receptor has a half-life of about 12 h in cultured fibroblasts (CASCIOLA et al. 1988), and on average, therefore, each receptor mediates about 60 endocytic cycles. LDL receptors are eventually degraded by unknown mechanisms and

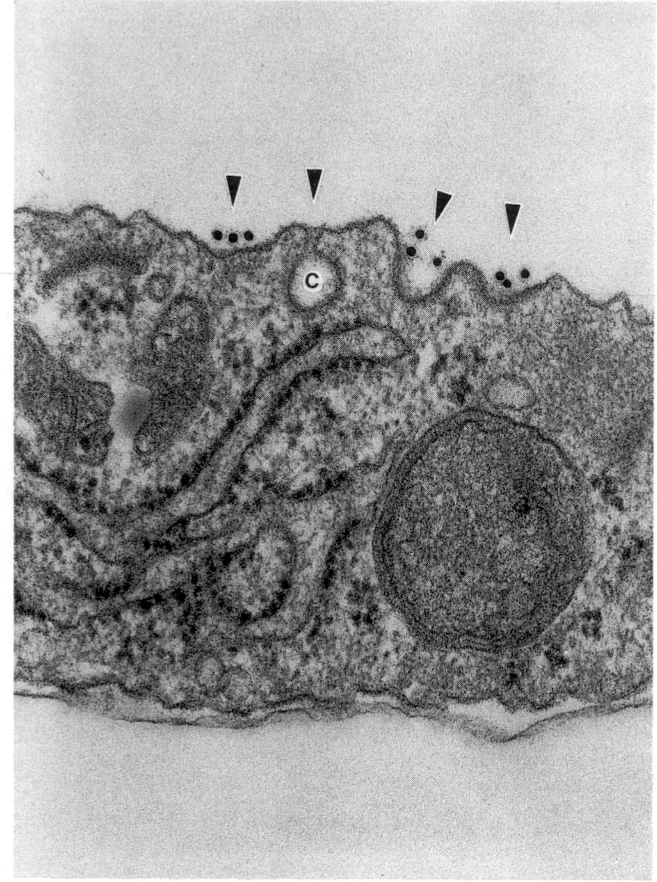

Fig. 2. *Continued*

at unknown sites that appear to be distinct from lysosomes (CASCIOLA et al. 1989).

IV. Low-Density Lipoprotein Receptor Mutations and Familial Hypercholesterolaemia

1. Classes of Low-Density Lipoprotein Receptor Mutations

FH is an autosomal dominant disorder caused by mutations in the LDL receptor gene. Many mutant alleles have been identified at the LDL receptor locus (reviewed by HOBBS et al. 1990) and more than 40 have been characterized to the extent that the DNA defect is known together with the LDL receptor protein malfunction. The human LDL receptor gene seems to be hypermutable and this is postulated to be due to (a) the presence at this locus of more than twice the average number of Alu repeats (reviewed by Hobbs et al. 1990), which increases the chances for recombination events, and (b) the presence of methylated CpG palindromes (Rideout et al. 1990), which increases the chances of cytosine to thymidine transitions via deamination of 5-methyl-cytosine. Of the 18 known point mutations, eight involve C to T transitions at CpG dinucleotides. Many families therefore have unique LDL receptor mutations. Exceptions to this general rule are found in certain inbred but expanded populations in which a high frequency of a particular LDL receptor mutation exists due to a founder effect. Such mutations are found among the French Canadians (Hobbs et al. 1987), South African Afrikaners (Leitersdorf et al. 1989b), Christian Lebanese (Lehrman et al. 1987), Finns (Aalto-Setala et al. 1989) and Ashkenazi Jews (Meiner et al. 1991).

Five classes of LDL receptor mutations have been defined on the basis of their phenotypic effects on the protein (Fig. 3). The mutations affect LDL receptor synthesis (class I), transport (class II), ligand binding (class III), internalization (class IV) and recycling (class V).

a) Class I Mutations (Null Alleles)

These result in a failure of LDL receptor synthesis for a number of different reasons. The FH French-Canadian-1 mutant allele has a deletion of the

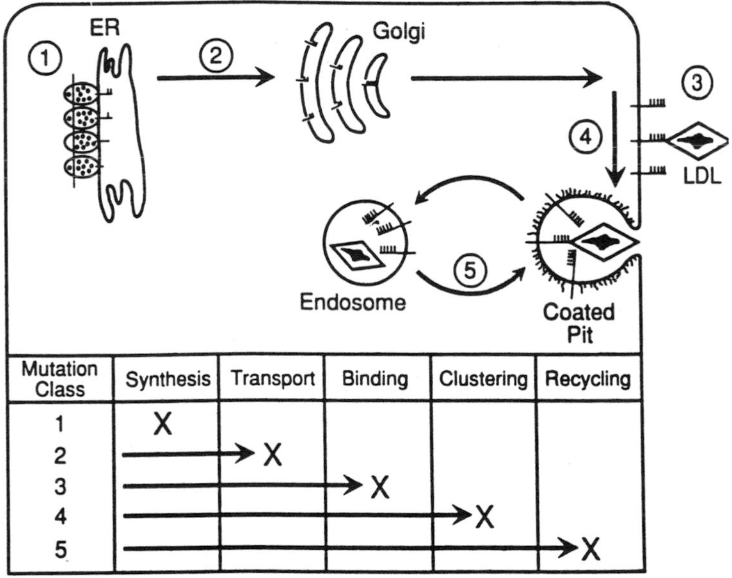

Fig. 3. Five classes of low-density lipoprotein (*LDL*) receptor mutations. These mutations affect receptor synthesis in the endoplasmic reticulum (*ER*), transport to the Golgi body, binding of ligands, clustering in coated pits and recycling in endosomes (from Hobbs et al. 1990)

promoter region and therefore fails to produce mRNA (HOBBS et al. 1987). In some cases, premature termination codons result in an absence of receptor, possibly due to unstable mRNA or enhanced protein degradation.

b) Class II Mutations (Transport-Defective Alleles)

These are the most common type and affect the processing and transport of receptors to the cell surface. In the so-called class IIA mutations, proteins are produced which fail completely to be transported. The truncated FH Lebanese LDL receptor (LEHRMAN et al. 1987) and the missense FH French-Canadian-2 mutation (LEITERSDORE et al. 1990) are examples. These mutations lie in the EGF precursor homology domain and presumably affect the correct folding of receptor molecules in such a way that the misfolded protein is detected and destroyed in the endoplasmic reticulum (ER). In class IIB defects, receptors exhibit slow but complete processing. Most of these mutations lie in the binding domain, including the founder FH Afrikaner-1 (LEITERSDORE et al. 1987b) and the FH Watanabe-heritable hyperlipidemic (WHHL) rabbit (YAMAMOTO et al. 1986) mutations. The ability of cells to recognize even small changes in protein conformation probably accounts for the high frequency of this type of defect.

c) Class III Mutations (Binding-Defective Alleles)

These lead to defective binding and usually occur in the binding domain, but also in the EGF precursor homology domain. The properties of these mutations have shed light on the structural requirements of the binding domain and the role of the different cysteine-rich repeats. The lower stringency for apoE binding accounts for certain mutant receptors being able to bind apoE-containing lipoproteins, but not LDL (see Sect. B.II.1).

d) Class IV Mutations (Internalization-Defective Alleles)

These prevent the clustering of LDL receptors into coated pits and consequently receptor and ligand uptake. These mutations all involve, through deletions or point mutations, the cytoplasmic tail of the receptor, which is known to carry signals necessary for internalization (CHEN et al. 1990).

e) Class V Mutations (Recycling-Defective Alleles)

These are all localized to the EGF precursor homology domain, which is necessary for both LDL binding and receptor recycling. These mutant receptors bind ligands efficiently but are unable to release them in response to the acidc conditions of the endosome. As a result, these receptors are unable to recycle back to the surface and are degraded rapidly (FOURIE et al. 1988; MIYAKE et al. 1989).

2. Clinical Variability in Familial Hypercholesterolaemia

To what extent can the different defects in mutant LDL receptors be correlated to the clinical severity of the disease? In general, receptor-negative status gives rise to more serious clinical features than does a receptor-defective phenotype. Patients with receptor alleles that produce some functional receptor activity are more responsive to therapy and show milder coronary atherosclerosis (GOLDSTEIN and BROWN 1989). The ability of some mutant receptors to bind apoE, but not apoB, may be one source of variability. These mutant receptors, although unable to bind LDL, can bing apoE-containing lipoproteins, including presumably IDL. As a result, they are associated with lower LDL levels than those found with receptors that bind neither apoB- nor apoE-containing ligands. The FH Denver-2 mutation is one such example and is associated with an absence of symptomatic coronary atherosclerosis (BILHEIMER et al. 1985). Clearly, other factors superimposed on the LDL receptor may affect the clinical phenotype, such as the proposed lipid-lowering gene (HOBBS et al. 1989) and the level (and isoform) of lipoprotein (a) (UTERMANN et al. 1989).

V. Low-Density Lipoprotein Receptor Regulation

1. Low-Density Lipoprotein Receptor Promoter

LDL receptor activity is regulated in response to the level of intracellular unesterified cholesterol (see BROWN and GOLDSTEIN 1986). Regulation is thought to occur solely through transcriptional control, which in turn controls the rate of receptor protein synthesis. There is no evidence for any mechanism of post-translational control of the LDL receptor, and conditions that cause marked changes in receptor expression also do not affect the rate of receptor degradation (CASCIOLA et al. 1988). Transcription of the LDL receptor gene is inhibited by free cholesterol or cholesterol delivered in lipoprotein particles. It seems as though as yet unidentified oxysterol products of cholesterol, and not cholesterol itself, mediate transcriptional control, but their exact nature is unclear.

The promoter region of the LDL receptor gene has been partially characterized by BROWN, GOLDSTEIN and coworkers (SÜDHOF et al. 1987; DAWSON et al. 1988; SMITH et al. 1990). All detectable *cis*-acting DNA sequences necessary for basal promoter activity as well as sterol regulation are located in a small region upstream of the transcription start site (Fig. 4). There are three direct imperfect repeats, each 16 base pairs in length, and a TATA-like sequence; each of these elements is essential for full promoter activity. Repeats 1 and 3 contain binding sites for the abundant positive transcription factor, Sp1. Repeat 2, which does not bind Sp1, contains an octamer sequence (5′-CACCCCAC-3) referred to as the sterol-regulatory element (SRE-1). A similar element is present in the promoters of HMG-

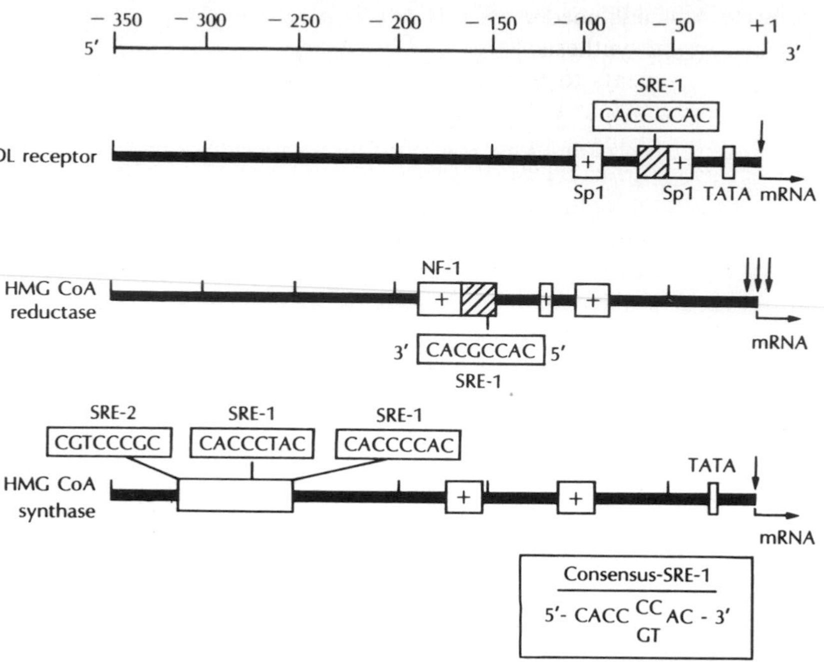

Fig. 4. Organization of transcription elements in the low-densitylipoprotein (*LDL*) receptor promoter and two other sterol-regulated promoters. The position of the sterol-regulatory element (*SRE-1*) in the LDL receptor is shown relative to the two *Sp1* binding sites. *HMG-CoA*, 3-hydroxymethylglutaryl coenzyme A (from Smith et al. 1988)

CoA reductase and HMG-CoA synthase, two other sterol-regulated genes (Smith et al. 1988).

How do these different elements promote and control gene expression? Evidence indicates that SRE-1 functions as a necessary positive element, in addition to repeats 1 and 3, but that sterols inactivate this element. The transacting factors that bind and regulate these *cis* DNA elements are unknown, although a cytosolic, oxysterol-binding protein has been identified (Taylor et al. 1984) and cloned (Dawson et al. 1989).

2. Control in the Liver

Numerous physiological conditions have been shown to affect LDL receptor activity (reviewed by Soutar and Knight 1990), and it has been suggested that LDL receptor transcription can be altered in ways that are not dependent on changes in the intracellular pool of regulatory sterols. LDL receptor regulation in the liver is of particular interest given the key role the liver plays in LDL clearance and its more complex requirements for cholesterol, which is needed for bile acid and lipoprotein production as well as for membrane synthesis.

The liver appears to be refractory to the suppression of LDL receptor activity by extracellular LDL and cholesterol (EDGE et al. 1986; HAVEKES et al. 1987), and changes in cholesterol synthesis appear to be the major response in liver to changing cholesterol requirements (SPADY et al. 1985). This suggests that either a separate intracellular regulatory pool of cholesterol exists or that sterol-independent mechanisms for receptor modulation are involved. Sterol-independent modulation of receptors has been indicated by a number of studies on cultured cells (GOLOS et al. 1987; FILIPOVIC and BUDDECKE 1986; CUTHBERT et al. 1989). In HepG2 cells, insulin increases LDL receptor mRNA levels and overcomes the suppressive effect of exogenous LDL (WADE et al. 1989). It is therefore possible that receptor expression is under the control both of intracellular cholesterol metabolites and of external signals operating through second-messenger systems. Dibutyryl cyclic AMP reduced LDL receptor activity in HepG2 cells (MAZIERE et al. 1988), although this effect may be regulated through free sterol levels, since cyclic AMP can stimulate neutral cholesterol esterase activity.

VI. Low-Density Lipoprotein Receptor and Therapy

1. Drug Therapy

Pharmaceutical agents that raise LDL receptor activity by increasing hepatic LDL receptor production should increase LDL clearance and lower plasma LDL cholesterol levels. Two classes of drugs have been developed which achieve this effect by increasing hepatic demands for cholesterol.

Bile acid sequestrants bind bile acids in the intestine and prevent their reabsorption. The subsequent increased conversion of cholesterol to bile acids depletes hepatic cells of cholesterol. This results in increased LDL receptor transcription and translation in hepatic cells, allowing hepatic cells to take up additional cholesterol. Plasma LDL levels decrease, but the effects are not profound (as is also found with the physiologically equivalent procedure of ileal bypass surgery) because the increase in cholesterol synthesis through induction of HMG-CoA reductase partially offsets the hepatic demand for cholesterol. Bile acid sequestrant therapy therefore lowers plasma LDL levels by only 10%–20% (GOLDSTEIN and BROWN 1989).

The HMG-CoA reductase inhibitors competitively inhibit this rate-limiting enzyme in the sterol biosynthetic pathway. The inhibition of hepatic cholesterol synthesis results in increased LDL receptor production as well as increased amounts of HMG-CoA reductase. Total body cholesterol is thus only slightly reduced, but plasma LDL cholesterol levels are lowered by 30%–40% because of the increase in LDL receptors (BILHEIMER et al.

1983). The combination of the two drug classes is more effective than either drug alone and reduces LDL cholesterol levels by 50%–60% (Mabuchi et al. 1983). These drugs may additionally reduce the hepatic production of cholesterol-rich lipoprotein precursors of LDL, which would also contribute to the lowering of plasma LDL levels.

2. Liver Transplantation

In FH heterozygotes, the drugs described earlier stimulate the expression of the LDL receptor from the normal receptor allele. This therapeutic approach is not effective in the treatment of FH homozygotes, especially those that show no functional LDL receptor activity. Portacaval anastomosis and chronic plasmapheresis are only partially effective in lowering plasma cholesterol levels in these patients, and plasmapheresis is technically difficult, time-consuming and costly (Goldstein and Brown 1989). A more direct approach is liver transplantation from a donor that expresses normal LDL receptor levels. Two FH homozygote patients that underwent this procedure subsequently showed normalization of their hyperlipidaemia (Bilheimer et al. 1984; Hoeg et al. 1987).

3. Gene Therapy

A potentially more effective and less morbid approach to the treatment of homozygous FH may be the replacement of LDL receptor function in vivo by somatic cell gene transfer (Wilson and Chowdhury 1990). Preliminary results from animal studies support the feasibility and efficacy of this approach. Overexpression of the LDL receptor in transgenic mice, for example, protects these animals from diet-induced hypercholesterolaemia (Yokode et al. 1990). The WHHL rabbit strain (Watanabe 1980) provides an experimental system to evaluate gene replacement therapy in vivo. Wilson et al. (1990) transfected recombinant retroviruses containing the normal LDL receptor gene into WHHL rabbit hepatocytes. Significant, although transient, reductions in plasma cholesterol levels occurred when the transfected hepatocytes were infused into the portal veins of the rabbits. Preliminary results in the development of gene transfer systems for the treatment of homozygous FH are therefore encouraging.

C. Chylomicron Remnant Receptor

Chylomicrons are the lipoprotein particles responsible for the transport of the bulk of dietary lipid. These large triglyceride-rich particles are secreted by the intestine and are then partially lipolysed to form remnant particles by the action of lipoprotein lipase in the capillary beds of the circulatory system. Remnants are cleared rapidly, primarily by receptors in the liver, and evidence now suggests that a surface receptor known as the LRP mediates this clearance.

I. Receptor-Mediated Clearance of Chylomicron Remnants

Chylomicron remnants are normally rapidly cleared from plasma, their half-life being in the order of a few minutes (GRUNDY and MOK 1976). This clearance occurs mainly in the liver and is dependent on the apolipoprotein composition of the particle (WINDLER et al. 1980) and on specific apoE isoforms (HAVEL et al. 1980). Although the liver is known to contain high-affinity binding sites for chylomicron remnants (SHERRILL et al. 1980), the identity of the receptor responsible for remnant uptake into hepatocytes has, until recently, remained elusive.

Chylomicron remnants can bind to the LDL-receptor. However, several lines of evidence have indicated that this receptor is not responsible for chylomicron clearance. The clearance of remnants in both the WHHL rabbit (KITA et al. 1982) and in persons homozygous for defective LDL receptors (RUBINSZTEIN et al. 1990) was found to be relatively normal, despite grossly impaired LDL clearance. In the FH homozygotes investigated, various LDL receptor mutations were present and these exhibited different degrees of functional activity when studied in cultured cells. Despite these variations in receptor activity, remnant clearance was found to be similar in the various FH patients. This argues for a non-LDL-receptor clearance mechanism and against the notion that the normal remnant clearance in those patients can be explained by an alternate clearance pathway that is employed by cells only in the absence of functional LDL receptors.

II. Low-Density Lipoprotein Receptor Related Protein: Candidate Chylomicron Remnant Receptor

The LRP is an integral membrane protein found in many animal cells and has been implicated as the chylomicron remnant receptor. This protein was identified by HERZ et al. (1988) during a search for proteins which contained a cysteine-rich motif found in certain complement proteins and in the LDL receptor. LRP, in fact, resembles a gigantic relative of the LDL receptor in which the extracellular domain of the LDL receptor is repeated approximately four times. Several properties of LRP, particularly its ability to bind apoE-containing lipoproteins, point to it being the chylomicron remnant receptor. It was realized more recently that LRP is identical to a protein previously characterized as being a receptor for another plasma protein, activated α_2-macroglobulin (STRICKLAND et al. 1990; KRISTENSEN et al. 1990). Possibly LRP has a dual function in being a receptor for both α_2-macroglobulin and apoE-containing lipoproteins.

III. Structure of Low-Density Lipoprotein Receptor Related Protein

LRP is encoded by a separate gene to that for the LDL receptor (HERZ et al. 1988). Although there is a high level of protein sequence similarity between LRP and the LDL receptor, less homology exists between these proteins at the nucleotide level. LRP is synthesized as a large, approximately 600-kDa polypeptide. The mRNA for LRP is approximately 15 kb in length and codes for 4544 amino acids. A 19 amino acid hydrophobic sequence at the N terminus corresponds to a cleavable signal peptide, while a hydrophobic region near the C terminus corresponds to the membrane-spanning segment of the polypeptide.

LRP has four approximate copies of the extracellular domain of the LDL receptor (Fig. 5). Like the LDL receptor, LRP is characterized by clusters of different types of cysteine-rich, short, repeating segments. The first of these corresponds to the binding-type repeats of the LDL receptor which are also found in a number of complement proteins (see Sect. B.II.1). Compared to the LDL receptor, which has seven such binding repeats, LRP has 31 in four separate clusters. The second type of cysteine-rich repeat is the growth factor repeat also found in the precursor molecule of EGF; three copies are present in LDL, whereas LRP has 22. A third type of repeat, not found in the LDL receptor, is an EGF-like repeat which is similar to the sequence of EGF itself. LRP contains six such repeats located just outside

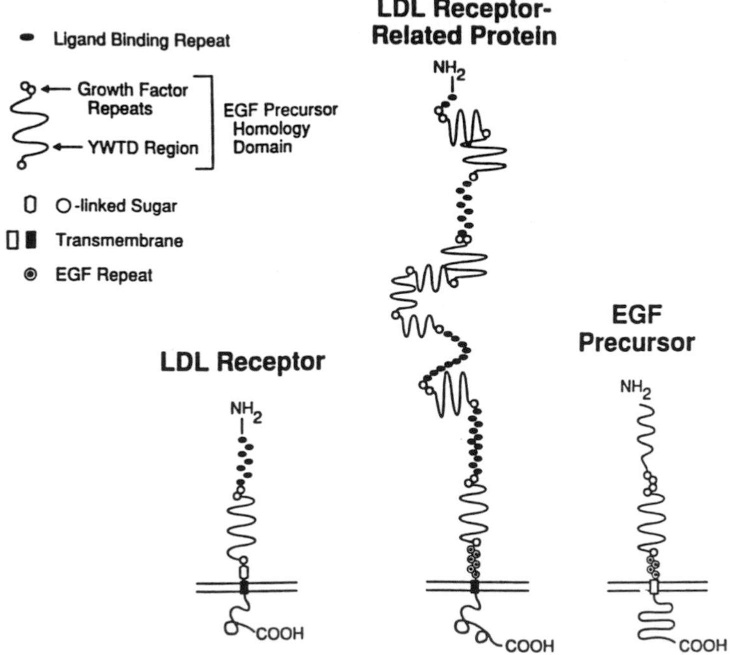

Fig. 5. Schematic representation of several members of the low-density lipoprotein (*LDL*) receptor supergene family, showing similarities in shared exons and transmembrane orientation. *EGF*, epidermal growth factor; *COOH*, C terminal; *YWTD*, a short conserved motif repeated once each 40–60 amino acids. (Modified from HOBBS et al. 1990)

the cell membrane in the place of the O-linked sugar domain of the LDL receptor.

The cytoplasmic tail of LRP is 100 amino acids in length, (about twice the size of the LDL receptor tail) and contains two copies of the sequence previously shown to be necessary for the clustering of LDL receptors into coated pits (CHEN et al. 1990). The high conservation of this region suggests that LRP functions in a receptor-mediated endocytic process. LRP is synthesized as a transmembrane glycoprotein referred to as LRP-600 on account of its apparent molecular weight on sodium dodecyl sulphide polyacrylamide gel electrophoresis (SDS PAGE). During its transport to the cell surface, the protein is proteolytically cleaved in the Golgi body into two subunits with apparent molecular masses of approximately 515 kDa and 85 kDa, respectively (HERZ et al. 1990a). The cysteine residues of the various cysteine-rich repeats are disulphide linked within each repeat. The proteolytic cleavage site is the peptide bond between the second arginine of a pair of arginines and a glutamine. This is a typical clipping site for proteins that are cleaved during secretion. The 515-kDa subunit lacks the transmembrane region, but remains attached to the cell surface through a stable association with the membrane-spanning 85-kDa subunit.

Recently, STIFANI et al. (1991) discovered that chickens synthesize two different proteins akin to mammalian LRP; one is found in the liver and corresponds closely to its mammalian counterpart, and the other is a smaller protein which is exclusively expressed in ovarian follicles. Chickens also have somatic and oocyte forms of the LDL receptor (BARBER et al. 1991). It therefore seems that, in oviparous species, at least four LDL-receptor-like proteins may have evolved to mediate lipoprotein clearance in somatic cells as well as lipid deposition in oocytes.

IV. Properties of Low-Density Lipoprotein Receptor Related Protein

1. Binding of Ligands

LRP interacts with lipoproteins enriched in apoE. This property is consistent with its proposed role in chylomicron remnant clearance, a process known to depend on apoE. KOWAL et al. (1990) demonstrated the LRP mediated the cellular uptake and lysosomal hydrolysis of cholesteryl esters from lipoproteins (β-VLDL) enriched in apoE. The binding of apoE (in apoE liposomes) to LRP on the surface of cells of a human hepatocyte line was also demonstrated by cross-linking experiments (BEISIEGEL et al. 1989). Like the LDL receptor, LRP binds calcium with high affinity and this divalent cation is required for ligand binding (KOWAL et al. 1990). The exact location of the ligand-binding site(s) on LRP is not known.

LRP is able to internalize bound ligands in a manner similar to the LDL receptor and other transport-type receptors (HERZ et al. 1990b) and is present in all the compartments involved in endocytosis (LUND et al. 1989).

The binding domain of LRP, like the corresponding domain in the LDL receptor, undergoes an acid-dependent conformational change in endosomes which releases the bound ligand for subsequent delivery to lysosomes and allows LRP to recycle back to the cell surface (HERZ et al. 1990b).

2. Regulation of Ligand Binding

Further evidence suggesting that LRP functions as the chylomicron remnant receptor is that the C apolipoproteins inhibit the binding of apoE-enriched β-VLDL to LRP (HUSSAIN et al. 1989). ApoC-I is known to inhibit the hepatic binding and uptake of triglyceride-rich or remnant lipoproteins (WINDLER et al. 1980) and exerts its effect at least partly by displacing apoE from lipoprotein particles (WEISGRABER et al. 1990).

Recent studies have identified additional potential regulators of LRP activity. The enzyme lipoprotein lipase (LPL), which is responsible for chylomicron lipolysis, has been shown to markedly increase the binding of chylomicrons to LRP (BEISIEGEL et al. 1991). This effect is not secondary to an enzymatic hydrolysis of lipoprotein particles by LPL. Another protein, 39 kDa in molecular mass, is known to associate with the 515-kDa subunit of LRP immediately after its synthesis and remains bound to the surface by virtue of this association. The 39-kDa has been shown to modulate the binding and uptake of ligand (HERZ et al. 1991), suggesting a potential role for this protein in vivo.

Modulation of receptor activity may be brought about by changes in lipoprotein ligands. The binding site on apoE for receptors is the α-helical region between amino acids 130–150, which is rich in positively charged amino acids (WILSON et al. 1991). Mutations in this region lead to dominantly inherited dysbetalipoproteinemia (also known as Type III hyperlipoproteinemia) and is characterized by the accumulation in plasma of chylomicron remnants (reviewed by MAHLEY et al. 1990). Mutations in apoE just outside this region sometimes also affect binding to receptors. The receptor-binding activity of the common apoE-2 isoform (Arg-158-Cys) is not always defective and can be modulated by a variety of conditions. As a result, this recessively inherited mutation therefore displays limited penetrance with respect to Type III hyperlipoproteinaemia, and the defective phenotype is revealed only under certain conditions (MAHLEY et al. 1990).

D. High-Density Lipoprotein Receptor

The HDL receptor is an HDL-binding cell surface protein postulated to mediate efflux of cholesterol from certain cells and endocytosis of HDL into other cells.

The HDL receptor is thought to play a pivotal role in reverse cholesterol transport. The process of reversible cholesterol transport, originally conceptualized by GLOMSET (1968), is responsible for the removal of excess cholesterol from extra-hepatic cells and for its subsequent transport to the liver for excretion from the body. In this process, HDL receptors have been proposed to mediate cholesterol movement between HDL and both extra-hepatic and hepatic cells. The importance of HDL in cholesterol homeostasis in the body is indicated by the inverse correlation between plasma HDL levels and risk for atherosclerosis (STEINBERG 1978), and it is generally accepted that HDL protects against overaccumulation of cholesterol in the arterial wall (REICHL and MILLER 1986).

The following lines of evidence support the existence of HDL receptors (reviewed by JOHNSON et al. 1991; ORAM 1990):

1. High-affinity binding sites for HDL exist on many cell types. Binding shows specificity for apoproteins A-I, A-II and A-IV, is inhibited by cross-linking of apoproteins and is independent of the LDL receptor. Fluctuations in cellular unesterified cholesterol have been shown to regulate the number of these sites. In extra-hepatic cells, HDL binding does not lead to cellular accumulation of HDL since the ligand is released, either by dissociation from the cell surface or following retro-endocytosis of internalized ligand. In liver cells, at least some of the bound HDL is endocytosed.
2. Several membrane proteins that bind HDL or apoA-I have been identified, but none of these have been characterized in detail. These proteins range from 58 to 120 kDa in size and have been reported in a variety of cell types. In contrast to these relatively high molecular weight proteins, the functional binding entities have been shown by the use of ionizing radiation to be molecules of very low molecular weights (10–16 kDa) (MENDEL et al. 1988).

I. Cholesterol Efflux

Cholesterol efflux from cells involves three steps: (1) hydrolysis of stored cholesteryl esters, (2) translocation of cholesterol from intracellular membranes to the surface plasma membrane and (3) transfer of cholesterol from the plasma membrane to acceptors such as HDL. The evidence for the physiological role of HDL receptors in mediating these events is, however, tenuous. It is generally accepted that the last step, the transfer from plasma membrane to HDL, does not depend on the binding of HDL to the cell surface (St CLAIR and LEIGHT 1983; MENDEL and KUNITAKE 1988; ORAM et al. 1991) and that cholesterol can diffuse passively through the unstirred water layer (PHILIPS et al. 1987). The participation of HDL binding in the second step, the translocation of cholesterol from intracellular to plasma membrane, is controversial. In some studies, the interaction of HDL with cells reportedly stimulated the translocation of intracellular unesterified cholesterol to the plasma membrane (SLOTTE et al. 1987; ORAM et al. 1991), and protein kinase C was implicated as a mediator of this process (MENDEZ et al. 1991). However, in other studies, the efflux of lysosomal cholesterol could not be linked to a specific binding of HDL to cells (Mahlberg et al. 1991). The translocation of intracellular cholesterol might possibly occur via different mechanisms under different conditions and in different cell types.

II. High-Density Lipoprotein Endocytosis

HDL receptors are thought to mediate the endocytosis of HDL in certain tissues such as the liver and gonads, but here again the identity and properties of such high-affinity receptors are unclear. HDL may either be taken up as a whole particle and its components degraded or its cholesteryl esters may be preferentially taken up by cells via a direct, non-endocytic transfer mechanism (Pitman et al. 1987a,b). In the former process, receptor-mediated HDL uptake and degradation might occur via a pronase-insensitive "binding site" (Bachorik et al. 1982), but the role of a specific HDL receptor remains undefined. In the latter process, specific HDL receptors appear not to be involved and cholesteryl esters are simply transferred passively from HDL into or through the plasma membrane of cells (Pitmann et al. 1987b).

It is likely that the clearance of HDL and its constituents, such as the apolipoproteins, involves a receptor-mediated process and that other lipoprotein receptors may play a role. For example, HDL that has acquired apoE can be cleared by the LDL receptor, while HDL that has been modified could be cleared by the scavenger receptor (see Sect. E). The questions that remain unanswered, therefore, are whether specific HDL receptors with a physiological role exist and whether the high-affinity binding sites identified represent these receptors.

E. The Scavenger Receptor

Although several risk factors have been implicated in the development of atherosclerosis, no unifying theory has emerged. The discovery and characterization of a "scavenger receptor" which is able to mediate the cellular uptake of certain modified lipoproteins may be the catalyst for the formulation of such a simple unifying theory for atherogenesis. This would incorporate causal factors as divergent as hyperlipidaemia, cigarette smoke, hypertension, age and hereditary factors.

I. The Modified Low-Density Lipoprotein Hypothesis

Scavenger receptors were discovered in 1979 when Brown, Goldstein and colleagues studied the accumulation of LDL cholesterol in macrophages in the atherosclerotic plaques of LDL receptor negative patients with FH. Macrophages from FH subjects fail to take up native LDL, but avidly take up LDL which had been acetylated in vitro. Because of the macrophage's inability to dispose of excess cholesterol, the chemically or modified LDL accumulates as foam droplets in the cytoplasm.

The cell surface receptors responsible for macrophage foam cell formation recognize the increased negative charge on the acetylated LDL protein. Other chemical modifications that increase the net negative charge of LDL also convert the lipoprotein into a ligand for the acetyl-LDL receptor. Because of this wide ligand specificity, the acetyl-LDL receptor soon became known as the scavenger receptor (GOLDSTEIN et al. 1979b; BROWN and GOLDSTEIN 1983). The search for a physiological ligand for the scavenger receptor uncovered an analogous biological modification when STEINBERG and co-workers (1989) discovered that oxidized LDL is recognized by this receptor.

Evidence has accumulated for the importance of oxidized LDL in atherogenesis. Oxidatively modified LDL is present in atherosclerotic lesions, but not in normal areas of aorta (HABERLAND et al. 1988). LDL extracted from aortic lesions shows advanced oxidative changes and is a ligand for scavenger receptors (YLÄ-HERTTUALA et al. 1989). Prevention of LDL oxidation by probucol, an antioxidant, decreases the rate at which LDL is taken up by macrophages and also effectively prevents atheroma formation in WHHL rabbits (KITA et al. 1987; CAREW et al. 1987).

The recent cloning and expression of a bovine lung scavenger receptor (KODAMA et al. 1990; ROHRER et al. 1990) therefore represents a major advance. Knowledge of the structure and function of macrophage scavenger receptors helps in understanding the diversity of the receptor's binding specificity.

II. Scavenger Receptor Structure

The scavenger receptor is a trimeric integral membrane glycoprotein composed of three 77-kDa subunits. It localizes to the cell surface and also to cells which take up modified LDL in vivo. Expression of the 220-kDa protein increased dramatically after the differentiation of receptor-inactive monocytes into receptor-active macrophages (VIA et al. 1985; KODAMA et al. 1988). Two closely related cDNAs were cloned from a bovine lung library (KODAMA et al. 1990; ROHRER et al. 1990). Each cloned receptor, when expressed in transfected cells, bound acetyl-LDL with similar affinity and showed the expected polyanion-binding specificity. Two types of cDNAs for human (MATSUMOTO et al. 1990) and mouse (KURIHARA et al. 1991) scavenger receptors have now also been cloned. Human type I and type II scavenger receptors are homologous (73% and 71%, respectively) to their bovine counterparts.

The amino acid sequence predicts six domains, of which domains I-V are common to both type I and II receptors (Fig. 6). Each highly glycosylated monomer of 453 amino acids (type I receptor) or 349 amino acids (type II receptor) possesses a single membrane-spanning region and

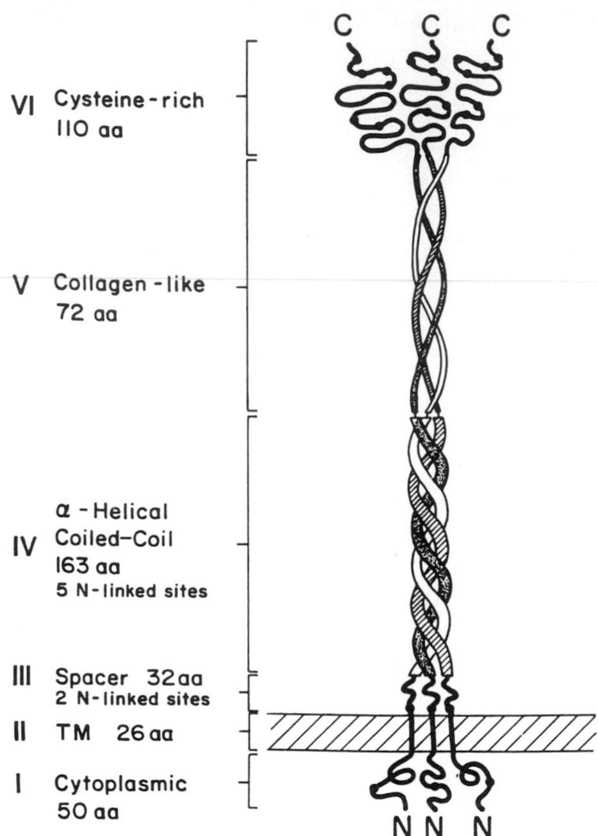

Fig. 6. Schematic model of the predicted trimeric structure of the type I bovine scavenger receptor. *aa*, amino acids; *TM*, transmembrane; *C*, C terminal; *N*, N terminal (from Kodama et al. 1990, with permission)

has an intracytoplasmic N-terminal "tail" of 50 amino acids and an extracellular C terminus.

There is no N-terminal signal peptide sequence among the 50 amino acids comprising the N-terminal cytoplasmic domain (domain I). One potential protein kinase substrate site, Arg-X-X-Ser/Thr (where X is another amino acid), of unknown functional significance, is conserved between species in the middle of the cytoplasmic domain (Kurihara et al. 1991). Although a receptor internalization sequence, Asn-Pro-X-Tyr (Chen et al. 1990), is absent in the scavenger receptor, a conserved tight-turn sequence, Try-X-Arg-Phe, near the N terminus is the structural motif necessary for endocytosis of both the transferrin receptor (Collawn et al. 1990) and the scavenger receptor (Kurihara et al. 1991).

The transmembrane region (domain II) consists of a single hydrophobic stretch of 26–28 amino acids, and the short proline-containing spacer (domain III) of 32 amino acids connects the transmembrane domain with extracellular domain IV. The spacer also contains two possible N-glycosylation sites.

Domain IV (the α-helical coiled-coil domain), consisting of 163 amino acids and five N-glycosylation sites, contains as many as 23 seven-amino-acid "heptad" repeats. The protein sequence of these repeats predicts an α-helical coiled-coil structure held together by an interhelical hydrophobic core of aliphatic residues. This structure plays an important role in the interaction and assembly of the functional trimeric receptor (KODAMA et al. 1988). Truncated C-terminal receptors containing only domains I–IV form trimeric structures on the cell surface, but are unable to bind either acetyl-LDL or oxidized LDL (KURIHARA et al. 1991). Scavenger receptors show a pH dependence in releasing their ligands in acidic intracellular compartments (KODAMA et al. 1990) and the presence of histidines (known to change their charges under acidic conditions) may influence the conformational stability of the receptor in acidic pH.

Domain V (the collagen-like domain) contains 24 (bovine and mouse) or 23 (human) uninterrupted Gly-X-Y (where X and Y are other amino acids) tripeptide repeats which form a collagenous triple helix (KODAMA et al. 1990). The common occurrence of proline or lysine as residue Y in the triplet is similar to collagen, where these residues are post-translationally hydroxylated (with ascorbic acid as essential cofactor) to stabilize the trimeric structure. Scavenger receptors are the first cell surface proteins reported to have collagen-like repeats. Studies of truncated mutant receptors indicate that a cluster of basic amino acids in the C-terminal region of domain V is essential for the binding of both acetyl-LDL and oxidized LDL (KODAMA et al. 1991). The C-terminal region of the collagen-like domain is thus essential for ligand recognition by scavenger receptors.

The two types of scavenger receptors differ only in the composition of domain VI (the carboxy-terminal type-specific domain). Domain VI of the type I receptor consists of 110 amino acids and is also termed the scavenger receptor cysteine-rich (SRCR) domain because of the presence of six cysteines which could be involved in intrachain and/or interchain disulphide bonds. Protein database analysis reveals that homologous SRCR domains are found in diverse membrane proteins from mammals (e.g., lymphocyte surface markers CD5 and Ly-1, complement factor I) and sea urchins (speract receptor; FREEMAN et al. 1990). This sequence therefore defines a family of highly conserved domains found in the extracellular portions of membrane proteins and also secreted proteins. Its function in the scavenger receptor is currently unknown, although its similarity to CD5 raises the possibility of a role for the scavenger receptor in the immune system, specifically in the early stages of the body's defence against pathogens (CASALI and NOTKINS 1989).

Type II scavenger receptors differ from type I receptors in having a very abbreviated domain VI which has no cysteines and only 17 amino acids (human) or six amino acids (mouse and bovine). Type I and II receptors show similar ligand-binding properties when expressed in COS cells , and this suggests that domain VI is not essential for ligand binding (Rohrer et al. 1990).

III. Scavenger Receptor Function

Dimeric and monomeric forms of the bovine receptor are unable to bind ligands, and scavenger receptors are therefore considered to function as trimers (Kodama et al. 1988). The presence of triplet repeats in domain V supports the proposed oligomerization of this domain into a collagenous triple helix (Kodama et al. 1990). Both type I and type II receptors are present on macrophages, and four possible combinations of subunits may make up either homo- or heterotrimers on the cell surface (Kurihara et al. 1991). During their synthesis, the receptors are co-translationally modified with high mannose N-linked sugars in the ER and then transported to the medial Golgi compartment, where the N-linked chains are converted into complex-type oligosaccarides (Penman et al. 1991).

Proteolytic removal of part of the extracellular cysteine-rich C terminus converts some mature type I receptors to resemble type II receptors (Penman et al. 1991); the functional significance of this post-translational cleavage has not yet been determined. Inhibition of the collagen-modifying enzymes prolyl and lysyl hydroxylases interfered with normal receptor processing, indicating that the collagenous domains probably do fold into collagen triple helices. Scavenger receptors function as noncovalently associated trimers. The α-helical coiled-coil domain may also contribute significantly to the trimerization of receptor subunits. Higher-order oligomerization, which would contain multiple collagenous triple helices, cannot be excluded.

The existence of more than one class of scavenger receptor has been proposed to explain apparent differences in acetyl-LDL and oxidized LDL binding to macrophages (Arai et al. 1989; Sparrow et al. 1989), Kupffer and endothelial liver cells (Van Berkel et al. 1991). Since both type I and II receptors are co-expressed on macrophages, multiple forms of the trimeric receptor may have different conformations which account for differing ligand specificities. However, studies with transfected receptors argue against this explanation. Transfected Type I and type II receptors recognize both acetyl-LDL and oxidized LDL, but show non-reciprocal cross-competition (Freeman et al. 1991). Acetyl-LDL efficiently competes for both its own endocytosis and for that of oxidized LDL; in contrast, oxidized LDL competes well against itself, but only poorly against acetylated LDL. This phenomenon may be ascribed to ligand binding to different, but interacting sites on a single receptor.

IV. Human Scavenger Receptor Gene

A single gene, located on chromosome 8 close to the lipoprotein lipase gene (MATSUMOTO et al. 1990), codes for the two scavenger receptors. The gene spans 80 kb and is composed of 11 exons. Alternative splicing results in two different mRNAs encoding type I and type II receptors, respectively (ASOAKA et al. 1991). The promoter region contains a TATA box of 28 base pairs (bp) upstream from the transcriptional initiation site. The promoter region from −817 bp to +50 bp relative to the transcription initiation site has been shown to exhibit functional promoter activity (MOULTON et al. 1991).

V. Expression and Regulation

The scavenger receptor is functionally expressed on macrophages, endothelial cells and Kupfer cells of the liver (NAGELKERKE et al. 1983; PITAS et al. 1985; STEIN and STEIN 1980; KUME et al. 1991). Both type I and II scavenger receptor mRNAs are highly expressed by bovine alveolar macrophages and liver (KODAMA et al. 1990; ROHRER et al. 1990) and human type I and II mRNAs have been detected in the lung, liver, placenta and in brain perivascular macrophages (MATSUMOTO et al. 1990). In contrast to the LDL receptor, the expression of scavenger receptors is not regulated by intracellular cholesterol. Continued delivery of lipoproteins to macrophages by this pathway can thus result in substantial intracellular cholesterol accumulation.

Monocytes do not normally have scavenger receptor mRNA, protein or cellular activity. Phorbol ester (MATSUMOTO et al. 1990; HAYASHI et al. 1991) and 1,25-dihydroxy-vitamin D3 (JOUNI and MCNAMARA 1991) induce scavenger receptor activity in human monocytic and tumour cell lines. This induction of activity may involve protein kinase C activation (AKESON et al. 1991). Phorbol esters also upregulate expression of the scavenger receptor in rabbit fibroblasts and smooth muscle cells (PITAS 1990). The enhanced expression of the scavenger receptor is unlikely to be related to the cholesterol status of these cells. Platelet secretory products are more likely regulatory candidates, since serum and secretory products from thrombin-stimulated platelets enhance scavenger receptor expression (PITAS 1990). The active protein-like factor originates from platelet alpha granules (FUHRMAN et al. 1991). The close proximity of smooth muscle cells, macrophages and aggregating platelets in the atherosclerotic lesion may contribute to the upregulation of scavenger receptor expression.

VI. Scavenger Receptor and Atherogenesis

Macrophage scavenger receptors are present in lipid-rich atherosclerotic lesions (MATSUMOTO et al. 1990). YLÄ-HERTTUALA et al. (1991) showed abundant scavenger receptor mRNA, but no detectable LDL receptor mRNA in human fatty streaks and atherosclerotic lesions. Expression of the scavenger receptor appears to be sufficient for foam cell formation, since the expression of either type of receptor in Chinese hamster ovary (CHO) cells causes the accumulation of lipid droplets (Freeman et al. 1991). The expression of 15-lipoxygenase mRNA and protein was seen in the same human lesions in which scavenger receptor mRNA colocalized with oxidation specific lipid-protein adducts (Yla-Herttuala et al. 1991). This would support the hypothesis that atherosclerosis is caused by the scavenger receptor-mediated uptake of 15-lipoxygenase oxidized LDL.

The release of nitric oxide by endothelial cells regulates vascular tone, and diseased atherosclerotic arteries respond less to endothelium-dependent vasodilatation (Rosenfeld 1991). Oxidized LDL also inhibits endothelium-dependent dilatation in coronary arteries and may do this by activating endothelial cell scavenger receptors in such a way as to decrease nitric oxide formation (Tanner et al. 1991). The resulting vasospasm and platelet aggregation further enhance atherogenesis.

The broad binding specificity of macrophage scavenger receptors suggests a multifunctional scavenging function in macrophage-associated immune responses and inflammation (Kodama et al. 1990; Freeman et al. 1990). The scavenger receptor may be part of an important defence mechanism that enables macrophages to degrade oxidized or chemically altered proteins. Liver scavenger receptors clear bacterial endotoxin (a lipopolysaccharide) from the circulation, possibly to prevent endotoxic shock (Hampton et al. 1991). Hepatic sinusoidal endothelial scavenger receptors might therefore filter oxidized proteins or pathogenic agents from the bloodstream as part of a detoxification process.

The structural resemblance between collagen and the scavenger receptor raises questions about possible functional similarities. Collagen also has a trimeric structure and a wide binding specificity. Oxidized LDL in collagen-rich tissues such as tendons and atherosclerotic plaques binds to collagen (Hoover et al. 1988). Collagen fibrils may concentrate potentially cytotoxic oxidized products like oxidized LDL for presentation to macrophages for scavenger-receptor-mediated detoxification.

The otherwise generally protective role of scavenger receptors could be damaging in the atherosclerotic plaque. The diverse atherogenic risk factors are known to be associated with oxidative changes in proteins. For example, oxidized proteins accumulate during the normal ageing process (Oliver et al. 1987) and cigarette smoke is known to oxidize LDL (Yokode et al. 1988). Hypertension also increases the collagen content of arteries. These conditions would favour the receptor-mediated uptake of trapped oxidized lipoproteins, resulting in foam cell formation. The macrophages are triggered to secrete cytokines and growth factors such as platelet-derived growth factor (Malden et al. 1991). These factors stimulate collagen production and smooth muscle proliferation. Further cell-mediated oxidation of LDL sets up a vicious circle culminating in the formation of a complex plaque. In this scenario, therefore, scavenger receptors are central to a unifying theory explaining the atherogenic effects of cholesterol, cigarette smoke, hypertension, ageing and heritable factors.

F. Conclusion

Knowledge gained from structural–functional analyses of the LDL receptor has enhanced our understanding of the hypercholesterolaemic syndrome caused by its dysfunction. Pharmacological manipulation of LDL receptor activity has been successful as a therapeutic strategy in lowering the risk of coronary heart disease in hyperlipidaemic patients, and gene therapy for homozygous FH patients is an exciting prospect for the future. The characterization of LRP and the scavenger receptor opens up new vistas for a better understanding of lipoprotein metabolism and atherogenesis and for the possible pharmacological manipulation of these receptors.

References

Aalto-Setala K, Helve E, Kovanen PT, Kontula K (1989) Finnish type of low density lipoprotein receptor gene mutation (FH-Helsinki) deletes exons encoding the carboxy-terminal part of the receptor and creates an internalization-defective phenotype. J Clin Invest 84:499–505

Akeson AL, Schroeder K, Woods C, Schmidt CJ, Jones WD (1991) Suppression of interleukin-1β and LDL scavenger receptor expression in macrophages by a selective protein kinase C inhibitor. J Lipid Res 32:1699–1707

Arai H, Kita T, Yokode M, Narumiya S, Kawai C (1989) Multiple receptors for modified low density lipoproteins in mouse peritoneal macrophages: different uptake mechanisms for acetylated and oxidized low density lipoproteins. Biochem Biophys Res Commun 159:1375–1382

Asaoka H, Emi M, Mukai T et al. (1991) Human scavenger receptor gene: its genomic structure and promoter region. Circulation 84(4) [Suppl 2]:II 230 (A914)

Bachorik PS, Franklin FA, Virgil DG, Kwiterovich Jr PO (1982) High-affinity uptake and degradation of apolipoprotein E free high-density lipoprotein and low-density lipoprotein in cultured porcine hepatocytes. Biochemistry 21:5675–5684

Barber DL, Sanders EJ, Aebersold R, Schneider WJ (1991) The receptor for yolk lipoprotein deposition in the chicken oocyte. J Biol Chem 266:18761–18770

Basu SK, Goldstein JL, Anderson RGW, Brown MS (1981) Monensin interrupts the recycling of low density lipoprotein receptors in human fibroblasts. Cell 24:493–502

Beisiegel U, Weber W, Ihrke G, Herz J, Stanley KK (1989) The LDL-receptor-related protein, LRP, is an apolipoprotein E-binding protein. Nature 341: 162–164

Beisiegel U, Weber W, Bengtsson-Olivecrona G (1991) Lipoprotein lipase enhances the binding of chylomicrons to low density lipoprotein receptor-related protein. Proc Natl Acad Sci USA 88:8342–8346

Bilheimer DW, Goldstein JL, Grundy SM, Brown MS (1975) Reduction in cholesterol and low density lipoprotein synthesis after portacaval shunt surgery in a patient with homozygous familial hypercholesterolemia. J Clin Invest 56:1420

Bilheimer DW, Grundy SM, Brown MS, Goldstein JL (1983) Mevinolin stimulates receptor-mediated clearance of low density lipoprotein from plasma in familial hypercholesterolaemia heterozygotes. Proc Natl Acad Sci USA 80:4124

Bilheimer DW, Goldstein JL, Grundy SM, Starzl TE, Brown MS (1984) Liver transplantation to provide low-density-lipoprotein receptors and lower plasma cholesterol in a child with homozygous familial hypercholestrertolemia. N Engl J Med 311:1658–1664

Bilheimer DW, East C, Grundy SM, Nora JJ (1985) II. Clinical studies in a kindred with a kinetic LDL receptor mutation causing familial hypercholesterolemia. Am J Med Genet 22:593–598

Brown MS, Goldstein JL (1983) Lipoprotein metabolism in the macrophage: implications for cholesterol deposition in atherosclerosis. Annu Rev Biochem 52:223–261

Brown MS, Goldstein JL (1986) A receptor-mediated pathway for cholesterol homeostasis. Science 232:34–37

Brown MS, Herz J, Kowal RC, Goldstein JL (1991) The low-density lipoprotein receptor-related protein: double agent or decoy? Curr Opin Lipidol 2:65–72

Carew TE, Schwenke DC, Steinberg D (1987) Antiatherogenic effect of probucol unrelated to its hypocholesterolemic effect: evidence that antioxidants in vivo can selectively inhibit low density lipoprotein degradation in macrophage-rich fatty streaks and slow the progression of atherosclerosis in the Watanabe heritable hyperlipidemic rabbit. Proc Natl Acad Sci USA 84:7725–7729

Casali P, Notkins AL (1989) CD5+ B lymphocytes, polyreactive antibodies and the human B-cell repertoire. Immunol Today 10:364–368

Casciola LAF, van der Westhuyzen DR, Gevers W, Coetzee GA (1988) Low density lipoprotein receptor degradation is influenced by a mediator protein(s) with a rapid turnover rate, but is unaffected by receptor up- or down-regulation. J Lipid Res 29:1481–1488

Casciola LAF, Grant KI, Gevers W, Coetzee GA, van der Westhuyzen DR (1987) Low-density-lipoprotein receptors in human fibroblasts are not degraded in lysosomes. Biochem J 262:681–683

Chen WJ, Goldstein JL, Brown MS (1990) NPXY, a sequence often found in cytoplasmic tails, is required for coated pit-mediated internalization of the low density lipoprotein receptor. J Biol Chem 265:3116–3123

Collawn JF, Stangel M, Kuhn LA, Esekogwu V, Jing S, Trowbridge IS, Tainer JA (1990) Transferrin receptor internalization sequence YXRF implicates a tight turn as the structural recognition motif for endocytosis. Cell 63:1061–1072

Cummings RD, Kornfeld S, Schneider WJ, Hobgood KK, Tolleshaug H, Brown MS, Goldstein JL (1983) Biosynthesis of the N- and O-linked oligosaccharides of the low density lipoprotein receptor. J Biol Chem 258:15261–15273

Cuthbert JA, Russell DW, Lipsky PE (1989) Regulation of low density lipoprotein receptor gene expression in human lymphocytes. J Biol Chem 264:1298–1304

Davis CG, Goldstein JL, Südhof TC, Anderson RGW, Russell DW, Brown MS (1987a) Acid-dependent ligand dissociation and recycling of LDL receptor mediated by growth factor homology region. Nature 326:760–765

Davis CG, Van Driel IR, Russell DW, Brown MS, Goldstein JL (1987b) The low density lipoprotein receptor. Identification of amino acids in cytoplasmic domain required for rapid endocytosis. J Biol Chem 262:4075–4082

Dawson PA, Hofmann SL, van der Westhuyzen DR, Südhof TC, Brown MS, Goldstein JL (1988) Sterol-dependent repression of low density lipoprotein receptor promoter mediated by 16-base pair sequence adjacent to binding site for transcription factor Sp1. J Biol Chem 263:3372–3379

Dawson PA, van der Westhuyzen DR, Goldstein JL, Brown MS (1989) Purification of oxysterol binding protein from hamster liver cytosol. J Biol Chem 264:9046–9052

Edge SB, Hoeg JM, Triche T, Schneider PD, Brewer HB (1986) Cultured human hepatocytes. Evidence for metabolism of low density lipoproteins by a pathway independent of the classical low density lipoprotein receptor. J Biol Chem 261:3800–3806

Esser VB, Limbird LE, Brown MS, Goldstein JL, Russell DW (1988) Mutational analysis of the ligand binding domain of the low density lipoprotein receptor. J Biol Chem 263:13282–13290

Filipovic I, Buddecke E (1986) Calmodulin antagonists stimulate LDL receptor synthesis in human skin fibroblasts. Biochim Biophys Acta 876:124–132

Fourie AM, Coetzee GA, Gevers W, van der Westhuyzen DR (1988) Two mutant low-density-lipoprotein receptors in Afrikaners slowly processed to surface forms exhibiting rapid degradation or functional heterogeneity. Biochem J 255:411–415

Freeman M, Ashkenas J, Rees DJG, Kingsley DM, Copeland NG, Jenkins NA, Krieger M (1990) An ancient, highly conserved family of cysteine-rich protein domains revealed by cloning type I and type II murine macrophage scavenger receptors. Proc Natl Acad Sic USA 87:8810–8814

Freeman M, Ekkel Y, Rohrer L, Penman M, Freedman NJ, Chisolm GM, Krieger M (1991) Expression of type I and type II bovine scavenger receptors in Chinese hamster ovary cells: Lipid droplet accumulation and nonreciprocal cross competition by acetylated and oxidized low density lipoprotein. Proc Natl Acad Sci USA 88:4931–4935

Fuhrman B, Brook GJ, Aviram M (1991) Activated platelets secrete a protein-like factor that stimulates oxidized-LDL receptor activity in macrophages. J Lipid Res 32:1113–1123

Glomset JA (1968) The plasma lecithin: cholesterol acyltransferase reaction. J Lipid Res 9:155–167

Goldstein JL, Brown MS (1989) Familial hypercholesterolemia. In: Scriver CR, Beaudet AL, Sly WS, Valle D (eds) The metabolic basis of inherited disease. McGraw-Hill, New York, pp 1215–1250

Goldstein JL, Anderson RGW, Brown MS (1979a) Coated pits, coated vesicles, and receptor-mediated endocytosis. Nature 279:679–685

Goldstein JL, Ho YK, Basu SK, Brown MS (1979b) Binding site on macrophages that mediates uptake and degradation of acetylated low density lipoprotein, producing massive cholesterol deposition. Proc Natl Acad Sci USA 76:333–337

Goldstein JL, Brown MS, Anderson RGW, Russell DW, Schneider WJ (1985) Receptor-mediated endocytosis: concepts emerging from the LDL receptor system. Annu Rev Cell Biol 1:1–39

Golos TG, Strauss JF, Miller WL (1987) Regulation of low density lipoprotein receptor and cytochrome P-450scc mRNA levels in human granulosa cells. J Steroid Biochem 27:767–773

Grant KI, Casciola LAF, Coetzee GA, Sanan DA, Gevers W, van der Westhuyzen DR (1990) Ammonium chloride causes reversible inhibition of low density lipoprotein receptor recycling and accelerates receptor degradation. J Biol Chem 265:4041–4047

Grundy SM, Mok HYI (1976) Chylomicron clearance in normal and hyperlipidemic man. Metab Clin Exp 25:1225–1239

Haberland ME, Fong D, Cheng L (1988) Malondialdehyde-altered protein occurs in atheroma of Watanabe heritable hyperlipidemic rabbits. Science 241:215–218

Hampton RY, Golenbock DT, Penman M, Krieger M, Raetz CRH (1991) Recognition and plasma clearance of endotoxin by scavenger receptors. Nature 352:342–344

Havekes LM, de Wit ECM, Princen HMG (1987) Cellular free cholesterol in Hep G2 cells is only partially available for down-regulation of low-density-lipoprotein receptor activity. Biochem J 247:739–746

Havel RJ, Chao Y-S, Windler EE, Kotite L, Guo LSS (1980) Isoprotein specificity in the hepatic uptake of apolipoprotein E and the pathogenesis of familial dysbetalipoproteinemia. Proc Natl Acad Sci USA 77:4349–4353

Hayashi K, Dojo S, Hirata Y, Ohtani H, Nakashima K, Nishio E, Kurushima H, Saeki M, Kajiyama G (1991) Metabolic changes in LDL receptors and an appearance of scavenger receptors after phorbol ester-induced differentiation of U937 cells. Biochim Biophys Acta 1082:152–160

Herz J, Hamann U, Rogne S, Myklebost O, Gausepohl H, Stanley KK (1988) Surface location and high affinity for calcium of a 500 kD liver membrane protein closely related to the LDL-receptor suggest a physiological role as lipoprotein receptor. EMBO J 7:4119–4127

Herz J, Kowal RC, Goldstein JL, Brown MS (1990a) Proteolytic processing of the 600 kD low density lipoprotein receptor related protein (LRP) occurs in a trans-Golgi compartment. EMBO J 9:1769–1776

Herz J, Kowal RC, Ho YK, Brown MS, Goldstein JL (1990b) Low density lipoprotein receptor-related protein mediates endocytosis of monoclonal antibodies in cultured cells and rabbit liver. J Biol Chem 265:21355–21362

Herz J, Goldstein JL, Strickland DK, Ho YK, Brown MS (1991) 39-kDa protein modulates binding of ligands to low density lipoprotein receptor-related protein/ α_2-macroglobulin receptor. J Biol Chem 266:21232–21238

Hobbs HH, Brown MS, Russell DW, Davignon J, Goldstein JL (1987) Deletion in the gene for the low-density-lipoprotein receptor in a majority of French Canadians with familial hypercholesterolemia. N Engl J Med 317:734–737

Hobbs HH, Leitersdorf E, Leffert CC, Cryer DR, Brown MS, Goldstein JL (1989) Evidence for a dominant gene that suppresses hypercholesterolemia in a family with defective low density lipoprotein receptors. J Clin Invest 84:656–664

Hobbs HH, Russell DW, Brown MS, Goldstein JL (1990) The LDL receptor locus in familial hypercholesterolemia. Mutational analysis of a membrane protein. Annu Rev Genet 24:133–170

Hoeg JM, Starzl TE, Brewer HB Jr (1987) Liver transplantation for treatment of cardiovascular disease: comparison with medication and plasma exchange in homozygous familial hypercholesterolemia. Am J Cardiol 59:705–707

Hoover GA, McCormick S, Kalant N (1988) Interaction of native and cell-modified low density lipoprotein with collagen gel. Arteriosclerosis 8:525–534

Hussain MM, Mahley RW, Boyles JK, Fainaru M, Brecht WJ, Lindquist P (1989) Chylomicron–chylomicron remnant clearance by liver and bone marrow in rabbits. Factors that modify tissue-specific uptake. J Biol Chem 264:9571–9582

Innerarity TL (1991) The low-density lipoprotein receptor. Curr Opin Lipidol 2: 156–161

Johnson WJ, Mahlberg FH, Rothblat GH, Phillips MC (1991) Cholesterol transport between cells and high-density lipoproteins. Biochim Biophys Acta 1085: 273–298

Jouni ZE, McNamara DJ (1991) Lipoprotein receptors of HL-60 macrophages: effect of differentiation with tetramyristic phorbol acetate and 1,25-dihydroxyvitamin D_3. Arterioscler Thromb 11:995–1006

Kaplan J (1981) Polypeptide-binding membrane receptors: analysis and classification. Science 212:14–20

Kita T, Goldstein JL, Brown MS, Watanabe Y, Hornick CA, Havel RJ (1982) Hepatic uptake of chylomicron remnants in WHHL Rabbits: a mechanism genetically distinct from the low density lipoprotein receptor. Proc Natl Acad Sci USA 79:3623–3627

Kita T, Nagano Y, Yokode M, Ishii K, Kume N, Ooshima A, Yoshida H, Kawai C (1987) Probucol prevents the progression of atherosclerosis in Watanabe heritable hyperlipidemic rabbit, an animal model for familial hypercholesterolemia. Proc Natl Acad Sci USA 84:5928–5931

Kodama T, Reddy P, Kishimoto C, Krieger M (1988) Purification and characterization of a bovine acetyl low density lipoprotein receptor. Proc Natl Acad Sci USA 85:9238–9242

Kodama T, Freeman M, Rohrer L, Zabrecky J, Matsudaira P, Krieger M (1990) Type I macrophage scavenger receptor contains α-helical and collagen-like coiled coils. Nature 343:531–535

Kodama T, Doi T, Matsumoto A et al. (1991) The C-terminal region of a collagen-like domain containing a cluster of basic amino acids mediates binding of modified low density lipoproteins by macrophage scavenger receptors. Circulation 84(4) [Suppl 2]:II 229 (A912)

Kowal RC, Herz J, Weisgraber KH, Mahley RW, Brown MS, Goldstein JL (1990) Opposing effects of apolipoproteins E and C on lipoprotein binding to low density lipoprotein receptor-related protein. J Biol Chem 265:10771–10779

Kristensen T, Moestrup SK, Gliemann J, Bendtsen L, Sand O, Sottrup-Hensen L (1990) Evidence that the newly cloned low-density-lipoprotein receptor related protein (LRP) is the α_2-macroglobulin receptor. FEBS Lett 276:151–155

Kume N, Arai H, Kawai C, Kita T (1991) Receptors for modified low-density lipoproteins on human endothelial cells: different recognition for acetylated low-density lipoprotein and oxidized low-density lipoprotein. Biochim Biophys Acta 1091:63–67

Kurihara Y, Matsumoto A, Itakura H, Kodama T (1991) Macrophage scavenger receptors. Curr Opin Lipidol 2:295–300

Kuwano M, Seguchi T, Ono M (1991) Glycosylation mutations of serine/threonine-linked oligosaccharides in low-density lipoprotein receptor: indispensable roles of O-glycosylation. J Cell Sci 98:131–134

Lehrman MA, Schneider WJ, Brown MS, Davis CG, Elhammer A et al. (1987) The Lebanese allele at the low density lipoprotein receptor locus: nonsense mutation produces truncated receptor that is retained in endoplasmic reticulum. J Biol Chem 262:401–410

Leitersdorf E, Chakravarti A, Hobbs HH (1989a) Polymorphic DNA haplotypes at the LDL receptor locus. Am J Hum Genet 44:409–421

Leitersdorf E, van der Westhuyzen DR, Coetzee GA, Hobbs HH (1989b) Two common low density lipoprotein receptor gene mutations cause familial hypercholesterolemia in Afrikaners. J Clin Invest 84:954–961

Leitersdorf E, Tobin EJ, Davignon J, Hobbs HH (1990) Common low-density lipoprotein receptor gene mutations in the French Canadian population. J Clin Invest 85:1014–1023

Lund H, Takahashi K, Hamilton RL, Havel RJ (1989) Lipoprotein binding and endosomal itinerary of the low density lipoprotein receptor-related protein in rat liver. Proc Natl Acad Sci USA 86:9318–9322

Mabuchi H, Sakai T, Sakai Y, Yoshimura A, Watanabe A, Bakasugi T, Koizumi J, Takeda R (1983) Reduction of serum cholesterol in heterozygous proteins with familial hypercholesterolemia: additive effects of compactin and cholestyramine. N Engl J Med 308:609

Mahlberg FH, Glick JM, Lund-Katz S, Rothblat GH (1991) Influence of apolipoproteins AI, AII, and C on the metabolism of membrane and lysosomal cholesterol in macrophages. J Biol Chem 266:19930–19937

Mahley RW, Hussain MM (1991) Chylomicron and chylomicron remnant catabolism. Curr Opin Lipidol 2:170–176

Mahley RW, Innerarity TL, Rall SC Jr, Weisgraber KH, Taylor JM (1990) Apolipoprotein E: genetic variants provide insights into its structure and function. Curr Opin Lipidol 1:87–95

Malden LT, Chait A, Raines EW, Ross R (1991) The influence of oxidatively modified low density lipoproteins on expression of platelet-derived growth factor by human monocyte-derived macrophages. J Biol Chem 266:13901–13907

Marazziti D, Eggertsen G, Fey GH, Stanley KK (1988) Relationships between the gene and protein structure in human complement component C9. Biochemistry 27:6529–6534

Matsumoto A, Naito M, Itakura H et al. (1990) Human scavenger receptors: primary structure, expression and localization in atherosclerotic lesions. Proc Natl Acad Sci USA 87:9133–9137

Mazière C, Mazière JC, Salmon S et al. (1988) Cyclic AMP decreases LDL catabolism and cholesterol synthesis in the human hepatoma line Hep G2. Biochem Biophys Res Commun 156:424–431

Mehta KD, Chen WJ, Goldstein JL, Brown MS (1991a) The low density lipoprotein receptor in Xenopus laevis: I. Five domains that resemble the human receptor. J Biol Chem 266:10406–10414

Mehta KD, Brown MS, Bilheimer DW, Goldstein JL (1991b) The low density lipoprotein receptor in Xenopus laevis: II. Feedback repression mediated by conserved sterol regulatory element. J Biol Chem 266:10415–10419

Meiner V, Landsberger D, Berkman N, Reshef A, Segal P, Seftel HC, van der Westhuyzen DR, Jeenah MS, Coetzee GA, Leitersdorf E (1991) A common Lithuanian mutation causing familial hypercholesterolemia in Ashkenazi Jews. Am J Hum Genet 49:443–449

Mendel CM, Kunitake ST (1988) Cell-surface binding sites for high density lipoproteins do not mediate efflux of cholesterol from human fibroblasts in tissue culture. J Lipid Res 29:1171–1178

Mendel CM, Kunitake ST, Kane JP, Kempner ES (1988) Radiation inactivation of binding sites for high density lipoproteins in human fibroblast membranes. J Biol Chem 263:1314–1319

Mendez AJ, Oram JF, Bierman EL (1991) Protein kinase C as a mediator of high density lipoprotein receptor-dependent efflux of intracellular cholesterol. J Biol Chem 266:10104–10111

Miyake Y, Tajima S, Funahashi T, Yamamoto A (1989) Analysis of a recycling-impaired mutant of low density lipoprotein receptor in familial hypercholesterolemia. J Biol Chem 264:16584–16590

Moulton KS, Wu H, Parthasarathy S, Glass CK (1991) Isolation and characterization of the acetylated-LDL receptor promoter. Circulation 84(4) [Suppl 2]:II 230 (A913)

Nagelkerke JF, Barto KP, van Berkel TJC (1983) In vivo and in vitro uptake and degradation of acetylated low density lipoprotein by rat liver endothelial, Kupffer, and parenchymal cells. J Biol Chem 258:12221–12227

Oliver CN, Ahn BW, Moerman EJ, Goldstein S, Stadtman ER (1987) Age-related changes in oxidized proteins. J Biol Chem 262:5488–5491

Oram JF (1990) Cholesterol trafficking in cells. Curr Opin Lipidol 1:416–421

Oram JF, Mendez AJ, Slotte JP, Johnson TF (1991) High density lipoprotein apolipoproteins mediate removal of sterol from intracellular pools but not from plasma membranes of cholesterol-loaded fibroblasts. Arterioscler Thromb 11:403–414

Packard CJ, Demant T, Shepherd J (1990) Genetics and apolipoprotein B metabolism. In: Lenfant C et al. (eds) Biotechnology of dyslipoproteinemias: applications in diagnosis and control. Raven, New York, pp 19–25

Pearse BMF (1988) Receptors compete for adaptors found in plasma membrane coated pits. EMBO J 7:3331–3336

Penman M, Lux A, Freedman NJ, Rohrer L, Ekkel Y, McKinstry H, Resnick D, Krieger M (1991) The Type I and Type II bovine scavenger receptors expressed in Chinese hamster ovary cells are trimeric proteins with collagenous triple helical domains comprising noncovalently associated monomers and cys^{83}-disulfide-linked dimers. J Biol Chem 266:23985–23993

Philips NC, Johnson WJ, Rothblat GH (1987) Mechanisms and consequences of cellular cholesterol exchange and transfer. Biochim Biophys Acta 906:223–276

Pitas RE (1990) Expression of the acetyl low density lipoprotein receptor by rabbit fibroblasts and smooth muscle cells: up-regulation by phorbol esters. J Biol Chem 265:12722–12727

Pitas RE, Innerarity TL, Arnold KS, Mahley RW (1979) Rate and equilibrium constants for binding of apo-E HDL_c (a cholesterol-induced lipoprotein) and low density lipoproteins to human fibroblasts: evidence for multiple receptor binding of apo-E HDL_c. Proc Natl Acad Sci USA 76:2311–2315

Pitas RE, Boyles J, Mahley RW, Bissell DM (1985) Uptake of chemically modified low density lipoproteins in vivo is mediated by specific endothelial cells. J Cell Biol 100:103–117

Pittman RC, Knecht TP, Rosenbaum MS, Taylor CA Jr (1987a) A nonendocytotic mechanism for the selective uptake of high density lipoprotein-associated cholesterol esters. J Biol Chem 262:2443–2450

Pittman RC, Glass CK, Atkinson D, Small DM (1987b) Synthetic high density lipoprotein particles: application to studies of the apoprotein specificity for selective uptake of cholesterol esters. J Biol Chem 262:2435–2442

Raychowdhury R, Niles JL, McCluskey RT, Smith JA (1989) Autoimmune target in Heymann nephritis is a glycoprotein with homology to the LDL receptor. Science 244:1163–1166

Reagan JW Jr, Miller LR, Clair RWS (1990) In vivo clearance of low density lipoprotein in pigeons occurs by a receptor-like mechanism that is not down-regulated by cholesterol feeding. J Biol Chem 265:9381–9391

Reichl D, Miller NE (1986) The anatomy and physiology of reverse cholesterol transport. Clin Sci 70:221–231

Rideout WM, Coetzee GA, Olumi AF, Jones PA (1990) 5-Methylcytosine as an endogenous mutagen in the human LDL receptor and p53 genes. Science 249:1288–1290

Rohrer L, Freeman M, Kodama T, Penman M, Krieger M (1990) Coiled-coil fibrous domains mediate ligand binding by macrophage scavenger receptor type II. Nature 343:570–572

Rosenfeld ME (1991) Oxidized LDL affects multiple atherogenic cellular responses. Circulation 83:2137–2140

Rubinsztein DC, Cohen JC, Berger GM, van der Westhuyzen DR, Coetzee GA, Gevers W (1990) Chylomicron remnant clearance from the plasma is normal in familial hypercholesterolemic homozygotes with defined receptor defects. J Clin Invest 86:1306–1312

Russell DW, Brown MS, Goldstein JL (1989) Different combinations of cysteine-rich repeats mediate binding of low density lipoprotein receptor to two different proteins. J Biol Chem 264:21682–21688

Sanan DA, van der Westhuyzen DR, Gevers W, Coetzee GA (1987) The surface distribution of low density lipoprotein receptors on cultured fibroblasts and endothelial cells: ultrastructural evidence for dispersed receptors. Histochem 86:517–523

Seguchi T, Merkle RK, Ono M, Kuwano M, Cummings RD (1991) The dysfunctional LDL receptor in a monensin-resistant mutant of Chinese hamster ovary cells lacks selected O-linked oligosaccharides. Arch Biochem Biophys 284:245–256

Sherrill BC, Innerarity TL, Mahley RW (1980) Rapid hepatic clearance of the canine lipoproteins containing only the E apoprotein by a high affinity receptor: identity with the chylomicron remnant transport process. J Biol Chem 255: 1804–1807

Simons LA, Reichl D, Myant NB, Mancini M (1975) The metabolism of the apoprotein of plasma low density lipoprotein in familial hyperbetalipoproteinemia in the homozygous form. Atherosclerosis 21:283

Slotte JP, Oram JF, Bierman EL (1987) Binding of high density lipoproteins to cell receptors promotes translocation of cholesterol from intracellular membranes to the cell surface. J Biol Chem 262:12904–12907

Smith JR, Osborne TF, Brown MS, Goldstein JL, Gil G (1988) Multiple sterol regulatory elements in promoter for hamster 3-hydroxy-3-methylglutaryl-coenzyme A synthase. J Biol Chem 263:18480–18487

Smith JR, Osborne TF, Goldstein JL, Brown MS (1990) Identification of nucleotides responsible for enhancer activity of sterol regulatory element in low density lipoprotein receptor gene. J Biol Chem 265:2306–2310

Soutar AK, Knight BL (1990) Structure and regulation of the LDL-receptor and its gene. Br Med Bull 46:891–916

Spady SK, Turley SD, Dietschy JM (1985) Rate of low density lipoprotein uptake and cholesterol synthesis are regulated independently in the liver. J Lipid Res 26:465–472

Sparrow CP, Parthasarathy S, Steinberg D (1989) A macrophage receptor that recognizes oxidized low-density lipoprotein but not acetylated low-density lipoprotein. J Biol Chem 264:2599–2604

St Clair RW, Leight MA (1983) Cholesterol efflux from cells enriched with cholesteryl esters by incubation with hypercholesterolemic monkey low density lipoprotein. J Lipid Res 24:183–191

Stein O, Stein Y (1980) Bovine aortic endothelial cells display macrophage-like properties towards acetylated ^{125}I-labelled low density lipoprotein. Biochim Biophys Acta 620:631–635

Steinberg D (1978) The rediscovery of high density lipoprotein; a negative risk factor in atherosclerosis. Eur J Clin Invest 8:107–109

Steinberg D, Parthasarathy S, Carew TE, Khoo JC, Witztum JL (1989) Beyond cholesterol: modifications of low-density lipoprotein that increases its atherogenicity. N Engl J Med 320:915–924

Stifani S, Barber DL, Aebersold R, Steyrer E, Shen X, Nimpf J, Schneider WJ (1991) The laying hen expresses two different low density lipoprotein receptor-related proteins. J Biol Chem 266:19079–19087

Strickland DK, Ashcom JD, Williams S, Burgess WH, Migliorini M, Argraves WS (1990) Sequence identity between the α_2-macroglobulin receptor and low density lipoprotein receptor-related protein suggests that this molecule is a multifunctional receptor. J Biol Chem 265:17401–17404

Südhof TC, Goldstein JL, Brown MS, Russell DW (1985a) The LDL receptor gene: a mosaic of exons shared with different proteins. Science 228:815–822

Südhof TC, Russell DW, Goldstein JL, Brown MS, Sanchez-Pescador R, Bell GI (1985b) Cassette of eight exons shared by genes for LDL receptor and EGF precursor. Science 228:893–895

Südhof TC, van der Westhuyzen DR, Goldstein JL, Brown MS, Russell DW (1987) Three direct repeats and a TATA-like sequence are required for regulated expression of the human low density lipoprotein receptor gene. J Biol Chem 262:10773–10779

Tanner FC, Noll G, Boulanger CM, Lüscher TF (1991) Oxidized low density lipoproteins inhibit relaxations of porcine coronary arteries: role of scavenger receptor and endothelium-derived nitric oxide. Circulation 83:2012–2020

Taylor RF, Saucier SE, Shown EP, Parish EJ, Kandutsch AA (1984) Correlation between oxysterol binding to a cytosolic binding protein and potency in the repression of hydroxymethylglutaryl coenzyme A reductase. J Biol Chem 259:10773–10779

Utermann G, Hoppichler F, Dieplinger H, Seed M, Thompson G, Boerwinkle E (1989) Defects in the low density lipoprotein receptor gene affect lipoprotein (a) levels: multiplicative interaction of two gene loci associated with premature atherosclerosis. Proc Natl Acad Sci USA 86:4171–4174

Van Berkel TJC, De Rijke YB, Kruijt JK (1991) Different fate in vivo of oxidatively modified low density lipoprotein and acetylated low density lipoprotein in rats: recognition by various scavenger receptors on Kupffer and endothelial liver cells. J Biol Chem 266:2282–2289

Van Driel IR, Goldstein JL, Südhof TC, Brown MS (1987a) First cysteine-rich repeat in ligand-binding domain of low density lipoprotein receptor binds Ca^{2+} and monoclonal antibodies, but not lipoproteins. J Biol Chem 262:17443–17449

Van Driel IR, Davis CG, Goldstein JL, Brown MS (1987b) Self-association of the low density lipoprotein receptor mediated by the cytoplasmic domain. J Biol Chem 262:16127–16134

Via DP, Dresel HA, Cheng SL, Gotto AM Jr (1985) Murine macrophage tumors are a source of a 260,000-Dalton acetyl-low density lipoprotein receptor. J Biol Chem 260:7379–7386

Wade DP, Knight BL, Soutar AK (1989) Regulation of low-density-lipoprotein-receptor mRNA by insulin in human hepatoma Hep G2 cells. Eur J Biochem 181:727–731

Watanabe Y (1980) Serial inbreeding of rabbits with hereditary hyperlipidemia (WHHL-rabbit). Incidence and development of atherosclerosis and xanthoma. Atherosclerosis 36:261–268

Weisgraber KH, Mahley RW, Kowal RC, Herz J, Goldstein JL, Brown MS (1990) Apolipoprotein C-I modulates the interaction of apolipoprotein E with β-migrating very low density lipoproteins (β-VLDL) and inhibits binding of β-VLDL to low density lipoprotein receptor-related protein. J Biol Chem 265: 22453–22459

Wilson JM, Chowdhury JR (1990) Prospects for gene therapy of familial hypercholesterolemia. Mol Biol Med 7:223–232

Wilson JM, Chowdhury NR, Grossman M, Wajsman R, Epstein A, Mulligan RC, Chowdhury JR (1990) Temporary amelioration of hyperlipidemia in low density lipoprotein receptor-deficient rabbits transplanted with genetically midified hepatocytes. Proc Natl Acad Sci USA 87:8437–8441

Wilson C, Wardell MR, Weisgraber KH, Mahley RW, Agard DA (1991) Three-dimensional structure of the LDL receptor-binding domain of human apolipoprotein E. Science 252:1817–1822

Windler E, Chao Y-S, Havel RJ (1980) Determinants of hepatic uptake of triglyceride-rich lipoproteins and their remnants in the rat. J Biol Chem 255: 5475–5480

Yamamoto T, Davis CG, Brown MS, Schneider WJ, Casey ML, Goldstein JL, Russell DW (1984) The human LDL receptor: a cysteine-rich protein with multiple Alu sequences in its mRNA. Cell 39:27–38

Yamamoto T, Bishop RW, Brown MS, Goldstein JL, Russell DW (1986) Deletion in cysteine-rich region of LDL receptor impedes transport to cell surface in WHHL rabbit. Science 232:1230–1237

Ylä-Herttuala S, Palinski W, Rosenfeld ME, Parthasarathy S, Carew TE, Butler S, Witztum JL, Steinberg D (1989) Evidence for the presence of oxidatively modified low density lipoprotein in atherosclerotic lesions of rabbit and man. J Clin Invest 84:1086–1095

Ylä-Herttuala S, Rosenfeld ME, Parthasarathy S, Sigal E, Sarkioja T, Witztum JL, Steinberg D (1991) Gene expression in macrophage-rich human atherosclerotic lesions: 15-lipoxygenase and acetyl low density lipoprotein receptor messenger RNA colocalize with oxidation specific lipid–protein adducts. J Clin Invest 87:1146–1152

Yokode M, Kita T, Arai H, Kawai C, Narumiya S, Fujiwara M (1988) Cholesteryl ester accumulation in macrophages incubated with low density lipoprotein pretreated with cigarette smoke extract. Proc Natl Acad Sci USA 85:2344–2348

Yokode M, Hammer RE, Ishibashi S, Brown MS, Goldstein JL (1990) Diet-induced hypercholesterolemia in mice: prevention by overexpression of LDL receptors. Science 250:1273–1275

CHAPTER 4

Genetic Disorders of Lipoprotein Metabolism

G. UTERMANN and H.J. MENZEL

A. Introduction

The transport of lipids in human blood plasma is a dynamic and complex process which requires the proper and coordinate action of many genes (HAVEL and KANE 1989; KANE and HAVEL 1989). Owing to their poor solubility in water, lipids are complexed with structural proteins called apolipoproteins and transported in the form of lipoproteins in human plasma. Apolipoproteins also have functions above and beyond their role as structural elements of lipoprotein particles. They act as cofactors for enzymes (apoA-I, apoC-II), modulators of enzyme activity or substrate accessibility (apoC-III), and as ligands for lipoprotein receptors (apoB, apoE; LI et al. 1988). Lipoproteins are assembled in parenchymal cells of the liver and in duodenal enterocytes. They are secreted into the space of Disse or the lymph, from where they enter the bloodstream (HAVEL and KANE, 1989; KANE and HAVEL, 1989). Extensive remodeling, including hydrolysis and removal of core lipids and exchange of lipids and surface apolipoproteins between lipoproteins, occurs within the vessel. These processes are mediated by enzymes (lipoprotein lipase, LPL; hepatic triglyceride lipase, HTGL; lecithin cholesterol acyltransferas, LCAT) and transfer proteins (cholesterolester transfer protein, CETP; HAVEL and KANE 1989; KANE and HAVEL 1989). Finally, lipoproteins are removed from plasma and taken up by tissues, a process which includes receptor-mediated endocytosis of lipoproteins by cells (BROWN and GOLDSTEIN 1986). In addition to their function in plasma lipid transport, some lipoproteins and apolipoproteins function in tissue regeneration and immunoregulation (HAJJAR et al. 1989).

All major genes coding for human apolipoproteins have been cloned and sequenced and their chromosomal localization is known. Likewise, genes for enzymes and transfer proteins which regulate the intravascular metabolism of plasma lipoproteins and for receptors that clear lipoproteins from the circulation have been characterized (Table 1). Mutations in most of these genes have been described. Many of them cause specific forms of dyslipoproteinemia and lipoprotein deficiency syndromes (Tables 2, 3) and some result in hyperlipidemia. Many of the genes involved in the intracellular regulation of lipoprotein biosynthesis are not yet known, but presumably mutations in such genes will also be important for some forms of

Table 1. Chromosomal localization of apolipoproteins, lipolytic enzymes, transfer proteins and lipoprotein receptors (adapted from LUSIS 1988 and LI et al. 1988)

	Amino Acids	Chromosomal localization
Apo A-I	243	11q23
Apo A-II	77	1p21–p23
Apo A-IV	377	11q23
Apo B (B-100 and B-48)	4536 or 2152	2p23–24
Apo C-II	79	19q13.2
Apo C-III	79	11q23
Apo E	299	19q13.2
CETP	476	16q13
LCAT	416	16q22.1
LPL	448	8p22
HTGL	477	15q21
LDL-Receptor	839	19p13.2
LRP	4525	?

CETP, cholesterolester transferase; LCAT, lecithin cholesterol acyltransferase; LPL, lipoprotein lipase; HTGL, hepatic triglyceride lipase; LRP, low-density lipoprotein receptor-related protein.

hyperlipidemia which are characterized by overproduction of lipoproteins.

It is also important to note that not all forms of genetic dys- or hyperlipidemia are monogenic. Rather, allelic variation at certain gene loci may have moderate effects on lipoprotein levels, and certain alleles may contribute to polygenic or multifactorial forms of hyperlipidemia. A good example is apolipoprotein E, where three common alleles (ε2, ε3, ε4) determine a polymorphism and affect apoE, apoB, low-density lipoprotein (LDL) cholesterol, and total cholesterol concentrations in the population (BOERWINKLE and UTERMANN 1988; HALLMAN et al. 1991; DAVIGNON et al. 1988). Moreover, they are associated with hyperlipidemic states (UTERMANN 1987a). Homozygotes for the ε2 allele inevitably develop a recessive form of dyslipidemia termed primary dysbetalipoproteinemia (UTERMANN et al. 1977), and some rare apoE mutations result in dominant type III hyperlipoproteinemia (RALL and MAHLEY 1992; MAHLEY et al. 1990). Accordingly, the classification and definition of genetic hyperlipidemics is not straightforward. Some hyperlipidemias are clearly monogenic following mendelian inheritance. These will be described here in detail. In most of the common hyperlipidemias, genetic factors also contribute to a variable extent. Only some well-documented examples of this will be given in this chapter, and the reader is referred to review articles covering this topic (UTERMANN 1987b; LUSIS 1988; BERG 1983).

Hyperlipidemias may be classified according to the principal lipids or lipoproteins which accumulate, e.g., hypercholesterolemia, hypertriglyceridemia, mixed hyperlipidemia (where both major lipids are elevated; Table 3), by the defects that cause them, e.g., receptor or ligand defects,

Table 2. Genetic dyslipoproteinemias and the affected genes and mutations

Genetic dyslipoproteinemias[a]	Affected gene/mutations
Autosomal recessive forms	
Familial hyperchylomicronemia	a) LPL b) ApoC-II
Hepatic lipase deficiency	Hepatic lipase
Cholesterol ester storage disease	Lysosomal acid cholesterol ester hydrolase
LCAT deficiency	LCAT
Fish-eye disease	LCAT/apoA-I
Familial hyperalphalipoproteinemia	CETP
Tangier disease	unknown
HDL Deficiency with planar xanthomatosis	ApoA-I
Familial ApoA-I/C-III Deficiencies types I and II	ApoA-I/C-III/A-IV gene cluster
Primary dysbetalipoproteinemia	a) ApoE (Arg158 → Cys) b) ApoE (Trp210 → Stop)
Abetalipoproteinemia	Microsomal triglyceride Transfer protein?
Anderson disease (chlyomicron retention disease)	Unknown
Sitosterolemia with xanthomatosis	HMG CoA-reductase deficiency
Cerebrotendinous xanthomatosis	??? Bile acid synthesis ????
Authosomal dominant forms	
Familial hypercholesterolemia	LDL receptor
FDB	ApoB (Arg3500 → Gln)
Familial combined hyperlipidemia	Heterogeneous: LPL, apoA-I/C-III/A-IV gene cluster and others
Familial Lp(a) hyperlipoproteinemia[b]	Apo(a)
Familial hypobetalipoproteinemia	ApoB
Familial hypoalpha-lipoproteinemia type Milano and some other apoA-I mutants	ApoA-I
Hyperlipoproteinemia type III dominant forms	ApoE (rare mutant forms)
Homozygous for autosomal dominant forms	
Homozygous familial hypercholesterolemia	LDL Receptor
Homozygous FDB	ApoB
Homozygous hypobetalipoproteinemia	ApoB
Multifactorial forms/genetics unclear	
Polygenic hypercholesterolemia	Multiple, apoE-4 (Cys112 → Arg)
Familial hypertriglyceridemia	Unknown (LPL ? and C-III?)
Hyperlipoproteinemia type III	ApoE (Arg158 → Cys) plus others factors
Familial type V hyperlipidemia	Unknown
Familial hypoalphalipoproteinemia with normal apoA-I	Unknown
Familial hyperalphalipoproteinemia with normal CETP	Unknown

LCAT, lecithin cholesterol acyltransferase; FDB, familial defective apoB-100; CETP, cholesterolester transfer protein; LPL, lipoprotein lipase; HMG CoA, hydroxymethylglutaryl coenzyme A; LDL, low-density lipoprotein.

[a] UTERMANN (1990)

[b] May be caused by one dominant or the additive effects of two codominant alleles.

Table 3. Major clincial and laboratory findings in the recessive dyslipoproteinemias[a]

Form	Lipid and lipoprotein abnormalities	Characteristic clinical and laboratory findings	Premature coronary heart disease
Familial hyperchylomicronemia	Triglycerides grossly elevated; low total plasma cholesterol; severe hyperchylomicronemia; low LDL and HDL levels	Eruptive xanthomas; episodic abdominal pain, pancreatitis, hepatosplenomegaly	None
Hepatic lipase deficiency	Similar to HLP type III; presence of β-VLDL; triglyceride-rich LDL and HDL	Eruptive skin xanthomas; linear yellow discoloration of the palmar creases, obesity	Unclear
Lecithin cholesterol acyltransferase deficiency (classical form)	Deficiency of cholesterolesters in plasma; abnormal chemical and physical properties of all major lipoproteins; HDL contains discoidal particles	Corneal opacities; anemia proteinuria, late nephropathy; Sea-blue histiocytes in spleen and bone marrow	Present in many patients
Fish-eye disease	Reduced levels of HDL; mild hypertriglyceridemia	Corneal opacities	Late atherosclerosis
Cholesterolester transfer protein deficiency	HDL_2 markedly elevated, abnormal chemical and physical properties of all major lipoproteins	None	Decreased risk
Tangier disease	Low plasma total cholesterol; deficiency of HDL, apoA-I, and apoA-II; abnormal LDL; abnormal chylomicron remnants	Hypoplastic orange tonsils; splenomegaly, storage of cholesterol esters in RES; relapsing neuropathy; corneal opacities	Unclear
HDL deficiency with planar xanthomas	Severe HDL deficiency; absent apoA-I	Planar xanthomas; diffuse corneal opacities	Severe

Familial apoA-I/C-III/A-IV deficiency types I and type II	Severe HDL deficiency; absence of apoA-I and C-III (type I) and apoA-I, C-III, and A-IV (type II)	Planar xanthomas (type I); mild diffuse corneal opacities (types I and II)	Severe
Primary dysbetalipoproteinemia	Low or normal plasma total cholesterol; mild hypertriglyceridemia; cholesterol-rich VLDL; presence of β-VLDL; low LDL	ApoE (Arg158 → Cys)	Decreased risk
Abetalipoproteinemia	Extremely low plasma total cholesterol, triglyceride, B-100 and B-48; absence of chylomicron, VLDL, and LDL	Fat malabsorbtion, retinis pigmentosa, acanthocytosis, ataxic neuropathy	None
Homozygous hypobetalipoproteinenemia	Extremely low plasma total cholesterol, triglyceride, B-100 and sometimes presence of truncated apoB species; absence of chylomicrons, VLDL, and LDL	Fat malabsorbtion, retinis pigmentosa, acanthocytosis, ataxic neuropathy	None
Anderson disease (chylomicron retention disease)	Absence of chylomicrons and B-48	Fat malabsorbtion, retinis pigmentosa, ataxic neuropathy	None
Sitosterolemia with xanthomatosis	Increased plant sterols (sitosterol, camposterol) in LDL and HDL; Facultative hypercholesterolemia	Tendon and tuberous xanthomas abnormal red blood cells	Severe
Cerebrotendinous xanthomatosis	Low HDL of abnormal composition cholesterol increased in plasma lipoproteins	Progressive neurological dysfunction (dementia, spinalcord paresis, cerebellar ataxia), tendon xanthomas, cataracts	Severe
Cholesterolester storage disease	LDL, elevated, type IIa or IIb	Hepatosplenomegaly	Moderate
X-linked recessive form			
X-linked ichthyosis (steroid sulphatase deficiency)	LDL and VLDL with cholesterol sulfate and increased mobility in electrophoresis	Ichthyosis, corneal opacities	Unknown

HDL, high-density lipoprotein; LDL, low-density lipoprotein; VLDL, very low density lipoprotein; RES, reticuloendothelial system; HLP, hyperlipoproteinemia.
[a] UTERMANN 1990.

Table 4. Classification of hyperlipidemias with known defects according to affected lipoprotein class

Form	Common designation	Mutant gene
HDL hyperlipoproteinemia	Hyperalphalipoproteinemia	CETP
LDL hyperlipoproteinemia	a) Familial hypercholesterolemia b) Familial defective apoB	a) LDL receptor b) ApoB-100
Hyperchylomicronemia and VLDL hyperlipoproteinemia	Type I hyperlipoproteinemia	a) LPL b) ApoC-II
Remnant hyperlipoproteinemia	Type III hyperlipoproteinemia, dysbetalipoproteinemia	a) ApoE b) HTGL
Lp(a) hyperlipoproteinemia	–	Apo(a)
Multiple phenotype hyperlipoproteinemia	Familial combined hyperlipoproteinemia	a) LPL b) ApoA-I/C-III/A-IV?

HDL, high-density lipoprotein; LDL, low-density lipoprotein; LDL, very low density lipoprotein; Lp(a), lipoprotein (a); CETP, cholesterolester transfer protein; HTGL, hepatic triglyceride lipase.

enzyme deficiencies, or apolipoprotein mutations, or by their mode of inheritance (Table 2). A phenotypic classification considering the etiological heterogeneity of hyperlipidemias will be used here. Mutations in receptors and their ligands (LDL receptor, apoB) and in enzymes and their cofactors may result in similar phenotypes (Table 4), though there are exceptions. Defects in the gene for the LDL receptor result in elevation of LDL and familial hypercholesterolemia (FH; Brown and Goldstein 1986), whereas those in apoE, one of the ligands for this receptor, may cause elevation of remnants (type III hyperlipoproteinemia; Mahley 1988; Utermann 1987a). Some situations have been described where mutations in different genes which each result in monogenic hyperlipidemia coexist in a single patient or family and determine a unique phenotype (Stalenhoef et al. 1981; Emi et al. 1991; Rauh et al. 1991).

The clinical importance of hyperlipidemias relates to the association of this class of inborn errors of metabolism with atherosclerotic vascular disease (Goldstein et al. 1973). Not all mutations in apolipoprotein genes or lipolytic enzymes result in hyperlipidemia, and not all forms of hyperlipidemia are associated with atherosclerosis and coronary heart disease (CHD; Table 3). Many mutations in the apoB gene result in truncated forms of this giant protein (512 kDa), hypobetalipoproteinemia (a condition characterized by low apoB, LDL, and total cholesterol in heterozygotes), and grossly reduced or absent apoB and LDL in homozygotes (Young 1990).

Defective LPL or apoC-II both cause massive hypertriglyceridemia (hyperchylomicronemia, type I hyperlipoproteinemia), but are apparently not associated with increased CHD risk in homozygotes or compound heterozygotes. Surprisingly though, heterozygotes for LPL deficiency may present with a lipid abnormality resembling familial combined hyperlipoproteinemia (FCH), a condition which is believed to increase the risk for CHD dramatically (WILLIAMS et al. 1991). In FCH due to LPL deficiency, hyperlipidemia is only moderate, but lipoprotein particles accumulate which are subnormal in concentration in homozygous LPL deficiency. Thus, the risk for CHD is more dependent on the type of elevated lipoprotein particles than on the extent of lipid elevation per se.

Hypolipidemic conditions may also predispose to premature CHD. Examples are gene defects which result in low or absent high-density lipoprotein (HDL; Table 3); these include some mutations in the gene for apoA-I and LCAT (ASSMANN et al. 1991). Familial LCAT deficiency is an example of an inborn error of lipoprotein metabolism where classification into hypo- or hyperlipidemia is not straightforward. All such patients are dyslipidemic from birth, and the condition is primarily associated with low HDL and hypolipidemia. With time, abnormal LDL particles may precipitate kidney disease, and "secondary" hyperlipidemia may develop in some patients. These patients may also develop atherosclerosis as a consequence of their kidney disease (NORUM et al. 1989). Multiple mutations in the LCAT gene, which either result in the classical clinical syndrome of LCAT deficiency (hypoalphalipoproteinemia, corneal clouding, anemia, proteinuria) or in fish-eye disease (corneal clouding) have been described (ASSMANN et al. 1991). In the following, the major hyperlipidemias will be described in detail.

B. Familial Hyperchylomicronemia

This clinical entity was originally described by BÜRGER and GRÜTZ in 1932 as a familial lipoidosis with hepatosplenomegaly and xanthomatosis (BÜRGER and GRÜTZ 1932) and was named after these authors. Later, the disease was renamed type I hyperlipoproteinemia or FH. The condition is inherited as an autosomal recessive trait and is characterized by a massive accumulation of chylomicrons and very low density lipoproteins (VLDI) in patients' plasma (BRUNZELL 1989), which has a milky appearance; triglyceride concentrations may exceed 5000 mg/dl and the concentrations of all other lipoproteins are subnormal. Both LDL and HDL concentrations are low and Lp(a) levels are also lower than in matched controls (SANDHOLZER et al. 1992). Thus, the typical lipid profile in FH is marked hypertriglyceridemia but low total cholesterol. Only when triglyceride elevation is excessive will cholesterol also be elevated. Upon agarose gel electrophoresis, practically all lipoproteins remain at the origin (type I hyperlipoproteinemia). Sometimes patients may exhibit a type V pattern, which is characterized by the presence of chylomicrons which remain at the origin and of pre-β-lipoproteins (VLDL).

Most of the lipoproteins from type I plasma float at a density of 1006 g/ml and can be obtained by short-term ultracentrifugation. When plasma from a type I patient is stored at +4°C overnight, a fatty white layer forms at the surface and the plasma underneath clears. The protein composition of the triglyceride-rich lipoproteins is characterized by the presence of apoB, apoE, apoC-I, C-II, and C-III, and traces of apoA-IV and other proteins. Both forms of apoB–apoB-48, which in humans is mainly of intestinal origin, and apoB-100, which is secreted from the liver in the form of VLDL – are present. The VLDL particles in type I hyperlipoproteinemia are usually very large and triglyceride rich, which makes it difficult to distinguish them from chylomicrons. The presence of apoB-100 and the lack of retinylester labeling demonstrates the hepatic origin of a fraction of the triglyceride-rich lipoproteins. This is consistent with the metabolic defect underlying the lipid disorder (Fig. 1). Chylomicrons and VLDL are both substrates for LPL (triacylglyceroprotein acylhydrolase, EC 3.134), which hydrolyzes the core triglycerides carried by the particles. This enzyme is located at the endothelial layer of blood vessels. It is anchored at the endothelial cell surface by binding to glycosaminoglycans (heparansulfate) and can be released from the cell surface by intraveneous heparin injection. This allows the measurement of LPL activity as a fraction of the so-called postheparin lipolytic

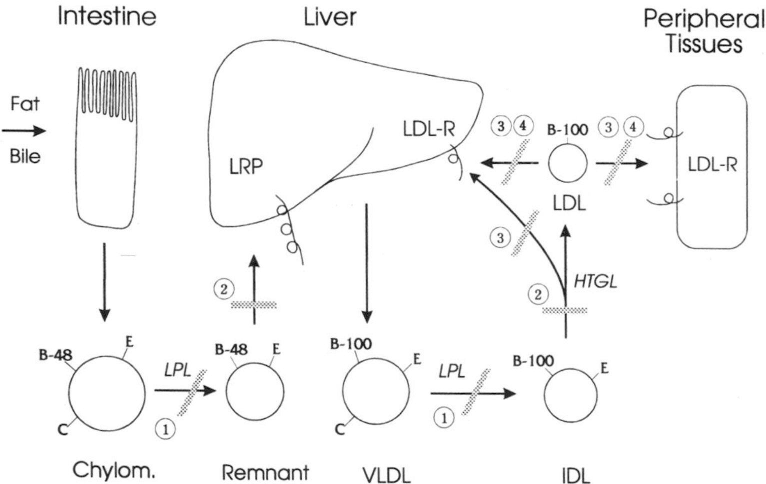

Fig. 1. Metabolic pathways and inborn errors of metabolism of apoB-containing lipoproteins. ①, Metabolic block due to LPL or apoC-II deficiency or malfunction; ②, block in the uptake of lipoprotein remnants due to apoE-2 mutations (see text); ③, block of LDL uptake by the LDL receptor due to deficient or defective LDL receptor; ④, block of lipoprotein uptake by the LDL receptor due to defective apoB-100 (Arg3500 → Gln). *B-100*, *B-48*, *E*, and *C* denote apolipoproteins B-100, B-48, E, and C. *LDL*, low-density lipoprotein; *LDL-R*, LDL receptor; *LRP*, LDL receptor-related protein; *Chylom.*, chylomicron; *VLDL*, very low density lipoprotein; *IDL*, intermediate-density lipoprotein; *HTGL*, hepatic triglyceride lipase

activity (BRUNZELL, 1989). LPL is activated by apoC-II, which is present at the surface of the lipoprotein substrate. Mutations in LPL and in the activator apoC-II both may result in familial hyperchylomicronemia.

I. Lipoprotein Lipase Deficiency

The LPL gene spans 30 kb at chromosome 8 and includes ten exons. It codes for a protein of 475 amino acid residues, including the leader sequence of 27 amino acids. Functional LPL is a homodimeric glycoprotein of 448 amino acids. Numerous molecular variants have been detected by analysis of the LPL gene in patients with LPL deficiency (LALOUEL et al. 1992; FAUSTINELLA et al. 1991; MONSALVE et al. 1990; DICHEK et al. 1991; HENDERSON et al. 1991; Henderson et al. 1990). Using functional and immunological criteria, LPL mutations have been grouped into three categories (AUWERX et al. 1989). Class I is defined as a null phenotype; neither LPL activity nor protein are present. Class II includes patients with reduced enzyme activity, but normal protein. Finally, class III subjects are characterized by the existence of LPL immunoreactive material in plasma. However, their LPL variants do not bind to heparin Sepharose. As for most other genetic disorders, mutations include large deletions and insertions, frameshift mutations, splicing defects, and nonsense and missense mutations (reviewed in LALOUEL et al. 1992).

LPL displays a high degree of structural similarity with human pancreatic lipase. The three-dimensional structure of human pancreatic lipase is known. This has made it possible to deduce the functional significance of LPL mutations by reference to the structure of pancreatic lipase (Fig. 2). Missense mutations in LPL cluster in the central exons 4, 5, and 6 of the gene (LALOUEL et al. 1992) and all missense mutations resulting in inactive enzyme are located here. These exons correspond to a doubly wound, mostly parallel ß-sheet structure which forms the amino-terminal domain of LPL and includes the catalytic triad Asp^{156}, His^{241}, Ser^{132}. Mutations at position 156 which directly affect the catalytic triad have been described (FAUSTINELLA et al. 1991). Other mutations affect conserved residues, but several occur at less conserved positions, including a common substitution at residue 188 (MONSALVE et al. 1990).

Several missense mutations impair the secretion rate or reduce the heparin-binding properties of LPL, but the mechanisms are largely unknown. Missense mutations may destroy the active center of the serine hydrolase (as is the case for mutations in position 156), impair the binding to membrane-bound glycosaminoglycans, which is mediated by the heparin-binding site, disturb the interaction with apoC-II or lipids, and destroy the site promoting interaction with the other glycoprotein domain to form an active dimer. As for other genes, mutations may also result in instable protein, truncated forms, or gross alterations without protein product (LALOUEL et al. 1992).

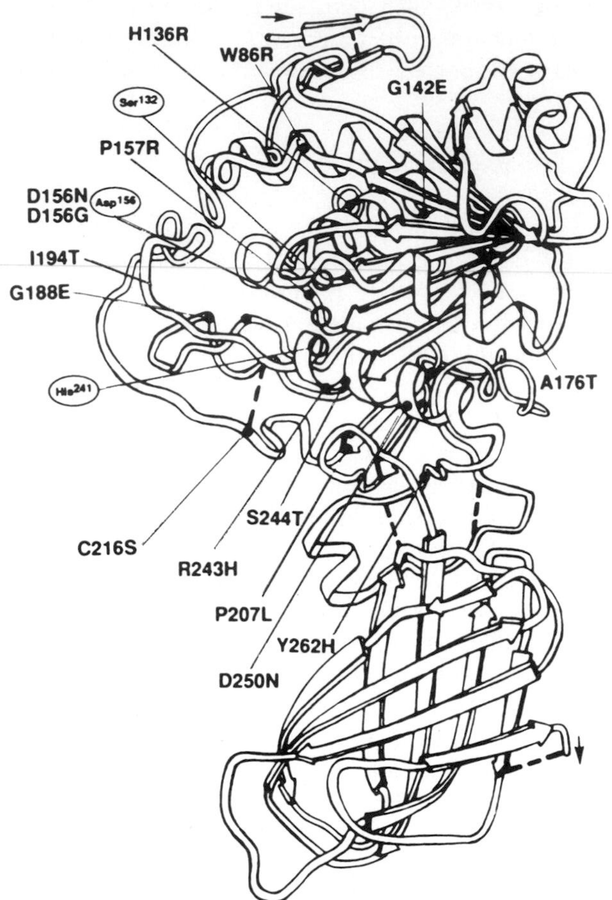

Fig. 2. Model of the three-dimensional structure of human pancreatic lipase published by WINKLER et al. (1990). Missense mutations of lipoprotein lipase (LPL) are superimposed on this structure. Because this representation relies upon homology between human LPL and human pancreatic lipase, the location of these residues may be only approximate. (From LALOUEL et al. 1992, with permission)

The clinical syndrome of LPL deficiency is characterized by intermittent abdominal pain, pancreatitis, eruptive xanthoma, hepatosplenomegaly, lipemia retinalis, and failure to thrive (SANDHOLZER et al. 1992). Patients with abdominal pain or pancreatitis and milky plasma should be suspicious of having LPL deficiency. Many patients underwent abdominal surgery because the metabolic disease was not recognized.

LPL deficiency occurs in most populations, but is generally a rare disorder. The frequency estimate of at least 1:1 million is based on patients whose condition was clinically ascertained. A significantly increased frequency of LPL deficiency has been found in French Canadians from certain regions of Quebec. In the Charlevoix and Saguenay-Lac St. Jean counties, the frequency of affected subjects ranges from 1:5000 to 1:10000, with carrier frequencies of about 1:40 (SANDHOLZER et al. 1992). This is the highest known frequency worldwide. Sequence analysis of exon 5 from 37 French Canadian patients with LPL deficiency revealed a missense substitution of leucine (CTG) for proline (CCG) at residue 207 on 54 of the mutant alleles. This mutation is the most frequent among French Canadians (73%) and is probably due to a founder effect (MA et al. 1991). It has been suggested that the $Pro^{207} \rightarrow$ Leu mutation was introduced into the province of Quebec by French immigrants in the seventeenth and eighteenth centuries. Studies of the origin of the patients have revealed the region of Perche, which is located between Paris and Normandy, as the likely center of diffusion of this major mutation in the Quebec population.

An identical missense mutation within exon 5 has been detected on 21 alleles from 13 unrelated affected individuals with LPL deficiency from different populations (MONSALVE et al. 1990). The mutation, which is an amino acid substitution of glutamic acid for glycin at position 188, is present in patients of French Canadian, English, Polish, German, Dutch, and East Indian ancestry and has also been detected in other populations, including Welsh (EMI et al. 1990) and Austrians (PAULWEBER et al. 1991). The mutation was also detected in nine South African individuals from four families which were all of Indian descent and have their origins in villages close to Bombay (HENDERSON et al. 1992). Many mutations in the LPL gene, including the frequent French Canadian mutation, have resulted in an altered restriction site, enabling a simple diagnosis of the disease and of the heterozygous carrier state in families and certain populations.

II. Apolipoprotein C-II Deficiency

ApoC-II is located on chromosome 19 in tandem array with apoC-I and apoE. The gene is 4 kb in size and includes four exons. ApoC-II is synthesized as a 101-amino acid preproprotein which undergoes cotranslational cleavage of a 22-amino acid prepeptide (FOJO et al. 1986).

Mutations in the LPL activator protein apoC-II may either result in absence of the protein or, rarely, in loss of its function but presence in plasma (SPRECHER et al. 1988). The underlying molecular defects that lead to apoC-II deficiency have been identified in several kindreds. All mutations identified in the apoC-II gene to date are single point mutations which result in the introduction of a premature stop codon, the substitution of the initiation codon for methionine, or the introduction of frameshift or donor splice site mutations (reviewed in CULOTTA and KOSHLAND 1992). This explains the severe deficiency or total absence of apoC-II in plasma from most patients.

It is not clear whether the clinical consequences of apoC-II deficiency are identical to those of LPL deficiency. It has been concluded that the extent of hypertriglyceridemia and the severity of clinical symptoms in patients with apoC-II deficiency appear to be less than in patients with LPL deficiency (SPRECHER et al. 1988; FOJO and BREWER 1992). However, recurrent pancreatitis was observed in German patients with apoC-II deficiency (BEIL et al. 1992).

It is presently not known how frequent complications and in particular pancreatitis occur in either condition. There are no systematic population-based studies on the spectrum of phenotypic expression in either LPL or apoC-II deficiency. All the information available is on clinical material, e.g., on patients that came to medical attention because of milky plasma or clinical symptoms. Thus, any data on the prevalence of clinical symptoms in type I patients, whether caused by apoC-II or LPL deficiency, are necessarily biased. In both patient groups, a low-fat diet is recommended as the only feasible long-term feasible therapy. Treatment of the acute phase of the disease, e.g., in patients with pancreatitis, is by starvation and electrolyte infusion.

C. Familial Hypercholesterolemia

FH exists in two forms: the heterozygous form, which has a frequency of about one in 500 in most populations, and the rare homozygous form, which is present in roughly one in every 1 million newborn babies (GOLDSTEIN and BROWN 1989). Considerably higher frequencies of FH have been noted in Afrikaners from South Africa (GEVERS et al. 1987), French Canadians from Quebec (HOBBS et al. 1992; BÉTARD et al. 1992), and some other populations (Table 5). FH is the best example of a monogenic inborn error of metabolism where the mutation produces both hypercholesterolemia and atherosclerosis.

I. Heterozygous Familial Hypercholesterolemia

The heterozygous form of FH was first recognized as a dominantly inherited syndrome by the Norwegian physician Müller (1938) and is characterized by hypercholesterolemia, xanthomatosis, and CHD. The classical lipoprotein abnormality is elevated LDL cholesterol, which is usually above the 95th percentile of the population. Average plasma total cholesterol concentrations are about 350 mg/dl. LDL cholesterol is usually elevated from birth, and the elevation persists throughout life. Agarose gel electrophoresis followed by lipid staining results in a pattern which is characterized by an increase in β-lipoproteins (type II hyperlipoproteinemia). Concomitant with the elevated LDL concentrations, apoB, which is the only LDL apolipoprotein, is also elevated.

Hypertriglyceridemia is present in some patients; however, this is probably not related to the primary defect, but rather reflects the prevalence of hypertriglyceridemia in the general population. Whether HDL is lower than

Table 5. Low-density lipoprotein receptor mutations in founder populations and isolates (HOBBS et al. 1992 and LEITERSDORF et al. 1993)

Ethnic group	Geographical location	Frequency of FH heterozygotes in populations	Mutant allele	Mutation	Proportion of FH heterozygotes with mutant allel (%)
Afrikaners	South Africa	1/100	FH Afrikaner-1	Gly206 → Glu	65–70
			FH Afrikaner-2	Val408 → Met	20–25
			FH Afrikaner-3	Asp154 → Asn	5–10
Ashkenazi Jews of Lithuanian ancestery	Israel South Africa	Not known	FH Lithuania	Δ Gly 197	80
Sephardic Jews	Israel	Not known	FH Sephardic	Asp147 → His	10
Christian Lebanese	Lebanon, Syria, and Israel	1/170	FH Lebanese	Stop 660	90
Finnish	Finland	~1/500	FH Helsinki	Δ Exon 16–18	30–40
			FH Karelia	Fs 287	34
French Canadians	Quebec Province	1/270	FH French Canadian-1	Δ Exon 1	60
			FH French Canadian-2	Cys646 → Tyr	5
			FH French Canadian-3	Glu207 → Lys	2
			FH French Canadian-4	Trp66 → Gly	7
			FH French Canadian-5	Δ Exon 2 and 3	5

Δ, Deletion; Fs, frameshift; FH, familial hypercholesterolemia.

in the general population is unclear. The only other consistent lipoprotein abnormality in FH patients is elevated Lp(a) (Utermann et al. 1989). High Lp(a) has been considered part of the clinical syndrome of FH. Though average Lp(a) levels are two- to threefold higher in FH patients than in controls, they are not high in all FH patients; they are, however, higher than expected for carriers of the respective apo(a) isoforms (see Sect. D). The reason for the elevation of Lp(a) in FH is at present still unclear.

In healthy humans, about 80% of LDL is cleared from plasma by receptor-mediated endocytosis (Brown and Goldstein 1986). A high-affinity LDL receptor is present on hepatic and most extrahepatic tissues (Fig. 1). This receptor, which has two ligands, apoB-100 in the LDL particle and apoE in β-VLDL (intermediate-density lipoprotein, IDL), is functionally deficient or absent in homozygous FH (see Sect C.II; Brown and Goldstein 1986). Heterozygotes for FH have one normal allele and one abnormal allele at the LDL receptor gene locus. Presence of only half the number of functionally normal receptors results in impaired binding, uptake, and degradation of LDL by cells and in an accumulation of LDL in the plasma compartment. FH is transmitted as an autosomal dominant trait with the rare occurrence of homozygotes. Considering either total cholesterol or LDL cholesterol as a marker for the disease, the dominant gene is highly penetrant at all ages (>0.9). A bimodal distribution of cholesterol and LDL cholesterol has been described in FH families but some overlap exists between normal and affected family members (Schrott et al. 1972). Some authors did not find the bimodal distribution of lipids in their families. An excellent example of the confounding effect of other genes that may obscure the bimodal distribution of cholesterol in FH families has been described in a large FH family from Puerto Rico (Hobbs et al. 1989). In this kindred, a dominant "cholesterol-lowering gene" segregated and family members who had inherited both a mutant LDL receptor gene and the cholesterol-lowering gene had normal LDL cholesterol levels. Measurement of LDL receptor activity on fibroblasts or lymphocytes also shows a bimodal distribution (Bilheimer et al. 1978; Steyn et al. 1989).

1. Low-Density Lipoprotein

The gene for the human LDL receptor encompasses 45 kb of DNA on the short arm of chromosome 19 (Franke et al. 1984). Over 150 mutations causing FH have been described in the LDL receptor gene to date (Hobbs et al. 1992). A total of 45 of the described mutations are large deletions and insertions. By measuring the binding, uptake, and degradation of LDL in fibroblast cultures, LDL receptor mutations were originally classified according to functional criteria into receptor-negative, receptor-defective, and internalization-defective mutations. Receptor-negative subjects have virtually no receptors at the cell surface. In contrast, receptor-defective is defined by a residual binding activity of approximately 2%–10% of controls.

Internalization-defective receptors bind, but do not internalize, LDL. The present classification system of LDL receptor mutations is based on the detailed knowledge of structure–function relationships in the LDL receptor which was obtained by characterization of the many natural mutations in the gene and of mutations produced by in vitro mutagenesis (Brown and Goldstein 1986; Russell et al. 1987; Hobbs et al. 1992). To understand the consequences and classification of LDL receptor mutations, one has to consider the physiology of the receptor in relationship to its structural domains (Fig. 3). The LDL receptor is synthesized in the endoplasmic reticulum (ER) as a precursor protein, processed to the mature form in the Golgi apparatus, and transported to the plasma membrane, where it binds LDL, clusters in coated pits, and is internalized together with its ligand in coated vesicles. By loss of their clathrin coat, endosomes are formed. Before endosomes fuse with lysosomes, where LDL is degraded, receptors and LDL are separated in the compartment of uncoupling of receptor and ligand

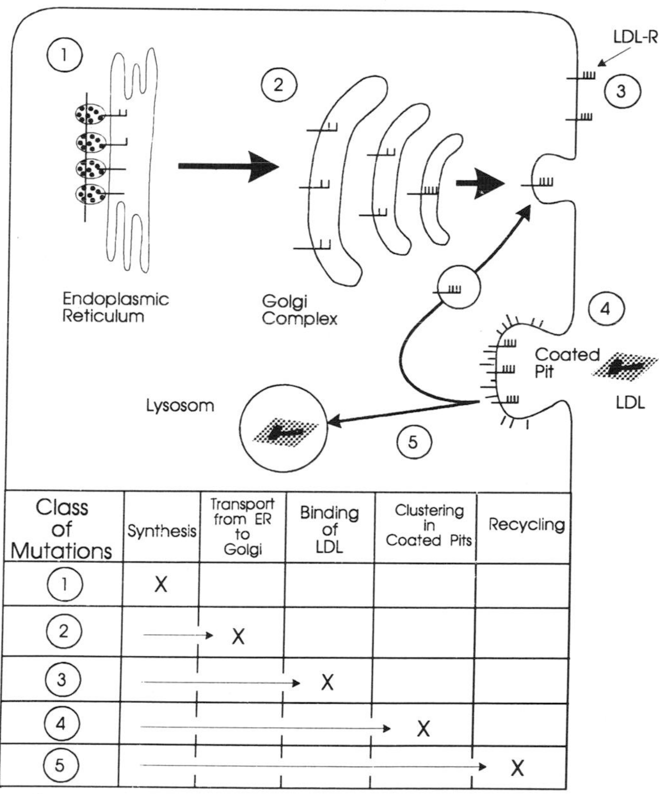

Class of Mutations	Synthesis	Transport from ER to Golgi	Binding of LDL	Clustering in Coated Pits	Recycling
1	X				
2	→	X			
3	→	→	X		
4	→	→	→	X	
5	→	→	→	→	X

Fig. 3. Classification of low-density lipoprotein (*LDL*) receptor (*LDL-R*) mutations. See text for explanation. *ER*, endoplasmic reticulum. (Adopted from Goldstein and Brown 1989)

(CURL) and LDL receptors recycle to the cell surface, from where they can undergo the next round of LDL binding and internalization.

Functional studies in cell culture have allowed LDL receptor gene mutations to be grouped into five major classes (Fig. 3). Class I mutants do not synthesize receptors; the null alleles result from large deletions, nonsense, frame-shift, or splicing mutations. Class II mutants are defined by defects in the processing and transport of the receptors from the ER to the Golgi compartment; few or no receptors occur at the cell surface. These are the most frequent mutations at the LDL receptor locus. Class III mutations are characterized by receptors that occur at the cell surface, but are defective in binding their ligand LDL. Receptors with class IV mutations occur at the cell surface and bind LDL normally, but are unable to cluster in coated pits and to internalize the bound LDL. Class V mutations are distinguished by the inability of the receptors to recycle to the cell surface, resulting in a deficiency of functional receptors at the plasma membrane.

A total of 127 mutations have been subjected to structure–function analysis, including analysis of binding and uptake of LDL, receptor biosynthesis, receptor transport, and gene structure. The mature LDL receptor protein is organized into five domains, which are represented by 17 exons in the gene (Fig. 4). The N-terminal ligand-binding domain contains seven cysteine-rich repeats which form a rather rigid structure. Each repeat forms

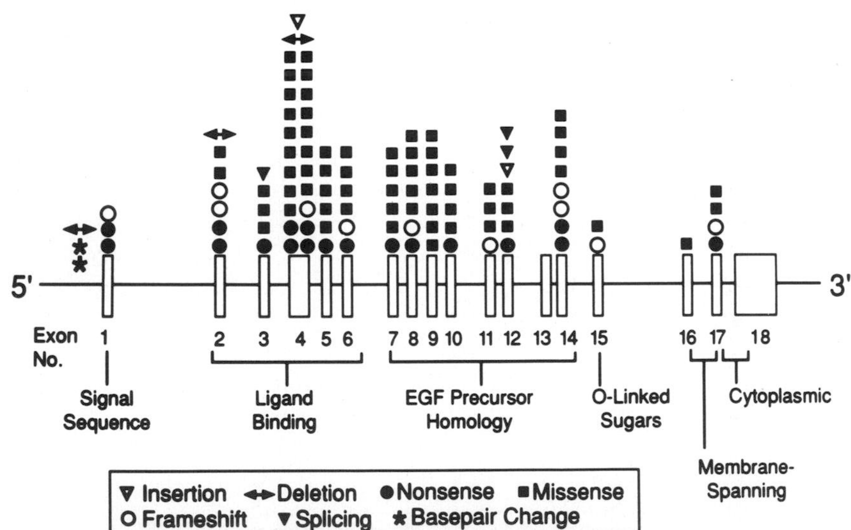

Fig. 4. Point mutations and small in-frame deletions/insertions (<25 bp) in the low-density lipoprotein (LDL) receptor gene in individuals with familial hypercholesterolemia (FH). Exons are shown as *vertical boxes* and introns as the *lines* connecting them. The map is drawn to approximate scale. Additional data for each mutation are given in Table 2 in HOBBS et al. (1992). *EGF*, epidermal growth factor. (From HOBBS et al. 1992, with permission)

a loop, which is stabilized by intrachain disulfide bridges. Repeats 2–7 are essential for LDL binding. Mutations which delete one or more of these repeats or missense mutations which effect critical residues (e.g., the conserved Cys residues) impair binding of LDL and result in binding-defective phenotypes (class III). These mutants are still able to bind IDL/β-VLDL. Repeat 5 in the binding domain is critical for β-VLDL binding. Deletion of this repeat results in the loss of apoE binding (RUSSELL et al. 1989). Mutations in the ligand-binding domain may also affect receptor transport (class II). Conversely, two thirds of class II mutations had mutation in the binding domain (HOBBS et al. 1992); many of these mutations directly or indirectly affect disulfide bond formation, preventing correct folding and passage of the quality control system of the ER.

Next follows a domain with homology to the epidermal growth factor (EGF) precursor. This domain is critical for the recycling of the receptor. The acid-dependent dissociation of receptor and LDL in the CURL is mediated by this domain. Loss of this function by mutation results in the cellular class V phenotype. All recycling-deficient alleles had mutations in the EGF precursor domain. Mutations in this domain may also effect the stability of the receptor (class II), but this is not specific and mutations in other parts of the receptor may also result in an unstable product. The domain where most 0-linked sugars are clustered is next to the outer plasma membrane. No class-specific effects of mutations in this domain have been reported.

The single hydrophobic transmembrane domain anchors the receptor in the lipid bilayer, but has no signal for clustering of receptors in coated pits, which is a prerequisite for internalization. Such a signal is contained in the cytoplasmic tail of the receptor. Tyr-807 is a critical residue in this region, facilitating interaction of the receptor with adaptin molecules in the clathrin coat of the pit. Mutations which result in substitutions of this residue for a nonaromatic amino acid result in a receptor which binds, but does not internalize, LDL (class IV). Likewise, C-terminal deletions, which include the membrane-spanning domain, result in receptors that are not internalized and moreover are shed from the cell surface due to loss of their membrane anchor.

It should also be noted that the classification is somewhat arbitrary, since several mutations affect more than one of the receptor functions, e.g., stability, transport, and ligand binding. The $Pro^{664} \rightarrow$ Leu mutation in the EGF precursor domain is an example for a mutation that affects both transport and ligand binding (SOUTAR et al. 1989). Three mutations in the LDL receptor are located in the promoter region of the gene.

The sterol-regulated expression of the LDL receptor gene is controlled by a short DNA sequence located within 200 bp of the initiator methionine codon. This sequence contains three imperfect direct repeats, two of which interact with transcription factor Sp1 (repeats 1 and 2). The second direct repeat contains a conditional-positive sterol regulatory element (SRE-1), which is required for high-level expression (GOLDSTEIN and BROWN 1990). All three promoter mutations affect one of the transcription factor Sp1-binding sites.

Several mutations in the LDL receptor gene have been observed in unrelated patients with different ethnic origin. Haplotype analysis demonstrated that the same mutation occurred on different chromosomes, suggesting an indepenent origin. Many of the recurrent mutations (16% of the point mutations) occur in CpG islands, as has also been observed in other genes (COOPER and YOUSSOUFIAN 1988).

A total of 45 large deletions and insertions have been reported in the LDL receptor gene. In ten of these, the deletion joints have been sequenced and nine were shown to possess Alu repeat elements at one or both sides of the joint. This strongly supports the hypothesis that Alu sequences are involved in genome rearrangements (BRITTEN et al. 1988).

2. Pathophysiology and Clinics of Familial Hypercholesterolemia

Whatever the class or mechanism of mutation, the consequences of all classes of LDL receptor mutations are similar: less LDL is delivered to the cells (Fig. 1). Most LDL receptors reside on liver cells, and the liver is the major site of LDL uptake and degradation in the human body. The impaired transport of LDL cholesterol from the plasma compartment to liver cells has several consequences for cholesterol metabolism. First, LDL concentration increases in plasma if the number of functional receptors is decreased at the liver cell surface. Second, decreased influx of cholesterol into liver cells results in an upregulation of the number of LDL receptors at the cell surface, which in part counteracts the loss of receptor activity caused by the mutation. Third, endogenous synthesis of cholesterol from its precursors is increased through activation of hydroxymethylglutaryl coenzyme A (HMG-CoA) reductase, and more lipoproteins are secreted from the liver. Together, this results in a new steady state with about two- to threefold elevated cholesterol levels in the plasma from heterozygotes and six- to ninefold elevation in homozygoes or compound heterozygotes. The fatal consequences of the elevated levels of LDL and their prolonged circulation in plasma are its deposition in the arterial wall, resulting in plaque formation, narrowing of the lumen of the vessels, and the clinical manifestations of atherosclerosis. The exact mechanism(s) by which LDL is deposited in the arterial wall is not yet completely understood, but uptake of oxidized or otherwise modified LDL by macrophages seems to be a crucial event (STEINBERG et al. 1989). Deposition of cholesterol in macrophages does also result in foam cell formation. These cells form the basis of the xanthomas frequently seen in FH patients. Clinical symptoms of the disorder are tendinous xanthomas, usually of the Achilles tendon and the extensors of the hands, xanthelasma, lipoid arcus of the cornea, and coronary atherosclerosis. The clinical course of heterozygous FH is that of a late onset disease. Though hypercholesterolemia, elevated LDL cholesterol, and high apoB are generally present from birth, the first clinical symptoms do not usually develop before the second decade. At this time, corneal arcus, xanthelasma, and tendon xanthomas may develop. By the third decade, corneal arcus and tendon xanthomas are present in about half of heterozygotes. They may be quite distinct and only be demonstrated by X-ray or ultrasound scanning of the Achilles tendon thickness (GOLDSTEIN and BROWN 1989). None of these symptoms is obligatory, even though as many as 80% of affected subjects ultimately develop xanthomas.

Progressive atherosclerosis and its fatal complications, myocardial infarction and stroke, are the most severe clinical signs of FH. The cumulative probabilty of CHD by the age of 60 years has been estimated to be 52% in affected males and 33% in affected females in a large study of over 1000 subjects from 116 kindreds with FH (STONE et al. 1974). The risk of CHD was about three to four times higher in affected subjects than in intrafamilial controls. The clinical course of FH is different in women from that in men, with all symptoms being delayed for about a decade in women. The mean age of onset of symptomatic CHD is about 45 years in affected males and 53 in females (HARLAN et al. 1966; GOLDSTEIN and BROWN 1989). The high risk of premature atherosclerosis in FH subjects is also evident from studies in survivors of myocardial infarction. Independent studies from Finland, Great Britain, and the USA have demonstrated a FH frequency of 3%–6% in unselected survivors of myocardial infarction compared to 0.1%–0.2% in the general population (GOLDSTEIN et al. 1973; NIKKILÄ and ARO 1973; PATTERSON and SLACK 1972). Though these studies may have included subjects with familial defective apoB-100 (FDB; see below), this does not invalidate the conclusions.

Not all patients with FH develop overt signs of CHD or die from myocardial infarction or stroke. It is a challange to find those factors which precipitate CHD, myocardial infarction, and stroke in FH patients – and those which protect against disease. Among the factors that accelerate atherosclerosis in FH patients are two which have also been identified as risk factors in the general population, but which may be of particular relevance in this high-risk group: one is low HDL, and the other is high Lp(a). As already discussed, many patients with FH have excessively high Lp(a) which is above the 90th percentile for the general population in about half to FH patients. Studies of patients from London and Sweden have shown that Lp(a) plasma level is the best predictor of CHD in FH patients (SEED et al. 1990; WIKLUND et al. 1990). Therefore, Lp(a) concentrations should be determined in all FH patients for risk assessment. The mechanism underlying the elevated Lp(a) levels in FH patients is at present unclear. Though Lp(a) contains apoB-100, which is a ligand for the LDL receptor, there is controversy as to whether Lp(a) removal from plasma is mediated by the LDL receptor. Together, the available data suggest that the LDL receptor pathway is not a major route for Lp(a) removal from human plasma in vivo.

Hence, the elevation of Lp(a) in FH patients must be something other than a direct consequence of the defective LDL receptor. Whether affected and nonaffected members from FH families differ in Lp(a) levels has not yet been ascertained either (SOUTAR et al. 1991; LEITERSDORF et al. 1991). One explanation for the conflicting results is that the elevation of Lp(a) in FH patients (and some of their non-FH family members) is due to selection bias. Because high Lp(a) is a strong risk factor for CHD is FH patients, Lp(a) levels may be elevated in patients who have been selected on the basis of clinical symptoms and also in some of their family members.

3. Diagnosis of Familial Hypercholesterolemia

Other forms of hypercholesterolemia have to be considered in the differential diagnosis of FH. Secondary hypercholesterolemia, polygenic hypercholesterolemia (the form of hypercholesterolemia present in one third of patients with FCH), and FDB all may be confused with FH. Given that the frequency of hypercholesterolemia in the population is one in 20 (taking the 95th percentile as the cut-off point), the chance of a pair of first-degree relative both being hypercholesterolemic is fairly high, and the chance of finding a second affected individual will increase with the number of family members tested. Thus, familial occurrence of hypercholesterolemia alone cannot be taken as evidence for the diagnosis FH (though the name does suggest this). The clinical diagnosis of FH heavily depends on the presence of xanthomas in a patient or first-degree relative. The diagnosis is presently accepted when at least two of three criteria are met: (1) plasma cholesterol with a percentile score of more than 95 in terms of the whole population; (2) presence of xanthomas in the patient; and (3) hypercholesterolemia and xanthomas in a first-degree relative. Except for FDB, patients with other forms of hypercholesterolemia rarely develop xanthomas and in particular do not develop xanthomas of the Achilles tendon. Xanthomas indistinguishable from those in FH may, however, develop in patients with FDB (Innerarity 1990) and in patients with autosomal recessive cerebrotendinous xanthomatosis. The latter patients have normal LDL cholesterol and may be distinguished by the presence of cataracts and mental deterioration. Xanthelasma and corneal arcus are also observed in normocholesterolemic subjects and may occur as a familial trait. It is impossible for any routine laboratory or even lipid clinic to make the diagnosis in an isolated patient with hypercholesterolemia, but no characteristic clinical symptoms. Some authors have developed sophisticated functional assays (LDL-binding assays), but the rate of false negatives or false positives of such assays is not known. Cloning and sequencing of the LDL receptor gene (RUSSELL et al. 1989; SÜDHOF et al. 1985) has allowed the direct diagnosis of FH by demonstrating the causal LDL receptor mutations (HOBBS et al. 1992). This approach presently still has practical limitations due to the large heterogeneity of LDL receptor mutations (HOBBS et al. 1992). However, in some isolated populations with founder effects, a high frequency of LDL receptor mutations, and prevalence of one or a few mutations, direct diagnosis of FH has been shown to be effective (see below) and even screening procedures may be feasible (AALTO-SETÄLÄ et al. 1988). Such populations include the Afrikaners from South Africa, French Canadians from Quebec, Askenazi Jews, Christian Lebanese, and Finns (GEVERS et al. 1987; HOBBS et al. 1992; BÉTARD et al. 1992; AALTO-SETÄLÄ et al. 1988; OPPENHEIM et al. 1991; RESHEF et al. 1992; HJELMS and STENDER, 1992). In all these populations, a few mutations represent the majority of defective LDL receptor alleles and in the Afrikaners, Christian Lebanese, and French Canadians, FH frequencies are much higher than in other populations (Table 5). A detailed analysis of LDL receptor mutations in Israel also demonstrated that DNA diagnosis of FH may be simplified if ethnicity and geographic origin are taken into account (RESHEF et al. 1992; OPPENHEIM et al. 1991; LEITERSDORF et al. 1993). Indirect diagnosis of FH by linkage in families has also been performed and may allow the correct classification of subjects that are borderline according to classical criteria (TAYLOR et al. 1988; TAYLOR et al. 1989; LEITERSDORF et al. 1989). This approach needs an informative family situation and, except in research, will rarely be applied for the diagnosis of FH heterozygotes in practical medicine. It may, however, be useful in the prenatal diagnosis of homozygous FH. Establishing the diagnosis of FH has a very practical consequence. Because of the autosomal dominant mode of transmission, every first-degree relative of a FH patient has a 50% risk of carrying the mutant allele and developing CHD. A family study is therefore indicated once the diagnosis of FH has been verified. This approach will make it possible to identify affected relatives before clinical complications have developed.

II. Homozygous Familial Hypercholesterolemia

The offspring of two heterozygous FH patients may be homozygous. Plasma LDL levels are elevated six- to ninefold, and total plasma cholesterol concentrations range from 600 mg/dl to over 1000 mg/dl in affected children. Gross hypercholesterolemia is present from birth and persists throughout life. Clinically, the condition is also more devastating than the heterozygous form of FH (GOLDSTEIN and BROWN 1989). There is massive tissue deposition of cholesterol, and cutaneous xanthomas that have a typical yellow–orange color may be present at birth. They have been found in every patient by the age of 4 years. A grayish corneal arcus and tendon xanthomas develop in childhood. Xanthomatous infiltrations are found throughout the arterial vascular system. A typical form may affect the aortic valve and may result in stenosis. Xanthomatous plaques and thickening of the endocardial surface of the mitral valve may cause regurgitation and stenosis. The cholesterol in the xanthoma is present within histiocytic foam cells mostly in the form of cytoplasmic droplets that are not bound to membranes. Typical atherosclerotic plaques are found in the arteries, including coronary arteries, thoracic and abdominal aorta, and also the major pulmonary arteries. In addition to severe atherosclerosis, there are intimal infiltrations of xanthomatous foam cells in the vessels which are similar in histologic appearance to those in the tuberous xanthomas. Other clinical and laboratory findings include painful and inflamed joints, cardiac murmurs, and a persistently elevated sedimentation rate. Myocardial infarction generally occurs as early as the first two decades of life, and the generalized atherosclerosis frequently results in death from myocardial infarction before 30 years of age (for review see GOLDSTEIN and BROWN 1989). However, environment seems to be important in precipitating clinical disease even in homozygotes. In a large series of FH patients from Japan, six survived beyond the third decade (MABUCHI et al. 1977) and the oldest reported living FH subject is a 57-year-old internalization-defective patient from Japan (KOMURO et al. 1987).

The parents of FH homozygous propositi are obligate heterozygotes and have the dominant form of FH, but not necessarily identical LDL receptor mutations. In face, most patients (60%) with clinically homozygous FH are genetic compounds. True homozygotes do occur in inbred populations where a large proportion of heterozygotes carry the same mutation (Table 5) or in consangineous marriages. The frequency of parental consanguinity was more than 30% in a study from Japan (MABUCHI et al. 1977). Parents of homozygous offspring occasionally have normal lipid levels even when identified as heterozygotes by receptor-binding assays or DNA studies. This probably reflects the influence of other genes (HOBBS et al. 1989) and environment.

III. Familial Defective ApoB-100

FDB was first recognized by kinetic studies in a subject with mild hypercholesterolemia (VEGA and GRUNDY 1986). The turnover of iodine-labeled LDL demonstrated a prolonged half-life of endogenous, but not foreign, LDL in the subject. LDL receptor function was normal. This suggested that a defect in the LDL rather than its receptor resulted in impaired uptake of LDL in this patient. Defective binding of LDL to the LDL receptor was subsequently demonstrated in fibroblast-binding assays. Using the monoclonal MB 19 antibody, which recognizes a biallelic polymorphism of apoB-100, two LDL populations were separated from subjects which were heterozygous for the MB 19 alleles and had FDB. One LDL fraction had normal binding to the receptor, whereas the other did not bind at all (INNERARITY et al. 1990). This is in agreement with the dominant transmission of FDB in families and suggested that FDB subjects are heterozygotes with one normal and one defective apoB allele. This was confirmed by sequencing the apoB genes from FDB patients and controls. All FDB patients had a G to A transition at position 10708 of the cDNA, resulting in the substitution of an Arg for a Gln at position 3500 of the protein. This mutation is in the region which has been identified as the receptor-binding domain of apoB-100 (SORIA et al. 1989).

Because LDL receptor function is normal in FDB subjects, the normal LDL present in the plasma from heterozygotes is taken up by receptor-mediated endocytosis, whereas LDL containing the mutant apoB-3500 are not. This not only results in an increase in total LDL, but also in a distorted ratio of mutant to normal LDL particles. About 60%–70% of the LDL from FDB patients contains the mutant apoB.

1. Population Genetics of Familial Defective ApoB-100

The apoB-3500 mutation has only been detected in Caucasian populations, with the notable exception of one Chinese subject. In all Caucasian FDB patients, the mutation is associated with the same apoB haplotype, suggesting that it represents an old mutation that has spread in Caucasian populations. The first FDB families identified in the USA were of Austrian ancestry (FRIEDL et al. 1991). Subsequently, patients with FDB were detected in Germany (RAUH et al. 1992), Denmark (TYBAERG-HANSEN et al. 1990), the Netherlands (see SCHUSTER et al. 1992a), and Great Britain (TYBAERG-HANSEN et al. 1990). The frequency of the mutation in North American Caucasian and European populations was estimated as being on the order of 1:600, making FDB one of the most common genetic causes of hypercholesterolemia. This frequency estimate was obtained by extrapolation from frequencies in mixed groups including patients with hypercholesterolemia and/or CHD. Such extrapolations are necessarily biased. The FDB mutation was not detected in Finland (HÄMÄLÄINEN et al. 1990).

2. Diagnosis of Familial Defective ApoB-100

Since FDB results from a single base substitution in the apoB gene, diagnosis is straightforward. Polymerase chain reaction (PCR) amplification followed by demonstration of the mutant allele is the method of choice. Originally, dot blot or slot blot/allele-specific oligonucleotide (ASO) was developed (SORIA et al. 1989). More recently, different variants of the amplification refractory mutation system technique have been applied (MOTTI et al. 1991; SCHUSTER et al. 1992b; MAMOTTE, VAN BOCKXMEER, 1993). These techniques are usually used on preselected patients that either have hypercholesterolemia, CHD, or both.

3. The Familial Defective ApoB-100 Phenotype

The original data on a few FDB patients and their families suggested that hypercholesterolemia is more moderate in FDB patients than in FH patients with LDL receptor defects and that the clinical consequences might be less severe. Average total cholesterol levels were 260 mg/dl, compared with 350 mg/dl in classical FH. None of the initial FDB patients had xanthomas.

This finding is in agreement with physiological considerations. Subsequently researchers from Denmark, Germany, and the USA have reported that their FDB patients have lipid levels and clinical symptoms indistinguishable from FH heterozygotes with defective LDL receptors. FDB patients have been identified who had high cholesterol, xanthomatosis, and CHD, as did their counterparts with FH, due to LDL receptor defects. These studies have demonstrated that overlap between the biochemical and clinical phenotypes of LDL receptor- and apoB-defective patients exists. This does not mean that the spectrum and frequencies of biochemical abnormalities and clinical findings are the same in the two groups.

In a population of patients diagnosed by classical criteria as having FH, 4%–6% were found to have the FDB mutation (TYBAERG-HANSEN et al. 1990; RAUH et al. 1992). This is lower than expected under the assumption that both diseases have a similar frequency and a similar phenotype. Either FDB is less frequent or the clinical phenotype is less severe. It is obvious that FDB patients identified among patients which fulfill the criteria for FH will themselves fulfill these criteria. In other words, they may not be significantly different from the group they are derived from.

In principal, a mutation resulting in a defective LDL receptor is expected to produce the same phenotype as one which results in defective binding of the ligand apoB. However, this does not take into account the fact that there is another ligand for the LDL receptor, apoE. LDL is generated in plasma from an apoE-rich precursor particle called VLDL remnant, β-VLDL, or IDL. This particle is also a ligand for the LDL receptor. Whereas in the situation of a defective LDL receptor, the removal of both LDL and IDL is impaired, mutations in apoB impair LDL, but not IDL, removal. Thus, the defect in FDB is expected to be less severe than the defect in classical FH (Fig. 1). This is supported by two other findings that at first glance seem surprising. One is that cholesterol and LDL cholesterol levels in recently described homozygotes for the apoB-3500 mutation are not significantly different from those in FDB heterozygotes (MÄRZ et al. 1992). The other is that FDB heterozygotes respond to therapy with HMG-CoA reductase inhibitors, though in theory one had expected they should not. One mechanism by which HMG-CoA reductase inhibitors lower cholesterol is by increasing the number of LDL receptors at the cell surface. Such an increase will have no effect on the concentration of a defective ligand.

Both phenomena may, however, be explained by the same physiological mechanism. Lack of uptake of defective LDL by the LDL receptor pathway in FDB homozygotes is expected to result in an upregulation of LDL receptors, because less cholesterol from LDL enters the cells. This will increase the uptake of apoE-containing LDL precursors (IDL, β-VLDL). Because the affinity of the LDL receptor for apoE is several times higher than for apoB, this might result in the effective removal of LDL precursors from plasma and hence decreased LDL production. By the same mechanism, HMG-CoA reductase inhibitors may lower cholesterol in FDB heterozygotes.

By increasing the uptake of LDL precursor lipoproteins (IDL) through upregulation of LDL receptor, less particles are channeled into the LDL pool. Mixed heterozygotes for FDB and FH were described, but had no higher cholesterol and LDL cholesterol levels than patients with either defect alone (RAUH et al. 1991).

D. Lp(a) Hyperlipoproteinemia

Elevated Lp(a) levels in plasma are a genetic risk factor for premature CHD and stroke (UTERMANN 1989; SCANU and FLESS 1990). The distribution of Lp(a) concentrations in Caucasian and Asian populations is very broad and highly skewed towards lower levels. Variation ranges from less than 0.2 mg/dl to over 120 mg/dl in unselected healthy subjects, but may exceed 200 mg/dl in certain disease conditions. Within healthy subjects, Lp(a) levels are genetically fixed and not significantly affected by diet, exercise, or even most lipid-lowering drugs. In agreement with this, genetic studies have shown that Lp(a) represents a continuous quantitative genetic trait (UTERMANN 1989). As for other lipoproteins or lipids, elevation is defined by levels exceeding the 95th percentile in the total population. For most Caucasian populations, this cut-off point is at approximately 50 mg Lp(a)/dl; Lp(a) hyperlipoproteinemia is therefore characterized by Lp(a) levels exceeding this value. Some factors that result in Lp(a) hyperlipoproteinemia have been identified, the most important being variation at the gene locus for apo(a), one of the two principal protein constituents of Lp(a) (SCANU and FLESS 1990). In most subjects, high Lp(a) levels are caused by alleles at the apo(a) locus. In order to understand the genetics of the Lp(a) trait and of Lp(a) hyperlipoproteinemia, it is essential of understand the genomic organization of and heterogeneity at the apo(a) locus.

I. Structure and Evolution

Lp(a) combines structural elements of the lipid-transport and of the blood-clotting systems. It is assembled from LDL – which contains apoB-100 and all the lipids of the complex – and from apo(a). Apo(a) is a member of the superfamily of trypsin-like proteinases and is closely related to plasminogen (PLG) from which it evolved during mammalian evolution (MCLEAN et al. 1987). PLG contains two different modules in its mature structure. One is represented by the protease domain. The other is a structure which is stabilized by three internal disulfide bonds and is called a kringle. Five such kringles exist in PLG with differing primary structure (see MCLEAN et al. 1987). Apo(a) has a protease domain with 80% homology to the PLG protease domain, though there are some notable differences. It is still controversial whether or not apo(a) functions as a protease in vivo. Further, human apo(a) contains two types of kringles. One has homology to kringle V of PLG and is present once. The other resembles PLG kringle IV and is present in multiple copies, several of which are 100% identical in DNA and amino acid sequence. Others show minor sequence variations. These "unique" kringles, which are represented by those numbered 30–37 in the cDNA sequence of McLEAN et al. (1987), may have distinct functional properties, including lysine binding (K 37) and involvement in interchain disulfide bridge formation with apoB-100 (K 36). Apo(a) does not contain kringle I-, II-, or III-like structures. Together, the structural data suggest a scenario in which apo(a) has evolved from PLG by duplication, deletion, gene conversion, and mutation around the time of divergence of Old and New World monkeys. This scenario is compatible with the presence of Lp(a) in primates and Old World, but not New World, monkeys and its absence in lower mammals. It does not, however, explain the presence of Lp(a) in the European hedgehog, where apo(a) may have evolved by convergent evolution.

II. Genetics

The apo(a) gene is located on human chromosome 6q2.6–q2.7 in close proximity to the PLG gene and is highly polymorphic due to differences in the number of K IV repeats in individual alleles of the gene (UTERMANN 1989; LACKNER et al. 1991; KRAFT et al. 1992). Between 11 and more than 40 PLG-like K IV repeats may be present in an apo allele. This heterogeneity is exceptional for a transcribed gene and is demonstrated by pulsed field gel electrophoresis of genomic DNA using restriction enzymes for digestion which do not cut in the K IV repeat domain of the gene. With several restriction enzymes the complete block of K IV repeats is obtained in one fragment, which varies in size depending on the enzyme used and the individual investigated. For practical reasons, KpnI has been used to investigate apo(a) alleles (LACKNER et al. 1991; KRAFT et al. 1992). With this enzyme, fragments ranging in size from 32 to more than 200 kb are generated. These differ in size by increases of 5.6 kb, which corresponds to one genomic kringle IV unit. This DNA polymorphism is reflected in the size variation of apo(a) mRNA (HIXSON et al. 1989; KOSCHINSKY et al. 1990) and explains the size polymorphism of apo(a) (UTERMANN et al. 1987). The protein polymorphism is detected by immunoblotting following size separation of apo(a) isoforms by sodium dodecyl sulfate (SDS) polyacrylamide or agarose gel electrophoresis (UTERMANN et al. 1987; KAMBOH et al. 1991). Historically, the protein polymorphism was the first to be recognized.

III. Genetics of Lp(a) and of Lp(a) Hyperlipoproteinemia

The apo(a) size polymorphism is associated with Lp(a) plasma levels. An inverse correlation exists between apo(a) size and Lp(a) concentration in

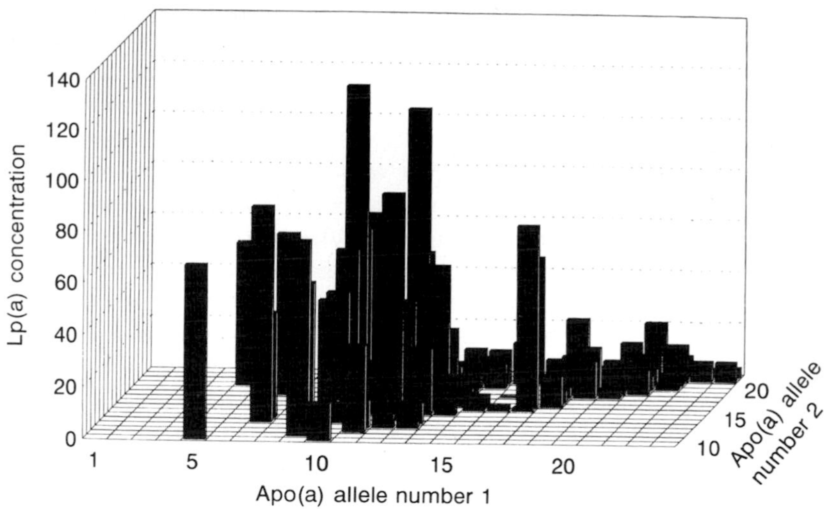

Fig. 5. Relation between apo(a) alleles and Lp(a) concentration. Graphical representation of the relationship between apo(a) Kpn I fragments and Lp(a) concentration in plasma. The two Kpn I fragments in a subject are denoted as allele number 1 and allele number 2. Note the decrease in Lp(a) concentration with the increase in allele number (= apo(a) Kpn I fragment size). (From KRAFT et al. 1992, with permission)

plasma. This has been demonstrated for protein isoforms and apo(a) DNA alleles (Fig. 5). In quantitative terms, the apo(a) size polymorphism explains about 45% of the variation in Lp(a) levels in Caucasians. Family studies have demonstrated the cosegregation of apo(a) alleles with Lp(a) concentration. Using a sib-pair linkage approach, it has been shown that more than 90% of the variability in Lp(a) concentrations within a population is explained by variation at the apo(a) gene locus (KRAFT et al. 1992; BOERWINKLE et al. 1992). According to a number of twin studies, Lp(a) levels are almost entirely under genetic control. Together, the data show that Lp(a) levels are almost entirely controlled by the apo(a) locus, but that only about half of the variation is explained by the size polymorphism. Lp(a) alleles appear to effect Lp(a) levels in an additive fashion. The Lp(a) concentration in an individual is the sum, rather than the mean, of the concentrations determined by each allele singly. In heterozygotes the two isoforms are present on different Lp(a) particles which occur with different concentrations in plasma. In other words, each apo(a) allele determines the concentration and physicochemical properties of the resulting Lp(a) particle. The Lp(a) concentrations determined by the two apo(a) alleles in a heterozygous subject may be similar, but may also be extremely different. In the latter case, this may result in apparent dominance. At the population level, there is, however, a continous distribution of "Lp(a) level alleles" (i.e., apo(a) alleles which determine a distinct Lp(a) concentration) with no obvious discontinuity. Therefore, Lp(a) concentrations are under the control of a large number of codominant alleles.

Lp(a) hyperlipoproteinemia may ensue when an individual inherits two alleles which both determine high Lp(a) concentrations or when a subject inherits one allele which determines a very high Lp(a) level. Although it is clear that Lp(a) hyperlipoproteinemia may be the result of single gene inheritance at the apo(a) locus, it is less clear whether and to what extent other factors may also contribute to the development of excessive Lp(a) concentrations. Lp(a) levels are about two- to threefold higher in patients with heterozygous FH than in controls. The patients have mutations in one LDL receptor gene, resulting in a reduction of functional LDL receptors at the cell surface. It has, therefore, been suggested that mutations in the LDL receptor gene may affect Lp(a) levels (UTERMANN et al. 1989), although this view has been challenged by recent family studies (SOUTAR et al. 1991). Possibly, the association of FH with high Lp(a) has other reasons, one being selection bias.

Other conditions which have been associated with Lp(a) hyperlipoproteinemia include end-stage renal disease and diabetes mellitus. Whereas reports on Lp(a) in diabetes are controversial and Lp(a) concentrations are certainly not dramatically increased in metabolically well-controlled type I or type II diabetics (CSÁSZÁR et al. 1993), the situation is different in kidney disease. Lp(a) levels are two- to threefold higher than normal in patients under hemodialysis or peritoneal dialysis (DIEPLINGER et al. 1993). Excessive Lp(a) hyperlipoproteinemia may be present in patients with nephrotic syndrome with values exceeding 300 or even 400 mg/dl. The mechanism of this dramatic increase in Lp(a) is unclear, but the high Lp(a) levels in kidney disease may explain part of the increased risk for atherosclerotic vascular disease in this patient group.

IV. Lp(a) Hyperlipoproteinemia and Atherosclerotic Vascular Disease

Most, though not all, studies have found an association of Lp(a) hyperlipoproteinemia with atherosclerotic vascular disease (SCANU and FLESS, 1990). An increased risk of CHD, stroke, and peripheral vascular disease (PVD) in subjects with elevated Lp(a) has been observed in several case control studies. Three prospective studies have confirmed the association of high Lp(a) with CHD (ROSENGREN et al. 1990; SIGURDSSON et al. 1992; DAHLEN et al. 1975), whereas one failed to find a significant risk associated with high Lp(a) (JAUHIAINEN et al. 1991). No prospective studies exist for stroke and PVD. In all positive studies where an appropriate analysis was performed, an increased risk was found for subjects with Lp(a) in the highest percentiles e.g., for subjects with Lp(a) hyperlipoproteinemia.

Lp(a) hyperlipoproteinemia is independent from other known risk factors, including elevated total cholesterol, LDL cholesterol, and HDL cholesterol levels, but a strong interaction between elevated Lp(a) and elevated LDL cholesterol exists. In subjects with elevated total cholesterol and Lp(a) hyperlipoproteinemia, the relative risk for CHD is about six times higher. In patients with FH that have LDL cholesterol levels above the 95th percentile of the population, Lp(a) levels are the best predictor for CHD. Many of these patients have excessive Lp(a) hyperlipoproteinemia (SEED et al. 1990; WIKLUND et al. 1990), and high Lp(a) has been included as part of the clinical syndrome of FH.

E. Hyperlipoproteinemia Type III

Hyperlipoproteinemia type III occurs in three genetically distinct forms: (1) a very rare recessive form which is due to apoE deficiency (SCHAEFER et al. 1986; LOHSE et al. 1992), (2) a rare dominant form (RALL and MAHLEY 1992; MAHLEY et al. 1990; DE KNIJFF et al. 1991), and (3) a common form that has a multifactorial etiology and is seen in more than 95% of Caucasian type III patients (UTERMANN 1987a). In all three forms, the major or sole defect is in the apoE gene.

I. ApoE Polymorphism

Prior to any discussion of type III hyperlipoproteinemia, it is necessary to introduce the polymorphism of apoE and the effect of common alleles at this gene locus on the normal variation of lipid levels. ApoE is a major constituent of plasma chylomicrons, VLDL, remnant lipoproteins, and of a subfraction of HDL (MAHLEY 1988). ApoE is a ligand for the LDL receptor, and lipoprotein particles enriched in apoE bind in vitro to the LDL receptor-related protein (LRP), which is believed to remove chylomicron remnants and β-VLDL from plasma (BROWN et al. 1991). ApoE exhibits a genetic polymorphism which was originally demonstrated by isoelectric focusing of delipidated, triglyceride-rich lipoproteins (UTERMANN et al. 1977). Today, the polymorphism is either investigated by electrofocusing of total plasma followed by western blotting (MENZEL and UTERMANN 1986) or by restriction analysis of PCR-amplified DNA (HIXSON and VERNIER 1990). The polymorphism is controlled by three common alleles designated $\varepsilon 2$, $\varepsilon 3$, and $\varepsilon 4$ at the apoE gene locus on chromosome 19. These alleles determine six phenotypes: apoE-4/2, apoE-4/3, apoE-4/2, apoE-3/3, apoE-3/2, and apoE-2/2. The three common gene products, apoE-4, apoE-3, and apoE-2, differ by single amino acid substitutions and have different functional properties (Table 6). In comparison to apoE-3, which is the most frequent isoform in all populations that have been studied so far, apoE-2 (Arg-158 $\rightarrow$ Cys) has less than 2% of the binding activity to the LDL receptor (SCHNEIDER et al. 1981; MAHLEY 1988). The plasma clearance of lipoproteins containing apoE-2 is therefore delayed. As a consequence, subjects which are homozygous for the binding-defective apoE-2 (Arg-158 $\rightarrow$ Cys) have a metabolic block

Table 6. Properties of common apolipoprotein E isoforms (from RALL and MAHLEY 1992)

	ApoE-4	ApoE-3	ApoE-2
LDL receptor binding	100%	100%	2%
Preferential lipoprotein association	VLDL	HDL	IDL + HDL
Sequence difference	Arg 112	Cys 112	Cys 112
	Arg 158	Arg 158	Cys 158

LDL, low-density lipoprotein; VLDL, very low density lipoprotein; HDL, high-density lipoprotein; IDL, intermediate-density lipoprotein.

that results in an accumulation of VLDL remnants (IDL or β-VLDL; Fig. 1). Chylomicron remnants which are believed to be cleared by a separate "remnant receptor" (LRP) also accumulate, but the mechanism for this is presently unclear. Surprisingly, most E-2 homozygotes have no hyperlipidemia, but rather are hypocholesterolemic (UTERMANN et al. 1977). This paradox is still not completely understood, but it has been hypothesized that a delayed clearance of remnant lipoproteins would result in a decreased cholesterol influx into the liver (especially from exogeneous lipoproteins), resulting in an upregulation of LDL receptors followed by a decrease in plasma LDL (UTERMANN 1985; BOERWINKLE and UTERMANN 1988). ApoE-2 (Arg-158 $\rightarrow$ Cys) also affects the LPL-mediated interconversion of β-VLDL/IDL to LDL. As a consequence, the homozygous apoE-2/2 phenotype, which has a frequency of about 1% in most populations, is associated with an inborn error of lipid metabolism termed primary dysbetalipoproteinemia (UTERMANN et al. 1977). This phenotype is characterized by subnormal total cholesterol concentrations (on the average 30–40 mg/dl below the mean of the respective population), an elevation of cholesterol-rich remnant particles, which may be demonstrated as so-called β-VLDL by agarose gel electrophoresis, and very low LDL concentrations.

Only occasionally do apoE-2/2 subjects develop hyperlipoproteinemia type III (see below). In contrast to apoE-2, apoE-4 exhibits enhanced in vivo catabolism (GREGG et al. 1986), which is probably mediated through preferential association of this isoform with triglyceride-rich lipoproteins. The functional differences between apoE isoforms are also expressed in heterozygotes and explain the significant effects of the apoE polymorphism on lipid and lipoprotein levels. In all populations, the average effect of the $\varepsilon 2$ allele is to lower plasma cholesterol by about 15 mg/dl. The average effect of the $\varepsilon 4$ allele is to increase plasma cholesterol by 5–10 mg/dl (Fig. 6). Whereas the effect of the $\varepsilon 2$ allele is believed to be the same across populations independent of ethnicity, dietary, and other possible environmental factors, it is not clear whether the effect of $\varepsilon 4$ is also identical among populations with different genetic and environmental backgrounds (UTERMANN 1987a; HALLMAN et al. 1991). Interactions are suggested by the obser-

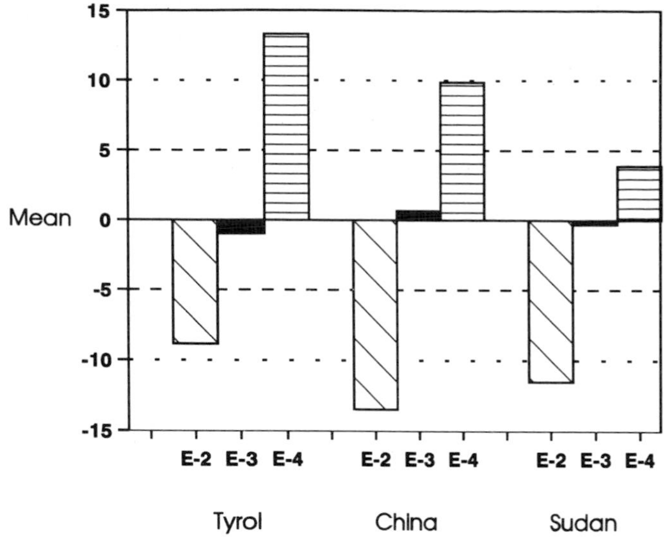

Fig. 6. Effect of apolipoprotein E alleles on total cholesterol concentrations in three populations. On the *y-axis*, the mean deviation of total cholesterol in mg/dl from the mean of the total cholesterol concentration of the respective population is given

vation that subjects with different apoE types differ in their responses to diet and to drug therapy from E-3 homozygotes (WEINTRAUB et al. 1987; DAVIGNON et al. 1988). Subjects which differ in apo-E genotype develop different forms of hyperlipoproteinemia if their lipoprotein removal system is challenged by other factors. Whereas apoE-2 homozygotes may develop type III hyperlipoproteinemia, apoE-4 homo- and heterozygosity are associated with hypercholesterolemia.

II. Multifactorial Type III Hyperlipoproteinemia

About one in 50 apoE-2 homozygotes develops severe type III hyperlipoproteinemia (broad beta disease, remnant-removal disorder; for review see MAHLEY and RALL 1989). The lipoprotein abnormality in type III hyperlipidemia is like the one in primary dysbetalipoproteinemia, with the only exception being that affected subjects have gross hyperlipidemia. Both cholesterol and triglycerides are elevated; cholesterol concentrations are usually over 300 mg/dl, and triglyceride levels tend to exceed those of cholesterol. The characteristic lipoprotein abnormality in type III hyperlipoproteinemia is the accumulation of cholesterylester-rich remnant lipoproteins of intestinal and endogenous origin. These lipoproteins are intermediate in density, size, chemical composition, and electrophoretic mobility between triglyceride-rich lipoproteins and LDL. Chylomicron-like particles which are rich in cholesterylester and represent chylomicron remnants may be present in fasting plasma. Agarose gel electrophoresis followed by lipid staining therefore results in a broad, unresolved band extending from the β- to the pre-β-region, and frequently "chylomicrons" are present at the origin. The abnormal lipoproteins float with the triglyceride-rich pre-β-lipoproteins upon ultracentrifugation, but exhibit β-mobility by agarose gel electrophoresis. Hence their name, "floating β-lipoproteins" or "β-VLDL." β-VLDL are depleted of apoC and enriched in apoE. LDL concentrations are typically subnormal, and HDL levels low in type III patients. Lipid levels and lipoprotein pattern are extremely sensitive to caloric intake, and consequently there is large variation in cholesterol and triglyceride levels, patients sometimes exhibiting a type V, rather than a type, III lipoprotein phenotype. Hyperlipoproteinemia type III is rarely seen before the age of 20 years, and there are only a few reported cases where the lipoprotein abnormality already existed in childhood. The most characteristic clinical manifestations of type III hyperlipoproteinemia, which is believed to be pathogenomic, are planar yellowish lipid deposits in the creases of the palm, which are called xanthoma striata palmaris. About 60% of patients reportedly have this type of xanthomas. Tuberous xanthomas are present in a fraction of type III subjects, and eruptive xanthomas and xanthelasma may occur. Tendon xanthomas and lipid arcus of the cornea, which characterize FH, are rare in type III. CHD occurs in about one third of patients, and one third has peripheral vascular disease. In a large series of patients at the National Institute of Health, ischemic vascular disease in men was diagnosed at a mean age of 39 years, but not until 10 years later in women. The frequencies of symptoms and age of onset of clinical disease do, however, depend on the mode of selection and vary between different studies (BORRIE 1969; MISHKEL et al. 1975; MORGANROTH et al. 1975; MOSER et al. 1974). All the data are probably severely biased by selection of patients. No data on unselected groups of E-2/2 homozygotes with type III hypercholesterolemia exist to date. However, there is no doubt that type III hyperlipoproteinemia represents an increased risk for premature CHD. In a study of survivors of myocardial infarction, 1% were E-2/2 homozygotes with type III hyperlipoproteinemia, whereas the frequency of type III hyperlipoproteinemia in the general population is one in 2000 at most (UTERMANN et al. 1984). The familial occurrence of type III hyperlipoproteinemia has been known since the pioneering work of FREDRICKSON et al. (1967), but the genetics of the lipid abnormality remained controlversial for years. It is now recognized that three genetic forms of the disease exist: (1) a frequent multifactorial forms, (2) a rare dominant form, and (3) a recessive form due to apoE deficiency. Abnormalities in apoE underly all forms of type III. Whereas mutations in some critical residues in the receptor-binding domain result in expression of type III hyperlipidemia in heterozygotes and dominant transmission in families, the etiology of the most common form of the disorder is more complex. Homozygosity for the apoE-2 (Arg-158 → Cys) mutation is a prerequisite, but is not sufficient for the development of type III hyperlipidemia. Other factors – genetic and/or nongenetic – are required to precipitate type III hyperlipoproteinemia in E-2 (Arg-158 → Cys) homozygotes (UTERMANN et al. 1979). Family studies of patients with severe type III hyperlipoproteinemia have demonstrated the occurrence of various forms of hyperlipidemia in about 50% of first-degree relatives. Some kindreds showed evidence of the coexistance of FCH. ApoE alleles segregated independently from hyperlipidemia in the families (UTERMANN et al. 1979).

Some authors have categorized this form as recessive type III hyperlipoproteinemia. This is, however not precise, unless one accepts a frequency of 1% for this lipid disorder. The inheritance of dysbetalipoproteinemia seen in virtually all E-2 (Arg-158 → Cys) homozygotes is indeed recessive, but hyperlipoproteinemia type III is not. Depending on the form of coinherited hyperlipidemia, the disorder may be dimeric in some and oligogenic or polygenic in others. Environmental and endogeneous factors also contribute to the phenotypic expression of the hyperlipoproteinemia. Among these are diet, drugs, and hormones. Hyperthyroidism and estrogen treatment can completely eliminate the hyperlipidemia – but not dyslipidemia – in type III subjects, whereas hypothyroidism markedly exaggerates the lipoprotein abnormality. Severe nephropathy may also precipitate type III hyperlipoproteinemia in an apoE-2 homozygote. In some patients, cholesterol may be lowered from over 500 mg/dl to less than 200 mg/dl solely by a stringent diet, but others may be refractory to diet and even fibrate therapy. Thus, different combinations of genetic and environmental factors cause the common form of type III hyperlipidemia that exists in over 95% of all Caucasian patients. The only factor in common is homozygosity for the binding-defective apoE-2 (Arg-158 → Cys) isoform, which acts as a recessive major gene defect. Because the common form of type III hyperlipidemia is caused by an interplay of different genes interacting with endogeneous and environmental factors, it is correctly defined as a multifactorial disorder. Due to the high frequency of the apoE-2 allele in the population and the prevalence of hyperlipidemia in type III in families, a vertical transmission of the lipid disorder has been observed in many kindreds. Here, the vertical transmission of the dyslipidemia represents pseudodominance (UTERMANN et al. 1977).

III. Dominant Type III Hyperlipoproteinemia

Several rare mutations in the apoE gene have been described which are also associated with type III hyperlipidemia. Among these are some which cause type III in heterozygotes and obviously do not require other factors for the development of hyperlipidemia (dominant type III hyperlipidemia). These include apoE-2 (Arg-145 → Cys) and E-2 (Lys-146 → Glu; Fig. 7, Table 7). One of the best documented cases of dominant type III hyperlipidemia is caused by the apoE-3 Leiden mutation, which is due to a seven-amino acid insertion in the apoE gene (Table 7).

Table 7. Genetic variants of apolipoprotein E (from RALL and MAHLEY 1992)

Designation or IEF position	Mutations	Binding activity	Association with hyperlipoproteinemia	Mode of inheritance of dysbetalipoproteinemia
E-3	"Wild type"	100%	No	
E-4*	Cys-112 → Arg	100%	No	–
E-2*	Arg-158 → Cys	<2%	Yes	Recessive
E-0	Trp-210 → Stop	–	Yes	Recessive
E-1	Arg-158 → Cys, Leu-252 → Glu	n.d.	Yes	–
E-1	Arg-158 → Cys, Gly-127 → Asp	4%	Yes	Recessive
E-1 Harrisburg	Lys-146 → Glu	Yes?	Yes	Dominant
E-2 Christchurch	Arg-136 → Ser	40%	Yes	Dominant
E-2	Arg-145 → Cys	45%	Yes	Dominant
E-2 Dunedin	Arg-228 → Cys	100%	Yes	–
E-2	Lys-146 → Gln	40%	Yes	Dominant
E-2	Arg-134 → Gln	n.d.	No	–
E-2	Val-236 → Gln	n.d.	Yes	–
E-3	Ala-99 → Thr, Ala-152 → Pro	n.d.	No	–
E-3	Arg-142 → Cys, Cys-112 → Arg	<4%	Yes	Dominant
E-3 Leiden	Cys-112 → Arg, 7 amino acid tandem insertion (AA 121–127)	25%	Yes	Dominant
E-3 Freiburg	Thr-42 → Ala	n.d.	No	–
E-4 Philadelphia	Cys-145 → Arg, Glu-13 → Lys	n.d.?	Yes	Dominant
E-4 Amsterdam	Ser-296 → Arg	n.d.	No	–
E-4 Freiburg	Leu-28 → Pro	n.d.	No	–
E-4 Doetinchem	Cys-112 → Arg, Arg-274 → His	n.d.	No	–
E-5	Glu-3 → Lys	>100%	Yes	–
E-5	Glu-13 → Lys	n.d.	No	–
E-5	Pro-84 → Arg	?	Yes	–
E-7 Suita	Glu-244 → Lys, Glu-245 → Lys	?	Yes	–

n.d., not determined; IEF, isoelectric focusing.

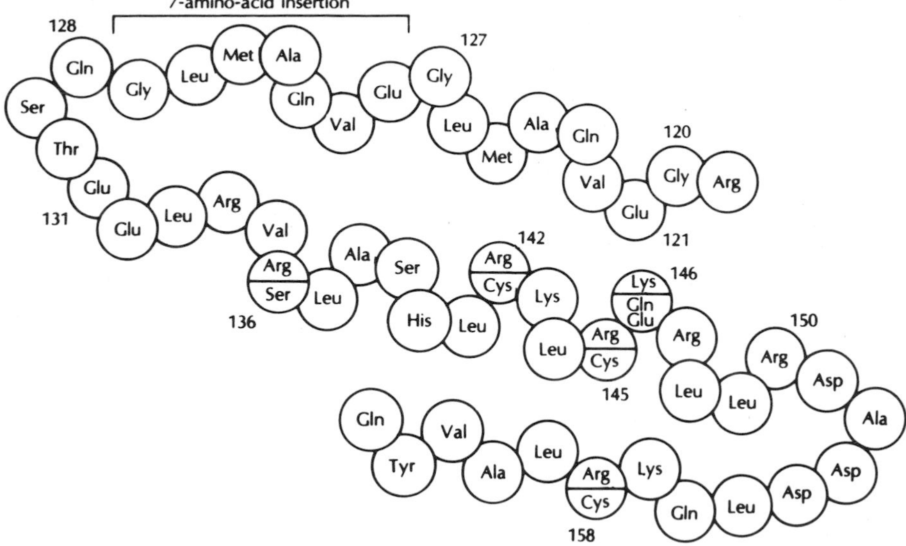

Fig. 7. The predicted secondary structure of the receptor-binding domain of apoE, indicating the location and identity of naturally occurring amino acid substitutions in this region of the protein. (In each substitution, the *bottom* amino acid substitutes for the *top* one, which represents the apoE-3 sequence.) (From MAHLEY et al. 1990, with permission)

IV. Pathophysiology of Type III Hyperlipoproteinemia

All mutations associated with type III hyperlipoproteinemia reported to date result in the absence of apoE from plasma or defective binding of apoE and apoE-containing lipoproteins to the LDL receptor (Table 7). Presumably, they also reduce uptake by the chylomicron remnant-removal system of the liver. As a result, all lipoproteins which are dependent on apoE for removal from the circulation accumulate in carriers of these mutants. Whereas for some rare mutations a single allele is sufficient to cause type III hyperlipidemia, even homozygosity for the binding-defective apoE-2 (Arg-158 → Cys) alone does not cause hyperlipoproteinemia type III. Clearly, the mechanisms by which the dominant mutations and the 158 mutation exert their effects must be different. Important insights into these mechanisms have recently been obtained by a combination of immunochemical, crystallographic, and biochemical analysis, but some aspects are still unclear. The apoE molecule has two domains, which may be separated by thrombin cleavage. The N-terminal 22-kDa fragment contains the receptor-binding region, which has been localized to the vicinity of residues 130–150 (MAHLEY 1988). In vitro binding assays have shown that the common E-2 isoform (apoE; Arg-158 → Cys) is the one most severely deficient in LDL receptor-binding activity (SCHNEIDER et al. 1981). At the same time, it is the only one where the binding defect is reversible upon chemical modification of the critical Cys-158 residue or by other manipulations, including dietary changes of the lipid composition of lipoprotein particles which carry the apoE variant (Mahley 1988). Further, residue 158 is outside the receptor-binding domain, and there is evidence that the effect of this mutation on receptor-binding activity is indirectly dependent on several modulating factors. The molecule may switch conformation from a severely binding-defective to an actively binding form by simply changing the diet of an apoE-2 (Arg-158 → Cys) homozygote. This has been taken as the explanation for why the 158 variant acts as a recessive mutation and why other factors are necessary to precipitate type III hyperlipidemia in homozygotes (Mahley et al. 1990).

In contrast to this situation, mutants which result in dominant type III hyperlipoproteinemia result from missense mutations in the LDL receptor-binding domain apoE-2 (Arg-142 → Cys), apoE-2 (Arg-146 → cys), or from mutations which result in an irreversible conformational change affecting the receptor-binding domain, as is the case for the apoE-3 Leiden variant (Wardell et al. 1989).

The high-affinity binding of apoE-containing lipoproteins to the LDL receptor requires the interaction of multiple apoE molecules on the particle with the receptor and may possibly involve more than one receptor (Mahley et al. 1990). The presence of mutant apoE molecules on the particles may interfere with this process, resulting in defective clearance of remnants. The available data explain the direct dominant effect of some apoE mutants and also why additional factors are required to precipitate type III hyperlipoproteinemia in carriers of the E-2 (Arg-158 → Cys) mutation. They do not, however, fully explain why homozygosity is required, nor why most homozygotes have hypocholesterolemic dysbetalipoproteinemia. Theoretically unfavorable conditions which arrest the mutant Cys-158 protein in a nonbinding form should result in dominant expression of type III hyperlipidemia. There might be further functional differences between the dominant mutation and the Arg-158 → Cys mutation. One may relate to the known preferences of apoE isoforms for lipoprotein association (Table 6). Others may relate to the uptake by the remnant-removal system which presumably requires interaction with heparan sulfate and LRP and/or with lipases. Some pieces in the E-2 (Arg-158 → Cys) puzzle are still missing.

F. Familial Combined Hyperlipoproteinemia

FCH (multiple-type hyperlipidemia, mixed hyperlipidemia) is the most common and least understood of all forms of genetic hyperlipidemias. This inborn error of metabolism is believed to be transmitted as an autosomal dominant trait, but this issue is controversial. FCH was not recognized as a distinct lipid disorder until 1973, when four groups independently described families in which the lipoprotein abnormality segregated (GOLDSTEIN et al. 1973; GLUECK et al. 1973; ROSE et al. 1973). Delineation had been difficult due to the pleomorphic manifestation of the disease. Affected subjects may be hyperlipidemic at one time and borderline or even normolipidemic at other times, suggesting environmental influences. About one third represents with pure hypercholesterolemia, pure hypertriglyceridemia, or elevation of both lipids, respectively, but the type of hyperlipidemia may change spontaneously. Accordingly, agaraose gel electrophoresis may result in a type IIa, IIb, IV, or (rarely) type V phenotype. In terms of lipoproteins, either VLDI, LDL, or both are elevated at a given time in an affected subject. Such a pattern is difficult to explain by a single gene defect. The lipoprotein abnormality rarely manifests itself before the middle of the second decade of life. The condition resembles hyperapobetaliporoteinemia, which is characterized by elevated apoB in LDL, even when LDL cholesterop is normal, and by the presence of small, dense LDL particles. Small, dense LDL have also been described in subjects with FCH (VEGA and GRUNDY 1986), and both conditions certainly have overlapping features and may be identical (SNIDERMAN et al. 1992). A relationship also exists to the so-called pattern B of LDL (AUSTIN et al. 1988). Kinetic studies suggest that the defect in FCH results in increased synthesis of lipoproteins containing apoB-100 (TENG et al. 1986; CHAIT et al. 1986).

The occurrence of phenotypically different lipoprotein patterns in the same family is believed to result from the expression of a single dominant gene in such kindreds, but locus heterogeneity has not been excluded and is suggested by some studies (WOJCIECHOWSKI et al. 1991; AUWERX et al. 1990). In one of the original studies (GOLDSTEIN et al. 1973), 47 families with FCH were identified among survivors of myocardial infarction. Half of the first-degree relatives above the age of 25 years had hyperlipidemia, which is consistent with a dominant mechanism. It has been criticized that the definition of the disorder and the identification of kindreds were based on the presence of further affected family members, which might have biased the subsequent genetic analysis. Multifactorial inheritance was suggested by reanalysis of the original data using complex segregation analysis (WILLIAMS et al. 1988). The strongest arguments for single gene defects underlying FCH was obtained from studies of families with LPL deficiency (AUWERX et al. 1990) and by a linkage study with candidate genes (WOJCIECHOWSKI et al. 1991). It had been noted for some time that some obligate heterozygotes for LPL deficiency may present with a milder form of hyperlipoproteinemia. A systematic study of LPL-deficient families in which heterozygotes were defined biochemically demonstrated an association of heterozygous LPL deficiency with hyperlipidemia. Affected individuals had multiple lipoprotein phenotypes fulfilling the criteria for FCH (AUWERX et al. 1990). This suggests that heterozygous LPL deficiency may be one cause of FCH and is consistent with a dominant gene effect. In one large family of a subject with homozygous LPL deficiency where heterozygotes were identified by linkage analysis, heterozygosity for LPL deficiency cosegregated with mild hypertriglyceridemia due to VLDL elevation, low HDL, and subnormal LDL (TASKINEN et al. 1990). The lipoprotein abnormality in this kindred was defined as familial hypertriglyceridemia rather than FCH.

A significant association of FCH with the X-2 allele at a polymorphic XmnI restriction site in the apoA-I/C-II/A-IV gene cluster (HAYDEN et al. 1987) prompted a linkage study in FCH families (WOJCIECHOWSKI et al. 1991). In the families of eight subjects with the X-2 allele, significant linkage (lod score 6.9 at a recombination fraction 0 = 0) was observed. This suggests that a mutation in or close to these genes contributes significantly to the development of FCH. The methodological approach used in the study was novel and appears to reduce potential bias. However, as is the rule in linkage studies of quantitative characters, even in this study several assumptions were made that have not yet been verified empirically. Until a mutation in the apoA-I/C-III/A-IV region has been identified in FCH subjects and has been proven to be causal, the reported linkage has to be interpreted with reservationl. Likewise, the heterozygous state of LPL deficiency as a causal factor in a subset of families with FCH has to be proven by larger studies. Though there is no doubt that heterozygous LPL deficiency is associated with hyperlipidemia, it is not clear whether this is indeed identical with FCH.

Results from a large Amish family where the lipid abnormality was defined as hyperapobetalipoproteinemia were also consistent with a dominant model (BEATY et al. 1986). Because the primary defect or defects in FCH are not known, no specific test – whether biochemical or by DNA analysis – is available to diagnose the disorder in an individual patient. The only way to make the diagnosis is still by family analysis.

Patients with FCH are frequently obese and have hyperinsulimenia and glucose intolerance. In contrast to FH or type III hyperlipoproteinemia, xanthomas are not part of the clinical syndrome. Patients with excessive hypertriglyceridemia occasionally develop pancreatitis, but this situation is probably not monocausal (BRUNZELL 1989). The one serious complication of FCH is coronary atherosclerosis and its clinical sequel. It has been estimated that 10%–20% of male survivors of myocardial infarction under the age of 60 years have FCH (GOLDSTEIN et al. 1973; NIKKILÄ and ARO 1973). A prevalence of 1%–2% in the general North American Caucasian population has been estimated. Due to the lack of a specific diagnostic test, however, this figure is rather uncertain. No data on the prevalence in different populations are yet available.

G. Polygenic Hypercholesterolemia

In a study of survivors of myocardial infarction, GOLDSTEIN et al. (1973) identified a group of patients with elevated plasma cholesterol levels, but no evidence of bimodality of plasma cholesterol concentrations was seen in the families of affected individuals. Instead, the distribution of cholesterol in these families was unimodal, but shifted towards a higher mean level. This form was designated polygenic hypercholesterolemia and by definition is caused by the clustering of several independent genes in one individual that all tend to moderately elevate plasma cholesterol levels. Attempts to identify these genes by the candidate gene approach have met with little success. Numerous studies have claimed associations between variation in candidate genes (mainly restriction fragment length polymorphs; RFLP) with cholesterol levels, but many of these were not reproduced or not consistent within or across populations (BOERWINKLE and HIXSON 1990). An XbaI allele (X-2) in the apoB gene was, however, consistently found to be associated with elevated cholesterol concentrations. The best example for an allele which is associated with hypercholesterolemia in patients and with elevated mean cholesterol levels in the general population is the $\varepsilon 4$ allele of the apoE polymorphism. In patients with primary hypercholesterolemia (none FH), the $\varepsilon 4$ allele is significantly increased. The allele is also associated with elevated LDL cholesterol in various ethnic groups around the world (HALLMAN et al. 1991; DAVIGNON et al. 1988). It may be one of several genes contributing to polygenic hypercholesterolemia (KOBAYASHI et al. 1989; UTERMANN 1988).

The classical clinical features of FH such as xanthomas and corneal arcus do not occur in polygenic hypercholesterolemia, but the disorder is likely to be associated with premature atherosclerosis. There are no studies on the prevalence of clinical symptoms in this group of patients. A frequency for polygenic hypercholesterolemia cannot be given, since it depends on the arbitrary definition of upper limits for normal cholesterol levels.

H. Familial Hypertriglyceridemia

Familial hypertriglyceridemia is an inborn error of lipid metabolism in adults. A moderate elevation of triglycerides in VLDL is characteristic of the disorder (GOLDSTEIN et al. 1973). Agarose gel electrophoresis exhibits a type IV or (rarely) type V pattern, but hyperchylomicronemia is rare. Xanthomas do not usually develop. Familial hypertriglyceridemia is associated with diabetes mellitus and obesity. Family studies on probands with familial hypertriglyceridemia and diabetes mellitus have provided evidence that the hyperlipidemia is transmitted independently from diabetes mellitus. However, patients that have both, familial hyperlipidemia and diabetes mellitus are more severely effected. A similar relationship may exist between hypertriglyceridemia and pancreatitis, which may occur in familial hypertriglyceridemia (BRUNZELL and SCHOTT 1973). Besides untreated diabetes, several other factors, including alcohol abuse and estrogen therapy, may lead to gross elevation of triglycerides, hyperchylomicronemia, and pancreatitis in patients with familial hypertriglyceridemia and the distinction between familial and acquired forms of hypertriglyceridemia is not always possible. Severe, acquired hypertriglyceridemia is sometimes the exacerbation of mild, preexisting familial hypertriglyceridemia. The role of triglyceride as an independent risk factor for atherosclerosis was long disputed, but is now generally accepted. The fact that 5% of unselected survivors of myocardial infarction under the age of 60 years have *familial* hypertriglyceridemia demonstrates that at least this form of hypertriglyceridemia may be associated with an increased risk of premature CHD (BRUNZELL et al. 1976).

Original family data on the inheritance of familial hypertriglyceridemia were considered compatible with an autosomal dominant mode of transmission (GOLDSTEIN et al. 1973). Subsequent studies suggested a multifactorial etiology of familial hypertriglyceridemia (WILLIAMS et al. 1988). The frequency of the disorder has been estimated as being on the order of two or three in 1000 (GOLDSTEIN et al. 1973). According to another study, this figure underestimates the frequency of familial hypertriglyceridemia (MOTULSKY and BOMAN 1975). The primary biochemical defects in familial hypertriglyceridemia are unknown. Overproduction of endogenous triglycerides as well as delayed catabolism of VLDL have both been observed in patients with familial hypertriglyceridemia. Familial hypertriglyceridemia is most probably a heterogeneous group of biochemically distinct disorders that all have in common an elevation of VLDL. The discovery of a large family where heterozygous LPL deficiency cosegregates with mild hypertriglyceridemia (WILSON et al. 1990) may be an example of such a specific biochemical entity out of the heterogeneous group collectively designated the familial hypertriglyceridemias. It does, however, also demonstrate the overlap of this condition with FCH. Isolated hypertriglyceridemia is also seen in one in three patients with FCH (GOLDSTEIN et al. 1973), and heterozygous LPL deficiency was also described as one defect causing FCH (BABIRAK et al. 1989). At present, no specific criteria exist to recognize familial hypertriglyceridemia. Differentation from other genetic or nongenetic forms of hypertriglyceridemia is not possible, except by extended family studies. If a patient presents with a combination of hypertriglyceridemia and premature vascular disease, the examining physician should consider familial hypertriglyceridemia as a possible diagnosis.

I. Familial Hyperalphalipoproteinemia and Familial Hypoalphalipoproteinemia

Genetic factors explain a large fraction of the variability in HDL cholesterol levels in the population. The extremes of the spectrum are called hyper- and hypoalphalipoproteinemia, respectively. A concentration of cholesterol in HDL exceeding 70 mg/100 ml is arbitrarily defined as hyperalphalipoproteinemia. GLUECK et al. (1973) first observed the familial aggregation of hyperalphalipoproteinemia and noticed that among blood relatives of propositi, about 50% again had elevated HDL cholesterol levels. Familial hyperalphalipoproteinemia is not a disorder, but rather the opposite. A significantly lower prevalence of coronary atherosclerosis and a higher mean age at death was noted in hyperalpha families, and it was concluded that hyperalphalipoproteinemia is a longevity syndrome (GLUECK et al. 1976). The etiology of familial hyperalphalipoproteinemia is heterogeneous, and different forms of hyperalphalipoproteinemia exist. Whereas hyperalphalipoproteinemia occurs as a familial trait with evidence of a major gene effect in Caucasians, most cases among black Americans are sporadic (SIERVOGEL et al. 1980). One form of hyperalphalipoproteinemia is frequent among Japanese and is caused by a deficiency of CETP activity. This condition is inherited as an autosomal recessive trait (KOIZUMI et al. 1991).

The molecular basis of this defect is a G → A point mutation in the slice donor site (position +1) in intron 14 of the CETP gene. Patients that are homozygous for this defect produce a CETP mRNA that is deficient in exon 9 sequence. This RNA translates in transient or stably transfected cells into a truncated protein that does not express CETP activity and is not secreted from these cells (INAZU et al. 1992). Deficiency of CETP activity in the homozygous patients results in very high levels of HDL and apoA-I. In heterozygotes, HDL is only moderately elevated. The increase in HDL results from accumulation of large HDL_2 particles which are rich in apoE and cholesterolesters. LDL is rich in triglycerides and unesterified cholesterol, and LDL concentration is low in the homozygous patients (SWENSON 1992).

Primary hypoalphalipoproteinemia has been difined as a condition in which HDL cholesterol levels are below the 10th percentile of the relevant age, sex, and race specific distribution and where no primary disease causing low HDL exists. Hypoalphalipoproteinemia has been described as a familial trait. Some rare inborn errors of lipid metabolism may cause hypoalphalipoproteinemia. Among these are LCAT deficiency and some mutations affecting apoA-I. Deficiency of LCAT results in a recessive form of hypoalphalipoproteinemia. Depending on the mutation in the LCAT gene, these may be clinically expressed as classical LCAT deficiency or as Fish-eye disease (ASSMANN et al. 1911; Table 2). The causes and genetics of the more common forms of low HDL are, however, unknown. In pedigrees ascertained through probands with low HDL cholesterol, dominant inheritance was suggested by segregation ratios of dichotomized levels (THIRD et al. 1984), but complex segregation analysis supported recessive inheritance (BYARD et al. 1984). A major gene for primary hypoalphalipoproteinemia has been suggested by BORECKI et al. (1986), and a dominant major locus resulting in low HDL cholesterol was revealed in two out of 55 Utah pedigrees selected through probands with early CHD, stroke, or hypertension (HASSTEDT et al. 1986). The condition is clearly associated with end points of atherosclerosis, e.g., myocardial infarction.

References

Aalto-Setälä K, Gylling H, Miettinen T, Kontula K (1988) Identification of a deletion in the LDL-receptor gene. FEBS-Letters 230:31–34

Assmann G, Von Eckardstein A, Funke H (1991) Lecithin: cholesterol acyltransferase deficiency and fish-eye disease. Curr Opin Lipid 2:110–117

Austin MA, Breslow JL, Hennekens CH, Buring JE, Willet WC, Krauss RM (1988) Low-density lipoprotein subclass phenotypes and the risk of myocardial infarction. JAMA 13:1917–1921

Auwerx JH, Babirak SP, Fujimoto WY, Iverius P-H, Brunzell JD (1989) Defective enzyme protein in lipoprotein lipase deficiency. Eur J Clin Invest 19:433–437

Auwerx JH, Babirak SP, Hokanson JE, Stahnke G, Will H, Deeb SS, Brunzell JD (1990) Coexistence of abnormalities of hepatic lipase and lipoprotein lipase in a large family. Am J Hum Genet 46:470–477

Babirak SP, Iverius P, Fujimoto A, Brunzell JD (1989) Detection and characterization of the heterozygotic state for lipoprotein lipase deficiency. Arteriosclerosis 9:326–334

Beaty TH, Kwiterovich PO Jr, Khoury MJ, Withe S, Bachorik PS, Smith HH, Teng B, Sniderman AD (1986) Genetic analysis of plasma sitosterol, apoprotein B and lipoproteins in a large Amish pedigree with sitosterolemia. Am J Hum Genet 38:492–504

Beil FU, Fojo SS, Brewer HB Jr, Greten H, Beisiegel U (1992) Apolipoprotein G-II deficiency syndrome due to apo C-II Hamburg: clinical and biochemical features and HphI restriction enzyme polymorphism. Eur J Clin Invest 22:88–95

Berg K (1983) Genetics of coronary heart disease. In: Steinberg AG, Bearn AG, Motulsky AR, Childs B (eds) Progress in medical genetics. Saunders, Philadelphia, p 35

Bilheimer DB, Ho YK, Brown MS, Anderson RGW, Goldstein JL (1978) Genetics of low density lipoprotein receptor: Diminished receptor activity in lymphocytes from heterozygotes with familial hypercholesterolemia. J Clin Invest 61:678–696

Boerwinkle E, Utermann G (1988) Simultaneous effects of the apolipoprotein E polymorphism on apolipoprotein E, apolipoprotein B and cholesterol metabolism. Am J Hum Genet 42:104–112

Boerwinkle E, Hixson JE (1990) Genes and normal lipid variation. Curr Opin Lipid 1:151–159

Boerwinkle E, Leffert CC, Lin J, Lackner C, Chiesa G, Hobbs HH (1992) Apolipoprotein(a) gene accounts for greater than 90% of the variation in plasma lipoprotein(a) concentrations. J Clin Invest 90:52–60

Borecki IB, Rao DC, Third JLHC, Laskarzewski PM, Glueck CJ (1986) A major gene for primary hypoalphalipoproteinemia. Am J Hum Genet 38:373–381

Borrie P (1969) Type III hyperlipoproteinemia. Br Med J 2:665–667

Britten RJ, Baron WF, Stout DB, Davidson EH (1988) Sources and evolution of human Alu repeated sequences. Proc Natl Acad Sci USA 85:4770–4774

Brown MS, Goldstein JL (1986) A receptor-mediated pathway for cholesterol homeostasis. Science 232:34–47

Brown MS, Herz J, Kowal RC, Goldstein JL (1991) The low density lipoprotein receptor – related protein: double agent or decoy. Curr Opin Lipid 2:65–72

Brunzell JD, Schott HG (1973) The interaction of familial and secondary causes of hypertriglyceridemia: role in pancreatitis. Trans Assoc Am Physicans 86:245–253

Brunzell JD, Schott HG, Motulsky AG, Bierman EL (1976) Myocardial infarction in the familial forms of hypertriglyceridemia. Metabolism 25:313–320

Brunzell JD (1989) Familial lipoprotein lipase deficiency and other causes of the chylomicronemia syndrome. In: Scriver CR, Beaudet AL, Sly WS, Valle D (eds) The metabolic basis of inherited disease, 6th edn. McGraw Hill, New York, p 1165

Byard PJ, Borecki IB, Glueck CJ, Laskarzewski PM, Third JLHC (1984) A genetic study of hypoalphalipoproteinemia. Epidemiology 1:43–51
Bétard Ch, Kessling AM, Chamberland A, Lussier-Cacan S, Davignon J (1992) Molecular genetic evidence for a founder effect in familial hypercholesterolemia among French Canadians. Hum Genet 88:529–536
Bürger M, Grütz O (1932) Über hepatosplenomegale Lipidose mit xanthomatösen Veränderungen in Haut und Schleimhaut. Arch Dermatol Syph 166:542
Chait A, Foster DM, Albers JJ, Failor A, Brunzell JD (1986) Low density lipoprotein metabolism in familial combined hyperlipidemia and familial hypercholesterolemia: kinetic analysis using an integrated model. Metabolism 35:697–704
Cooper DN, Youssoufian H (1988) The CpG dinucleotide and human genetic disease. Hum Genet 78:151–155
Császár A, Dieplinger H, Sandholzer Ch, Karádi I, Juhász E, Drexel H, Halmos T, Romics L, Patsch JR, Utermann G (1993) Plasma lipoprotein(a) concentration and phenotypes in diabetes mellitus. Diabetologia 36:47–51
Culotta E, Koshland DE Jr (1992) No news is good news. Science 258:1862–1865
Dahlen G, Berg K, Gillnas T, Ericson C (1975) Lp(a) lipoprotein/prebetal-lipoprotein in Swedish middle-aged males and in patients with coronary heart disease. Clin Genet 7:334
Davignon J, Gregg RE, Sing CF (1988) Apolipoprotein E polymorphism and atherosclerosis. Arteriosclerosis 8:1–21
De Knijff P, Van den Maagdenberg AMJM, Stalenhoef AFH, Leuven JAG, Demacker PNM, Kuyt LP, Frants RR, Havekes LM (1991) Familial dysbetalipoproteinemia associated with apolipoprotein E3-Leiden in an extended multigeneration pedigree. J Clin Invest 88:643–655
Dichek HL, Fojo SS, Beg OU, Skarlatos SI, Brunzell JD, Cutler GB Jr, Brewer HB Jr (1991) Identification of two separate allelic mutations in the lipoprotein lipase gene of a patient with the familial hyperchylomicronemia syndrome. J Biol Chem 266:473–477
Dieplinger H, Lackner C, Kronenberg F, Sandholzer Ch, Lhotta K, Hoppichler F, Graf H, König P (1993) Elevated plasma concentrations of lipoprotein(a) in patients with end-stage renal disease are not related to the size polymorphism of apolipoprotein(a). J Clin Invest 91:397–401
Emi M, Wilson DE, Iverius P-H, Wu L, Hata A, Hegele R, Williams RR, Lalouel J-M (1990) Missense mutation (Gly → Glu^{188}) of human lipoprotein lipase imparting functional deficiency. J Biol Chem 265:5910–5916
Emi M, Hegele RM, Hopkins PN, Wu LL, Plaetke R, Williams RR, Lalouel J-M (1991) Effects of three genetic loci in a pedigree with multiple lipoprotein phenotypes. Arterioscler Thromb 11:1349–1355
Faustinella F, Chang A, Van Biervliet JP, Rosseneu M, Vinaimont N, Smith LC, Chen S-H, Chan L (1991) Catalytic triad residue mutation (Asp^{156} → Gly) causing familial lipoprotein lipase deficiency. Co-inheritance with a nonsense mutation (Ser^{447} → Ter) in a Turkish family. J Biol Chem 266:14418–14424
Fojo SS, Taam T, Ronan R, Bishop C, Meng MS, Hoeg JM, Sprecher DL, Brewer BJr (1986) Human preproapolipoprotein C-II: analysis of major plasma isoforms. J Bill Chem 261:9591–9594
Fojo SS, Brewer HB (1992) Hypertriglyceridaemia due to genetic defects in lipoprotein lipase and apolipoprotein C-II. J Intern Med 231:669–677
Franke U, Brown MS, Goldsten JL (1984) Assignment of the human gene for the low density lipoprotein receptor to chromosome 19: synthesis of a receptor, a ligand, and a genetic disease. Proc Natl Acad Sci USA 81:2826–2830
Fredrickson DS, Levy RI, Lees RS (1967) Fat transport in lipoproteins – an integrated approach to mechanism and disorders. New Engl J Med 267:34–40
Friedl W, Ludwig EH, Balestra ME, Arnold KS, Paulweber B, Sandhofer F, McCarthy BJ, Innerarity TL (1991) Apolipoprotein B gene mutations in Austrian subjects with heart disease and their kindred. Arteriosclerosis 11:371–378

Gevers W, Casciola AF, Fourie AM, Sanan DA, Coetzee GA, Van der Westhuyzen DR (1987) Familial hypercholesterolemia in South Africa. Defective LDL receptors that are common in a large population. Hoppe-Seyler's Z Physiol Chem 368:1233–1243
Glueck CJ, Fallat RW, Millet F, Gartside P, Elston RC, Go RCP (1973) Familial hyperalphalipoproteinemia: studies in eighteen kindreds. Metabolism 22:1403–1428
Glueck CJ, Gartside P, Fallat RW, Sielski J, Steiner PM (1976) Longevity syndromes: familial hypobeta- and familial hyperalphalipoproteinemia. J Lab Clin Med 88:941–957
Goldstein JL, Schrott HG, Hazard WR, Bierman EL, Motulsky AG (1973) Hyperlipidemia in coronary heart disease. II. Genetic analysis of lipid levels in 176 families and delineation of a new inherited dysorder, combined hyperlipidemia. J Clin Invest 52:1544–1568
Goldstein JL, Brown MS (1989) Familial hypercholesterolemia. In: Scriver CR, Beaudet AL, Sly WS, Valle D (eds) The metabolic basis of inherited disease. McGraw-Hill, New York, p 1215
Goldstein JL, Brown MS (1990) Regulation of the mevalonate pathway. Nature 343:425–430
Gregg RE, Zech LA, Schaefer EJ, Stark D, Wilson D, Brewer HB Jr (1986) Abnormal in vivo metabolism of apolipoprotein E-4 in humans. J Clin Invest 78:815–821
Hajjar KA, Gavish D, Breslow JL, Nachman RL (1989) Lipoprotein(a) modulation of endothelial cell surface fibrinolysis and its potential role in atherosclerosis. Nature 339:303–305
Hallman DM, Boerwinkle E, Saha N, Sandholzer C, Jürgen Menzel H, Csázár A, Utermann G (1991) The appolipoprotein E polymorphism: a comparison of allele frequencies and effects in nine populations. Am J Hum Genet 49:338–349
Harlan WR, Graham JB, Estes H (1966) Familial hypercholesterolemia: a genetic and metabolic study. Medicine 45:77–110
Hasstedt SJ, Ashe OK, Williams RR (1986) A reexamination of major locus hypotheses for high density lipoprotein cholesterol level using 2170 persons screened in 55 Utah pedigrees. Am J Med Genet 24:57–67
Hasstedt SJ, Williams RR (1986) Three alleles for the quantitative Lp (a). Genet Epidemiol 3:53–55
Havel RJ, Kane JP (1989) Introduction: structure and metabolism of plasma lipoproteins. In: Scriver CR, Beaudet AL, Sly WS, Valle D (eds) The metabolic basis of inherited disease, 6th edn. McGraw Hill, New York, p 1129
Hayden MR, Kirk H, Clarl C, Frohlich J, Rapkin S, McLeod R, Hewitt J (1987) DNA polymorphism in and around the apo AI/CIII genes and genetic hyperlipidemias. Am J Hum Genet 40:421–430
Henderson HE, Devlin R, peterson J, Brunzell JD, Hayden MR (1990) Frameshift mutation in exon 3 of the lipoprotein lipase gene causes a premature stop codon and lipoprotein lipase deficiency. Mol Biol Med 7:511–517
Henderson HE, Ma Y, Hassan MF, Monsalve MV, Marais AD, Winkler F, Gubernator K, Peterson J, Brunzell JD, Hayden MR (1991) Amino acid substitution ($Ile^{194} \rightarrow$ Thr) in exon 5 of the lipoprotein lipase gene causes lipoprotein lipase deficiency in three unrelated probands. Support for a multicentric origin. J Clin Invest 87:2005–2011
Henderson HE, Hassan F, Berger GMB, Hayden MR (1992) The lipoprotein lipase $Gly^{188} \rightarrow$ Glu mutation in South Africans of Indian descent: evidence suggesting common origins and an increased frequency. J Med Genet 29:119–122
Hixson JE, Britten ML, Manis GS, Rainwater DL (1989) Apolipoprotein (a) glycoprotein isoforms result from size differences in apo(a) mRNA in baboons. J Biol Chem 264:6013–6016

Hixson JE, Vernier DT (1990) Restriction isotyping of human apolipoprotein E by gene amplification and cleavage with HhaI. J Lipid Res 31:545–548

Hjelms E, Stender S (1992) Accelerated cholesterol accumulation in homologous arterial transplants in cholesterol-fed rabbits: a surgical model to study transplantation atherosclerosis. Arterioscler Thromb 12:771–779

Hobbs HH, Leitersdorf E, Leffert CC, Cryer DR, Brown MS, Goldstein JL (1989) Evidence for a dominant gene that suppresses hypercholesterolemia in a family with defective low density lipoprotein receptors. J Clin Invest 84:656–664

Hobbs HH, Brown MS, Goldstein JL (1992) Molecular genetics of the LDL receptor gene in familial hypercholesterolemia. Hum Mutation 1:445–466

Hämäläinen T, Palotie A, Aalto-Setälä K, Kontula K, Tikkanen MJ (1990) Absence of familial defective apolipoprotein B-100 in Finnish patients with elevated serum cholesterol. Atherosclerosis 82:117–125

Inazu A, Quinet EM, Wang S, Brown ML, Stevenson S, Barr ML, Moulin P, Tall AR (1992) Alternative splicing of the mRNA encoding the human cholesteryl ester transfer protein. Biochemistry 31:2352–2358

Innerarity TL (1990) Familial hypobetalipoproteinemia and familial defective apolipoprotein B-100: genetic disorders associated with apolipoprotein B. Curr Opin Lipid 1:104–109

Innerarity TL, Mahley RW, Weisgraber KH, Bersot TP, Krauss RM, Vega GL, Grundy SM, Friedl W, Davignon J, McCarthy BJ (1990) Familial defective apolipoprotein B-100: a mutation of apolipoprotein B that causes hypercholesterolemia. J Lipid Res 31:1337–1349

Jauhiainen M, Koskinen P, Ehnholm C, Frick HM, Mänttäri M, Manninen V, Huttunen JK (1991) Lipoprotein(a) and coronary heart disease risk: a nested case-control study of the Helsinki Heart Study participants. Atherosclerosis 89:59–67

Kamboh MI, Ferrell RE, Kottke BA (1991) Expressed hypervariable polymorphism of apolipoprotein (a). Am J Hum Genet 49:1063–1074

Kane JP, Havel RJ (1989) Disorders of the biogenesis and secretion of lipoproteins containing the B Apolipoproteins. In: Scriver CR, Beaudet AL, Sly WS, Valle D (eds) The metabolic basis of inherited disease, 6th edn. McGraw Hill, New York, p 1139

Kobayashi J, Shirai K, Saito Y, Yoshida S (1989) Lipoprotein lipase with a defect in lipid interface recognition in a case with type I hyperlipidaemia. Eur J Clin Invest 19:424–432

Koizumi J, Inazu A, Yagi K, Koizumi I, Uno Y, Kajinami K, Miyamoto S, Moulin P, Tall AR, Mabuchi H, Takeda R (1991) Serum lipoprotein lipid concentration and composition in homozygous and heterozygous patients with cholesteryl ester transfer protein deficiency. Atherosclerosis 90:189–196

Komuro I, Kato H, Nakagawa T, Takahashi K, Mimori A, Takeuchi F, Nishida Y, Miyamoto T (1987) Case report: the longest-lived patient with homozygous familial hypercholesterolemia secondary to a defect in internalization of the LDL receptor. Am J Med Sci 294:341–345

Koschinsky ML, Beisiegel U, Henne-Bruns D, Eaton DL, Lawn RM (1990) Apolipoprotein(a) size heterogeneity is related to variable number of repeat sequences in its mRNA. Biochemistry 29:640–644

Kraft HG, Köchl S, Menzel HJ, Sandholzer C, Utermann G (1992) The apolipoprotein(a) gene: a transcribed hypervariable locus controlling plasma lipoprotein(a) concentration. Hum Genet 90:220–230

Lackner C, Boerwinkle E, Leffert CC, Rahmig T, Hobbs HH (1991) Molecular basis of apolipoprotein (a) isoform size heterogeneity as revealed by pulsed-field gel electrophoresis. J Clin Invest 87:2153–2161

Lalouel JM, Wilson DE, Iverius PH (1992) Lipoprotein lipase and hepatic triglyceride lipase: molecular and genetic aspects. Curr Opin Lipid 3:86–95

Leitersdorf E, Chakravarti A, Hobbs HH (1989) Polymorphic DNA haplotypes at the LDL receptor locus. Am J Hum Genet 44:409-421
Leitersdorf E, Friedlander Y, Bard J-M, Fruchart J-C, Eisenberg S, Stein Y (1991) Diverse effect of ethnicity on plasma lipoprotein[a] levels in heterozygote patients with familial hypercholesterolemia. J Lipid Res 32:1513–1519
Leitersdorf E, Reshef A, Meiner V, Dann EJ, Beigel Y, Graadt van Roggen F, Van der Westhuyzen DR, Coetzee GA (1993) A missense mutation in the low density lipoprotein receptor gene causes familial hypercholesterolemia in Sephardic Jews. Hum Genet 91:141–147
Li WH, Tanimura M, Luo CC, Datta S, Chan L (1988) The apolipoprotein multigene family: biosynthesis, structure, structure-function relationship, and evolution. J Lipid Res 29:245–271
Lohse P, Brewer HB, III, Meng MS, Skarlatos SI, LaRosa JC, Brewer HB Jr (1992) Familial apolipoprotein E deficiency and type III hyperlipoproteinemia due to a premature stop codon in the apolipoprotein E gene. J Lipid Res 33:1583–1590
Lusis AJ (1988) Genetic factors affecting blood lipoproteins: the candidate gene approach. J Lipid Res 29:397–428
Ma Y, Henderson HE, Ven Murthy MR, Roederer G, Monsalve MV, Clarke LA, Normand T, Julien P, Gagné C, Lambert M, Davignon J, Lupien PJ, Brunzell J, Hayden MR (1991) A mutation in the human lipoprotein lipase gene as the most common cause of familial chylomicronemia in French Canadians. N Engl J Med 324:1761–1766
Mabuchi H, Ito S, Haba T, Ueda K, Ueda R, Tatami R, Kametani T, Koizumi I, Ohta M, Miyamoto S, Takeda R, Takegoshi T (1977) Discriminiation of familial hypercholesterolemia and secondary hypercholesterolemia by Achille's tendon thickness. Atherosclerosis 28:61–68
Mahley RW (1988) Apolipoprotein E: cholesterol transport protein with expanding role in cell biology. Science 240:622–633
Mahley RW, Rall SC Jr (1989) Type III hyperlipoproteinemia (dysbetalipoproteinemia): the role of apolipoprotein E in normal and abnormal lipoprotein metabolism. In: Scriver CR, Beaudet AL, Sly WS, Valle D (eds) The metabolic basis of inherited disease, 6th edn. McGraw Hill, New York, p 1195
Mahley RW, Innerarity TL, Rall SC Jr, Weisgraber KH, Taylor JM (1990) Apolipoprotein E: genetic variants provide insights into its structure and function. Curr Opin Lipid 1:87–95
Mamotte CDS, Van Bockxmeer FM (1993) A robust strategy for screening and confirmation of familial defective apolipoprotein B-100. Clin Chem 39:118–121
McLean JW, Tomlinson JE, Kuang W-J, Eaton DL, Chen EY, Fless GM, Scanu AM, Lawn RM (1987) cDNA sequence of human apolipoprotein (a) is homolgous to plasminogen. Nature 300:132–137
Menzel HJ, Utermann G (1986) Apolipoprotein E phenotyping from serum by Western blotting. Electrophoresis 11:492–495
Mishkel MA, Nazir DI, Crowther S (1975) A longitudinal assessment of lipid ratios in the diagnosis of type III hyperlipoproteinemia. Clin Chim Acta 53:121–136
Monsalve MV, Henderson H, Roederer G, Julien P, Deeb S, Kastelein JJP, Peritz L, Devlin R, Bruin T, Murthy MRV, Cagne C, Davignon J, Lupien PJ, Brunzell JD, Hayden MR (1990) A missense mutation at codon 188 of the human lipoprotein lipase gene is a frequent cause of lipoprotein lipase dificiency in persons of different ancestries. J Clin Invest 86:728–734
Morganroth JR, Levy RY, Fredrickson DS (1975) The biochemical, clinical and genetic features of type III hyperlipoproteinemia. Ann Intern Med 82:158–174
Moser H, Slack J, Borrie P (1974) Type III hyperlipoproteinemia: a genetic study with an account of the risk of coronary death and first degree relative. In: Schettler G, Weizel A (eds) Atherosclerosis III. Springer, Berlin Heidelberg New York, p 854

Motti C, Funke H, Rust S, Dergunov A, Assmann G (1991) Using mutagenic polymerase chain reaction primers to detect carriers of familial defective apolipoprotein B-100. Clin Chem 37:1762-1766

Motulsky AP, Boman H (1975) Screening for the hyperlipidemias. In: Milunsky A (ed) The prevention of genetic disease and mental retardation. Saunders, Philadelphia, p 303

März W, Ruzicka C, Pohl T, Usadel KH, Gross W (1992) Familial defective apolipoprotein B-100: mild hypercholesterolaemia without atherosclerosis in a homozygous patient. Lancet 340:1362

Müller C (1938) Xanthoma, hypercholesterolemia, angina pectoris. Acta med Scand Suppl 89:75

Nikkilä EA, Aro A (1973) Family study of serum lipids and lipoproteins and coronary heart disease. Lancet 1:954–959

Norum KR, Gjone E, Glomset JA (1989) Familial lecithin: cholesterol acyltransferase deficiency, including fish-eye disease. In: Scriver CR, Beaudet AL, Sly WS, Valle D (eds) The metabolic basis of inherited disease, 6th edn. McGraw Hill, New York, p 1181

Oppenheim A, Friedlander Y, Dann EJ, Berkman N, Pressman Schwartz S, Leitersdorf E (1991) Hypercholesterolemia in five Israeli Christian-Arab kindreds is caused by the "Lebanese" allele at the low density lipoprotein receptor gene locus and by an additional independent major factor. Hum Genet 88:75–84

Patterson D, Slack J (1972) Lipid abnormalities in male and female survivors of myocardial infarction. Lancet 1:393

Paulweber B, Weibusch H, Miesenböck G, Funke H, Assmann G, Hölzl B, Sippl JM, Friedl W, Patsch JR, Sandhofer F (1991) Molecular basis of lipoprotein lipase deficiency in two Austrian families with type I hyperlipoproteinemia. Atherosclerosis 86:239–250

Rall SC Jr, Mahley RW (1992) The role of apolipoprotein E genetic variants in lipoprotein disorders. J Intern Med 231:653–569

Rauh G, Schuster H, Fischer J, Keller C, Wolfram G, Zöllner N (1991) Identification of a heterozygous compound individual with familial hypercholesterolemia and familial defective apolipoprotein B-100. Klin Wochenschr 69:320–324

Rauh G, Keller C, Kormann B, Spengel F, Schuster H, Wolfram G, Zöllner N (1992) Familial defective apolipoprotein B-100: clinical characteristics of 54 cases. Atherosclerosis 92:233–241

Reshef A, Meiner V, Dann EJ, Granat M, Leitersdorf E (1992) Prenatal diagnosis of familial hypercholesterolemia caused by the "Lebanese" mutation at the low density lipoprotein receptor locus. Hum Genet 89:237–239

Rose HG, Krantz P, Weinstock M, Juliano J, Haft JI (1973) Inheritance of combined hyperlipoproteinemia: evidence for new lipoprotein phenotypes. Am J Med 54:148–152

Rosengren A, Wilhelmsen L, Eriksson E, Risberg B, Wedel H (1990) Lipoprotein (a) and coronary heart disease: a prospective case-control study in a general population sample of middle aged men. Br Med J 301:1248–1251

Russell DW, Lehrman MA, Südhof TC, Yamamoto T, Davis CG, Hobbs HH, Brown MS, Goldstein JL (1987) The LDL receptor in familial hypercholesterolemia: use of human mutations to dissect a membrane protein. Cold Spring Harbor Symposia on Quantitative Biology 51:811–819

Russell DW, Brown MS, Goldstein JL (1989) Different combinations of cysteine-rich repeats mediate binding of low density lipoprotein receptor to two different proteins. J Biol Chem 264:21682–21688

Sandholzer C, Feussner G, Brunzell J, Utermann G (1992) Distribution of apolipoprotein (a) in the plasma from patients with lipoprotein lipase deficiency and with type III hyperlipoproteinemia. No evidence for a triglyceride-rich precursor of lipoprotein (a). J Clin Invest 90:1958–1965

Scanu AM, Fless GM (1990) Lipoprotein (a). Heterogeneity and biological relevance. J Clin Invest 85:1709–1715

Schaefer J, Gregg RE, Giselli G, Forte TN, Ordovas JM, Zech LA, Brewer HB Jr (1986) Familial apolipoprotein E deficiency. J Clin Invest 78:1206–1219

Schneider WJ, Kovanen PT, Brwon MS, Goldstein JL, Utermann G, Weber W, Havel RJ, Kotite L, Kane JP, Innerarity TL, Mahley RW (1981) Familial dysbetalipoproteinaemia abnormal binding of mutant apoprotein E to low density lipoprotein receptors of human fibroblasts and membranes from liver and adrenal of rats, rabbits and cows. J Clin Invest 68:1075–1085

Schrott HG, Goldstein JL, Hazzard WR, McGoodwin MM, Motulsky AG (1972) Familial hypercholesterolemia in a large kindred. Evidence for a monogenic mechanism. Ann Intern Med 76:711–720

Schuster H, Humphries S, Rauh G, Keller C (1992a) First international workshop on familial defective apo B-100. Klin Wochenschr 70:961–964

Schuster H, Rauh G, Müller S, Keller C, Wolfram G, Zöllner N (1992b) Allele-specific and asymmetric polymerase chain reaction amplification in combination: a one step polymerase chain reaction protocol for rapid diagnosis of familial defective apolipoprotein B-100. Anal Biochem 204:22–25

Seed M, Hoppichler F, Reaveley D, McCarthy S, Thompson GR, Boerwinkle E, Utermann G (1990) Relation of serum lipoprotein (a) concentration and apolipoprotein (a) phenotype to coronary heart disease in patients with familial hypercholesterolemia. N Engl J Med 322:1494–1499

Siervogel RM, Morrison JA, Kelly K, Mellies M, Gartside P, Glueck CJ (1980) Familial hyperalphalipoproteinemia in 26 kindreds. Clin Genet 17:13–25

Sigurdsson G, Baldursdottir A, Sigvaldason H, Agnarsson U, Thorgeirsson G, Sigfusson N (1992) Predictive value of apolipoproteins in a prospective survey of coronary artery disease in men. Am J Cardiol 69(16):1251–1254

Sniderman A, Brown BG, Stewart BF, Cianflone K (1992) From familial combined hyperlipidemia to hyperapo B: unravelling the overproduction of hepatic apolipoprotein B. Curr Opin Lipid 3:137–142

Soria LF, Ludwig EH, Clarke HRG, Vega GL, Grundy SM, McCarthy BJ (1989) Association between a specific apolipoprotein B mutation and familial defective apolipoprotein B-100. Proc Natl Acad Sci USA 86:587–591

Soutar AK, Knight BL, Patel DD (1989) Identification of a point mutation in growth factor repeat C of the low density lipoprotein receptor gene in a patient with homozygous familial hypercholesterolemia that affects ligand binding and intracellular movement of receptor. Proc Natl Acad Sci USA 86:4166–4170

Soutar AK, McCarthy SN, Seed M, Knight BL (1991) Relationship between apolipoprotein(a) phenotype, lipoprotein(a) concentration in plasma, and low density lipoprotein receptor function in a large kindred with familial hypercholesterolemia due to the $Pro^{664} \rightarrow$ Leu mutation in the LDL receptor gene. J Clin Invest 88:483–492

Sprecher DL, Taam L, Gregg RE, Fojo SS, Wilson DM, Kashyap ML, Brewer B Jr (1988) Identification of an apo C-II variant (apo C-II Bethesda) in a Kindred with apo C-II deficiency and type I hyperlipoproteinemia. J Lipid Res 29:273–278

Stalenhoef AFH, Casparie AF, Demacker PNM, Stouten JTJ, Lutterman JA, van't Laar A (1981) Combined deficiency of apolipoprotein C-II and lipoprotein lipase in familial hyperchylomicronemia. Metabolism 30(9):919–926

Steinberg D, Parthasarathy S, Carew TE, Khoo JC, Witztum JL (1989) Beyond cholesterol: modification of low density lipoprotein that increase its atherogenicity. New Engl J Med 320:915–924

Steyn K, Weight MJ, Dando BR, Chrstopher KJ, Rossouw BR (1989) The use of low density lipoprotein receptor activity of lymphocytes to determine the prevalence of familial hypercholesterolemia in a rural South African community. J Med Genet 26:32–36

Stone NJ, Levy RI, Fredrickson DS, Verter J (1974) Coronary artery disease in 116 kindreds with familial type II hyperlipoproteinemia. Circulation 49:476–488

Swenson TL (1992) Transfer proteins in reverse cholesterol transport. Curr Opin Lipid 3:67–74
Südhof TC, Goldstein JL, Brown MS, Russell DW (1985) The LDL receptor gene: a mosaic of exons shared with different proteins. Science 228:815–822
Taskinen M-R, Packard CJ, Shepherd J (1990) Effect of insulin therapy on metabolic fate of apolipoprotein B-containing lipoproteins in NIDDM. Diabetes 39:1017–1027
Taylor R, Jeenah M, Seed M, Humphries S (1988) Four DNA polymorphisms in the LDL receptor gene: their genetic relationship and use in the study of variation at the LDL receptor locus. J Med Genet 25:653–659
Taylor R, Bryant J, Gudnason V, Sigurdsson G, Humphries S (1989) A study of familial hypercholesterolemia in Iceland using RFLPs. J Med Genet 26:494–498
Teng B, Sniderman AD, Soutar AK, Thompson GR (1986) Metabolic basis of hyperapolipoproteinemia. Turnover of apolipoprotein B in low density lipoprotein and its precursors and subfractions compared with normal and familial hypercholesterolemia. J Clin Invest 77:663–672
Third JLHC, Montag J, Flynn M, Freidel J, Laskarzewski P, Glueck CJ (1984) Primary and familial hypoalipoproteinemia. Metabolism 33:136–146
Tybaerg-Hansen A, Gallagher J, Vincent J, Houlston R, Talmud P, Dunning AM, Seed M, Hamsten A, Humphries S, Myant NB (1990) Familial defective apolipoprotein B-100 detection in the United Kingdom and Scandinavia, and clinical characteristics of ten cases. Atherosclerosis 80:235–242
Utermann G, Hees M, Steinmetz A (1977) Polymorphism of apolipoprotein E and occurrence of dysbetalipoproteinemia in man. Nature 269:604–607
Utermann G, Vogelberg KH, Steinmetz A, Schoenborn W, Pruin N, Jaeschke M, Hees M, Canzler H (1979) Polymorphism of apolipoprotein E II: genetics of hyperlipoproteinaemia type III. Clin Genet 15:37–62
Utermann G, Hardewig A, Zimmer F (1984) Apolipoprotein E phenotypes in patients with myocardial infarction. Hum Genet 65:237–241
Utermann G (1985) Genetic polymorphism of apolipoprotein E – impact on plasma lipoprotein metabolism. In: Crepaldi G, Tiengo A, Baggio G (eds) Diabetes, obesity and hyperlipidemias III. Excerpta Medica, Amsterdam, p 1
Utermann G (1987a) Apolipoprotein E polymorphism in health and disease. Am Heart J 113:433–440
Utermann G (1987b) Apolipoproteins, quantitative lipoprotein traits, and multifactorial hyperlipidemia. In: Collins G (ed) Ciba Foundation Symposium Nr. 130: molecular approaches to human polygenic disease, 52nd end. Wiley, Chichester, p 52
Utermann G, Menzel HJ, Kraft HG, Duba HC, Kemmler HG, Seitz C (1987) Lp(a) glycoprotein phenotypes. Inheritance and relation to Lp(a)-lipoprotein concentrations in plasma. J Clin Invest 80:458–465
Utermann G (1988) Apolipoprotein polymorphism and multifactorial hyperlipidemia. J Inher Metab Dis 11 [Suppl 1]:74–86
Utermann G (1989) The mysteries of lipoprotein (a). Science 246:904–910
Utermann G, Hoppichler F, Dieplinger H, Seed M, Thompson G, Boerwinkle E (1989) Defects in the LDL receptor gene affect Lp (a) lipoprotein levels: multiplicative interaction of two gene loci associated with premature atherosclerosis. Proc Natl Acad Sci USA 86:4171–4174
Utermann G (1990) Coronary Heart Disease. In: Emery AEH, Rimoin DL (eds) Principles and practice of medical genetics, 2nd edn. Churchill Livingstone, Edinburgh, p 1239
Vega GL, Grundy SM (1986) In vivo evidence for reduced binding of low density lipoproteins to receptors as a cause of primary moderate hypercholesterolemia. J Clin Invest 78:110–114
Wardell MR, Weisgraber KH, Havekes LM, Rall SC Jr (1989) Apolipoprotein E 3-Leiden contains a seven amino acid insertion that is a tandem repeat of residues 121 to 127. J Biol Chem 264:21205–21210

Weintraub MS, Eisenberg S, Breslow JL (1987) Dietary fat clearance in normal subjects is regulated by genetic variation in apolipoprotein E. J Clin Invest 80:1571–1577

Wiklund O, Angelin B, Olofsson SO, Eriksson M, Fager G, Berglund L, Bondjers G (1990) Apolipoprotein(a) and ischaemic heart disease in familial hypercholesterolaemia. Lancet 335:1360–1363

Williams KJ, Petrie KA, Brocia RW, Swenson TL (1991) Lipoprotein lipase modulates net secretory output of apolipoprotein B in vitro. A possible pathophysiologic explanation for familial combined hyperlipidemia. J Clin Invest 88:1300–1306

Williams RR, Hunt SC, Hopkins PN, Stults BM, Wu LL, Hasstedt SJ, Barlow GK, Stephenson SH, Lalouel JM, Kuida A (1988) Familial dyslipidemic hypertension: evidence from 58 Utah-families for the syndrome present in approximately 12% of patients with essential hypertension. JAMA 259:3579–3586

Wilson DE, Emi M, Iverius P-H, Hata A, Wu LL, Hillas E, Williams RR, Lalouel J-M (1990) Phenotypic expression of heterozygous lipoprotein lipase deficiency in the extended pedigree of a proband homozygous for a missense mutation. J Clin Invest 86:735–750

Winkler FK, D'Arcy A, Hunziker W (1990) Structure of human pancreatic lipase. Nature 343:771–774

Wojciechowski AP, Farrall M, Cullen P, Wilson TME, Bayliss JD, Farren B, Griffin BA, Caslake MJ, Packard CJ, Shepherd J, Thakker R, Scott J (1991) Familial combined hyperlipidaemia linked to the apolipoprotein AI-CIII-AIV gene cluster on chromosome 11q23–q24. Nature 349:161–164

Young SG (1990) Recent progress in understanding apolipoprotein B. Circulation 82:1574–1594

CHAPTER 5

Interactions Between Lipoproteins and the Arterial Wall

A.J.R. HABENICHT, P.B. SALBACH, and U. JANSSEN-TIMMEN

A. The Molecular Mechanisms of Atherogenesis Are Not Known

Epidemiological evidence indicates that hypercholesterolemia, smoking, diabetes mellitus, and hypertension are major "risk factors" for cardiovascular disease. However, until now the molecular mechanisms of how these and other risks trigger the disease have escaped detection (SCROTT et al. 1972; GOLDSTEIN and BROWN 1977, 1989; HAVEL and KANE 1989; STEINBERG et al. 1989; HAUSS 1990; BROWN and GOLDSTEIN 1986, 1990; SCHWARTZ et al. 1991; CAMBIEN et al. 1992). Furthermore, unlike cancerogenesis with the discovery of oncogenes and tumor suppressor genes, no unifying hypothesis of atherogenesis has emerged (BISHOP 1987; HENDERSON et al. 1991; WEINBERG 1985, 1991). This lack of direction is not surprising, because atherosclerosis is a chronic, multicellular, and probably multistep disease that manifests itself in several distinguishable morphological forms as well as in different parts of the arterial tree depending on the risk factor (ROSS and GLOMSET 1974a,b; ROSS 1986). Nevertheless, significant progress has recently been made in our understanding of arterial wall biology. These results have shed new light on the potential molecular origins of the disease and in particular on mechanisms that are associated with hypercholesterolemia. In this chapter we will focus on the relation between atherogenesis and hypercholesterolemia.

The cellular constituents of "fatty streaks" and of "proliferative lesions," the early stages of atherosclerosis in response to experimental hypercholesterolemia, have been identified as functionally altered endothelial cells, differentiating macrophages, activated T lymphocytes, and proliferating smooth muscle cells (see ROSS 1986 and SCHWARTZ et al. 1991 for recent reviews). Each of these cell types is capable of forming distinct sets of biologically active molecules upon appropriate stimulation conditions in vitro. In theory, fatty streaks and proliferative lesions could therefore arise from a multitude of direct and indirect cellular interactions. The enormous potential complexity of the disease is illustrated by the number and striking biological activities of the mediators that can be formed by arterial wall cells. For example, macrophages are highly differentiated secretory cells that can be induced to form and secrete more than 100 biologically active products (NATHAN et al. 1980) under different stimulation conditions both in vitro and in vivo, and smooth muscle cells, endothelial cells, and lymphocytes are also producers of many potential mediators (ROSS 1986; MUNRO and COTRAN 1988; RUBIN et al. 1988, 1991; HANSSON et al. 1989, 1991; WITZTUM and STEINBERG 1991; INABA et al. 1992a,b). While much of the information on cells that form atherosclerotic lesions has been obtained with cultured cells, we are only just beginning to understand the alterations of the arterial wall cells phenotype in the intact artery. In vivo data have largely been obtained in normal animals, in particular in subhuman primates, rabbits, and pigs experimentally exposed to risk factors including cholesterol-rich diets or in animals with genetic defects of lipoprotein metabolism such as the Watanabe heritable hyperlipidemic rabbit, which lacks functional low-density lipoprotein (LDL) receptors (FAGGIOTO et al. 1984; reviewed in ROSS 1986; CYBULSKI and GIMBRONE 1991). While such animal models have been extensively exploited in attempts to mimic the human disease, the chronicity as well as the cellular and biochemical complexity of the induced changes in the arterial wall have limited the interpretation of the data, posing the problem of how to distinguish between causative, secondary, and defensive or even protective alterations in arterial wall cells using experiments that require months or years. However, a distinct sequence of morphological alterations has been deduced from cholesterol-fed animals that is likely to mirror the human disease. It appears that one of the first changes in the arterial wall is the adherence of mononuclear cells to the intact endothelium and their subsequent transendothelial migration into the intima of the arterial wall (HAUSS et al. 1965; GERRITY et al. 1979, 1985; GERRITY 1981; FAGGIOTO et al. 1984; ROSS 1986; JONASSON et al. 1986, 1988). Furthermore, mononuclear cells appear to become activated before or concomitant with their migration into the subendothelial space and apparently begin to form and secrete molecules with potent biological activities (see below). The resulting metabolic networks and pathways has led investigators to propose hypotheses of the pathogenesis of atherosclerosis that consider widely differing morphological, biochemical, genetic, and hemodynamic aspects. A major focus of current research that has the potential to link epidemiologically defined risk factors such as hypercholesterolemia to the biology of the arterial wall are the fields of cell adhesion molecules (CAMs), growth factors, differentiation factors, and cytokines (ROSS 1986; RUBIN et al. 1988; FERNS et al. 1991). For example, macrophages and endothelial cells produce the B-chain of the smooth muscle cell mitogen, platelet-derived growth factor (PDGF), the angiogenic factor, basic fibroblast growth factor (bFGF) as well as interleukin 1 (IL-1) in human carotid artery lesions and diseased coronary arteries, but to a much lesser extent in normal blood vessels (ROSS et al. 1990). Moreover, animal studies suggest that interference with growth factor action in vivo limits the formation of proliferative lesions after mechanical injury of the arterial wall (LINDNER and REIDY 1991; FERNS et al. 1991). While these recent data have paved the way to a better understanding of arterial wall biology, several fundamental and urgent questions remain unanswered: What are the initial molecular mechanisms and triggering links between risk factor action and early changes in the arterial wall such as the induction of adhesion molecules regulating leukocyte–endothelial interaction? Do different risk factors act through similar or distinct signaling mechanisms or metabolic pathways? What are the primary target cells of risk factors and in particular hypercholesterolemia? Why do risk factors act on arteries and within the arterial tree at distinct sites such as coronary arteries? It is obvious that answers to these questions will be crucial to the development of new therapeutic strategies of primary prevention.

B. The Low-Density Lipoprotein Receptor Pathway and the Pathogenesis of Atherosclerosis

The discovery of the LDL receptor by Brown and Goldstein is a milestone in lipoprotein research (reviewed in Brown and Goldstein 1986 and Goldstein and Brown 1989); it has influenced virtually every aspect of this research and has also greatly affected atherosclerosis research. The most convincing evidence for a close relation between atherogenesis and disorders of lipid metabolism is the striking incidence of coronary heart disease in patients afflicted with the LDL receptor negative phenotype of familial hypercholesterolemia (here collectively termed homozygous FH; Scrott et al. 1972; reviewed in Goldstein and Brown 1989). Many homozygous FH individuals have been studied in considerable detail at both the clinical and the molecular levels, and distinct subtypes of their LDL receptor mutations have been defined (also see Chap. III by de Villiers et al., this volume). Moreover, evidence has been obtained for a genetic linkage of an "atherogenic" lipoprotein phenotype to or close to the LDL receptor locus on chromosome 19 (Nishina et al. 1992 and references therein; reviewed by MacCluer 1992 and Lusis et al. 1992). The most severe forms of homozygous FH is almost invariably associated with early atherosclerosis, and only few patients survive beyond childhood. The mean age at death in individuals with less than 2% of normal LDL receptor activity is approximately 10 years, and most patients die from coronary heart disease rather than from other forms of cardiovascular disease. An FH patient who suffered multiple myocardial infarctions during early childhood and received a combined heart–liver transplant to replace her ailing heart and to provide a source of normal LDL receptors showed a striking reduction of plasma LDL levels immediately after surgery (Goldstein and Brown 1989). In addition, animal studies using transgenic mice revealed that the induced overexpression of normal LDL receptors eliminates LDL from the circulation (Hofmann et al. 1988). Furthermore, it prevented an increase in plasma LDL in response to cholesterol feeding (Yokode et al. 1990). Taken together, these studies demonstrate that the level of LDL receptor expression largely determines plasma LDL levels; they also document the inverse relation between normal LDL catabolism and atherosclerosis. They do not, however, provide evidence for a molecular link between high LDL levels and atherogenesis. Instead, the dominant association of homozygous FH with coronary heart disease raises several important new questions. At least three possibilities merit consideration and are focuses of current research: (a) FH patients are at risk of developing coronary heart disease, because the abnormally high concentration of native plasma LDL (that results mainly from the inability of the liver to catabolize LDL through receptor-mediated endocytosis and from higher than average LDL synthesis rates) injures the arterial wall and/or circulating blood cells directly and thereby triggers the disease (BROWN and GOLDSTEIN 1986); (b) native plasma LDL undergo some kind of modification within the arterial wall that is yet to be characterized and thus acquire atherogenic properties locally (STEINBERG et al. 1989); (c) the LDL receptor defect prevents the normal actions and metabolism of LDL-derived molecules or metabolites, for example, unsaturated fatty acids or antioxidants that – under normal circumstances – protect the arterial wall (HABENICHT et al. 1990b). It should be noted that these possibilities are not mutually exclusive. The discovery of the LDL receptor has led to the identification of a complex lipoprotein-dependent metabolic pathway, referred to as the LDL pathway (reviewed in BROWN and GOLDSTEIN 1986). It is initiated by high-affinity binding of several lipoprotein classes to the LDL receptor, endocytotic uptake of the lipoproteins in coated pits, the formation of secondary lysosomes, and the lysosomal degradation of the lipoprotein lipid and protein moieties (also see Chap. III by DE VILLIERS et al., this volume). Unesterified cholesterol, unesterified fatty acids, and other LDL-derived molecules are subsequently released from the lysosomal compartment. Among them, cholesterol (and/or an oxidation product of it), the best studied molecule in LDL, mediates negative feedback control on 3-hydroxy-3-methylglutary coenzyme A (HMG-GoA) reductase and other enzymes of the cholesterol biosynthetic pathway and the LDL receptor to achieve cellular cholesterol homeostasis (RIDGWAY et al. 1992). In contrast, the enzyme acyl CoA cholesterol acyltransferase (ACAT) is activated by cholesterol. The metabolism and functional roles of other molecules that are derived from lysosomal LDL catabolism, including unsaturated fatty acids and antioxidants, are less well understood (see below). The normal functioning of the LDL pathway implies that cellular cholesterol balance is maintained by either up- or downregulation of LDL receptors and the postreceptor-mediated regulation of enzyme activities involved in cholesterol synthesis and cholesterol esterification. Consequently, when the concentration of LDL in the extracellular milieu rises, LDL receptor expression is suppressed. However, there are several exceptions to this rule: (a) steroidogenic tissues, in particular the adrenal gland and the ovary, that use the LDL-derived cholesterol precursor for steroid hormone production have a high constitutive LDL receptor expression that is resistent to the down-regulatory effects of even high LDL levels (reviewed in SCHREIBER and WEINSTEIN 1986): (b) certain brain cells including sensory neurons in the cortex, hippocampal Ca1 and Ca2 neurons, and glial cells show a high constitutive LDL receptor expression (SWANSON et al. 1988); (c) rapidly dividing cells including neoplastic cells and activated T lymphocytes have a higher than average LDL receptor expression (HABENICHT et al. 1980; CUTHBERT et al. 1989; CUTHBERT and LIPSKY 1989). Nevertheless, the down-regulation of LDL receptors and of enzymes in the cholesterol biosynthetic pathway including HMG-CoA reductase very effectively prevent the over-accumulation of cellular cholesterol even at very high LDL concentrations. Therefore, the LDL receptor pathway plays a protective role in the sense that it maintains the cellular cholesterol balance. Are there other functions that have escaped detection? It is important to note that the LDL-mediated suppression of HMG-CoA reductase is incomplete and can be further enhanced by the addition of mevalonic acid, the product of the enzyme (reviewed in GOLDSTEIN and BROWN 1990). Research related to this phenomenon and into the metabolic routes of mevalonic acid and the biological roles of nonsteroid metabolites of mevalonic acid (HABENICHT et al. 1980; WOLDA and GLOMSET 1988) led to the identification of new types of post-translational modification of important cell proteins involving long chain, thioether-linked prenyl groups by Glomset and coworkers (reviewed in GLOMSET et al. 1990; GOLDSTEIN and BROWN 1990; GLOMSET 1994). These posttranslational modification reactions including farnesylation, geranyl-geranylation, and possibly other modifications apparently function to anchor proteins in specific membrane domains or receptors and are potential new drug targets (GIBBS 1991). Prenylation occurs on cysteine residues near the C terminus of the proteins (reviewed by GLOMSET et al. 1990). Prenylated proteins discovered so far have important biological activities. They include several members of the proto-oncogene family of ras proteins that act as GTP-binding and hydrolyzing proteins and are involved in signal transduction of multiple hormones acting through specific cell surface receptors. Furthermore, nuclear lamins – lamin B was the first prenylated protein ever described (WOLDA and GLOMSET 1988) – present in the inner core of the nuclear envelope are prenylated and appear to participate in nuclear envelope assembly (reviewed in GLOMSET et al. 1990; GOLDSTEIN and BROWN 1990). Other prenylated proteins are members of the smg p21B, rap, rho, rac, and rab families, some of which appear to regulate specific steps of intracellular transport (KAWATA et al. 1990; DIDSBURY et al. 1990; KINSELLA and MALTESE 1991; KINSELLA et al. 1991; YAMANE et al. 1991; reviewed in GLOMSET 1994). This new exciting and important area is likely to hold surprises for the future, because prenylation of some of the proteins has already been shown to be related to their biological function. Whether altered prenylation patterns of proteins are related to atherogenesis is, however, completely unknown.

C. Is Low-Density Lipoprotein Atherogenic?

The regulatory role of cholesterol on the LDL receptor at the cellular level is reflected by an inverse relation between the concentration of plasma LDL and LDL receptor expression in the general population (GOLDSTEIN and BROWN 1989). As a result, populations with high dietary cholesterol intake show suppressed LDL receptor activities when compared to populations with low dietary cholesterol intake. The existence of this inverse relationship between LDL levels and LDL receptor expression in the general population raises the important question of which of these two parameters trigger the disease, i.e., is there evidence for an atherogenic potential of LDL? The epidemiological evidence of an association between high LDL plasma levels and the incidence of cardiovascular disease consequently led to an intensive search for an "injurious" or "atherogenic" potential of native human plasma LDL. However, evidence that native LDL is atherogenic by directly affecting arterial wall biology is scarce, if not nonexistent. For example, enhanced LDL uptake by dividing cells such as rapidly proliferating smooth muscle cells or activated T lymphocytes may not be an "atherogenic," but rather a physiological response and may simply reflect the need of dividing cells for cholesterol and other LDL components such as essential long-chain polyunsaturated fatty acids (HABENICHT et al. 1980, 1986; CUTHBERT and LIPSKY 1989). In patients with high plasma LDL levels, this interaction is not qualitatively different: high plasma LDL levels reflect the low LDL receptor expression in the liver, which is limited in its capacity to catabolize LDL by receptor-mediated endocytosis. Is there evidence that cells that form atherosclerotic lesions behave differently, thus implying a mechanism for the striking accumulation of cholesteryl esters in arterial wall cells as seen in hypercholesterolemic patients afflicted with cardiovascular disease or in experimental animals fed a cholesterol-rich diet? Early studies performed with human blood-derived monocytes or monocytic cell lines using β-migrating very low density lipoproteins (β-VLDL), which have a very high affinity for the LDL receptor because they contain apolipoprotein E, indicated that LDL receptor downregulation was defective in these cells (reviewed in GOLDSTEIN and BROWN 1977). This data appeared to offer a potential clue for cellular cholesteryl ester accumulation and thus "foam cell" formation, believed to be important in the pathogenesis of atherosclerosis. However, later studies dismissed some of the early interpretations: there is now general agreement that endothelial cells, smooth muscle cells, macrophages, and lymphocytes express the LDL receptor at normal rates and that the regulatory functions associated with the LDL pathway in arterial wall cells are indistinguishable from the regulatory functions found in other cells (with the exceptions mentioned above). Finally, normal smooth muscle cells or monocytes in culture cannot be transformed into lipid-laden foam cells even upon incubation with very high concentrations of native LDL (MAHLEY 1988; HUSSAIN et al. 1989; STEINBERG et al.

1989; WITZTUM and STEINBERG 1991). Therefore, the interaction of native LDL with the LDL receptor expressed by cells of the arterial wall does not appear to have pathophysiological relevance. In fact, it is not the presence, but the absence of LDL receptors that is associated with lipid accumulation and thus foam cell formation and atherosclerosis.

D. The Low-Density Lipoprotein Receptor Is a Member of a Family of Lipoprotein Receptors: The Low-Density Lipoprotein Receptor Related Proteins, a Very Low-Density Lipoprotein Receptor, and Glycoprotein 330

In mammals the LDL receptor interacts with two protein ligands, apolipoprotein B_{100} and apolipoprotein E. These ligands bind to distinct pockets within the general binding domain of the LDL receptor (BARBER et al. 1991; reviewed in BROWN et al. 1991; STIFANI et al. 1991). However, there is no accumulation of apolipoprotein E-containing lipoproteins in the plasma of homozygous FH patients, and chylomicron remnant removal appears to be unaffected in Watanabe heritable hyperlipidemic rabbits, an animal model of homozygous FH (RUBINSTEIN et al. 1990). It has therefore been suspected for a number of years that there is an additional lipoprotein receptor that can bind apolipoprotein E (MAHLEY 1988). While early attempts to identify this putative "remnant" or "apolipoprotein E" receptor have failed, candidate receptors including one in man have recently been reported (HERZ et al. 1988). The first of these was isolated from a cDNA library of cytotoxic human T lymphocytes by HERZ and coworkers in attempts to chase for complement peptide-like sequences (the classical LDL receptor contains a cluster of seven cysteine-rich repeats with significant complement peptide sequence homology). Using cross-hybridization screening, a cDNA clone that encoded a lipoprotein cell surface receptor distinct but closely related to the classical LDL receptor was isolated. This receptor, termed the "LDL receptor related protein" (LRP) has several surprising and potentially important properties. It is a 600-kDa glycoprotein and one of the largest proteins ever described (see BROWN et al. 1991). It contains 32 cysteine-rich repeats of the complement type. Evidence gathered so far strongly suggests that LRP represents a "remnant receptor." It binds β-VLDL and apolipoprotein-E-containing liposomes in a Ca^{2+}-sensitive way (BEISIEGEL et al. 1991; STIFANI et al. 1991), and transfection experiments revealed that LRP can specifically bind VLDL when the medium is enriched in apolipoprotein E (BROWN et al. 1991). A recent observation by BEISIEGEL et al. (1991) added to the evidence that LRP represents an apolipoprotein E receptor. These researchers showed that lipoprotein lipase, apparently independently of its lipolytic activity, was capable of greatly enhancing the binding of "remnant-like" lipoproteins and apolipoprotein E-containing liposomes to LRP. EISENBERG et al. (1992) reported that lipoprotein lipase enhances binding of LDL and VLDL to heparan sulfate on cell surfaces and extracellular matrix. In this context it is of interest that YLÄ-HERTTUALA et al. (1991b) have recently shown in rabbit atherosclerotic lesions that lipoprotein lipase is expressed by macrophages in lipid-rich areas and that SALOMON et al. (1992) demonstrated with the help of polymerase chain reaction amplification the presence of increased apolipoprotein E transcripts in advanced human atheroma. Together with recent evidence that lipoprotein lipase enhances LDL and VLDL retention by the subendothelial cell matrix in vivo, the coexpression of LRP, lipoprotein lipase, and apolipoprotein E in the intima of the arterial wall could have pathophysiologic significance (SALOMON et al. 1992; SAXENA et al. 1992).

YAMADA et al. (1992) found that the administration of apolipoprotein E to Watanabe heritable hyperlipidemic rabbits prevented the progression of atherosclerosis in this LDL receptor deficient animal model of homozygous FH. In a recent study using transgenic mice overexpressing apolipoprotein E, SHIMANO et al. (1992a,b) demonstrated that the transgenes were resistant to cholesterol-rich diets in that they did not develop hypercholesterolemia. The transgenes had subnormal VLDL and LDL levels whereas the HDL concentrations were less affected. Moreover, mice generated by gene targeting that lacked apolipoprotein E developed hypercholesterolemia, fatty streak formation in their coronary arteries, and premature atherosclerosis (ZHANG et al. 1992; PLUMP et al. 1992). This recent data reveals that apolipoprotein E has antiatherogenic properties. However, the molecular mechanism of the antiatherogenic properties of apolipoprotein E remains to be determined.

A further potentially important observation, made independently by KRISTENSEN et al. (1990) and STRICKLAND et al. (1990), was the demonstration that LRP is identical to the α_2-macroglobulin receptor. Since α_2-macroglobulin is known to bind plasma lysosomal enzymes and growth factors such as PDGF and transforming growth factor β (TGFβ), LRP could conceivably function as a "remnant receptor" and as a "scavenger receptor" for ligands that bind to the plasma protein α_2-macroglobulin or for ligands that do not require prior binding to α_2-macroglobulin. This hypothesis is supported by a recent study by KOUNNAS et al. (1992) that demonstrates that the α_2-macroglobulin receptor binds and internalizes *Pseudomonas* exotoxin A. Moreover, recent studies demonstrated that LRP binds complexes of tissue-type and urokinase-type plasminogen activators with plasminogen activator inhibitor type 1 (BU et al. 1992; ORTH et al. 1992; HERZ et al. 1992) and pregnancy zone protein (GÅFVELS et al. 1992). The fact that LRP functions in vivo to remove chylomicrom remnants can be inferred from a recent study that demonstrated that lactoferrin (which binds to LRP with high affinity in vitro) specifically inhibited hepatic chylomicron remnant removal (HUETTINGER et al. 1992). An additional in vivo function of the LRP appears to be the implantation of the embryo, as shown by disruption of LRP gene expression in intact animals (HERZ et al. 1992). This novel function of LRP appears to be mediated by the binding of urokinase-type plasminogen activator–plasminogen activator inhibitor complexes (HERZ et al. 1992; see also GÅFVELS et al. 1992). If further studies substantiate the notion that LRP is a multifunctional receptor that can bind molecules as diverse as chylomicron remnants, α_2-macroglobulin, bacterial toxins, plasminogen activators, and pregnancy zone protein, additional questions will arise. These concern the functional relationships between the multiple ligands as well as the regulation of the expression of different members of the LRP family of lipoprotein receptors (LUND et al. 1989). Recent studies in the chicken may have important implications for the understanding of the biology of mammalian lipoprotein receptors. Thus, STIFANI et al. (1991) isolated two different types of LRP, one expressed by the liver and possibly other somatic cells, while another is exclusively expressed by the oocyte. Analagously, BARBER et al. (1991) identified an oocyte-specific LDL receptor and a somatic cell-specific LDL receptor. The identification of four distinct lipopoprotein receptors in the chicken is particularly intriguing in view of the absence of apolipopoprotein E in birds. All four chicken lipoprotein receptors have been shown by ligand-blotting experiments to be capable of binding Ca^{2+}, but they differentially bind vitellogenin, apolipoprotein E, α_2-macroglobulin, apolipoprotein B, and probably lipoprotein lipase (STEYRER et al. 1990; BEISIEGEL et al. 1991). These results are consistent with the possibility that different classes of LDL receptors exist that are differentially regulated in a tissue-specific way and that lipoproteins, α_2-macroglobulin, and other ligands bind to them in a regulated fashion, thus implying distinct biological functions. Despite rapid progress in LRP research, questions remain to be answered, in particular as far as the biological role of LRP in lipid metabolism in humans is concerned. Up to now no naturally transduced mutants of LRP have been isolated to test the possibility that chylomicron remnant removal in LDL receptor normal individuals is dependent on LRP. Recent studies by CHOI et al. using anti-LDL-receptor antibodies that do not show cross-reactivity against LRP and in vivo competition studies of chylomicron remnants and α_2-macroglobulin in mice speak in favor of this possibility (CHOI et al. 1991; HUSSAIN et al. 1991; but see VAN DIJK et al. 1992). What is the relation between the LRP and the pathogenesis of atherosclerosis? Unlike the LDL receptor, LRP apparently cannot be downregulated by exogenous cholesterol and cannot be upregulated by estrogens (BROWN et al. 1991). This indicates that LDL receptor and LRP expression are differentially regulated. Whether LRP mediates the overaccumulation of cholesteryl esters in appropriate target cells, including macrophages and smooth muscle cells within atherosclerotic lesions, deserves further study. While LRP could contribute to foam cell formation, believed to be important in "fatty streak" formation (ROSS 1986), the biological sequelae and significance of this cholesteryl ester accumulation for the biology of the arterial wall are anything but clear. Although foam cell formation has been widely assumed to represent a hallmark of atherogenesis, no atherogenic or other biological function of foam cells has been clearly delineated. Studies with cultured cells show that macrophages expressing LRP when incubated with β-VLDL acquire the appearance of lipid-laden cells (these lipoproteins contain apolipoproteins E, B_{48}, B_{100}, and apolipoprotein C). Indeed, ligand blots using biotinylated β-VLDL have shown that this lipoprotein class binds to the LRP and that apolipoprotein C inhibits binding (KOWAL et al. 1990). Since LRP can be expressed by cells of the arterial wall in vitro, it will be of interest to determine whether it is expressed by arterial wall cells in vivo during the early stages of atherogenesis and to search for a functional role of LRP in the biology of blood vessels.

VAN DIJK et al. (1992) recently provided evidence that β-VLDL are recognized by a receptor with properties distinct from LRP. In the rabbit, TAKAHASHI et al. (1992) recently isolated a cDNA clone from a heart library that displays very close sequence homology with the LDL receptor but which binds VLDL, intermediate-density lipoprotein, and β-VLDL with higher affinity than LDL. Transcripts of this receptor were particularly expressed in heart, muscle, brain, and adipose tissue, but not in liver. It will be of major interest to study the function of this lipoprotein receptor and search for homologous receptors in humans.

Two further molecules with significant sequence homology to the LDL receptor family have recently been identified. One of these, termed glycoprotein 330 (GP330), is the autoimmune target in Heymann nephritis (RAYCHOWDHURY et al. 1989) and another, termed perlecan, is a basement membrane heparan sulfate proteoglycan that also shows sequence homology with laminin A chain and the neural cell adhesion molecule (NOONAN et al. 1991). The function of these molecules in relation to lipoprotein metabolism is unknown, but their presence raises the possibility that there are family members in humans yet to be discovered. These, as well as the ones already identified, could conceivably play distinct physiological as well as pathophysiological roles in lipid metabolism.

E. The Low-Density Lipoprotein Receptor Dependent Arachidonic Acid Pathway

Cholesterol is a key regulator of LDL receptor expression, HMG-CoA reductase activity, and ACAT activity (Fig. 1). Are there molecules in LDL other than cholesterol that affect intracellular metabolic pathways or is cholesterol the only one? Following this line of thought, we have turned our attention to the fact that LDL and other lipoprotein classes that bind to the LDL receptor are carriers of essential long-chain polyunsaturated fatty acids; in fact, LDL receptor-binding lipoproteins are the major carriers of linoleic acid and arachidonic acid (eicosatetraenoic acid 20:4 ω6) in fasting and postprandial human plasma (J.A. GLOMSET, P.B. SALBACH, W.C. KING and A.J.R. HABENICHT, unpublished observation). These fatty acids are

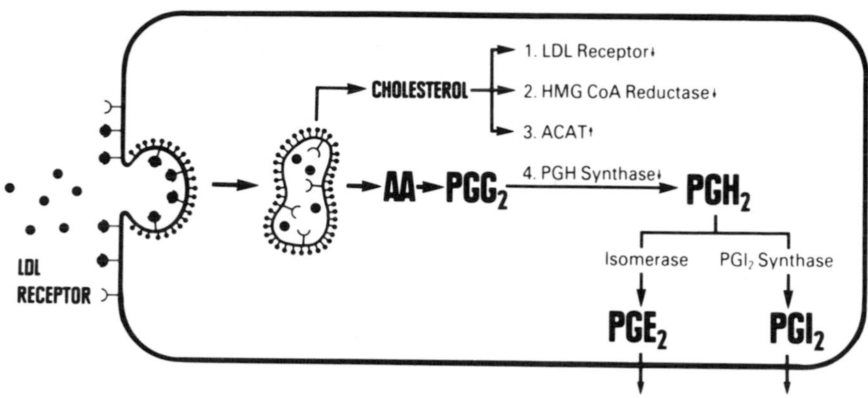

Fig. 1. Proposed relation between the classical low-density lipoprotein (LDL) receptor pathway and LDL-dependent arachidonic acid metabolism in platelet-derived growth factor (PDGF) stimulated fibroblasts. LDL and other lipoprotein classes that bind to the LDL receptor such as chylomicron remnants are taken up through receptor-mediated endocytosis and cholesteryl esters, and phospholipids are hydrolyzed in secondary lysosomes. Unesterified cholesterol and unsaturated fatty acids including arachidonic acid (*AA*) are subsequently released from the lysosomal compartment. While unesterified cholesterol regulates the activity of 3-hydroxy-3-methylglutaryl-coenzyme A (HMG-CoA) reductase, the expression of the LDL receptors, and the enzyme acyl CoA cholesterol acyltransferase (*ACAT*) (BROWN and GOLDSTEIN 1986), LDL-derived unesterified AA is first converted by the prostaglandin H (*PGH*) synthase reaction into the unstable endoperoxide PGH_2 and subsequently converted into prostacyclin (PGI_2) by prostacyclin synthase or PGE_2 by endoperoxide PGE_2 isomerase (see HABENICHT et al. 1990b for further details)

precursor molecules for a large group of oxidation products that exert potent biological activities in a variety of physiological and disease processes including cardiovascular disease (reviewed in NEEDLEMAN et al. 1986; SAMUELSSON et al. 1987). Using gas liquid chromatography analyses of free versus esterified arachidonic acid, we found that 96%–98% of the total fasting plasma arachidonic acid is present in the form of lipid esters within lipoproteins. In an attempt to establish a relation between unsaturated fatty acid metabolism and the LDL pathway (HABENICHT et al. 1986), we removed the native cholesteryl esters by heptane and reconstituted the LDL phospholipid apoprotein B shell with cholesteryl esters of radiolabeled arachidonic acid. Subsequent studies in cultured cells revealed that arachidonic acid present in the reconstituted lipoproteins was utilized by rapidly growing fibroblasts to form prostacyclin and prostaglandin E_2 (PGE_2), two biologically active metabolites of the fatty acid (HABENICHT et al. 1990b). Interestingly, PGE_2 exerts immunosuppressive effects on lymphocytes and also has potent inhibitory effects on macrophage activation (NEEDLEMAN et al. 1986). With the help of cultured FH cells, normal cells, and anti-LDL-receptor antibodies, we have further shown that this delivery is mediated by the

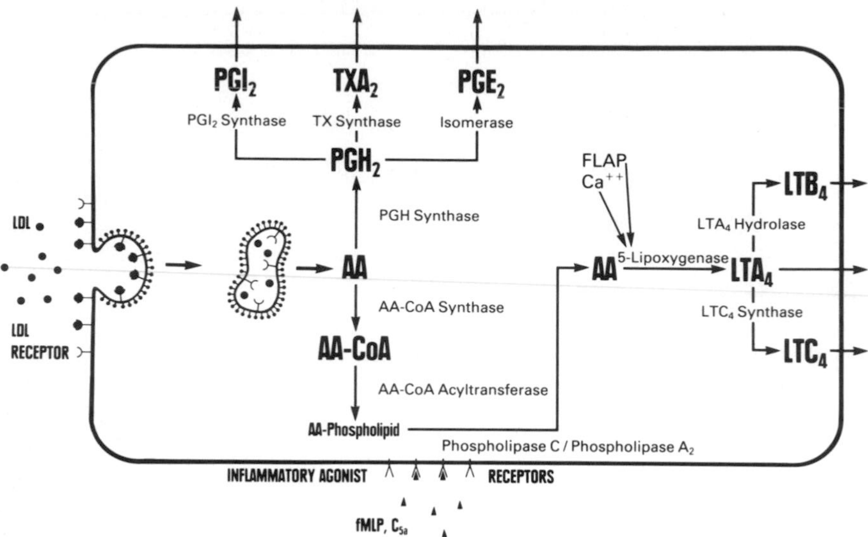

Fig. 2. Proposed low-density lipoprotein (*LDL*) receptor-dependent uptake and metabolism of arachidonic acid (*AA*) in human-blood-derived monocytes/macrophages. Unesterified AA, released from secondary lysosomes, is preferentially utilized by the prostaglandin H (*PGH*) synthase reaction in resting macrophages or incorporated into cellular phospholipids by the coupled actions of the AA-Coenzyme A (*CoA*) synthase and AA-CoA acyltransferase reactions. AA that is converted into the unstable endoperoxide PGH_2 by PGH synthase and further metabolized by three peripheral enzymes of this pathway: by prostacyclin synthase into PGI_2, by thromboxane (*TX*) synthase into TXA_2, and by endoperoxide PGE_2 isomerase into PGE_2. LDL-derived AA may also be metabolized by the 5-lipoxygenase pathway in response to the activation of signaling pathways generated by Ca^{2+} ionophore or the chemotactic peptide, fMeth-Leu-Phe (*fMLP*), the complement peptide C_{5a} or other activators of phospholipases, and the 5-lipoxygenase activating protein (*FLAP*) (see SALBACH et al. 1992 for further details)

classical LDL pathway of Brown and Goldstein and that arachidonic acid is converted into distinct sets of metabolites depending on the cell type and the stimulation conditions (SALBACH et al. 1992). For example, cultured fibroblasts mainly form prostacyclin and PGE_2 from LDL-derived arachidonic acid (Fig. 1), while cultured human-blood-derived monocytes form prostacyclin and PGE_2 as well as thromboxane, and they form leukotrienes from LDL when stimulated by inflammatory agonists (Fig. 2). The latter group of substances constitute potent mediators of proinflammatory tissue reactions (reviewed by SAMUELSSON et al. 1987). Taken together, these preliminary results indicate that the LDL receptor pathway has a regulatory role in the formation of several eicosanoids in addition to its known role in the maintenance of cellular cholesterol homeostasis. However, many questions regarding the LDL arachidonic acid pathway remain to be investigated, and further work is required to demonstrate the in vivo significance of this pathway as well as to elucidate the potential biological activities of the LDL-derived eicosanoids in the arterial wall and other target tissues.

F. The Oxidation and Scavenger Receptor Hypotheses

We have stated above that little evidence supports the concept that native LDL is atherogenic or toxic to cells, even at high concentrations. A hypothesis that addresses the puzzle of the relation between high concentrations of plasma LDL and atherogenesis, regardless of whether hypercholesterolemia is due to dietary and/or genetic factors, has been proposed by STEINBERG et al. (see STEINBERG et al. 1989; WITZTUM and STEINBERG 1991 for recent reviews). These researchers proposed that cells that form the proliferative lesions of atherosclerosis, in particular macrophages and endothelial cells, modify native plasma LDL locally within the arterial wall in such a way that it acquires atherogenic potential (PALINSKI et al. 1989). During the process of modification, native LDL loses its ability to bind to the classical LDL receptor. Moreover, the modification of LDL involves an alteration of apolipoprotein B_{100}, so that the modified versions of LDL bind to one or more classes of "scavenger receptors" (several distinct, but closely related, bovine and human acetyl-LDL receptors have recently been cloned, see below). The nature of this modification and the amino acid sequence within the modified sequence of apolipoprotein B_{100} required for binding to the acetyl-LDL receptor remains to be identified. In addition to its altered binding characteristics towards the classical LDL receptor and its acquired binding characteristics towards the acetyl-LDL receptor, modified LDL adopts striking biological properties both in vitro and in vivo that are consistent with an atherogenic role (Table 1). For example, modified LDL has been shown to act as a potent promoter of inflammatory tissue reactions: it induces the formation of cytokines, chemotactic peptides, and the synthesis of proinflammatory arachidonic acid metabolites including thromboxane and leukotrienes (Table 1). A recent observation that is inconsistent with a proinflammatory role of modified LDL indicates that oxidized LDL inhibits the expression of the principal growth factor for smooth muscle cells, PDGF, in cultured monocytes/macrophages (MALDEN et al. 1991). Given these in vitro activities (but see below), it is of interest that in vivo evidence has been obtained to indicate that portions of the arterial wall that are affected by the disease are recognized by antibodies directed against modified forms of LDL (HABERLAND et al. 1988; PALINSKI et al. 1989).

What is the nature of LDL modification and what are the potential triggering events? First, it should be recognized that both lipids and the apolipoprotein B of LDL can undergo modification and in particular oxidation (STEINBERG et al. 1989). The major type of LDL modification most likely occurs in the lipid moieties, however, and it is long-chain unsaturated fatty acid peroxidation that appears to be the initiating event, whereas protein modification is believed to be secondary (STEINBERG et al. 1989;

Table 1. Properties of modified low-density lipoproteins

	Reference
Cytotoxic towards arterial wall cells	reviewed in STEINBERG et al. (1989)
Chemotactic towards human monocytes through lysophospholipid content	reviewed in WITZTUM and STEINBERG (1991)
Reduced uptake through the LDL receptor	FREEMAN et al. (1991)
Uptake through acetyl LDL receptors	FREEMAN et al. (1991)
Decreased content of unsaturated fatty acids	ESTERBAUER et al. (1987)
Increased content of oxidized cholesterol	WITZTUM and STEINBERY (1991)
Induces prostaglandin formation in monocytes	HARTUNG et al. (1986)
Induces leukotriene formation in vivo	LEHR et al. (1991)
Induces formation of oxidized arachidonic acid metabolites in monocytes	YOKODE et al. (1988)
Reduces PDGF mRNA and protein in macrophages	MALDEN et al. (1991)
Induces monocyte adhesion to plastic surfaces	FROSTEGARD et al. (1990)
Induces major histocompatibility complex class II molecules and leuM3 antigen and decreases CD4 antigen in monocytes	FROSTEGARD et al. (1990)
Induces monocyte chemotactic protein (MCP) 1 in endothelial and smooth muscle cells	KELLEY et al. (1988); CUSHING et al. (1990)

LDL, low-density lipoprotein; PDGF, platelet-derived growth factor.

ESTERBAUER et al. 1987, 1992). Both the cholesterol moiety and the long-chain polyunsaturated fatty acids, in particular linoleic acid, seem to be involved (KÜHN et al. 1992). For apolipoprotein B, these reactions include acetylation, acetoacetylation, and malondialdehyde conjugation, as shown in in vitro studies (STEINBERG et al. 1989: WITZTUM and STEINBERG 1991). Oxidation reactions in vitro have been induced by metal ions such as Cu^{2+}, and this type of LDL modification was shown to occur after the lipid soluble antioxidants present in LDL such as α-tocopherol and β-carotene were used up through the oxidation reaction (ESTERBAUER et al. 1987). Once LDL contains small amounts of fatty acid peroxides, the number of free radicals greatly increases and leads to fatty acid breakdown products including several short-chain aldehydes and ketones (ESTERBAUER et al. 1987) and, as a consequence of the presence of reactive oxygen species, may secondarily mediate protein modification (reviewed in HALLIWELL and GUTTERIDGE 1990). Furthermore, it was found that endothelial cells, macrophages, and smooth muscle cells were also capable of modifying LDL in such a way that it binds to acetyl-LDL receptors (HENRIKSON et al. 1981; reviewed in WITZTUM and STEINBERG 1991). It is not clear which enzyme is reponsible for cell-mediated LDL oxidation, and much needs to be learned about the triggers of its activation as well as the molecular structure of some of the products (KÜHN et al. 1992). It seems to be clear, however, that lipid peroxides and activated oxygen species have potent biological activities including alteration in the gene expression patterns of cells and possibly alter the gene activation program of the intact artery (STAAL et al. 1990; XANTHOUDAKIS and CURREN 1992 and references therein). Halliwell and Gutteridge have proposed several mechanisms to explain how the chain reaction of LDL oxidation could be triggered in vivo (see HALLIWELL and GUTTERIDGE 1990 for a recent review). In earlier studies it has been suggested that superoxide anion would be the critical mediator. However, studies utilizing superoxide anion dismutase added to cell cultures of macrophages have shown little effects of the enzyme on cell-mediated lipid oxidation in one study (RANKIN et al. 1991), but strong effects in another (McNALLY et al. 1990). Which molecules are responsible for cell-mediated LDL oxidation reactions therefore remains an unresolved question. A consistent finding of several investigators was, however, that inhibitors of lipoxygenases including the 5- and 15-lipoxygenase almost completely blocked cell-mediated LDL oxidation (RANKIN et al. 1991). The specificity of these inhibitors and their mode of action remain to be fully characterized and only limited information can be gained from pharmacological studies of this kind. A stronger argument for the involvement of a 15-lipoxygenase in cell-mediated LDL oxidation is the demonstration of 15-lipoxygenase mRNA by in situ hybridization and of protein in lipid-rich areas of the atherosclerotic lesion in rabbits, the absence of this enzyme in unaffected arteries, and the fact that 15-lipoxygenase colocalizes with epitopes of oxidized LDL and mRNA of acetyl-LDL receptors (YLÄ-HERTTUALA et al. 1991a). No evidence for the presence of 5-lipoxygenase in atherosclerotic lesions could be obtained in this study. Moreover, the cytokine interleukin (IL) 4 is a potent inducer of 15-lipoxygenase in human monocytes in vitro (CONRAD et al. 1992). It is also possible, however, that LDL oxidation is achieved by an oxidase yet to be characterized. This view is supported by a recent pharmacological study of endothelial cells and macrophages that compares the efficiencies of different lipoxygenase inhibitors with their ability to block cell-mediated LDL oxidation (SPARROW and OLSZEWSKI 1992). In this in vitro study, the effect of lipoxygenase inhibitors to block hydroxyoctadecadienoic acid (HODE) production (no absolute values were presented and the cells were resting) could be distinguished from the cells ability to modify LDL. Regardless of what the precise molecular mechanisms of LDL oxidation in the arterial wall are, it is of interest that antioxidants such as α-tocopherol, β-carotene, and ascorbic acid are all able to inhibit LDL oxidation in vitro, β-carotene being the most potent on a molar basis (JIALAL et al. 1991; JIALAL and GRUNDY 1991). In this regard it is of further interest that lipid-soluble pharmacological antioxidants of the probucol family have been shown to reduce the development of atherosclerotic lesions in the Watanabe heritable hyperlipidemic rabbit (CAREW et al. 1987). This latter data does not necessarily imply cellular lipoxygenases in the oxidation process of LDL, because lipid-soluble antioxidants could conceivably act in the extracellular space by their ability to scavenge reactive oxygen species and thus act as lipid peroxidation chain-breaking reagents or could exert biological activities yet to be described. LEHR et al. (1991) injected high concentrations of oxidized LDL into control hamsters and hamsters pretreated with the specific 5-lipoxygenase inhibitor MK-886. Treatment of animals with the inhibitor completely prevented leukocyte adherence to the endothelium of arterioles induced by oxidized LDL, implicating and involvement of 5-lipoxygenase in the recruitment of leukocytes in this experimental system. Again, the interpretation of such studies (here the introduction of fairly high concentrations of oxidized LDL into the circulation) is limited, but the studies support the notion that different enzymes of oxidative arachidonic acid metabolism, including the 5-lipoxygenase, could be involved in leukocyte endothelial interaction. These unanswered questions led KÜHN et al. (1992) to analyze the oxidized lipids of aortic lesions of patients afflicted with severe atherosclerosis. Interestingly, using chiral-phase high-pressure liquid chromatogaphy to separate the enantiomers (i.e., optic isomers of a given positional isomer) of lipid oxidation products in human atherosclerotic plaques, these researchers demonstrated that the bulk of the oxidized lipid species were racemic mixtures rather than one or another enantiomer. This finding is of major interest because it makes an involvement of a lipoxygenase or other enzyme-mediated reactions that show strict optical stereospecificity for the generation of the bulk of the oxidized lipids in the arterial wall unlikely. However, the possibility remains that one of the known oxigenases or one yet to be identified trigger an oxidation reaction that leads to a pathological chain reaction culminating in LDL oxidation in the extracellular space. Other studies concern the cholesterol-fed rabbit. Diseased or unaffected aortic tissue strips were incubated with labeled unesterified linoleic acid, the major fatty acid in LDL, and the products were analyzed by high-pressure liquid chromatography (SIMON et al. 1989). The results indicated that there was formation of 13-HODE in hypercholesterolemic, but much less in control animals. The enantiomers of 13-HODE in this study were not separated, and an involvement of a lipoxygenase can, therefore, not be inferred with certainty.

At least two types of modified forms of LDL, oxidized LDL and acetylated LDL, are recognized by the type I and type II acetyl-LDL receptors; their complementary DNA sequences were isolated from a bovine lung library (KODAMA et al. 1990) and subsequently the human forms were isolated by cross-hybridization screening using the bovine probes (MATSUMOTO et al. 1990). SPARROW et al. (1989), on the basis of cross-competition binding studies using acetylated LDL and oxidized LDL, proposed that there is an oxidized LDL receptor that does not recognize acetyl-LDL, but evidence for its presence at a molecular level has not been presented. On the other hand, transfection of cells with types I or II acetyl LDL receptor expression vectors revealed nonreciprocal cross-competition between acetyl-LDL and oxidized LDL (FREEMAN et al. 1991). It therefore remains to be clarified whether these binding characteristics towards the acetyl-LDL receptors explains the data of SPARROW et al. or whether there is a separate scavenger receptor that specifically binds oxidized LDL and is inactive towards acetyl-LDL. STANTON et al. (1992) recently obtained strong evidence for such a receptor and identified it as a macrophage Fc receptor for IgG. Since it has recently been shown that the acetylated LDL receptors are regulated by TGFβ (BOTTALICO et al. 1991), interferon-γ (GENG and HANSSON 1992), and protein kinase C activating phorbol esters (MOULTON et al. 1992), it is likely that more information on the role and mechanisms of regulation of this class of lipoprotein receptors will be available soon. Studies into the function of scavenger receptors will help to understand their role, if there is one, in arterial wall biology. In this regard, it is of interest – and in contrast to suggestions originally made by KODAMA et al. (1990) – that VIA et al. (1992) recently showed for the murine acetylated LDL receptor that its 80-kDa subunit is as active in binding the acetylated lipoprotein as the trimeric form. The finding by HAMPTON et al. (1991) that acetyl-LDL receptors are endotoxin-binding receptors in the liver is of interest in this respect and may indicate that acetyl-LDL receptors are multifunctional molecules, as the LRP class of receptors are.

A further potentially significant aspect of the oxidation hypothesis of the pathogenesis of atherosclerosis was revealed by the finding that high-density lipoproteins (HDL) can also undergo oxidation and that the modified HDL loses its ability to stimulate cholesterol efflux from cultured cells (NAGANO et al. 1991 and see below). Before LDL loses its ability to bind to the LDL receptor during cell- or chemical-mediated oxidation in vitro, the lipoproteins undergo a process that has been termed "minimal oxidation" (BERLINER et al. 1990). This type of modification implies that "minimally oxidized lipoproteins," while still binding to the classical LDL receptor, exert biological activities both in vitro and in vivo that are very different from those of native LDL and similar to those of oxidized LDL. For example, it has been shown that "minimally oxidized LDL" generated by mild iron oxidation or prolonged storage of native LDL induce macrophage-colony stimulating factor (M-CSF) mRNA and activity in cultured endothelial cells and of monocyte chemotactic protein-1 (MCP-1) (CUSHING et al. 1990) (Table 1). Moreover, when minimally oxidized LDL was injected into the circulation of mice, M-CSF activity was induced (RAJAVASHISTH et al. 1990; LIAO et al. 1991). These studies indicate a potential link between LDL metabolism and the response to injury hypothesis in atherogenesis (see below).

While there is ample evidence that oxidized LDL is present in atherosclerotic lesions, it will be a major challenge to demonstrate that these lipoproteins are pathogenetically important either by helping to trigger changes in the arterial wall or to propagate important chain reactions that support injurious changes in the arterial wall.

G. High-Density Lipoproteins and the Reverse Cholesterol Transport Hypothesis

Epidemiological studies indicate an inverse relationship between the concentration of plasma HDL and the incidence of cardiovascular disease (BRESLOW 1989a; TALL 1990). As is the case for LDL, the molecular basis for this relationship is not clear. GLOMSET (1968) suggested that one of the potential physiological roles of HDL could be its ability to take up cholesterol from peripheral cells and carry it back to the liver, thus mediating an antiatherogenic effect. This "reverse cholesterol transport hypothesis" has attracted general attention and considerable experimental support for it has been accumulating. Several lines of evidence suggest that reverse cholesterol transport involves distinctly regulated steps. The first step is presumed to be the esterification reaction of unesterified cholesterol on the surface of nascent HDL particles catalyzed by the enzyme lecithin cholesterol acyltransferase (LCAT), a circulating plasma enzyme that GLOMSET (1968) had discovered earlier. LCAT is particularly active towards cholesterol-poor nascent HDL precursors, disk-shaped lipoproteins that mainly consist of apolipoproteins and phospholipids. The esterification reaction removes unesterified cholesterol from the HDL disk's surface and relocates the newly formed cholesteryl esters into the HDL's hydrophobic core (HARA and YOKOYAMA 1991; JOHNSON et al. 1990, 1991). Thus, during the course of the LCAT reaction, HDL precursors become increasingly enriched in cholesteryl esters. This is presumed to create a concentration gradient for unesterified cholesterol and phospholipids between cells and the lipoprotein surface that can now serve as a sink for these lipids, resulting in net transfer of cholesterol from various sources including cells of the arterial wall to HDL (MENDEZ et al. 1991). How is the transfer of cholesterol to HDL achieved? It has been suggested that it involves a putative HDL receptor that either acts at the cell surface or that is recycled back to the surface after being transiently endocytosed and its ligand loaded with unesterified cholesterol intracellularly (GRAHAM and ORAM 1987; ORAM et al. 1987; STEINMETZ et al. 1990; MORRISON et al. 1991). While the HDL receptor has not yet been characterized in molecular terms, several lines of evidence speak in favor of the presence of a plasma membrane component that specifically binds HDL through its apolipoprotein A-I moiety. Recently, MORRISON et al. (1991) demonstrated high-affinity binding of apolipoprotein A-I cyanogen bromide fragments complexed with liposomes to hepatocyte membranes, and ligand blots revealed the presence of a protein doublet with a molecular mass of around 94 kDa. In a study indirectly supporting the HDL receptor concept, RUBIN et al. (1991) used transgenic mice overexpressing human apolipoprotein A-I and showed that the transgenes were protected from cholesterol accumulation in arteries and that plasma accumulation of apolipoprotein A-I was associated with increased formation of small HDL species similar to those that have been presumed to be nascent HDL particles. According to a recent suggestion by

MENDEZ et al. (1991), the HDL receptor activates a hypothetical intracellular signal transduction pathway that leads to the translocation of intracellular cholesterol stores to the plasma membrane. The reverse cholesterol transfer scenario then implies that cholesteryl ester transfer protein (CETP) removes these esters from HDL and carries them over to other lipoproteins, including intermediate density lipoproteins and liver membranes (BROWN et al. 1989; MORTON 1988). The lipoprotein acceptors are converted into LDL in the circulation or directly taken up through the LDL receptor or the LRP in the liver. However, there is also reason to believe that CETP is not necessary and/or not important in reverse cholesterol transport (FRANCONE et al. 1991). In addition, HDL species that contain apolipoprotein E can act as cholesteryl ester acceptors (BRESLOW 1989a,b).

From the above it is apparent that much needs to be learned about reverse cholesterol transport, and we are far away from a detailed understanding at the molecular level: the hypothesis as a whole and many of the steps described above remain speculative, because they are based on indirect, pharmacological or in vitro evidence. However, several inherited diseases of lipid metabolism, LCAT deficiency (reviewed in NORUM et al. 1989; ASSMANN et al. 1991), CETP deficiency (DRAYNA et al. 1984; BROWN et al. 1989; reviewed in SWENSON 1992), and other HDL deficiency syndromes (reviewed in BRESLOW 1989b) have been identified and provide in vivo systems to test the involvement of several components of reverse cholesterol transport.

H. Arterial Wall Cells Form Biologically Active Mediators of Inflammation in Response to Cholesterol Feeding

Because cholesterol feeding in animal models of atherogenesis leads to rapidly inducible alterations of arterial wall morphology that are similar to atherosclerotic lesions observed in humans, it has long been suspected that the direct interaction of lipoproteins with arterial wall cells causes injury of endothelial cells. HAUSS et al. studied this possibility in considerable detail in several animal models of atherosclerosis (HAUSS et al. 1965; reviewed in HAUSS 1990). ROSS and GLOMSET used an approach that combined in vivo morphological studies of arteries of cholesterol-fed monkeys with the study of arterial wall cells in culture of the same animals (ROSS and GLOMSET 1974a,b and references therein). On the basis of the results that they had obtained with cultured arterial smooth muscle cells and scanning electron microscopy, they proposed a detailed hypothesis of the pathogenesis of atherosclerosis, later termed the "response to injury hypothesis." This hypothesis was recently updated by ROSS (1986). It presumes that risk factors such as hypercholesterolemia initially lead to an injury of the endothelial cell monolayer and that, as a result of this injury, circulating white blood cells adhere to the endothelium. While it was suspected earlier that platelets were involved during this early stage of the disease, investigators more recently focused on monocytes, because GERRITY (1981) had demonstrated by scanning electron microscopy and immunohistochemistry that monocytes, rather than platelets, were among the earliest detectable cell types after the initiation of diet-induced fatty streaks in the aortic arch of the pig. T lymphocytes were not identified until recently and were shown by JONASSON et al. (1986, 1988) to be present in monocyte-rich areas as well as the adventitia. The identification of monocytes and lymphocytes (which are present in close physical proximity) in arterial wall lesions was important in this connection because it stimulated research into immunological phenomena in the pathogenesis of the disease (reviewed by MUNRO and COTRAN 1988). Recently, HANSSON et al. (1991) provided evidence for a protective role of T lymphocytes in atherogenesis. Using monoclonal antibodies that eliminated T lymphocytes from the circulation of rats, it was found that T-cell-depleted animals developed larger arterial lesions than control rats upon balloon injury. In another animal model, the same investigators further demonstrated that rats that are constitutively deficient in T cells develop larger intimal lesions than their normal counterparts. Since γ-interferon, a T-cell product, inhibits smooth muscle cell proliferation, these data raise the interesting possibility that T lymphocytes protect rather than promote the development of intimal lesions. If these events proved to be the in vivo situation during the course of the disease in humans, the accumulation of T lymphocytes in the atherosclerotic lesions would reflect an attack of the immune system by atherogenic agents yet to be identified (see JUNE 1991 for a recent review).

Whatever the initial injurious agents are, both cell types, apparently after or concomitant with their adhering to the endothelium and during their transendothelial migration into the intima (for review see BUTCHER 1991), undergo a differentiation and activation pathway: while blood-borne monocytes become metabolically highly activated tissue macrophages, lymphocytes are transformed into immunologically active T cells. At least one of the chemotactic activities that are generated during cholesterol feeding experiments appears to be the chemotactic factor MCP-1 (see Tables 1, 2). The triggering molecules of the differentiation pathway of monocytes and of the activation pathway of T lymphocytes during the early stages of atherogenesis as a result of hypercholesterolemia are unknown, but several candidates can be envisaged. Among them, adhesion molecules (see below) are of special interest, together with a growing number of hematopoietic growth factors, differentiation factors, and cytokines that tightly regulate the adhesive properties, the proliferative capacity as well as the state of differentiation of arterial wall cells: M-CSF, granulocyte macrophage-colony stimulating factor (GM-CSF), and the interleukins trigger the formation and secretion of other blood cell and smooth muscle cell products with potent biological activities (reviewed in METCALF 1991). Since macrophages, lymphocytes, and smooth muscle cells can be identified in close physical association in the intima of the arterial wall, these events resemble both an exaggerated immune response (see JUNE 1991 for review) and a chronic inflammatory tissue reaction (MUNRO and COTRAN 1985). The development of culture conditions for all four cell types that form the atherosclerotic plaque led to the characterization of the differentiated phenotype of these cells in vitro and in vivo (WANG et al. 1989). A major result of this was the identification and characterization of biologically active molecules that

Table 2. Identification of biologically active molecules in atherosclerotic lesions

	Reference
Mononuclear leukocyte adhesion molecule (ATHERO-ELAM-1)	CYBULSKI and GIMBRONE (1991)
PDGF A-chain (sequence homology with B chain)	References, see below
PDGF B chain (identical with *c-sis* proto-oncogene)	RUBIN et al. (1988); MAJESKY et al. (1988); WILCOX et al. (1988); WANG et al. (1989); GOLDEN et al. (1991)
PDGF-β receptor	WILCOX et al. (1988); ROSS et al. (1990)
TGFβ	ROSS et al. (1990); NIKOL et al. (1992)
M-CSF receptor (identical with *c-fms* proto-oncogene)	SHERR et al. (1985); ROSS et al. (1990); SALOMON et al. (1992); INABA et al. (1992a,b)
bFGF	LINDNER and REIDY (1991); reviewed in FOLKMAN and KLAGSBRUNN (1987)
c-myb	SIMONS et al. (1992)
IL-1	WANG et al. (1989); ROSS et al. (1990)
15-Lipoxygenase	YLÄ-HERTTUALA et al. (1991c)
Oxidized and malondialdehyde-modified LDL	HABERLAND et al. (1988); PALINSKI et al. (1989)
Racemic mixture of 13-hydroxyoctadecadienoic acids	KÜHN et al. (1992)
Malondialdehyde altered protein	HABERLAND et al. (1988)
Monocyte chemoattractant protein (MCP) 1	YLÄ-HERTTUALA et al. (1991b); YU et al. (1992)
Lipoprotein lipase	YLÄ-HERTTUALA et al. (1991a)
Apolipoprotein E	SALOMON et al. (1992)
Stromelysin	HENNEY et al. (1991)
Macrophage and smooth muscle cell scavenger receptors	MATSUMOTO et al. (1990); INABA et al. (1992b)
Plasminogen activator inhibitor (PAI) 1 mRNA	SCHNEIDERMAN et al. (1992)
Thrombin receptors	NELKEN et al. (1992)

PDGF, platelet-derived growth factor; TGF, transforming growth factor; M-CSF, macrophage colony-stimulating factor; bFGF, basic fibroblast growth factor; IL, interleukin; LDL, low-density lipoprotein.

are secreted by the cells. Among those are growth factors including PDGF (HELDIN and WESTERMARK 1990), angiogenesis factors such as bFGF (FOLKMAN and KLAGSBRUNN 1987; FOLKMAN and SHING 1992), differentiation factors (METCALF 1991), cytokines, and proteinases such as the stromelysins that are capable of degrading extracellular matrix proteins (Tables 1, 2; HENNEY et al. 1991). Moreover, in most cases, the receptors of these factors have been identified and cloned (YARDEN et al. 1986; reviewed in LEUTZ and GRAF 1990; Table 2). Of special interest is the finding that the receptor for M-CSF, the proto-oncogene *c-fms*, is expressed by smooth muscle cells of atherosclerotic plaques (INABA et al. 1992a) and that PDGF induces the *c-fms* proto-oncogene and the scavenger receptor in vascular smooth muscle cells (INABA et al. 1992b). Recently, FERNS et al. (1991) found in a nude rat model of angioplasty of the carotid artery that antibodies directed against PDGF reduced the thickening of the intima, and similar results were obtained in rat carotid arteries after balloon injury when antibodies directed against bFGF were administered (LINDNER and REIDY 1991). A close relationship between the angiogenic properties of bFGF and the proliferation of smooth muscle cells in arteries was also reported in an in vivo model of angiogenesis (EDELMAN et al. 1992).

Moreover, the identification of bFGF receptors specifically on proliferating smooth muscle cells may lead to specific drug-targeting approaches that are promising as far as therapeutic intervention is concerned (CASSCELLS et al. 1992; LINDNER et al. 1992). Conversely, when PDGF was infused into rats, intima thickening resulting from carotid injury was greatly increased (JAWIEN et al. 1992). These studies, together with earlier data obtained by the same investigators, provide the first evidence that growth factors are not only expressed in the arterial wall of affected patients or induced in appropriate animal models of atherogenesis, but that interference with growth factor action has the potential to reduce intimal thickening.

I. Adhesion Molecules Are Involved in Leukocyte–Endothelial Cell Interactions

Using cholesterol feeding experiments and the Watanabe heritable hyperlipidemic rabbit model, CYBULSKI and GIMBRONE (1991) demonstrated that endothelial cell adhesion molecules (ELAMs) are expressed during the early stages of the development of fatty streaks. Interestingly, monocytes and lymphocytes adhere to the intact endothelium apparently as a function of the expression of an adhesion molecule, termed ATHERO-ELAM-1. Moreover, ATHERO-ELAM-1 was specifically expressed by endothelial cells that covered atherosclerotic lesions of cholesterol-fed rabbits, while histologically normal vessel walls did not. A close relation between the LDL pathway and expression of the ATHERO-ELAM-1 was provided by the finding that the LDL-receptor-deficient Watanabe heritable hyperlipidemic rabbit also expressed ATHERO-ELAM-1 during the early stages of the disease while, as stated above, normal rabbits expressed it after cholesterol feeding. In vitro studies added to this field by the demonstration that antibodies directed against the ELAM prevented leukocyte–endothelial interactions. These findings may represent a major step forward, because they demonstrate a direct relationship between the expression of an early inducible molecule with potent biological activity as a function of LDL receptor deficiency (Watanabe rabbit), hypercholesterolemia (cholesterol-fed normal rabbit), and atherosclerosis. It will now be of major interest to search for human homologs of rabbit ATHERO-ELAM-1. One of these molecules may have already been identified: the N-terminal amino acid sequence of ATHERO-ELAM-1 shows high homology with human VCAM-1 (venous cell adhesion molecule-1). VCAM-1 mediates the adhesion of circulating lymphocytes to endothelial cells through the interaction with the leukocyte $\beta 1$ integrin receptor very late activation antigen-4 (also referred to as CD49d/CD29; KOVACH et al. 1992). It will therefore be of major interest to determine whether antibodies directed against the human adhesion molecule(s) or their leukocyte receptors can exert a protective effect on the development of proliferative lesions. A better understanding of the role of adhesion molecules in experimental animals will clarify the question as to whether leukocyte adhesion is a protective or an injurious event in hypercholesterolemia, and whether it is therefore promising for patients that undergo angioplasty of coronary arteries showing high rates of reocclusion due to proliferative lesions. It should be noted that leukocyte–endothelial-cell recognition can be divided into at least three distinct steps: primary and often transient adhesion, leukocyte activation, and activation-dependent binding (reviewed in BUTCHER 1991). Together with recent findings that lymphocyte adherence is closely associated with lymphocyte activation and activation-dependent binding (reviewed in VAN SEVENTER et al. 1991), ELAMs may be located ideally to mediate the recruitment of blood cells into the arterial wall. Therefore, adhesion is closely associated with activation and irreversible, i.e., activation-dependent binding, provided that there are signals that overcome the transient form of blood-cell binding to the endothelium. Conversely, both activation and the differentiation phenotype of white blood cells including lymphocytes can promote binding (SHIMIZU et al. 1991 and references therein). Which of these steps are primarily affected by disorders of lipid metabolism and LDL receptor deficiency is not clear at all. Therefore, further studies of the molecular events of leukocyte–endothelial-cell interaction in relation to hypercholesterolemia are needed. In particular it will be important to answer the question as to whether distinct steps of the leukocyte adhesion process are triggered from within the arterial wall or whether circulating factors can be identified that affect these events before there is detectable adherence of mononuclear cells to the arterial wall. Answers to these questions are critical for the development of new therapeutic strategies to fight the disease.

J. Proto-oncogenes Are Expressed in Atherosclerotic Lesions

The independent discovery by WATERFIELD et al. (1983) and DOOLITTLE et al. (1983) that the B chain of PDGF is encoded by the *c-sis* proto-oncogene and that this proto-oncogene is expressed in the diseased, but not normal, arterial wall has greatly affected atherosclerosis research. Until now, aproximately 45–50 members of the oncogene family of peptides have been identified, and many of their normal counterparts, the proto-oncogenes, occupy strategic positions in physiological pathways of proliferation and differentiation at the plasma membrane, the cytoplasm, and the nucleus. While mutations or deregulated expression of proto-oncogenes (JOHNSON et al. 1985; LEAL et al. 1985) can result in uncontrolled cell growth and/or a block in the differentiation program of cells, the expression of proto-oncogenes is highly regulated during physiological processes such as embryogenesis, cell cycle traverse, and differentiation (BISHOP 1987; MAJESKI et al. 1988). In addition to the role of oncogenes in cancerous growth and the role of proto-oncogenes in physiology, the possibility exists that proto-oncogenes can act as mediators of disease other than cancer. For example, selected proto-oncogenes are induced in a variety of pathological states such as inflammation, atherogenesis, and brain seizures (JONAT et al. 1990; ABATE and CURRAN 1990). The functional role of the expression of a single proto-oncogene or a combination of them in disease states is not entirely clear, but their occurence indicates that they may play important roles in tissue response towards injurious agents (PAULSSON et al. 1987; reviewed in AARONSON 1991). Do they act as mediators of disease or do they counteract the injurious agent? Probably, proto-oncogenes are capable of doing both. Some may support the progression of the disease, while others may counteract it (ABATE and CURRAN 1990; JONAT et al. 1990; NEPVEU et al. 1987; REILLY et al. 1989; SIMONS and ROSENBERG 1992). For example, *c-fos* expression is dramatically increased in phorbol-ester-mediated skin inflammation and evidence indicates that anti-inflammatory glucocorticosteroids exert their pharmacological properties (both anti-inflammatory and anti-cancerous) by interfering with *c-fos*/*c-jun*-dependent binding to specific DNA sequences (ABATE and CURRAN 1990). However, the expression of a proto-oncogene in one cell type may promote the disease while the expression of the same proto-oncogene in another cell type may represent a repair response. Thus, *c-fos* induction has been shown to be important in connective tissue cell proliferation and in PDGF-stimulated mitogenesis as well as in lymphocyte and macrophage differentiation (see ABATE and CURRAN 1990 for review). Recently, several members of the proto-oncogene family have been identified in surgical specimens of atherosclerotic lesions of patients and animals in response to cholesterol feeding. The *c-sis* proto-oncogene (encoding the B chain of PDGF) has been identified by in situ hybridization in both endothelial cells and macrophages of atherosclerotic lesions (WILCOX et al. 1988), while the *c-fms* proto-oncogene encoding the receptor for M-CSF (SHERR et al. 1985) has been shown to be present in lesion macrophages (ROSS et al. 1990). In view of the known presence of activated macrophages and proliferating smooth muscle cells in atherosclerotic lesions, these results are, of course, not surprising, and it can be predicted that more members of the proto-oncogene family will be identified in the diseased arterial wall in the near future. Among them are growth factor receptors that share sequence homology to several proto-oncogenes such as the PDGF receptors (*c-fms* and *c-kit*; YARDEN et al. 1986), vascular endothelial growth factor receptor (*c-fms*), and other growth-factor-related oncogenes (reviewed in LEUTZ and GRAF 1990). In the rat carotid artery angioplasty model of atherogenesis, it has recently been shown that the *c-myb* gene is strongly induced in the area of blood vessel injury. Furthermore, treatment of these injured segments using a stable *c-myb* antisense oligonucleotide applied as a gel immediately after injury significantly reduced intimal thickening after 2 months (SIMONS et al. 1992; SIMONS and ROSENBERG 1992). As the list of the proto-oncogenes that are expressed in the diseased arterial wall grows and as we learn more about the sequence of proto-oncogene expression during the course of atherogenesis, together with a better understanding of the molecular mechanisms of proto-oncogene action in general and in the arterial wall in particular, we will understand better the pathogenesis of atherosclerosis.

Acknowledgements. This work was supported by the Deutsche Forschungsgemeinschaft, Bonn. We wish to thank Alois Dresel and Karl Lackner for advice in preparing this manuscript.

References

Aaronson S (1991) Growth factors and cancer. Science 254:1146–1153
Abate C, Curran T (1990) Encounters with fos and jun on the road to AP-1. Semin. Cancer Biol 1:19–26
Assmann G, v Eckardstein A, Funke H (1991) Lecithin: cholesterol acytransferase deficiency and fish-eye disease. Curr Opin Lipidol 2:110–117
Barber DL, Sanders, EJ, Aebersold R, Schneider WJ (1991) The receptor for yolk lipoprotein deposition in the chicken oocyte. J Biol Chem 266:18761–18770
Beisiegel U, Weber W, Bengtsson-Olivecrona G (1991) Lipoprotein lipase enhances the binding of chylomicrons to low density lipoprotein receptor-related protein. Proc Natl Acad Sci USA 88:8342–8346
Berliner JA, Territo MC, Sevanian A, Ramin S, Kim JA, Bamshad B, Esterson, M., Fogelman, A.M. (1990) Minimally modified low density lipoprotein stimulates monocyte endothelial interactions. J Clin Invest 85:1260–1266
Bishop JM (1987) The molecular genetics of cancer. Science 235:305–311
Bottalico LA, Wager RE, Agellon LB, Assoian RK, Tabas I (1991) Transforming growth factor-β1 inhibits scavenger receptor activity in THP-1 human macrophages. J Biol Chem 266:22866–22871
Breslow JL (1989a) Genetic basis of lipoprotein disorders. J Clin Invest 84:373–380
Breslow JL (1989b) Familial disorders of high density lipoprotein metabolism. In: Scriver, CR, Baudet AL, Shy WS, Valee D (eds) The metabolic basis of

inherited disease 1. McGraw-Hill Information Services, New York, pp 1251–1266
Brown ML, Inazu A, Hesler CB (1989) Molecular basis of lipid transfer protein deficiency in a family with increased high-density lipoproteins. Nature 342:448–451
Brown MS, Goldstein JL (1986) A receptor-mediated pathway for cholesterol homeostasis. Science 232:334–347
Brown MS, Goldstein JL (1990) Scavenging for receptors. Nature 343:508–509
Brown MS, Herz J, Kowal RC, Goldstein JL (1991) The low-density lipoprotein receptor-related protein: double agent or decoy? Curr Opin Lipidol 2:65–72
Bu G, Williams S, Strickland DK, Schwartz AL (1992) Low density lipoprotein receptor-related protein/α_2-macroglobulin receptor is an hepatic receptor for tissue-type plasminogen activator. Proc Natl Acad Sci USA 89:7427–7431
Butcher EC (1991) Leukocyte-endothelial cell recognition: three (or more) steps to specificity and diversity. Cell 67:1033–1036
Cambien F, Poirier O, Lecerf L, Evans A, Cambou J-P, Arveiler D, Luc G, Bard J-M, Bara L, Ricard S, Tiret L, Amouyel P, Alhenc-Geelas, F, Soubrier, F (1992) Deletion polymorphism in the gene for angiotensin-converting enzyme is a potent risk factor for myocardial infarction. Nature 359:641–644
Carew TE, Schwenke DC, Steinberg D (1987) Antiatherogenic effect of probucol unrelated to its hypocholesterolemic effect: evidence that antioxidants in vivo can selectively inhibit low density lipoprotein degradation in macrophage-rich fatty streaks and slow the progression of atherosclerosis in the Watanabe heritable hyperlipidemic rabbit. Proc Natl Acad Sci USA 84:7725–7729
Cascells W, Lappi DA, Olwin BB, Wai C, Siegman M, Speir EH, Sasse J, Baird A (1992) Elimination of smooth muscle cells in experimental restenosis: targeting of fibroblast growth factor receptors. Proc Natl Acad Sci USA 89:7159-7163
Choi SY, Fong LG, Kirven MJ, Cooper AD (1991) Use of an anti-low density lipoprotein receptor antibody to quantify the role of the LDL receptor in the removal of chylomicron remnants in the mouse in vivo. J Clin Invest 88:1173–1181
Conrad DJ, Kühn H, Mulkins M, Highland E, Sigal E (1992) Specific inflammatory cytokines regulate the expression of human monocyte 15-lipoxygenase. Proc Natl Acad Sci USA 89:217–221
Cushing SD, Berliner JA, Valente AJ, Territo MC, Navab M, Parhami F, Gerrity R, Schwartz CJ, Fogelman AM (1990) Minimally modified low density lipoprotein induces monocyte chemotactic protein 1 in human endothelial cells and smooth muscle cells. Proc Natl Acad Sci USA 87:51134–51138
Cuthbert JA, Lipsky PE (1989) Lipoproteins may provide fatty acids necessary for human lymphocyte proliferation by both low density lipoprotein receptor-dependent and -independent mechanisms. J Biol Chem 264:13468–13474
Cuthbert JA, Russell DW, Lipsky PE (1989) Regulation of low density lipoprotein receptor gene expression in human lymphocytes. J Biol Chem 264:1298–1304
Cybulski MI, Gimbrone MA (1991) Endothelial expression of a mononuclear leukocyte adhesion molecule during atherogenesis. Science 251:788–791
Didsbury JR, Uhing RJ, Snyderman R (1990) Isoprenylation of the low molecular mass GTP-binding proteins Rac1 and Rac2: possible role in membrane localization. Biochem Biophys Res Commun 171:804–812
Doolittle RF, Hunkapiller MW, Hood LE, Devare SG, Robbins KC, Aaronson SA, Antoniades HN (1983) Simian sarcoma virus onc gene, v-sis, is derived from the gene (or genes) encoding a platelet-derived growth factor. Science 221:275–277
Drayna D, Jarnagin AS, McLean J, Henzel W, Kohr W, Fielding C, Lawn R (1984) Cloning and sequencing of human cholesteryl ester transfer protein cDNA. Nature 327:632–634
Edelman ER, Nugent MA, Smith LT, Karnowsky MJ (1992) Basic fibroblast growth factor enhances the coupling of intimal hyperplasia and proliferation of vasa vasorum in injured rat arteries. J Clin Invest 89:465–473

Eisenberg S, Sehayek E, Olivecrona T, Vlodavsky I (1992) Lipoprotein lipase enhances binding of lipoproteins to heparan sulfate on cell surfaces and extracellular matrix. J Clin Invest 90:2013–2021

Esterbauer H, Jürgens G, Quehenberger O, Koller E (1987) Autooxidation of human low density lipoprotein: loss of polyunsaturated fatty acids and vitamin E and generation of aldehydes. J Lipid Res 28:495–509

Esterbauer H, Gebicki J, Puhl H, Jürgens G (1992) The role of lipid peroxidation and antioxidants in oxidative modification of LDL. Free Radic Biol Med 13:341–390

Faggioto A, Ross R, Harker L (1984) Studies of hypercholesterolemia in the nonhuman primate: I. Changes that lead to fatty streak formation. Arteriosclerosis 4:323–340

Ferns GAA, Raines EW, Sprugel KH, Motani AS, Reidy MA, Ross R (1991) Inhibition of neointimal smooth muscle accumulation after angioplasty by an antibody to PDGF. Science 253:1129–1132

Folkman J, Klagsbrun M (1987) Angiogenic factors. Science 235:442–447

Folkman J, Shing Y (1992) Angiogenesis. J Biol Chem 267:10931–10934

Francone OL, Fielding PE, Fielding CJ (1991) The distribution of cell-derived cholesterol between plasma lipoproteins – a comparison between three techniques. J Lipid Res 31:2195–2200

Freeman M, Ekkel Y, Rohrer L, Penman M, Freedman NJ, Chisolm GM, Krieger M (1991) Expression of type I and type II bovine scavenger receptors in chinese hamster ovary cells: lipid droplet accumulation and nonreciprocal cross competition by acetylated and oxidized low density lipoprotein. Proc Natl Acad Sci USA 88:4931–4935

Frostegard J, Nilsson J, Haegerstrand A, Hamsten A, Wigzell H, Gidlund M (1990) Oxidized low density lipoprotein induces differentiation and adhesion of human monocytes and the monocytic cell line U937. Proc Natl Acad Sci USA 87:904–908

Gåfvels ME, Coukos G, Sayegh R, Coutifaris C, Strickland DK, Strauss III JF (1992) Regulated expression of the trophoblast α_2-macroglobulin receptor/low density lipoprotein receptor-related protein. J Biol Chem 267:21230–21234

Geng Y-J, Hansson GK (1992) Interferon-γ inhibits scavenger receptor expression and foam cell formation in human monocyte-derived macrophages. J Clin Invest 89:1322–1330

Gerrity RG (1981) The role of the monocyte in atherogenesis: I.Transition of blood-borne monocytes into foam cells in fatty lesions. Am J Pathol 103:181–190

Gerrity RG, Naito HK, Richardson M, Schwartz CJ (1979) Dietary induced atherogenesis in swine. Am J Pathol 95:775–792

Gerrity RG, Goss J, Soby L (1985) Control of monocyte recruitment by chemotactic factor(s) in lesion-prone areas of swine aorta. Arteriosclerosis 5:55–65

Gibbs JB (1991) Ras C-terminal processing enzymes – new drug targets? Cell 65:1–4

Glomset JA (1968) The plasma lecithin: cholesterol acyltransferase reaction. J Lipid Res. 9:155–167

Glomset JA (1994) Ann Rev Cell Biol, in press

Glomset JA, Gelb MH, Farnsworth CC (1990) Prenyl proteins in eukaryotic cells: a new type of membrane anchor. Trends Biochem Sci 15:139–142

Golden MA, Tina Au YP, Kirkman TR, Wilcox JN, Raines EW, Ross R, Clowes AW (1991) Platelet-derived growth factor activity and mRNA expression in healing vascular grafts in baboons, association in vivo of platelet-derived growth factor mRNA and protein with cellular proliferation. J Clin Invest 87:406–414

Goldstein JL, Brown MS (1977) The low density lipoprotein pathway and its relation to atherosclerosis. Annu Rev Biochem 46:897–930

Goldstein JL, Brown MS (1989) Familial hypercholesterolaemia. In: Scriver CR, Baudet AL, Shy WS, Valee D (eds) The metabolic basis of inherited diseases 1. McGraw Hill Information Services, New York, pp 1215–1250

Goldstein JL, Brown MS (1990) Regulation of the mevalonate pathway. Nature 343:425–433

Graham DL, Oram JF (1987) Identification and characterization of a high density lipoprotein-binding protein in cell membranes by ligand blotting. J Biol Chem 262:7439–7442

Habenicht AJR, Glomset JA, Ross R (1980) The relation between cholesterol and mevalonic acid metabolism to the cell cycle in smooth muscle and Swiss 3T3 cells stimulated to divide by platelet-derived growth factor. J Biol Chem 255, 5134–5140

Habenicht AJR, Dresel HA, Goerig M, Weber JA, Stoehr M, Glomset JA, Ross R, Schettler G (1986) Low density lipoprotein receptor-dependent prostaglandin synthesis in Swiss 3T3 cells stimulated by platelet-derived growth factor. Proc Natl Acad Sci USA 83:1344–1348

Habenicht AJR, Salbach, P, Blattner C, Janßen-Timmen U (1990a) Platelet-derived growth factor, formation and biological activities. In: Habenicht, A (ed.) Growth factors, differentiation factors, and cytokines. Springer, Habenicht, A. (ed) Springer, Berlin Heidelberg New York, pp 31–42

Habenicht AJR, Salbach, P, Goerig M, Zeh W, Janßen-Timmen, U, Blattner C, King, WC, Glomset, JA (1990b) The LDL receptor pathway delivers arachidonic acid for eicosanoid formation in cells stimulated by platelet-derived growth factor. Nature 345:634–636

Haberland ME, Fong D, Cheng L (1988) Malondialdehyde-altered protein occurs in atheroma of Watanabe heritable hyperlipidemic rabbit. Science 241:215–218

Halliwell B, Gutteridge, JM (1990) Role of free radicals and catalytic metal ions in human disease: an overview. Methods Enzymol 186:1–85

Hampton RY, Golenbock GT, Penman M, Krieger M, Raetz CRH (1991) Recognition and plasma clearance of endotoxin by scavenger receptors. Nature 352: 342–344

Hansson GK, Jonasson L, Seifert, PS, Stemme S (1989) Immune mechanisms of atherosclerosis. Arteriosclerosis 9:567–571

Hansson GK, Holm, J, Holm S, Fotev Z, Hedrich H-J, Fingerle J (1991) T lymphocytes inhibit the vascular response to injury. Proc Natl Acad Sci USA 88: 10530–10534

Hara HJ, Yokoyama S (1991) Interaction of free apolipoproteins with macrophages. Formation of high density lipoprotein-like lipoprotein and reduction of cellular cholesterol. J Biol Chem 266:3080–3086

Hartung H-P, Kladetzky RG, Melnik, B, Hennerici M (1986) Stimulation of the scavenger receptor on monocytes-macrophages evokes release of arachidonic acid metabolites and reduced oxygen species. Lab Invest 55:209–216

Hauss WH (1990) Die Arteriosklerose. Pathogenese der arteriosklerotischen, der rheumatischen und weiterer reaktiver, chronischer Mesenchymerkrankungen. Steinkopff, Darmstadt

Hauss WH, Junge-Hülsing G, Matthes KJ, Wirth W (1965) Über den Einfluß von Schock und Hyperlipidämie auf den Lipidgehalt, die Lipidsynthese und Mucopolysaccharidsynthese der Gefäßwand. J Atheroscler Res 5:451–465

Havel RJ, Kane JP (1989) Structure and metabolism of plasma lipoproteins. In: Scriver CR, Baudet AL, Shy WS, Valee D (eds) The metabolic basis of inherited diseases 1. McGraw Hill Information Services, New York, pp 1129–1138

Heldin C-H, Westermark, B (1990) Autocrine stimulation of growth of normal and transformed cells. In: Habenicht A (ed) Growth factors, differentiation factors, and cytokines. Springer, Berlin Heidelberg New York, pp 267–279

Henderson BE, Ross RK, Pike MC (1991) Toward the primary prevention of cancer. Science 254:1131–1138

Henney AM, Wakeley, PR, Davies MJ, Foster K, Hembry R, Murphy G, Humphries S (1991) Localization of stromelysin mRNA expression in atherosclerotic plaques by in situ hybridization. Proc Natl Acad Sci USA 88:8154–8158

Henrikson T, Mahoney EM, Steinberg D (1981) Enhanced macrophage degradation of low density lipoprotein previously incubated with cultured endothelial cells: recognition by receptor for acetylated low density lipoproteins Proc Natl Acad Sci USA 78:6499–6503

Herz J, Hammann U, Rogne, S, Myklebost O, Gausepohl H, Stanley KK (1988) Surface location and high affinity for calcium of a 500 kDa liver membrane protein closely related to the LDL receptor suggest a physiological role as lipoprotein receptor. EMBO J 7:4119–4127

Herz J, Clouthier DE, Hammer RE (1992) LDL receptor-related protein internalizes and degrades uPA-PAI-1 complexes and is essential for embryo implantation. Cell 71:411–421

Hofmann SL, Russell DW, Brown MS, Goldstein JL, Hammer RE (1988) Overexpression of low density lipoprotein (LDL) receptor eliminates LDL from plasma in transgenic mice. Science 239:1277–1281

Huettinger M, Retzek H, Hermann M, Goldenberg H (1992) Lactoferrin specifically inhibits endocytosis of chylomicron remnants but not α-macroglobulin. J Biol Chem 267:18551–18557

Hussain MM, Mahley RW, Boyles JK, Fainaru M, Brecht WJ, Lindquist P (1989) Chylomicron-chylomicron remnant clearance by liver and bone marrow in rabbits. Factors that modify tissue-specific uptake. J Biol Chem 264:9571–9582

Hussain MM, Maxfield FR, Mas-Oliva J, Tabas I, Ji Z-S, Innerarity TL, Mahley, RW (1991) Clearance of chylomicron remnants by the low-density lipoprotein receptor-related protein/α_2-macroglobulin receptor. J Biol Chem 266: 13936–13940

Inaba T, Yamada N, Gotada T, Shimano H, Shimada M, Momomura K, Kadowaki T, Motoyoshi K, Tsukada T, Morisaki N, Saito Y, Yoshida S, Takaku F, Yazaki Y (1992a) Expression of M-CSF receptor encoded by c-fms on smooth muscle cells derived from arteriosclerotic lesion. J Biol Chem 267:5693–5699

Inaba T, Gotoda T, Shimano H, Shimada M, Harada K, Kozaki K, Watanabe Y, Hoh E, Motoyoshi K, Yazaki Y, Yamada N (1992b) Platelet-derived growth factor induces c-fms and scavenger receptor genes in vascular smooth muscle cells. J Biol Chem 267:13107–13112

Jawien A, Bowen-Pope DF, Lindner V, Schwartz SM, Clowes AW (1992) Platelet-derived growth factor promotes smooth muscle migration and intimal thickening in a rat model of balloon angioplasty. J Clin Invest 89:507–511

Jialal I, Grundy SM (1991) Preservation of the endogenous antioxidants in low density lipoprotein by ascorbate but not probucol during oxidative modification. J Clin Invest 87:597–601

Jialal I, Norkus EP, Cristol L, Grundy SM (1991) β-Carotene inhibits the oxidative modification of low-density lipoprotein. Biochim. Biophys. Acta 1086:134–138

Johnson WJ, Chacko GK, Phillips MC, Rothblat, GH (1990) The efflux of lysosomal cholesterol from cells. J Biol Chem 265:5546–5553

Johnson, WJ, Mahlberg, FH, Rothblat, GH, Phillips, MC (1991) Cholesterol transport between cells and high density lipoproteins. Biochim Biophys Acta 1085:273–298

Johnsson A, Betsholtz C, Heldin C-H, Westermark B (1985) Antibodies against platelet-derived growth factor inhibit acute transformation by simian sarcoma virus. Nature 317:438–440

Jonasson L, Holm J, Skalli O, Bonders G, Hansson GK (1986) Regional accumulation of T-cells, macrophages, and smooth muscle cells in the human atherosclerotic plaque. Arteriosclerosis 6:131–138

Jonasson L, Holm J, Hansson GK (1988) Cyclosporin A inhibits smooth muscle cell proliferation in the vascular response to injury. Proc Natl Acad Sci USA 85:2302–2306

Jonat C, Rahmsdorf HJ, Park KK, Cato ACB, Gebel S, Ponta H, Herrlich P (1990) Antitumor promotion and antiinflammation: downmodulation of AP-1 (fos/jun) activity by glucocorticoid hormone. Cell 62:1189–1204

June CH (1991) Signal transduction in T cells. Curr Opin Immunol 3:287–293
Kawata M, Farnsworth CC, Yoshida Y, Gelb MH, Glomset JA, Takai Y (1990) Posttranslationally processed structure of the human platelet protein smg p21B: evidence for geranylgeranylation and carboxymethylation of the C terminal cysteine. Proc Natl Acad Sci USA 87:8960–8964
Kelley JL, Rozek MM, Suenram CA, Schwartz CJ (1988) Activation of human peripheral blood monocytes by lipoproteins. Am J Pathol 130:223–231
Kinsella BT, Maltese WA (1991) Rab GTP-binding proteins implicated in vesicular transport are isoprenylated in vitro at cysteines within a novel carboxy-terminal motif. J Biol Chem 266:8540–8544
Kinsella BT, Erdman RA, Maltese WA (1991) Carboxyl-terminal isoprenylation of ras-related GTP-binding proteins encoded by rac1, rac2, and ralA. J Biol Chem 266:9786–9794
Kodama T, Freeman M, Rohrer L, Zabricky J, Matsudaira P, Krieger, M (1990) Type I scavenger receptor contains α-helical and collagen-like coils. Nature 343:531–535
Kounnas MZ, Morris RE, Thompson MR, FitzGerald DJ, Strickland, DK, Saelinger CB (1992) The α_2-macroglobulin receptor/low density lipoprotein receptor-related protein binds and internalizes pseudomonas exotoxin A. J Biol Chem 267:12420–12423
Kovach NL, Carlos TM, Yee, E, Harlan JM (1992) A monoclonal antibody to ß1 integrin (CD29) stimulates VLA-dependent adherence of leukocytes to human umbilical vein endothelial cells and matrix components. J Cell Biol 116:499–509
Kowal RC, Herz J, Weisgraber KH, Mahley RW, Brown MS, Goldstein JL (1990) Opposing effects of apolipoprotein E and C on lipoprotein binding to low-density lipoprotein receptor-related protein. J Biol Chem 265:10771–10779
Kristensen T, Moestrup SK, Glieman J, Bendtsen L, Sand O, Sottrup-Jensen L (1990) Evidence that the newly cloned low-density lipoprotein receptor-related protein (LRP) is the α_2-macroglobulin receptor. FEBS Lett 276:151–155
Kühn H, Belkner J, Wiesner R, Schewe T, Lankin VZ, Tikhaza AK (1992) Structure elucidation of oxygenated lipids in human atherosclerotic lesions. Eicosanoids 5:17–22
Leal F, Williams, LT, Robbins KC, Aaronson, SA (1985) Evidence that the v-sis gene product transforms by interaction with the receptor for platelet-derived growth factor. Science 230:327–330
Lehr HA, Hübner C, Finckh B, Angermüller S, Nolte D, Beisiegel U, Kohlschütter A, Messmer K (1991) Role of leukotrienes in leukocyte adhesion following systemic administration of oxidatively modified human low density lipoprotein in hamsters. J Clin Invest 88:9–14
Leutz A, Graf, T (1990) Relationships between oncogenes and growth control. In: Sporn MB, Roberts AB (eds) Peptide growth factors and their receptors. Springer, Berlin Heidelberg New York, pp 655–703 (Handbook of pharmacology, vol 95/2)
Liao FL, Berliner JA, Mehrabian M, Navab M, Demer LL, Lusis AJ, Fogelman AM (1991) Minimally modified low density lipoprotein is biologically active in vivo in mice. J Clin Invest 87:2253–2257
Lindner V, Reidy MA (1991) Proliferation of smooth muscle cells after vascular injury is inhibited by an antibody against basic fibroblast growth factor. Proc Natl Acad Sci USA 88:3739–3743
Lindner V, Olson NE, Clowes AW, Reidy MA (1992) Inhibition of smooth muscle cell proliferation in injured rat arteries. J Clin Invest 90:2044–2049
Lund H, Takahashi K, Hamilton RL, Havel RJ (1989) Lipoprotein binding and endosomal itinerary of the low-density lipoprotein receptor-related protein in rat liver. Proc Natl Acad Sci USA 86:9318–9322
Lusis AJ, Castellani LW, Fisler JS (1992) Fitting pieces from studies of animal models into the puzzle of atherosclerosis. Curr Opin Lipidol 3:143–150

MacCluer JW (1992) Biometrical studies to detect new genes with major effects on quantitative risk factors for atherosclerosis. Curr Opin Lipidol 3:114–121

Mahley RW (1988) Apolipoprotein E: cholesterol transport protein with expanding role in cell biology. Science 240:622–630

Majesky MW, Benditt EP, Schwartz SM (1988) Expression and developmental control of platelet-derived growth factor A-chain and B-chain/sis genes in rat aortic smooth muscle cells. Proc Nat Acad Sci USA 85:1524–1528

Malden LT, Chait A, Raines EW, Ross R (1991) The influence of oxidatively modified low density lipoproteins on expression of platelet-derived growth factor by human monocyte-derived macrophages. J Biol Chem 266:13901–13907

Matsumoto A, Naito M, Itakura H, Ikemoto S, Asaoka H, Hayakawa I, Kanamori H, Aburatani H, Takaku F, Suzuki H, Kobari Y, Miyai T, Takahashi K, Cohen EH, Wydrow R, Housman DE, Kodama T (1990) Human macrophage scavenger receptors: primary structure, expression, and localization in atherosclerotic lesions. Proc Natl Acad Sci USA 87:9133–9137

McNally AK, Chisolm III GM, Morel DW, Cathcart MK (1990) Activated human monocytes oxidize low-density lipoprotein by a lipoxygenase-dependent pathway. J Immunol 145:254–259

Mendez AJ, Oram JF, Bierman EL (1991) Protein kinase C as a mediator of high density lipoprotein receptor-dependent efflux of intracellular cholesterol. J Biol Chem 266:10104–10111

Metcalf D (1991) Control of granulocytes and macrophages: molecular, cellular, and clinical aspects. Science 254:529–533

Morrison J, Fidge NH, Tozuka M (1991) Determination of the structural domain of apoAI recognized by high density lipoprotein receptors. J Biol Chem 266: 18780–18785

Morton RE (1988) Free cholesterol is a potent regulator of lipid transfer protein. J Biol Chem 263:12235–12241

Moulton KS, Wu H, Barnett J, Parthasarathy S, Glass, CK (1992) Regulated expression of the human acetylated low density lipoprotein receptor gene and isolation of promoter sequences. Proc Natl Acad Sci USA 89:8102–8106

Munro JM, Cotran RS (1988) Biology of disease, the pathogenesis of atherosclerosis: atherogenesis and inflammation. Lab Invest 58:249–259

Nagano Y, Arai H, Kita, T (1991) High density lipoprotein loses its effect to stimulate efflux of cholesterol from foam cells after oxidative modification. Proc Natl Acad Sci USA 88:6457–6461

Nathan CF, Murray HW, Cohn Z (1980) Current concepts: the macrophage as an effector cell. N Engl J Med 303:622–626

Needleman P, Turk J, Jakschik BA, Morrison AR (1986) Arachidonic acid metabolism. Annu Rev Biochem 55:69–102

Nelken NA, Soifer, SJ, O'Keefe J, Vu T-KH, Charo, IF, Coughlin, SR (1992) Thrombin receptor expression in normal and atherosclerotic human arteries. J Clin Invest 90:1614–1621

Nepveu A, Levine RA, Campisi J, Greenberg ME, Ziff EB, Marcu KB (1987) Alternative modes of c-myc regulation in growth factor-stimulated and differentiating cells. Oncogene 1:243–250

Nikol S, Isner JM, Pickering JG, Kearny M, Leclerc G, Weir L (1992) Expression of transforming growth factor-β1 is increased in human vascular restenosis lesions. J Clin Invest 90:1582–1592

Nishina PM, Johnson JP, Naggert JK, Krauss RM (1992) Linkage of atherogenic lipoprotein phenotype to the low density lipoprotein receptor locus on the short arm of chromosome 19. Proc Natl Acad Sci USA 89:708–712

Noonan DM, Fulle A, Valente P, Cai S, Horigan E, Sasaki M, Yamada Y, Hassell JR (1991) The complete sequence of perlican, a basement membrane heparan sulfate proteoglycan, reveals extensive similarity with laminin A chain, low density lipoprotein receptor, and the neural cell adhesion molecule. J Biol Chem 266:22939–22947

Norum KR, Gjone E, Glomset JA (1989) Familial lecithin: cholesterol acyltransferase deficiency, including fish eye disease. In: Scriver CR, Baudet, AL, Shy, WS, Valee, D (eds) The metabolic basis of inherited diseases 1. McGraw-Hill Information Services, New York, pp 1181–1194

Oram JF, Johnson CJ, Aulinskas Brown T (1987) Interaction of high density lipoprotein with its receptor on cultured fibroblasts and macrophages. J Biol Chem 262:2405–2410

Orth K, Madison EL, Gething M-J, Sambrook JF, Herz J (1992) Complexes of tissue-type plasminogen activator and its serpin inhibitor plasminogen-activator inhibitor type 1 are internalized by means of the low density lipoprotein receptor-related protein/α_2-macroglobulin receptor. Proc Natl Acad Sci USA 89: 7422–7426

Palinski W, Rosenfeld ME Ylä-Herttuala S, Gurtner GC Socher SS Butler SW, Parthasarathy S, Carew TE Steinberg D Witztum JL (1989) LDL undergoes oxidative modification in vivo. Proc Natl Acad Sci USA 86:1372–1376

Paulsson Y, Hammacher A, Heldin C-H, Westermark, B (1987) Possible positive autocrine feedback in the prereplicative phase of human fibroblasts. Nature 328:715–717

Plump AS, Smith JD, Hayek T, Aalto-Setälä K, Walsh A, Verstuyft JG, Rubin EM, Breslow JL (1992) Severe hypercholesterolemia and atherosclerosis in apolipoprotein E-deficient mice created by homologous recombination in ES cells. Cell 71:343–353

Rajavashisth TB, Andalibi A Territo MC, Berliner JA, Navab M, Fogelman AM, Lusis AJ (1990) Induction of endothelial cell expression of granulocyte and macrophage colony-stimulating factors by modified low density lipoproteins. Nature 344:254–257

Rankin SM, Parthasarathy A, Steinberg D (1991) Evidence for a dominant role of lipoxygenase(s) in the oxidation of LDL by mouse peritoneal macrophages. J Lipid Res 32:449–456

Raychowdhury R, Niles JL, Mc Cluskey RT, Smith, JA (1989) Autoimmune target in Heymann nephritis is a glycoprotein with homology to the LDL receptor. Science 244:1163–1166

Reilly CF, Kindy MS, Brown KE, Rosenberg RD, Sonenshein GE (1989) Heparin prevents vascular smooth muscle cell progression through the G1 phase of the cell cycle. J Biol Chem 264:6990–6995

Ridgway ND, Dawson PA, Ho YK Brown MS, Goldstein, JL (1992) Translocation of oxysterol binding protein to Golgi apparatus triggered by ligand binding. J Cell Biol 116:307–319

Ross R (1986) The pathogenesis of atherosclerosis – an update. N Engl J Med 314:488–500

Ross R, Glomset JA (1974a) The pathogenesis of atherosclerosis. N Engl J Med 295:369–377

Ross R Glomset JA (1974b) The pathogenesis of atherosclerosis. N Engl J Med 295:420–425

Ross R, Masuda J, Raines EW, Gown AM, Katsuda S, Sasahara M Malden LT, Masuko H, Sato H (1990) Localization of PDGF-B protein in macrophages in all stages of atherosclerosis. Science 248:1009–1012

Rubin EM, Krauss RM, Spangler EA, Verstuft JG, Clift SM (1991) Inhibition of early atherogenesis in transgenic mice by human apolipoprotein A-I. Nature 353:265–266

Rubin K, Tingstrom A, Hansson GK Larsson E Ronnstrand L Klareskog L, Claesson-Welsh, L Heldin, C-H Fellstrom, B Terracio, L (1988) Induction of B-type PDGF receptors for platelet-derived growth factor in vascular inflammation: possible implications for development of vascular proliferative lesions. Lancet 1:1353–1356

Rubinstein DC Cohen JC Berger GM, Van Der Westhuyzen, DR, Coetzee, GA Gevers W (1990) Chylomicron remnant clearance from the plasma is normal in

familial hypercholesterolemic homozygotes with defined receptor defects. J Clin Invest 86:1306–1312
Salbach PB, Specht E, Hodenberg Ev, Kossmann J, Janßen-Timmen U, Schneider WJ, Hugger P, King W, Glomset JA, Habenicht AJR (1992) Differential LDL receptor-dependent eicosanoid production in human blood-derived monocytes. Proc Natl Acad Sci USA 89:2439–3443
Salomon RN, Underwood R, Doyle MV, Wang A, Libby P (1992) Increased apolipoprotein E and c-fms gene expression without elevated interleukin 1 or 6 mRNA levels indicates selective activation of macrophage functions in advanced human atheroma. Proc Natl Acad Sci USA 89:2814–2818
Samuelsson B, Dahlén DA, Lindgren CA, Rouzer CA, Serhan CN (1987) Leukotrienes and lipoxins: structures, biosynthesis, and biological effects. Science 237: 1171–1176
Saxena U, Klein MG, Vanni TM, Goldberg IJ (1992) Lipoprotein lipase increases low density lipoprotein retention by subendothelial cell matrix J Clin Invest 89:373–380
Schneiderman J, Sawdey MS, Keeton MR, Bordin GM, Bernstein EF, Dilley RB, Loshutoff DJ (1993) Increased type 1 plasminogen activator inhibitor gene expression in atherosclerotic human arteries. Proc Natl Acad Sci USA 89:6998–7002
Schreiber JR, Weinstein DB (1986) Lipoprotein receptors in steroidogenesis. In: Scanu AM, Spector AA (eds) Biochemistry and biology of plasma lipoproteins. Dekker, New York
Schwartz CJ, Valente AJ, Sprague EA, Kelley JL, Nerem RM (1991) The pathogenesis of atherosclerosis: an overview. Clin Cardiol 14:1–16
Scrott HG, Goldstein JL, Hazzard WR, Mc Goodwin MM, Motulsky AG (1972) Familial hypercholesterolemia in a large kindred. Evidence for a monogenic mechanism. Ann Intern Med 67:711–718
Sherr CJ, Rettenmier CW, Sacca R, Roussel MF, Look AT, Stanley ER (1985) The c-fms proto-oncogene product is related to the receptor for the mononuclear phagocyte growth factor, CSF-1. Cell 41:665–676
Shimano H, Yamada N, Katsuki M, Shimada M, Gotoda T, Harada K, Murase T Fukazawa C, Takaku F, Yazaki Y (1992a) Overexpression of apolipoprotien E in transgenic mice: marked reduction in plasma lipoproteins except high density lipoprotein and resistance against diet-induced hypercholesterolemia. Proc Natl Acad Sci USA 89:1750–1745
Shimano H, Yamada N, Katsuki M, Yamamoto K, Gotoda T, Harada K, Shimada M, Yazaki Y (1992b) Plasma lipoprotein metabolism in transgenic mice overexpressing apolipoprotein E accelerated clearance of lipoproteins containing apolipoprotein B. J Clin Invest 90:2084–2091
Shimizu Y, Newman W, Gopal TV, Horgan KJ, Graber N, Dawson Beall L, van Seventer GA, Shaw, S (1991) Four molecular pathways of T cell adhesion to endothelial cells: roles of LFA-1, VCAM-1, and ELAM-1 and changes in pathway hierarchy under different activation conditions. J Cell Biol 113: 1203–1212
Simon TC, Makheja AN, Bailey JM (1989) The induced lipoxygenase in atherosclerotic aorta converts linoleic acid to the chemorepellant factor 13-HODE. Thromb Res 55:171–178
Simons M, Rosenberg RD (1992) Antisense nonmuscle myosin heavy chain and c-myb oligonucleotides suppress smooth muscle cell proliferation in vitro. Circ Res 70:835–843
Simons M, Edelman ER, DeKeyser J-L, Langer R, Rosenberg RD (1992) Antisense c-myb oligonucleotides inhibit intimal arterial smooth muscle cell accumulation in vivo. Nature 359:67–70
Sparrow CP, Olszewski J (1992) Cellular oxidative modification of low density lipoprotein does not require lipoxygenases. Proc Natl Acad Sci USA 89:128–131

Sparrow CP, Parthasarathy S, Steinberg D (1989) A macrophage receptor that recognizes oxidized low-density lipoprotein but not acetylated low-density lipoprotein. J Biol Chem 264:2599–2604

Staal FJT, Roederer M, Herzenberg LA, Herzenberg LA (1990) Intracellular thiols regulate activation of nuclear factor kB and transcription of human immunodeficiency virus. Proc Natl Acad Sci USA 87:9943–9947

Stanton LW, White RT, Bryant CM, Protter AA, Endemann G (1992) A macrophage Fc receptor for IgG is also a receptor for oxidized low density lipoprotein. J Biol Chem 267:22446–22451

Steinberg D, Parthasarathy S, Carew TE, Khoo JC, Witztum JL (1989) Beyond cholesterol: modification of low-density lipoproteins that increase their atherogenicity. N Engl J Med 320:915–923

Steinmetz A, Barabas R, Ghalim N, Claveyx V, Fruchart J-C, Ailhaud G (1990) Human apolipoprotein A IV binds to apolipoprotein A-I/A-II receptor sites and promotes cholesterol efflux from adipose cells. J Biol Chem 265:7859–7863

Steyrer E, Barber DL, Schneider WJ (1990) Evolution of lipoprotein receptors. The chicken oocyte receptor for very low density lipoprotein and vitelogenin binds the mammalian ligand apolipoprotein E. J Biol Chem 265:19575–19581

Stifani S, Barber DL, Aebersold R, Steyrer E, Shen, X Nimpf J, Schneider WJ (1991) The laying hen expresses two different low density lipoprotein receptor-related proteins. J Biol Chem 266:19079–19087

Strickland DK, Ashcom JD, Williams S, Burgess WH, Mirgliorini M, Argraves WS (1990) Sequence identity between the α_2-macroglobulin receptor and low-density lipoprotein receptor-related protein suggests that this molecule is a multifunctional receptor. J Biol Chem 265:17401–17404

Swanson LW, Simmons DM, Hofmann SL, Goldstein JL, Brown MS (1988) Localization of mRNA for low density lipoprotein receptor and a cholesterol synthetic enzyme in rabbit nervous system by in situ hybridization. Proc Natl Acad Sci USA 85:9821–9825

Swenson TL (1992) Transfer proteins in reverse cholesterol transport. Curr Opin Lipidol 3:67–74

Takahashi S, Kawarabayashi Y, Nakai T, Sakai J, Yamamoto T (1992) Rabbit very low density lipoprotien receptor: a low density lipoprotein receptor-like protein with distinct ligand specificity. Proc Natl Acad Sci USA 89:9252–9556

Tall AR (1990) Plasma high density lipoproteins – metabolism and relationship to atherogenesis. J Clin Invest 86:379–384

Van Dijk MCM, Kruijt JK, Boers W, Linthorst C, van Berkel TJC (1992) Distinct properties of the recognition sites for β-very low density lipoprotein (remnant receptor) and α_2-macroglobulin (low density lipoprotein receptor-related protein) on rat parenchymal cells. J Biol Chem 267:17732–17737

Van Seventer GA, Shimizu Y, Shaw S (1991) Roles of multiple accessory molecules in T-cell activation. Curr Opin Immunol 3:294–303

Via DP, Kempner ES, Pons L, Fanslow AE, Vignale S, Smith LC Gotto AM Jr Dresel HA (1992) Mouse macrophage receptor for acetylated low density lipoprotein: demonstration of a fully functional subunit in the membrane and with purified receptor. Proc Natl Acad Sci USA 89:6780–6784

Wang W, Doyle MV, Mark DF (1989) Quantitation of mRNA by polymerase chain reaction. Proc Natl Acad Sci USA 86:9717–9721

Waterfield MD, Scrace GT, Whittle N, Stroobant P, Johnsson A, Wasteson A, Westermark B, Heldin C-H Huang JS, Deuel TF (1983) Platelet-derived growth factor is structurally related to the putative transforming protein $p28^{sis}$ of simian sarcoma virus. Nature 304:35–39

Weinberg RA (1985) The action of oncogenes in the cytoplasm and nucleus. Science 230:770–776

Weinberg, RA (1991) Tumor suppressor genes. Science 254:1138–1146

Wilcox JN, Smith KM, Williams LT, Schwartz SM, Gordon D (1988) Platelet-derived growth factor mRNA detection in human atherosclerotic plaques by in situ hybridization. J Clin Invest 82:1134–1143
Witztum JL, Steinberg D (1991) Role of oxidized low density lipoprotein in atherogenesis. J Clin Invest 88:1785–1792
Wolda SL, Glomset JA (1988) Evidence for modification of lamin B by a product of mevalonic acid. J Biol Chem 263:5997–6000
Xanthoudakis S, Curran T (1992) Identification of ref-1, a nuclear protein that facilitates AP-1 DNA-binding activity. EMBO (Eur Mol Biol Org) J 11:635–665
Yamane HK, Farnsworth CC, Xie H, Evans T, Howald WN, Gelb MH, Glomset JA, Clarke S, Fung BK (1991) Membrane binding domain of the small G protein G25K contains an S-(all-trans-geranyl-geranyl)cysteine methyl ester at its carboxyl terminus. Proc Natl Acad Sci USA 88:286–290
Yamada N, Inoue I, Kawamura M, Harada K, Watanabe Y, Shimano H, Gotoda T, Shimada M, Kohzaki K, Tsukada T, Shiomi M, Watanabe Y, Yazaki Y (1992) Apolipoproein E prevents the progression of atherosclerosis in Watanabe heritable hyperlipidemic rabbits. J Clin Invest 89:706–711
Yarden Y, Escobedo JA, Kuang W-J, Yang-Feng TL, Daniel TO, Tremble PM, Chen EY, Ando ME, Harkins RN, Francke U, Fried VA, Ullrich A, Williams LT (1986) Structure of the receptor for platelet-derived growth factor helps define a family of closely related growth factor receptors. Nature 323:226–232
Ylä-Herttuala S, Lipton BA, Rosenfeld ME, Goldberg IJ, Steinberg D, Witztum JL (1991a) Macrophages and smooth muscle cells express lipoprotein lipase in human and rabbit atherosclerotic lesions. Proc Natl Acad Sci USA 88:10143–10147
Ylä-Herttuala S, Lipton BA, Rosenfeld ME, Särkioja T, Yoshimura T, Leonhard EJ, Witztum JL, Steinberg D (1991b) Expression of monocyte chemoattractant protein 1 in macrophage-rich areas of human and rabbit atherosclerotic lesions. Proc Natl Acad Sci USA 88:5252–5256
Ylä-Herttuala S, Rosenfeld ME, Parthasarathy S, Sigal E, Särkioja Witztum JL, Steinberg D (1991c) Gene expression in macrogphage-rich human atherosclerotic lesions, 15-lipoxygenase and acetyl low density lipoprotein receptor messenger RNA colocalize with oxidation specific lipid–protein adducts. J Clin Invest 87:1146–1152
Yokode M, Kita T, Kikawa Y, Ogorochi T, Narumiya S, Kawai C (1988) Stimulated arachidonate metabolism during foam cell transformation of mouse peritoneal macrophages with oxidized low density lipoprotein. J Clin Invest 81:720–729
Yokode M, Hammer RE, Ishibashi S, Brown MS, Goldstein JL (1990) Diet-induced hypercholesterolemia in mice: prevention by overexpression of LDL receptors. Science 250:1273–1275
Yu X, Dluz S, Graves DT, Zhang L, Antoniades HN, Hollander W, Prusty S, Valente AJ, Schwartz CJ, Sonenshein (1992) Elevated expression of monocyte chemoattractant protein 1 by vascular smooth muscle cells in hypercholesterolemic primates. Proc Natl Acad Sci USA 89:6953–6957
Zhang SH, Reddick RL, Piedrahita JA, Maeda N (1992) Spontaneous hypercholesterolemia and arterial lesions in mice lacking apolipoprotein E. Science 258:468–471

Note Added in Proof

While this chapter was in preparation, the following important articles relating to this chapter have appeared:

1. Ross R (1993) The pathogenesis of atherosclerosis: A perspective for the 1990s. Nature 362:801–809
2. Bates P, Young JAT, Varmus HE (1993) A receptor for subgroup A rous sarcoma virus is related to the low density lipoprotein receptor. Cell 74:1043–1051
3. Fischer DG, Tal N, Novick D, Barak S, Rubinstein M (1993) An antiviral form of the LDL receptor induced by interferon. Science 262:250–253

In addition the conclusions of the manuscript by Herz et al. 1993 have been retracted.

CHAPTER 6

Lipoprotein Lipase and Hepatic Lipase

T. Olivecrona and G. Olivecrona

A. Introduction

Lipoprotein metabolism involves two major steps (Eisenberg 1990). First, triglyceride (TG) rich lipoproteins bind transiently to lipoprotein lipase (LPL) at the vascular endothelium (Fig. 1). The enzyme rapidly hydrolyzes triglycerides, which accomplishes two things (Eckel 1989): it enables the tissues to utilize fatty acids from the lipoproteins, and it transforms large TG-rich lipoproteins (chylomicrons and very low density lipoproteins, VLDL) into cholesterol-rich remnant lipoproteins. This process is completed within the space of a few minutes to a few hours after the lipoproteins have entered circulation. Some of the remnants are rapidly removed from the circulation by receptor-mediated endocytosis, but some are transformed into low-density lipoproteins (LDL) and high-density lipoproteins (HDL). Subjects with genetic deficiency of LPL demonstrate that the enzyme is indeed necessary for these reactions; there is massive accumulation of TG-rich lipoproteins in plasma and low levels of LDL and HDL (Brunzell et al. 1986).

While the role of LPL in delipidation of TG-rich lipoproteins is fairly well defined, the role of hepatic lipase (HL) has remained somewhat enigmatic. In HL deficiency in humans, VLDL remnants and TG-rich HDL accumulate in plasma (Breckenridge 1987; Auwerx et al. 1989b). Studies on lipoprotein kinetics in patients with HL deficiency (Demant et al. 1989) and studies on the effects of adding purified HL to plasma (Clay et al. 1989) also indicate that HL acts on HDL and LDL and that it may be involved in the terminal stages of delipidation of chylomicron and VLDL remnants. Hence, LPL and HL appear to have partially overlapping roles in lipoprotein metabolism.

LPL action can be viewed as the bulk reaction in lipoprotein metabolism, whereas HL is involved in further degradation and remodelling of the remnants formed by LPL and in the metabolism of LDL and HDL. Table 1 shows lipase activities in postheparin plasma of a number of species. These data indicate that LPL activity is high in most species. This is logical, since animals must be able to rapidly catabolize TG-rich lipoproteins after lipid meals. In contrast, HL activity varies widely between species (Jansen and Hülsmann 1985). It is high in humans and in rats, moderately high in

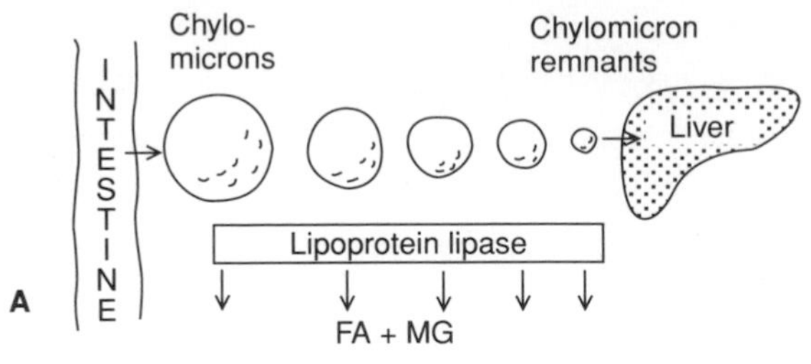

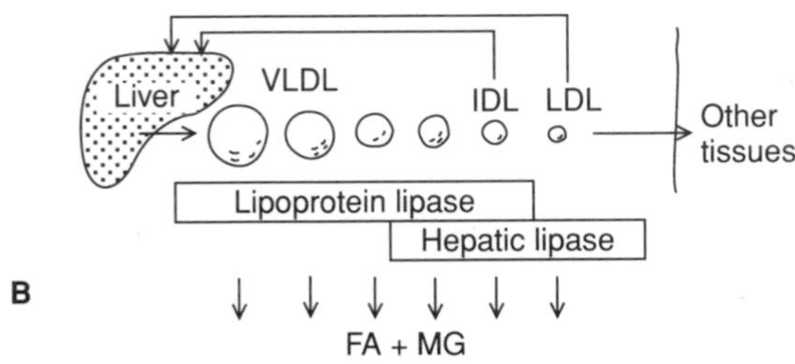

Fig. 1. A,B. Delipidation of triglyceride-rich lipoproteins by the lipases. **A** Exogenous pathway. **B** Endogenous pathway. *VLDL*, very low density lipoprotein; *IDL*, intermediate-density lipoprotein; *LDL*, low-density lipoprotein; *FA*, fatty acid; *MG*, monoglyceride

dogs, low in guinea pigs, and virtually nonexistent in calves. Similar observations have been made for other factors involved in the metabolism of the cholesterol-rich lipoproteins (BROWN et al. 1990). For instance, lipid transfer activity in plasma differs widely between species. The concentrations of LDL and HDL in plasma also vary widely between animals. For instance, both LDL and HDL are relatively high in humans, whereas dogs have low LDL, but high HDL concentrations, and guinea pigs have high LDL and low HDL concentrations. It appears that LPL carries out an indispensable initial reaction in lipoprotein metabolism and is consistently high in animal species, whereas there is more variance in how animals are disposed to handle the metabolism of remnant particles and of LDL and HDL.

B. Gene Structures

From their cDNA sequences it became apparent that LPL, HL, and pancreatic lipase are structurally related and belong to a gene family (for a list of references on sequences, see HIDE et al. 1992). Amino acid and cDNA sequences are known for LPL from five species (man, cow, mouse, guinea

Table 1. Lipase activities in postheparin plasma of some species

	LPL (mU/ml)	HL (mU/ml)
Man	350	370
Rat	440	700
Dog	630	170
Guinea pig	790	70
Calf	530	<5

Heparin (100 IU/kg b.w., about 0.65 mg) was injected i.v. and a plasma sample was taken 10 or 15 min later. Hepatic lipase (HL) was assayed using a gum arabic stabilized triglyceride emulsion in the presence of 1 *M* NaCl to suppress low-density lipoprotein (LPL) activity. LPL was assayed with a phosphatidylcholine-stabilized triglyceride emulsion using rat serum as source of activator. For assay of LPL activity in human and rat plasma, HL was suppressed by immunoinhibition using anti-HL serum. For dog, guinea pig, and calf plasma, LPL represents the difference in activity between samples treated with anti-LPL or control serum.

pig, and chicken) and for HL from four species (human, rat, rabbit, and mouse). The gene structures are known for human and guinea pig LPL (Kirchgessner et al. 1989a; Enerbäck and Bjursell 1989) and for human HL (Cai et al. 1989). The LPL gene (human and mouse) has been mapped to chromosome 8, while the human HL gene is on chromosome 15. The evolutionary relationships between the lipases have been discussed in several recent articles based on gene structures and degrees of conservation (Hide et al. 1992; Kirchgessner et al. 1989a; Persson et al. 1989, 1991; Mickel et al. 1989; Kern 1991; Bensadoun 1991; Derewenda and Cambillau 1991; Giller et al. 1992). The pancreatic lipase gene (canine) contains 13 exons, while the human LPL gene has ten exons and the human HL gene has nine exons. The exon structures of LPL and HL are similar, but LPL contains an additional exon which codes for a long 3′-untranslated region. It seems likely that a first gene duplication gave rise to a digestive lipase and a lipoprotein-metabolizing enzyme. A second gene duplication allowed this to evolve into LPL and HL, which are specialized for different, but overlapping, roles in lipoprotein metabolism. As far as is known, there is only one variant (one gene) for LPL and HL in each species.

The lipase family has a fourth member in some yolk proteins from *Drosophila* (for references, see Bownes 1992). These proteins have no catalytic activity, but they bind fatty acid ecdysteroid conjugates and are transported in the hemolymph to the oocyte, where they are taken up by receptor-mediated endocytosis. Thus, the lipid transport theme is also relevant for this distant family member.

C. Molecular Structure

Mature human LPL contains 448 amino acid residues, and two N-linked oligosaccharide chains, giving a total molecular mass of about 55 kDa. HL has 476 amino acid residues and also has N-linked oligosaccharide chains. The three-dimensional structures for pancreatic lipase and its complex with colipase have been resolved (Winkler et al. 1990; Van Tilbeurgh et al. 1992). Human pancreatic lipase, which has 449 amino acid residues, is composed of two independently folded domains (Fig. 2). The aminoterminal domain spans residues 1–335 and contains the residues of the active site triad (Ser_{152}, Asp_{176}, and His_{263}). The active site is covered by a short loop between the disculfide-linked residues 237 and 261. Similar loops, sometimes called "flap" or "lid," are found in some fungal lipases (Brady et al. 1990; Schrag et al. 1991). The loop has to be moved to enable entrance of substrate molecules into the active site (Lawson et al. 1992). The loop structure has a hydrophobic side which in water is turned against the active site. On interfacial binding the loop may be repositioned so that its inside becomes part of a hydrophobic, lipid-binding region around the entrance of the active site.

In a central homology region spanning residues 125–230 in pancreatic lipase, there is 33% residue identity between all known lipases of this protein family, whereas in neighboring segments on either side the identity is only 15%. Based on structure prediction analyses, it has been concluded that this region is similarly folded in all members of the family (Hide et al. 1992; Derewenda and Cambillau 1991; Persson et al. 1991) and that it contains the active site. Natural and synthetic mutants affecting Ser_{132}, Asp_{156}, or His_{241} in LPL (Emmerich et al. 1992; Faustinella et al. 1991a,b, 1992; Ma et al. 1992) or Ser_{147} in rat HL (Davis et al. 1990) have no catalytic activity, which supports the hypothesis that these residues participate in the catalytic triad, as do corresponding residues in pancreatic lipase.

In contrast to the strong sequence similarities in the central homology region, there is not a single residue identity across all three lipases in the loop which covers the active site in pancreatic lipase (Persson et al. 1991). The disulfide bond which closes the loop is, however, conserved. An earlier study showed that trypsin causes selective cleavage of LPL at Arg_{230} (in bovine LPL), indicating that this bond is particularly exposed (Bengtsson-Olivecrona et al. 1986). This corresponds to the tip of the loop in pancreatic lipase and hence supports the hypothesis that LPL has a similar loop. The catalytic properties were dramatically changed in the nicked enzyme, indicating that the cleavage had disturbed the finely tuned motions of the putative loop. In a recent study by Faustinella et al. (1992), the loop in LPL was partly or fully replaced by the loop sequence from HL. This lowered the LPL activity by at the most 40%. The predicted secondary structure of the loop is similar for LPL and for HL, and they may, there-

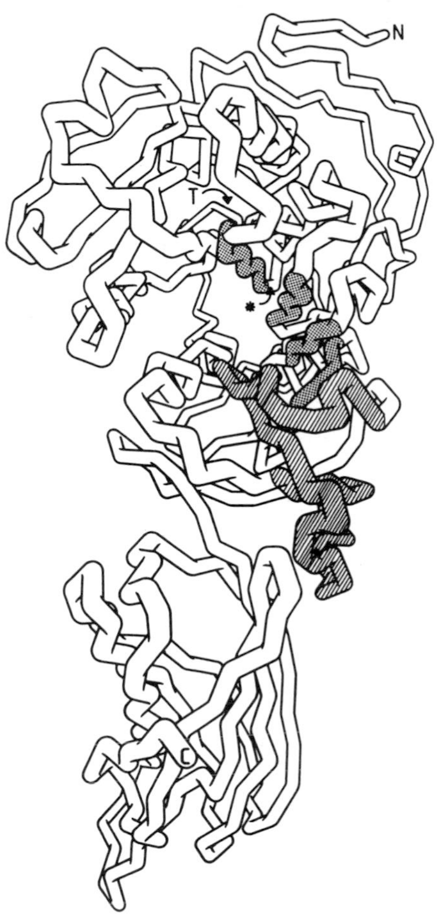

Fig. 2. Three-dimensional structure of pancreatic lipase. Drawing by Hans Nilsson representing an artist's interpretation of the three-dimensional figure in WINKLER et al. (1990). *Stippled segments* show the four β-sheets which are conserved in both the lipases and the insect yolk proteins. The *hatched region* corresponds to the segment postulated to be heparin binding in LPL. The *arrow* with *asterisk* marks active site Ser-152. *N* and *C* indicate the corresponding termini. From PERSSON et al. (1991) with permission

fore, be functionally almost equivalent. In contrast, when the loop was partly deleted, or replaced by the loop structure from pancreatic lipase, the expressed LPL was completely inactive.

Another area of special interest in the LPL molecule is the site for binding to heparin (Fig. 2). The physiological function is to anchor the lipase to the glycosaminoglycan chains of heparan sulfate proteoglycans. The heparin-binding site has been localized to the region which in pancreatic lipase forms a surface loop between the two folding units on the back side of the molecule (Persson et al. 1991). Within this segment–residues 260–306 in human LPL, residues 281–330 in human pancreatic lipase according to the alignment in Hide et al. (1992) – there are 14 positively charged residues in LPL, nine in HL, and four in pancreatic lipase. The segment of human LPL contains the sequence KVRAKRSSK that carries a high density of positive charge (five out of nine residues) and is similar to the consensus sequence found in other heparin-binding proteins such as apolipoproteins B and E. It is likely that the three-dimensional arrangement of the charge is also important for the binding affinity (Bengtsson-Olivecrona and Olivecrona 1985).

Active LPL is a homodimer. The dimer appears to be in rapid equilibrium with "active" monomers, but these are prone to undergo an essentially irreversible change in conformation with loss of activity (Osborne et al. 1985). Hence, the dimer state is important for conformational stability. This may represent a built-in mechanism to self-destruct which would limit the lifetime of the enzyme after it has been released from the lipase-producing cells. Monomeric, inactive LPL (and active HL) binds with lower affinity to heparin than the LPL dimer does (Bengtsson-Olivecrona and Olivecrona 1985; Liu et al. 1992a). The stronger binding of the dimer may be because heparin-binding sites on both subunits cooperate (Clarke et al. 1983) or because the dimer has a more favorable arrangement of charges in the heparin-binding region. Some mutations in regions that are not considered to be involved in heparin binding lower the heparin affinity (Fojo 1992). This is probably due to structural rearrangements in the molecule, possibly leading to monomerization. It is not known which parts of LPL are involved in the subunit interaction.

The carboxy-terminal domains of the lipase family members show little sequence similarity. Out of 114 residues, only ten are conserved. However, structure prediction calculations lead to the view that all three lipases have a similar arrangement (β-sandwich) in this part as well (Derewenda and Cambillau 1991). One piece of evidence for this is that bovine LPL is readily cleaved by chymotrypsin at Phe_{390} and Trp_{392} (A. Lookene and G. Bengtsson-Olivecrona, 1993). In comparison with the pancreatic lipase structure, this site would be in the middle of the putative carboxyterminal folding unit in a surface loop connecting the $\beta5$ and the $\beta6$ strands. In pancreatic lipase, binding of colipase occurs to a preformed site in this region (van Tilbeurgh et al. 1992). A similar binding of apoC-II to LPL is less likely, since studies of LPL-HL chimers in two different laboratories (Wong et al. 1991; Fojo 1992) indicate that binding of apoC-II is within the 314 amino-terminal residues, i.e., to the amino-terminal folding unit. Furthermore, the chymotrypsin-truncated LPL has the same affinity for apoC-II as the intact LPL (A. Lookene and G. Bengtsson-Olivecrona, 1993). Monoclonal antibodies directed to the carboxy-terminal region of LPL inhibit the activity against emulsified lipid substrates (Wong et al. 1991; Fojo 1992) and the chymotryptic cleavage made LPL less active against emulsified lipid substrates (Bengtsson-Olivecrona et al. 1986). These data indicate that the carboxy-terminal region of LPL contains structures of importance for proper arrangement of the enzyme at certain interfaces (a secondary lipid-binding site).

D. Effects of Apolipoproteins on Lipase Action

The activity of LPL in its natural environment is dependent on apolipoprotein C-II. Patients with apoC-II deficiency have symptoms similar to patients with LPL deficiency (Fojo 1992). In model systems, however, the effect of C-II on LPL varies (Olivecrona and Bengtsson-Olivecrona 1987). The effect is specific; C-II does not activate any other lipase. Pancreatic lipase also has a protein activator, colipase. There is no sequence or structural homology between apoC-II and colipase. Colipase (101 amino acid residues) contains five disulfide bonds which stabilize three finger-like structures responsible for binding to lipid (van Tilbeurgh et al. 1992). ApoC-II is a member of the apolipoprotein family, has no disulfides, but contains flexible regions which bind to lipid by forming amphipathic helices (for review, see Wang et al. 1992). The main role of colipase is to assist pancreatic lipase in binding to lipid–water interfaces that are covered by bile salts and proteins. LPL, on the other hand, binds to most interfaces, even in the absence of apoC-II. Therefore, the activator has its main effect on LPL that is already bound to the interface. It can only be assumed that the interaction causes structural changes in the enzyme which make it more efficient in catalysis. A logical mechanism would be to facilitate the movement of the loop covering the active site. This is, however, less likely, since the LPL/HL chimer with the loop sequence from HL was stimulated by apoC-II (Faustinella et al. 1992). The lipid-binding region of apoC-II has been localized to the amino-terminal two-thirds of the molecule, while the structures involved in binding to LPL and activation are in the carboxy-terminal third (Smith and Pownall 1984). From studies with synthetic fragments of apoC-II, it was concluded that the lipid-binding regions were not necessary for activation of the lipase. This may not be true for synthetic interfaces at high surface pressures (Jackson et al. 1986a) or with biologically packaged lipids such as chylomicrons (G. Bengtsson-Olivecrona and U. Beisiegel, unpublished). All mutations identified thus far in patients with clinically manifest apoC-II deficiency lead to disturbed expression or to expression of prematurely truncated forms of the protein (Fojo 1992). There is no example of a point mutation of an individual amino acid residue which leads to loss of function.

In vitro studies with synthetic emulsions have shown that other apolipoproteins, e.g., apoC-III (Jackson et al. 1986b) and apoE (McConathy and Wang 1989), can inhibit LPL. The mechanism may be nonspecific, since many unrelated lipid-binding proteins have the same effect (Olivercrona and Bengtsson-Olivecrona 1987). The likely mechanism is crowding of the lipid–water interface, making it more difficult for the enzyme to adsorb. In contrast to this view, MCCONATHY et al. (1992) could localize inhibitory activity to the N-terminal part of apoC-III, which does not bind to lipid. There is no firm evidence that apolipoproteins other than C-II modulate LPL action in the physiological milieu. For instance, JACKSON et al. (1986b), working with lipoproteins from LPL-deficient patients and with model emulsions, estimated that it would take apoC-III to apoC-II ratios about 20-fold higher than normal to significantly affect lipolysis.

There are also reports on the effects of apolipoproteins on HL action. It is clear that HL is not activated by apoC-II and has no obligatory need for any other apolipoprotein. It is also clear that HL action on TG emulsions is impeded by any of a number of lipid-binding proteins and that this can be related to covering of the lipid–water interface (KUBO et al. 1982). Whether any of these effects have physiological relevance remains to be established.

E. Lipoprotein Lipase and Hepatic Lipase – What Are the Functional Differences?

The two enzymes are structurally similar, have similar active sites, and hydrolyze the same lipids. Yet they carry out quite different functions in vivo. What is the molecular basis for this? We will first turn our attention to the question of how they select lipoproteins as substrates. A generally accepted theory for lipase action states that there are two major steps (VERGER and DEHAAS 1976). First, the enzyme absorbs to the lipid particle. Then the enzyme seeks out a single substrate molecule at the interface. Thermodynamic considerations suggest that the lipid-binding site is separate from the active site. This prediction has been experimentally verified for pancreatic phospholipase A_2, the best studied lipase.

The fact that LPL and HL select different types of lipoproteins as their substrate is evident from a number of in vitro and in vivo observations. For instance, when the lipases are added to whole plasma, LPL hydrolyzes lipids mainly in VLDL, whereas HL acts mainly on HDL (CLAY et al. 1989). This particle selection probably reflects differences in binding affinity for the lipoprotein particles, i.e., differences in the "lipid-binding" or "interface-recognition" sites on the two lipases. Direct demonstrations of substrate selection have come from competition experiments. For instance, in a study by Bengtsson et al., HL was found to efficiently hydrolyze TG in phospholipid (PL)-stabilized emulsion droplets when these were the only lipid particles in the system (BENGTSSON and OLIVECRONA 1980a). When HDL was added, hydrolysis of the emulsified TG was suppressed and the enzyme acted on HDL instead. In contrast, addition of HDL to such a system did not impede the action of LPL, which preferred to stay on the emulsion droplets.

Another aspect of substrate selection is how the enzymes seek out and hydrolyze individual lipid molecules at the interface. Model studies have shown that LPL and HL are relatively nonspecific enzymes. They hydrolyze TG, diglycerides, monoglycerides, PL, and a variety of model substrates such as paranitrophenylbutyrate (OLIVECRONA and BENGTSSON-OLIVECRONA 1987). An important exception is that they do not hydrolyze cholesteryl esters. In lipoproteins, the two major substrates are therefore TG and PL. TG are largely confined to the core of the lipoproteins. HAMILTON and SMALL (1981) have shown, by model experiments, that the molar ratio of PL to TG in the surface film is more than 30. SCOW and EGELRUD (1976) were the first to show that LPL hydrolyzes not only TG, but also PL in rat chylomicrons. Subsequent studies, mainly in model systems, have led to the view that HL is a more potent phospholipase than LPL is. This was directly tested in a recent study on the action of the two lipases on PL and TG in isolated lipoproteins (DECKELBAUM et al. 1992). For each type of lipoprotein used, HL always hydrolyzed more PL for a given amount of TG hydrolysis. Rojas et al. addressed the same question with TG-containing PL liposomes and PL-enveloped TG droplets as model substrates. They reached the same conclusion as DECKELBAUM et al., i.e., HL always hydrolyzed more PL molecules for each TG molecule hydrolyzed than LPL did (ROJAS et al. 1991). There are several other studies in the literature which lead to similar conclusions. For instance, Ikeda et al., using LPL and HL purified from human postheparin plasma, compared product profiles during hydrolysis of TG in a TG–PL emulsion by the two enzymes (IKEDA et al. 1989). As has been demonstrated before, LPL hydrolyzed TG mainly to monoglycerides and free fatty acids. Under identical conditions, HL hydrolyzed more monoglycerides, yielding primarily free fatty acids as products. Another piece of evidence comes from observations by ÅKE NILSSON and his coworkers (1987). They noted that LPL has relatively low activity against ester bonds involving fatty acids with double bonds close to the carbonyl. This is probably due to steric hindrance, as has been shown for pancreatic lipase. HL shows relatively higher activity on these ester bonds than LPL does. In fact, addition of HL to reaction mixtures with LPL will drive hydrolysis of such glycerides towards completion in a mixture of natural TG (MELIN et al. 1991).

In summary, there are differences between LPL and HL, both in terms of which lipoprotein particles they prefer and how they select substrate molecules at the interface.

F. Actions of Lipoprotein Lipase

LPL is an efficient enzyme. Its turnover number is about 1000/s for TG hydrolysis under "physiological conditions"(OLIVECRONA and BENGTSSON-OLIVECRONA 1987). This raises questions on how substrate molecules enter and products leave the active site. To approach this, Rojas et al. (1991) compared hydrolysis of TG in liposomes and in emulsion droplets made up of the same PL and TG (Rojas et al. 1991). The relations of rate vs. surface area were similar for liposomes and emulsion droplets, indicating that the lipid-binding sites on the lipase did not detect any marked difference between the surface of the two types of particles. If each hydrolytic event was followed by dissociation of the lipase into the aqueous phase, one would expect to see the same ratio of PL:TG hydrolysis and similar kinetics with liposomes and emulsion droplets. The ratios of TG:PL hydrolysis were, however, quite different. With liposomes the enzyme hydrolyzed more PL than TG, whereas with emulsion droplets TG hydrolysis was much faster. Furthermore, the maximal rate of TG hydrolysis was more than 40 times higher with the emulsion droplets than with the liposomes. These results indicate that when LPL binds to the substrate particle, it stays for several rounds of lipolysis, i.e., the action is processive. Furthermore, the results suggest that when products leave the active site, it is easier for a new substrate molecule to come in from below, i.e., the core, than from the side, i.e., from the surface. This would seem to fit with the X-ray structure of pancreatic lipase, which shows the active site at the bottom of a hydrophobic pocket (Winkler et al. 1990).

This picture of lipolysis fits the action of LPL (and HL) in vivo. Degradation of chylomicrons is a rapid process in which many TG and a few PL molecules are hydrolyzed. A typical chylomicron may contain 2 million TG molecules. Hydrolysis of 90% of these TG to form a remnant particle requires 3.6 million hydrolytic events. Hence, several lipase molecules must act simultaneously on the particle and the lipase molecules must stay attached to the particle for many rounds of lipolysis. It is clear that the process can not be accomplished by the chylomicron hopping from one lipase molecule to another for each hydrolytic event. On the other hand, cooperative binding to many lipase molecules might lock the particle at the endothelial site (Fig. 3). There is a built-in property in the lipase which balances this. In vitro experiments have demonstrated that LPL binds fatty acids and that the resulting complexes have much reduced affinities for lipid–water interfaces, for apoC-II, and for heparan sulfate (Bengtsson and Olivecrona 1980b; Saxena and Goldberg 1990). Saxena et al. (1989) have shown that physiological concentrations of free fatty acids dissociate LPL from cultured endothelial cells. Thus, the following sequence of events is suggested (Peterson et al. 1990; Karpe et al. 1992a): fatty acids are sometimes released by LPL at endothelial sites more rapidly than the underlying tissue can transport and metabolize the fatty acids, which, in turn, results in spillage of fatty acids into the blood and to the formation of LPL–fatty-acid complexes with inhibition of continued hydrolysis and dissociation of LPL from endothelial heparan sulfate. The implication is that LPL may either sometimes, often, or even usually be present in excess and the limiting factor is transport and metabolism of fatty acids by the underlying tissue.

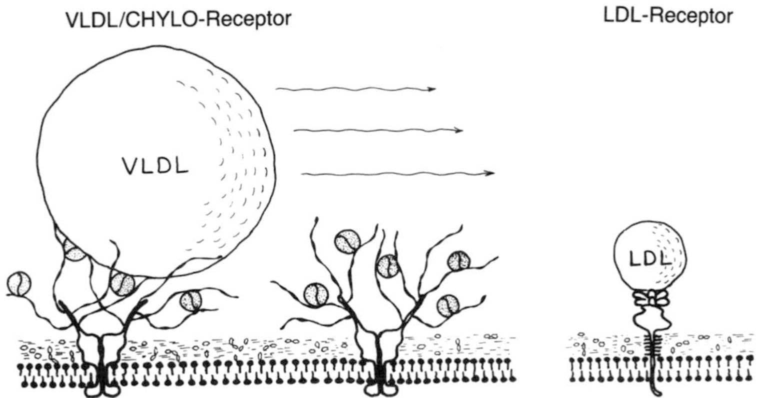

Fig. 3. Endothelial lipolysis sites. *Left panel*: heparan sulfate proteoglycans are intercalated in the plasma membrane of an endothelial cell. Each core protein carries several polysaccharide chains. The attached lipoprotein lipase (LPL) molecules (*small spheres*) are in a position where they can freely interact with lipoproteins, illustrated here by a very low density lipoprotein (VLDL) particle. Chylomicrons are much larger than VLDL and probably interact simultaneously with several heparan sulfate proteoglycans. *Right panel*: the low density lipoprotein (*LDL*) receptor also has its ligand-binding domain outside the glycocalyx. This is accomplished by a domain with O-linked glycan chains, which spans the glycocalyx. From OLIVECRONA and BENGTSSON-OLIVECRONA (1987), with permission

G. Sites of Synthesis

LPL acts at, but is not made in, vascular endothelial cells. Other cells produce and release the lipase for transfer to binding sites at the endothelium. Studies on mRNA levels and by in situ hybridization (GOLDBERG et al. 1989; YACOUB et al. 1990; SEMENKOVICH et al. 1991; KIRCHGESSNER et al. 1989b; CAMPS et al. 1990b) have shown that LPL is made in wide variety of tissues and cells.

Quantitatively, the dominant tissues for LPL production are adipose and muscles (Table 2). There are large differences between adipose tissue from different anatomical locations (ECKEL 1987; ARNER et al. 1992); red muscles tend to have higher LPL activity than white (TAN et al. 1977). In adipose tissue, the enzyme is made both in preadipocytes and in mature adipocytes (ECKEL 1987). For both skeletal and heart muscle, there is evidence that LPL is synthesized in myocytes (BORENSZTAJN 1987; BLANCHETTE-MACKIE et al. 1989; SERVERSON et al. 1988; CAMPS et al. 1990b). The Steins and their coworkers have proposed that in heart, LPL is also made in a nonbeating cell type (FRIEDMAN et al. 1991).

During lactation, there is high LPL activity in the mammary gland; this serves to direct plasma TG fatty acids to synthesis of milk lipids (SCOW and CHERNICK 1987). Opinion differs as to whether the enzyme is made in the

Table 2. Sources of lipaproein lipase in some guinea pig tissues

Tissue	Cell type	LPL activity (mU/g tissue)
Adipose tissue	Adipocytes, preadipocytes	4000
Skeletal muscle	Myocytes	500
Heart muscle	Myocytes	3300
Mammary gland	Milk-producing cells	6000
Lung	Macrophages	1000
Spleen	Macrophages	90
Ovary	Cells producing	140
Adrenal	steroid hormones	90
Liver (adult)	Kupffer cells	240
Liver (fetal)	Mainly hepatocytes	Data not available
Nervous system	Certain neurons	Data not available

Data on cellular sources of lipoprotein lipase (LPL) are based on in situ hybridization and immunofluorescence (CAMPS et al. 1990b, 1991). Data on LPL activity are in part from the same studies, and in part previously unpublished data.

mammary alveolar, i.e., the milk-producing, cells (CAMPS et al. 1990b) or in adipocyte-like cells (JENSEN et al. 1991).

LPL is synthesized in several other cell types (Table 2). This includes macrophages in several tissues (KHOO et al. 1981; CAMPS et al. 1991) and hormone-producing cells in the adrenals (GÅFVELS et al. 1991) and ovaries (CAMPS et al. 1990a). An interesting observation is that LPL is made in certain neuronal cells in the nervous system (VILARÓ et al. 1990; GOLDBERG et al. 1989). In hepatocytes, LPL is made during the fetal state (VILARÓ et al. 1987), but becomes suppressed soon after birth (STAELS and AUWERX 1992).

H. Maturation into Active Lipoprotein Lipase

Early studies suggested that adipocytes contained inactive forms of LPL which could be recruited and released into the medium as active LPL on incubation of the cells with appropriate signals (e.g., insulin) under conditions when synthesis of new enzyme molecules was blocked (STEWART and SCHOTZ 1974; PRADINES-FIGUÈRES et al. 1988). It now appears that this was due to an underestimation of cellular LPL because of assay problems (PRADINES-FIGUÈRES et al. 1990). There is no firm evidence for LPL in storage vesicles of a regulated secretion pathway. Recent studies indicate that newly synthesized LPL travels by the so-called default pathway and is either rapidly released from the cells or is degraded within the cells (BENSADOUN 1991; BRAUN and SEVERSON 1992). LPL has been demonstrated in secretory vesicles in adipocytes (VANNIER et al. 1986) and in heart myocytes (BLANCHETTE-MACKIE et al. 1989). Cells differ in the fraction of newly synthesized LPL that is spontaneously released to the outside. Examples of cells which show little or no spontaneous release are the nonbeating heart cells studied by the Steins (Chajek et al. 1977), cardiac myocytes (Severson et al. 1988), and Ob17 adipocytes (Vannier et al. 1989). High rates of LPL release have been reported with preadipocytes (Chajek-Shaul et al. 1985a) and 3T3-F442A adipocytes (Vannier et al. 1989). A particularly dramatic example are Ob17 cells. In the presence of heparin, these cells appeared to transfer all newly synthesized LPL to the medium, whereas in the absence of heparin no release was observed (Vannier et al. 1985). Cellular LPL activity remained constant, indicating a steady state where newly synthesized LPL was soon degraded. This was directly demonstrated by Semb and Olivecrona (1987) in studies with guinea pig adipocytes. Without heparin, 16% of pulse-labeled LPL appeared in the medium during a 1-h chase; with heparin, this figure increased to 49%. The difference did not shown up as increased LPL retained in the cells, but was made up by increased degradation. These data suggest that LPL can be diverted towards release or towards degradation by secretagogues acting outside the cell. Cisar et al. (1989) have suggested that newly synthesized LPL is transported to the cell surface where it binds to heparan sulfate proteoglycans. The enzyme can then be released into the medium or be internalized and either degraded or recycled back to the cell surface. Model experiments have shown that the binding of LPL to heparan sulfate gets tighter at lower pH values and that the enzyme retains catalytic activity when bound to heparin/heparan sulfate, even though in solution it becomes very unstable at low pH values (Bengtsson-Olivecrona and Olivecrona 1985; Sivaram et al. 1992). Hence, the enzyme should be able to survive passage through acidified vesicles if bound to heparan sulfate.

Several investigators have proposed that a main site of regulation lies in the channeling of LPL towards release. Nilsson-Ehle et al. (1976) found that adipocytes from rats given insulin released more LPL to the medium than adipocytes from control rats. In 3T3-L1 adipocytes deprived of insulin for several days, only 7% of the LPL could be rapidly released by heparin, i.e., was located at the cell surface. Insulin treatment of these cells for 4 h caused a small increase in total LPL activity, but increased the heparin-releasable fraction to 36% (Olivecrona et al. 1987). When fat pads from fasted rats were incubated with ^{35}S-methionine, radioactivity in LPL leveled off within 1 h, suggesting a steady state where LPL was degraded as rapidly as it was formed. In contrast, with fat pads from fed rats, radioactivity in LPL increased continuously for the 3-h incubation period, indicating that the lipase was released from the adipocytes and thereby rescued from intracellular degradation (Olivecrona et al. 1987). This was studied in more detail by Doolittle et al. (1990). When rats were fasted overnight, LPL catalytic activity in adipose tissue decreased by 50%, while LPL mRNA levels and rates of LPL synthesis increased nearly twofold. Clearly, enzyme activity was controlled in part by posttranscriptional mechanisms. Pulse-chase experiments indicated that the main difference was that in the fasted state a much larger fraction of the enzyme was diverted from the Golgi towards degradation. Taken together, these data indicate that there are mechanisms to modulate the fraction of newly synthesized LPL molecules that are released from adipocytes. The nature of this putative regulatory system remains unclear.

One aspect that has aroused much interest is the role of the oligosaccharide structure for assembly of active LPL (Braun and Severson 1992). This goes back to studies on fat pads by the Robinson group (Parkin et al. 1980) and studies on 3T3-L1 adipocytes by Spooner et al. (1979). The Steins and their collaborators proposed that there is a regulated, critical step in the processing of the oligosaccharide chains which determines the activity of the enzyme (Chajek-Shaul et al. 1985a). Vannier and Ailhaud (1989) concluded from experiments with murine Ob17 and 3T3-F442A cells that core glycosylated but catalytically inactive LPL monomers are transferred from the endoplasmic reticulum (ER) to Golgi, where their oligosaccharides are processed and the enzyme assembles into catalytically active dimers. Semb and Olivecrona (1989) studied the relation between glycosylation and activity of LPL in guinea pig adipocytes. Using lectin affinity chromatography, they isolated LPL containing only high-mannose oligosaccharides and showed that this form of the engyme was catalytically active. In the presence of methyldeoxynojirimycin or deoxymannojirimycin, which inhibit trimming and processing of the oligosaccharides, catalytically active high-mannose-type LPL molecules were released into the medium. These results indicate that core-glycosylated LPL can be catalytically active. Recently, Ben-Zeev et al. (1992) have shown that in COS cells incubated at 16°C to inhibit vesicular transport, LPL was retained in the ER in a high-mannose, catalytically active form. Furthermore, active LPL accumulated in the ER of COS cells transfected with an LPL cDNA construct that contained a tetrapeptide (KDEL) ER retention signal at its C-terminal end. All of these studies show that LPL can attain its catalytically active form already in the ER, while in a high-mannose form. This is in accord with the present view that most oligomeric proteins are assembled in the ER (Hurtley and Helenius 1989). In fact, it appears that correct assembly is often a prerequisite for transfer to the Golgi (Pelham 1989). Some recent studies indicate that, at least in rats and mice, removal of glucose residues from the core oligosaccharide(s) in the ER may be necessary before the enzyme can fold correctly (Masuno et al. 1992; Ben-Zeev et al. 1992; Carroll et al. 1992).

Chicken LPL is sulfated (Hoogewerf et al. 1991). It is not known if this is also true for mammalian LPL.

I. Transfer to the Endothelium

Blanchette-Mackie et al. (1989) studied the movement of LPL in mouse hearts by electron microscopic immunocytochemistry. Their results indicate that LPL moves along cell surfaces, crosses the endothelial cells via vesicles or intracellular channels, and then concentrates at the surface of luminal projections of the endothelium. Goldberg and coworkers demonstrated a saturable transport system for LPL in cultured bovine aortic endothelial cells (SAXENA et al. 1991). Interestingly, this transport required heparan sulfate proteoglycans and was impeded by addition of increasing ratios of oleic acid: albumin at the basal surface. This suggests that transport of LPL over the endothelial cells is metabolically controlled. The same group have shown that endothelial cells are polarized to accumulate LPL on their apical surface and that the enzyme recycles into and back out of the endothelial cells (SAXENA et al. 1990). Recently, they have demonstrated a 116-kDa protein which binds to LPL and also to heparan sulfate chains (SIVARAM et al. 1992). They suggest that this protein mediates binding, internalization, and recycling of LPL towards the apical surface of endothelial cells.

J. Transport in Blood

It appears that a main route for turnover of endothelial LPL is transport in blood to the liver, where the enzyme is avidly taken up and degraded (Fig. 4; OLIVECRONA et al. 1991b). When the liver was excluded from circulation in rats, plasma LPL activity rose continuously, directly demonstrating the importance of the liver for the clearing of plasma LPL (CHAJEK-SHAUL et al. 1985c). After injection of labeled LPL, the lipase rapidly disappeared from blood and about 40% was found in the liver (WALLINDER et al. 1984). During perfusion of rat livers with bovine LPL, VILARÓ et al. observed uptake of more than 75% within 5 min. CHAJEK-SHAUL et al. (1988b) perfused rat livers with LPL from rat preadipocytes and found continuous uptake of about 50% of the enzyme in single pass. The uptake in liver occurs for both active and inactive forms of LPL (WALLINDER et al. 1984). It is impeded, but not abolished, by heparin (LIU et al. 1991a).

Injected labeled LPL also binds in extrahepatic tissues (WALLINDER et al. 1984). This binding is dependent on the native, dimeric state of the enzyme and is competed out by heparin. In perfused rat hearts, exogenous LPL bound with displacement of endogenous LPL (CHAJEK-SHAUL et al. 1988a). The bound exogenous enzyme could act on perfused TG and most of it rapidly returned to the perfusate on addition of heparin. Hence, binding was probably to the normal binding sites. No degradation of the enzyme was detected during a 1-h perfusion. A slow release of lipase back to the medium occurred even without heparin, as had been reported previously by BAGBY (1983). These data are in accord with continuous release of LPL from the heart endothelium into blood. Hence, both release by an extrahepatic tissue and extraction and degradation by the liver have been directly demonstrated. Whether this plasma–liver pathway accounts for most or all of LPL turnover is not known.

Heparan sulfate proteoglycans are present throughout the vascular tree. Hence, transport of LPL in blood would be expected to result in spreading

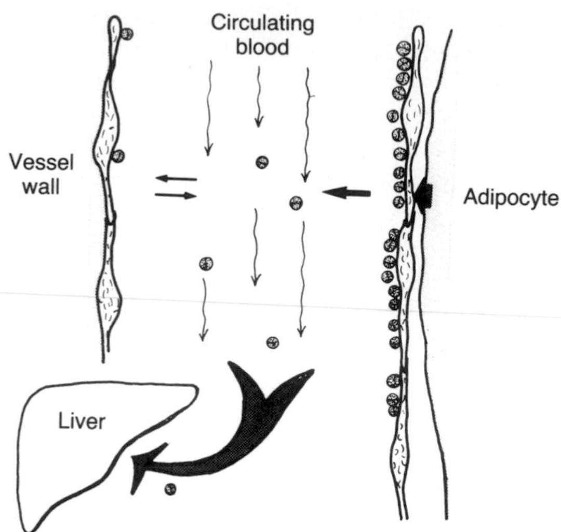

Fig. 4. Transport of lipoprotein lipase (LPL) in blood. The enzyme is produced in parenchymal cells, illustrated here by an adipocyte to the *right*. It is released from these cells and transferred to binding sites at the nearby vascular endothelium. From here LPL can dissociate into the circulating blood and bind to endothelial sites in other tissues, as illustrated in the *left* part of the figure. Avid uptake in the liver keeps the concentration of LPL low in the blood. From OLIVECRONA and BENGTSSON-OLIVECRONA (1987) with permission

of the enzyme to sites where there is no local synthesis. Evidence for such spreading was obtained in studies with immunofluorescence (CAMPS et al. 1990b). A particularly clear example were the glomeruli in the kidney. There are no adjacent LPL-producing cells, yet there was immunoreaction for LPL at the endothelial surface. Hence, LPL activity will depend not only on local production, but on the number and affinity of endothelial binding sites and on exchange with LPL in blood.

Another example of such transport comes from studies on HL. This enzyme is synthesized solely in the liver, but is found also in adrenals and ovaries. Careful studies have failed to demonstrate any synthesis of the enzyme or any mRNA for the enzyme in these tissues (DOOLITTLE et al. 1987; HIXENBAUGH et al. 1989). Hence, it is clear that HL must have been transported from the liver with blood and picked up by binding sites in the adrenals and ovaries.

K. Lipases in Pre- and Postheparin Plasma

Several studies have shown a correlation between plasma HL activity before and after heparin (OLIVECRONA 1990; ECKEL et al. 1988; OLIVECRONA et al. 1991a; YOUNG et al. 1992; GLASER et al. 1992). This suggests that plasma HL is in equilibrium with HL at tissue binding sites and that the effect of heparin is to shift this equilibrium towards soluble complexes in blood. In accord with this, a study where human HL was injected in mice showed that the enzyme rapidly became bound in the liver, from where it could be released again by heparin (PETERSON et al. 1986). Furthermore, HL is turned over relatively slowly; the $t_{1/2}$ has been estimated at 4.6-h in rats (SCHOONDERWOERD et al. 1981) and 3h in mice (PETERSON et al. 1986). Hence, relatively small amounts would be expected to turn over during the 15–60 min that are usually studied after heparin injection.

For LPL the situation is more complex. This enzyme is turned over more rapidly. The turnover time for endothelial LPL in heart has been estimated at 12 min (BAGBY 1983) and 40 min (CHAJEK-SHAUL et al. 1988a) in rats and 35–95 min in guinea pigs (LIU and OLIVECRONA 1992). There is production and net release of the enzyme from extrahepatic tissues and net extraction and degradation in the liver, and these processes are likely to be influenced by a number of physiological parameters. Therefore, the flow of LPL through plasma is not expected to be an equilibrium situation. In accord with this, most studies show no correlation between LPL activity in pre- and postheparin plasma (OLIVECRONA 1990; OLIVECRONA et al. 1991a; YOUNG et al. 1992; GLASER et al. 1992), although one early study claimed such a correlation (ECKEL et al. 1988).

Heart perfusions show a rapid burst of LPL activity emerging within the first 2 min after addition of heparin (BEN-ZEEV et al. 1981; BORENSZTAJN and ROBINSON 1970; CHAJEK-SHAUL et al. 1988a; LIU and OLIVECRONA 1992). This is presumably LPL that was directly exposed to the circulating blood. Continued perfusion with heparin was recently studied in guinea pigs (LIU and OLIVECRONA 1992). Over the next 20 min almost twice as much LPL activity emerged in the perfusate as had been released over the first 2 min. Hence, heparin was able to recruit relatively large amounts of LPL from sites not immediately exposed to blood. This may represent the pool of LPL located along cell surfaces in the tissue, as shown by electron microscopic immunostaining (BLANCHETTE-MACKIE et al. 1989), and perhaps LPL recycling in the endothelial cells (SAXENA et al. 1990). Translating this to the whole animal, we might expect to first see a quick rise of LPL activity after heparin injection, followed by a continued rise. This was seen in animals where the liver had been excluded from the circulation (EHNHOLM et al. 1980). In the intact animal, however, the liver extracts LPL even in the presence of heparin (WALLINDER et al. 1984). Hence postheparin LPL soon reaches a plateau value. This does not represent all the LPL that was exposed at endothelial sites, but represents a balance between release from peripheral tissues and extraction in liver.

Heparin is a polydisperse mixture of molecules varying in length and degree of sulfation. We have recently compared the effects of size-fractionated heparins on LPL (LIU et al. 1991b, 1992b). Shorter heparins were quite effective in releasing the enzyme from peripheral tissues, but relatively less efficient in retarding the uptake of the enzyme by the liver. Hence, plasma LPL activity fell off more rapidly after the shorter heparins. Preliminary studies indicate this can lead to a period of depletion of functional LPL, with impeded TG clearing ability.

L. Regulation

LPL is subject to regulation at several levels, and there is developmental regulation in many tissues (Chajek et al. 1977; Semenkovich et al. 1991; Kirchgessner et al. 1989a). For instance, LPL is expressed in fetal and neonatal liver (Vilaró et al. 1987), but then becomes suppressed (Staels and Auwerx 1992); LPL is an early marker during adipocyte development (Dani et al. 1990); macrophages express LPL on activation (Auwerx et al. 1989a); and in the mammary gland, LPL is normally low, but at parturition there is a dramatic increase (Scow and Chernick 1987).

There are also many examples of LPL regulation in response to physiological situations, e.g., feeding/fasting (Braun and Severson 1992), cold adaptation (Mitchell et al. 1992), exercise (Nikkilä 1987), in response to pathophysiological situations, e.g., trauma/shock/endotoxin (Bagby and Pekala 1987), and as part of disease processes, e.g., diabetes and obesity (Eckel 1989). In some of these instances, regulation is mainly at the mRNA level, e.g., cold adaptation (Mitchell et al. 1992), endotoxin/tumor necrosis factor (TNF; Fried and Zechner 1989; Cornelius et al. 1988). In most of the other situations, additional mechanisms seem to be at work. For adipose tissue there are changes in mRNA levels during feeding/fasting, but these changes do not fully explain the magnitude of the changes in LPL activity and they occur more slowly than the changes of LPL activity (for review see Braun and Severson 1992). An additional mechanism seems to be that newly synthesized LPL molecules are partitioned between transfer to the endothelium and intracellular degradation, as discussed above (see Sect. H). It is as if the cells often produce the enzyme in excess, e.g., during fasting, and then degrade most of the newly synthesized enzyme molecules. The purpose of this could be to enable the tissue to respond quickly to meals with temporary abundance of lipid.

Changes in tissue LPL activity are often most pronounced in the heparin-releasable fraction. This is assumed to be the fraction of the enzyme which is located at the capillary endothelium and serves to direct the flow of fatty acids from lipoprotein TG into the tissue (Eckel 1989; Braun and Severson 1992). This is often referred to as "functional LPL." Liu and Olivecrona (1992) have recently studied regulation of LPL activity in perfused guinea pig hearts. There was no difference in mRNA or in the rate of LPL synthesis, but about three times more LPL activity could be rapidly released by heparin in hearts from fasted compared to fed animals. Spontaneous release was also higher. Hence, in guinea pigs as in rats, functional LPL is

Table 3. Comparison of endogenous lipoprotein lipase activity in some rat tissues, with uptake of exogenous lipoprotein lipase by the tissues

	Uptake of 125I-LPL[a]	LPL activity (mU/g)
Liver	6.5	32
Heart	5.6	1200
Adipose tissue	0.45	1500

Previously unpublished data. ^{125}I-labeled lipoprotein lipase (LPL) was injected i.v. to anesthetized rats. Tissue radioactivity was measured 10 min later and was corrected for radioactivity in blood remaining in the tissue.
[a] Percent of dose per g tissue, 10 min after injection.

higher in the fasted state. When perfusion with heparin was continued, a steady level of release was obtained from about 20 min onward; this level did not differ between hearts from fed and fasted guinea pigs. In pulse-chase labeling experiments, there were no detectable differences in how newly synthesized LPL molecules were transported. The authors suggested that endothelial LPL activity in heart may be regulated by exchange with blood, rather than by changes in local production. Data from several laboratories indicate that plasma LPL activity is higher in the fasted than in the fed state (Benson et al. 1975; Chajek-Shaul et al. 1985c; Wallinder et al. 1984; Semb et al. 1987).

To study to what extent circulating LPL is bound by different tissues, labeled LPL has been injected (Wallinder et al. 1984; Hultin et al. 1992). Table 3 compares data on the uptake of labeled exogenous LPL by tissues to the endogenous LPL activity in these tissues. In the adipose tissue there was high LPL activity, but relatively low uptake of exogenous LPL. This is what we might expect from a tissue with high local synthesis of LPL, which is regulated, and a relatively slow blood flow. In the heart, on the other hand, there was high activity and high uptake. Here, the endothelial activity may be largely governed by exchange with the rapid flow of blood. The liver also showed avid uptake of exogenous LPL, but displayed very little LPL activity. The liver does not synthesize LPL, but extracts it from blood and degrades it (Vilaró et al. 1988; Chajek-Shaul et al. 1988b).

M. Lipoprotein Lipase as Ligand for Binding of Lipoproteins to Cells and Receptors

The model for the binding of LPL to heparan sulfate proteoglycans (Fig. 3) implies that the enzyme should be able to anchor lipoproteins to all cells that have heparan sulfate at their surface. This includes most cells in the body, except some blood cells. Interest has earlier focused mainly on the hydrolytic function. More recently, other aspects have been considered. The Steins and coworkers found that LPL dramatically enhanced transfer of cholesteryl esters to a variety of cells (FRIEDMAN et al. 1981). This did not depend on the hydrolytic action of the enzyme (STEIN et al. 1984), but required that the enzyme bound on one side to the cell surface and on the other side to the lipoprotein (CHAJEK-SHAUL et al. 1982). LPL has also been shown to mediate transfer of tocopherol from chylomicrons to cells (TRABER et al. 1985). More recently, direct evidence for LPL-mediated binding of lipoproteins to cell surface heparan sulfate has emerged from several laboratories (EISENBERG et al. 1992; MULDER et al. 1992; WILLIAMS et al. 1991, 1992). These actions occur with all types of lipoproteins (EISENBERG et al. 1992; MULDER et al. 1992) and even with liposomes prepared from nonhydrolyzable PL analogues (STEIN et al. 1984). Hence, the liganding function by which LPL links lipoprotein particles to cell surfaces appears to be independent of the hydrolytic function of the enzyme. It is not yet clear if this binding in itself leads to endocytosis and degradation of the LPL–lipoprotein complex, or if LPL mainly serves to guide the lipoprotein to the cell surface, where it then transfers to the appropriate receptors, e.g., the LDL receptor, for internalization.

BEISIEGEL et al. (1991) recently found that LPL also mediates binding of lipoproteins to the low-density lipoprotein related protein (LRP), the presumed receptor for chylomicron remnants (BROWN et al. 1991). This effect was not dependent on lipolysis, but appeared to be due to the lipase protein itself. By chemical cross-linking agents, direct interaction of the lipase and the receptor was demonstrated.

LPL thus appears to be a multifunctional protein, with properties of importance for several aspects of lipoprotein catabolism. The lipase is required for rapid unloading of TG from chylomicrons at endothelial binding/lipolysis sites. Lipase molecules adsorbed to the leaving particle may then signal that it is ready for receptor-mediated removal mechanisms.

N. Is Lipoprotein Lipase Rate-Limiting for Catabolism of Triglyceride-Rich Lipoproteins?

It is clear that LPL is needed for normal catabolism of TG-rich lipoproteins. The massive hypertriglyceridemia seen in patients deficient in LPL or in its activator, apolipoprotein C-II, attest to this (FOJO 1992). However, it is not so clear whether LPL is rate limiting in normal individuals.

Postheparin LPL activity is assumed to give a measure of LPL available at endothelial sites. A recent study showed an activity of 483 ± 180 mU/ml for a group of middle-aged, healthy, normolipidemic men (KARPE et al. 1992a). This assay was with a PL-stabilized TG emulsion at 25°C and pH 8.5. The activity would be higher at 37°C, but lower at pH 7.4; these factors roughly equal out. The activity is somewhat higher with (rat) chylomicrons than with the synthetic emulsion. Nonetheless, the value should give an estimate of the capacity for TG hydrolysis in vivo. For each TG molecule, two ester bonds are split by the enzyme. Therefore, the activity corresponds to clearing of roughly 250 nmol TG per ml and min, or about 40 g of TG per h if the plasma volume is 3 l. This is well above rates of TG transport after normal meals, suggesting that other processes are rate limiting for the entire process, e.g., lipid absorption. Even after peroral fat loads, this LPL activity should be enough for efficient clearing, but the margin is not large. Hence, we may expect to see a correlation between clearing capacity and postheparin LPL activity (or LPL activity measured in tissue biopsies). A few studies have shown an inverse relation between LPL activity and the magnitude of postprandial lipemia in healthy subjects (WEINTRAUB et al. 1987; PATSCH et al. 1987), but the correlations have been weak. A recent study on heterozygotes for LPL deficiency showed pronounced lipemia in response to a peroral fat load (MIESENBÖCK et al. 1992). In patients with coronary artery disease, on the other hand, LPL activity did not predict the lipemia (GROOT et al. 1991; KARPE et al. 1992a). Taken together, these studies indicate that low LPL activity can be a limiting factor in the clearing of large peroral fat loads, but is probably not limiting under normal conditions. The implication is that the amount of LPL, as a restriction, can be overridden by other factors. One possibility that has been raised is that TG transport may be limited more by the ability of the tissues to assimilate the products of lipolysis, than by the capacity for lipolysis as such (PETERSON et al. 1990; SNIDERMAN et al. 1992). A molecular coupling device could be the interaction of fatty acids with LPL (see above).

A related question is whether LPL is required for assimilation of lipids into extrahepatic tissues. Humans have very limited capacity for synthesizing fatty acids in adipose tissue. Hence, transport in lipoprotein TG and uptake through LPL is thought to be the major pathway by which adipose tissue acquires fatty acids. In rats, a strong correlation between LPL activity in adipose tissue and uptake of fatty acids from radiolabeled chylomicrons has been demonstrated (CRYER et al. 1976). However, BRUN et al. (1989) found no difference in total body fat content, or in the distribution of subcutaneous fat tissue, between patients with LPL deficiency and normal controls. The implication is that assimilation of fatty acids by adipose tissue depends less on the pathway of transport than on metabolic factors.

O. Impact on Lipoprotein Levels

Another aspect is the impact LPL activity has on the concentration and composition of lipoproteins in the postabsorptive and fasting states. Most studies on the relation between LPL activity and basal plasma TG levels have yielded weak or nonsignificant relations (TASKINEN 1987). In a recent study, 29 carriers of a defective LPL gene, defined by allele-specific probes, were compared to noncarriers in a pedigree of 126 members (WILSON et al.

1990). Carriers of the mutant allele often had a combination of increased plasma TG, VLDL cholesterol, and apoB, but decreased LDL and HDL cholesterol. The differences became significant only in individuals over 40 and in those with compouding diseases, e.g., obesity. Thus, almost all young carriers had normal TG levels, whereas two-thirds of carriers over 40 years of age had hypertriglyceridemia. In the general population, the most consistent correlation has been a positive relation between LPL activity and HDL cholesterol, particularly HDL_2 cholesterol (TASKINEN and KUUSI 1987). The molecular mechanisms behind this relation have been extensively discussed. One line of thought is based on the fact that surface components transfer from TG-rich lipoproteins to HDL as a result of LPL-mediated hydrolysis of core TG. Hence, efficient delipidation, as opposed to early receptor-mediated particle removal, would channel more PL, cholesterol, and apolipoproteins to HDL (TASKINEN and KUUSI 1987). Another line of thought stresses the role of cholesteryl ester transfer protein (CETP), which catalyzes homo- and heteroexchange of cholesteryl esters and TG between lipoproteins (BROWN et al. 1990; EISENBERG 1990; MIESENBÖCK and PATSCH 1992). Increased levels of TG-rich lipoproteins (basal or postprandial) would cause increased flow of cholesteryl esters from LDL and HDL into the TG-rich lipoproteins. On the other hand, TG transferred to LDL and HDL would be susceptible to hydrolysis by HL. The result would be a preponderance of small HDL (HDL_3) and small LDL (pattern B). Recent studies have in fact demonstrated an association between LPL activity and small dense LDL (MIESENBÖCK et al. 1992; KARPE et al. 1992b).

Many studies have demonstrated an inverse relation between HL activity and HDL cholesterol levels (TASKINEN and KUUSI 1987). The suggested underlying mechanisms follow the same lines of thought as for LPL. Based on the observation that HL acts on HDL, it has been argued that high HL activity leads to a generally enhanced catabolism of the particles (TASKINEN and KUUSI 1987). Alternatively, it has been suggested that the major impact of high HL activity is to enhance hydrolysis of TG transferred into the particles with CETP and hence drive depletion of HDL core lipids through exchange with TG-rich lipoproteins (DECKELBAUM et al. 1984; MIESENBÖCK and PATSCH 1992).

Whatever the mechanisms, many studies now indicate that low LPL activity and high HL activity alone, or in combination, can drive the lipoproteins towards an "atherogenic profile," probably imparting increased risk of cardiovascular disease for the individual. Both enzymes respond profoundly to physiological conditions such as hormones, diet, and exercise. It is becoming increasingly probable that imbalances in the lipases are at the root of many of the common derangements of lipoprotein composition and concentration.

Acknowledgement. Our studies on the lipases are supported by grant 13X-727 from the Swedish Medical Research Council.

References

Arner P, Lithell H, Wahrenberg H, Brönnegård M (1992) Expression of lipoprotein lipase in different human subcutaneous adipose tissue regions. J Lipid Res 32:423–429

Auwerx JH, Deeb SS, Brunzell JD, Wolfbauer G, Chait A (1989a) Lipoprotein lipase gene expression in THP-1 cells. Biochemistry 28:4563–4567

Auwerx JH, Marzetta CA, Hokanson JE, Brunzell JD (1989b) Large buoyant LDL-like particles in hepatic lipase deficiency. Arteriosclerosis 9:319–325

Bagby GJ (1983) Heparin-independent release of lipoprotein lipase from perfused rat hearts. Biochim Biophys Acta 753:47–52

Bagby GJ, Pekala PH (1987) Lipoprotein lipase in trauma and sepsis. In: Borensztajn J (ed) Lipoprotein lipase. Evener, Chicago, p 247

Beisiegel U, Weber W, Bengtsson-Olivecrona G (1991) Lipoprotein lipase enhances the binding of chylomicrons to low density lipoprotein receptor-related protein. Proc Natl Acad Sci USA 88:8342–8346

Ben-Zeev O, Schwalb H, Schotz MC (1981) Heparin-releasable and nonreleasable lipoprotein lipase in the perfused rat heart. J Biol Chem 256:10550–10554

Ben-Zeev O, Doolittle MH, Davis RC, Elovson J, Schotz MC (1992) Maturation of lipoprotein lipase. Expression of full catalytic activity requires glucose trimming but not translocation to the cis-Golgi compartment. J Biol Chem 267:6219–6227

Bengtsson G, Olivecrona T (1980a) The hepatic heparin-releasable lipase binds to high density lipoproteins. FEBS Lett 119:290–292

Bengtsson G, Olivecrona T (1980b) Lipoprotein lipase. Mechanism of product inhibition. Eur J Biochem 106:557–562

Bengtsson-Olivecrona G, Olivecrona T (1985) Binding of active and inactive forms of lipoprotein lipase to heparin: effects of pH. Biochem J 226:409–413

Bengtsson-Olivecrona G, Olivecrona T, Jörnvall H (1986) Lipoprotein lipase from cow, guinea-pig and man. Structural characterization and identification of protease-sensitive internal regions. Eur J Biochem 161:281–288

Bensadoun A (1991) Lipoprotein lipase. Annu Rev Nutr 11:217–237

Benson JD, Hearn V, Boyd T, Bensadoun A (1975) Triglyceride hydrolase of chicken and rat pre- and post-heparin plasma. Effects of fasting and comparison with adipose tissue lipoprotein lipase. Int J Biochem 6:727–734

Blanchette-Mackie EJ, Masuno H, Dwyer NK, Olivecrona T, Scow RO (1989) Lipoprotein lipase in myocytes and capillary endothelium of heart: immunocytochemical study. Am J Physiol 256:E818–E828

Borensztajn J (1987) Heart and skeletal muscle lipoprotein lipase. In: Borensztajn J (ed) Lipoprotein lipase. Evener, Chicago, p 133

Borensztajn J, Robinson DS (1970) The effect of fasting on the utilization of chylomicron triglyceride fatty acids in relation to clearing factor lipase (lipoprotein lipase) releasable by heparin in the perfused rat heart. J Lipid Res 11:111–117

Bownes M (1992) Why is there sequence similarity between insect yolk proteins and vertebrate lipases. J Lipid Res 33:777–790

Brady L, Brzozowski AM, Derewenda ZS, Dodson E, Dodson G, Tolley S, Turkenburg JP, Christiansen L, Huge-Jensen B, Norskov L, Thim L, Menge U (1990) A serine protease triad forms the catalytic centre of a triacylglycerol lipase. Nature 343:767–770

Braun JEA, Severson DL (1992) Regulation of the synthesis, processing and translocation of lipoprotein lipase. Biochem J 287:337–347

Breckenridge WC (1987) Deficiencies of plasma lipolytic activities. Am Heart J 113:567–573

Brown ML, Hesler C, Tall AR (1990) Plasma enzymes and transfer proteins in cholesterol metabolism. Curr Opin Lipidol 1:122–127

Brown MS, Herz J, Kowal RC, Goldstein JL (1991) The low-density lipoprotein receptor-related protein: double agent or decoy? Curr Opin Lipidol 2:65–72

Brun L-D, Gagné C, Julien P, Tremblay A, Moorjani S, Bouchard C, Lupien P-J (1989) Familial lipoprotein lipase-activity deficiency: study of total body fatness and subcutaneous fat tissue distribution. Metabolism 38:1005–1009

Brunzell JD, Iverius PH, Scheibel MS, Fujimoto WY, Hayden MR, McLeod R, Frolich J (1986) Primary lipoprotein lipase deficiency. Adv Exp Med Biol 201:227–239

Cai S-J, Wong DM, Chen S-H, Chan L (1989) Structure of the human hepatic triglyceride lipase gene. Biochemistry 28:8966–8971

Camps L, Gåfvels M, Reina M, Wallin C, Vilaró S, Olivecrona T (1990a) Expression of lipoprotein lipase in ovaries of the guinea pig. Biol Reprod 42:917–927

Camps L, Reina M, Llobera M, Vilaró S, Olivecrona T (1990b) Lipoprotein lipase: cellular origin and functional distribution. Am J Physiol 258:C673–C681

Camps L, Reina M, Llobera M, Bengtsson-Olivecrona G, Olivecrona T, Vilaró S (1991) Lipoprotein lipase in lungs, spleen, and liver: synthesis and distribution. J Lipid Res 32:1877–1888

Carroll R, Ben-Zeev O, Doolittle MH, Severson DL (1992) Activation of lipoprotein lipase in cardiac myocytes by glycosylation requires trimming of glucose residues in the endoplasmic reticulum. Biochem J 285:693–696

Chajek T, Stein O, Stein Y (1977) Rat heart in culture as a tool to elucidate the cellular origin of lipoprotein lipase. Biochim Biophys Acta 488:140–144

Chajek T, Stein O, Stein Y (1977) Pre- and post-natal development of lipoprotein lipase and hepatic triglyceride hydrolase activities in rat tissues. Atherosclerosis 26:549–561

Chajek-Shaul T, Friedman G, Stein O, Olivecrona T, Stein Y (1982) Binding of lipoprotein lipase to the cell surface is essential for the transmembrane transport of chylomicron cholesteryl ester. Biochim Biophys Acta 712:200–210

Chajek-Shaul T, Friedman G, Knobler H, Stein O, Etienne J, Stein Y (1985a) Importance of the different steps of glycosylation for the activity and secretion of lipoprotein lipase in rat preadipocytes studied with monensin and tunicamycin. Biochim Biophys Acta 837:123–134

Chajek-Shaul T, Friedman G, Stein O, Etienne J, Stein Y (1985c) Endogenous plasma lipoprotein lipase activity in fed and fasting rats may reflect the functional pool of endothelial lipoprotein lipase. Biochim Biophys Acta 837:271–278

Chajek-Shaul T, Bengtsson-Olivecrona G, Peterson J, Olivecrona T (1988a) Metabolic fate of rat heart endothelial lipoprotein lipase. Am J Physiol 255:E247–E254

Chajek-Shaul T, Friedman G, Ziv E, Bar-On H, Bengtsson-Olivecrona G (1988b) Fate of lipoprotein lipase taken up by the rat liver. Evidence for a conformational change with loss of catalytic activity. Biochim Biophys Acta 963:183–191

Cisar LA, Hoogewerf AJ, Cupp M, Rapport CA, Bensadoun A (1989) Secretion and degradation of lipoprotein lipase in cultured adipocytes. Binding of lipoprotein lipase to membrane heparan sulfate proteoglycans is necessary for degradation. J Biol Chem 264:1767–1774

Clarke AR, Luscombe M, Holbrook JJ (1983) The effect of the chain length of heparin on its interaction with lipoprotein lipase. Biochim Biophys Acta 747:130–137

Clay MA, Hopkins GJ, Ehnholm C, Barter PJ (1989) The rabbit as an animal model of hepatic lipase deficiency. Biochim Biophys Acta 1002:173–181

Cornelius, P. SE, Bjursell G, Olivecrona T, Pekala PH (1988) Regulation of lipoprotein lipase mRNA levels in 3T3-L1 cells by tumor necrosis factor. Biochem J 249:765–769

Cryer A, Riley SE, Williams ER, Robinson DS (1976) Effect of nutritional status on rat adipose tissue, muscle and post-heparin plasma clearing factor lipase activities: their relationship to triglyceride fatty acid uptake by fat-cells and to plasma insulin concentrations. Clin Sci Mol Med 50:213–221

Dani C, Amri E-Z, Bertrand B, Enerbäck S, Bjursell G, Grimaldi P, Ailhaud G (1990) Expression and regulation of pOb24 and lipoprotein lipase genes during adipose conversion. J Cell Biochem 43:103–110

Davis RC, Stahnke G, Wong H, Doolittle MH, Ameis D, Will H, Schotz MC (1990) Hepatic lipase: site directed mutagenesis of a serine residue important for catalytic activity. J Biol Chem 265:6291–6295

Deckelbaum RJ, Granot E, Oschry Y, Rose I, Eisenberg S (1984) Plasma triglyceride determines structure-composition in low and high density lipoproteins. Arteriosclerosis 4:225–231

Deckelbaum RJ, Ramakrishnan R, Eisenberg S, Olivecrona T, Bengtsson-Olivecrona G (1992) Triglyceride and phospholipid hydrolysis in human plasma lipoproteins: role of lipoprotein and hepatic lipase. Biochemistry 31:8544–8551

Demant T, Carlson LA, Holmquist L, Karpe F, Nilsson-Ehle P, Packard CJ, Shepherd J (1988) Lipoprotein metabolism in hepatic lipase deficiency: studies on the turnover of apolipoprotein B and on the effect of hepatic lipase on high density lipoprotein. J Lipid Res 29:1603–1611

Derewenda ZS, Cambillau C (1991) Effects of gene mutations in lipoprotein and hepatic lipases as interpreted by a molecular model of the pancreatic triglyceride lipase. J Biol Chem 266:23112–23119

Doolittle MH, Wong H, Davies RC, Schotz MC (1987) Synthesis of hepatic lipase in liver and extrahepatic tissues. J Lipid Res 28:1326–1334

Doolittle MH, Ben-Zeev O, Elovson J, Martin D, Kirchgessner TG (1990) The response of lipoprotein lipase to feeding and fasting. Evidence for posttranslational regulation. J Biol Chem 265:4570–4577

Eckel RH (1987) Adipose tissue lipoprotein lipase. In: Borensztajn J (ed) Lipoprotein lipase. Evener, Chicago, p 79

Eckel RH (1989) Lipoprotein lipase: a multifunctional enzyme relevant to common metabolic diseases. N Engl J Med 320:1060–1068

Eckel RH, Goldberg IJ, Steiner L, Yost J, Paterniti JR Jr (1988) Plasma lipolytic activity. Relationship to postheparin lipolytic activity and evidence for metabolic regulation. Diabetes 37:610–615

Ehnholm C, Schröder T, Kuusi T, Bång B, Kinnunen PKJ, Kahma K, Lempinen M (1980) Studies on the effect of hepatectomy on pig post-heparin plasma lipases. Biochim Biophys Acta 617:141–149

Eisenberg S (1990) Metabolism of apolipoproteins and lipoproteins. Curr Opin Lipidol 1:205–221

Eisenberg S, Sehayek E, Olivecrona T, Vlodavsky I (1992) Lipoprotein lipase enhances binding of lipoproteins to heparan sulfate on cell surfaces and extracellular matrix. J Clin Invest 90:2013–2021

Emmerich J, Beg OU, Peterson J, Previato L, Brunzell JD, Brewer HB Jr, Santamarina-Fojo S (1992) Human lipoprotein lipase. Analysis of the catalytic triad by site-directed mutagenesis of Ser-132, Asp-156, and His-241. J Biol Chem 267:4161–4165

Enerbäck S, Bjursell G (1989) Genomic organization of the region encoding guinea pig lipoprotein lipase; evidence for exon fusion and unconventional splicing. Gene 84:391–397

Faustinella F, Chang A, Van Biervliet JP, Rosseneu M, Vinaimont N, Smith LC, Chen S-H, Chan L (1991a) Catalytic triad residue mutation ($Asp^{156} \rightarrow Gly$) causing familial lipoprotein lipase deficiency. Co-inheritance with a nonsense mutation ($Ser^{447} \rightarrow Ter$) in a Turkish family. J Biol Chem 266:14418–14424

Faustinella F, Smith LC, Semenkovich CF, Chan L (1991b) Structural and functional roles of highly conserved serines in human lipoprotein lipase. Evidence that serine 132 is essential for enzyme catalysis. J Biol Chem 266:9481–9485

Faustinella F, Smith LC, Chan L (1992) Functional topology of a surface loop shielding the catalytic center in lipoprotein lipase. Biochemistry 31:7219–7223

Fojo SS (1992) Genetic dyslipoproteinemias: role of lipoprotein lipase and apolipoprotein C-II. Curr Opin Lipidol 3:186–195

Fried SK, Zechner R (1989) Cachectin/tumor necrosis factor decreases human adipose tissue lipoprotein lipase mRNA levels, synthesis, and activity. J Lipid Res 30:1917–1923

Friedman G, Chajek-Shaul T, Stein O, Olivecrona T, Stein Y (1981) The role of lipoprotein lipase in the assimilation of cholesteryl linoleyl ether by cultured cells incubated with labeled chylomicrons. Biochim Biophys Acta 666:156–164

Friedman G, Ben-Naim M, Halimi O, Etienne J, Stein O, Stein Y (1991) The expression of lipoprotein lipase activity and mRNA in mesenchymal rat heart cell cultures is modulated by bFGF. Biochim Biophys Acta Lipids Lipid Metab 1082:27–32

Gåfvels M, Vilaró S, Olivecrona T (1991) Lipoprotein lipase in guinea-pig adrenals: activity, mRNA, immunolocalization and regulation by ACTH. J Endocrinol 129:213–220

Giller T, Buchwald P, Blum-Kaelin D, Hunziker W (1992) Two novel human pancreatic lipase related proteins, hPLRP1 and hPLRP2. Differences in colipase dependence and in lipase activity. J Biol Chem 267:16509-16516

Glaser DS, Yost TJ, Eckel RH (1992) Preheparin lipoprotein lipolytic activities: relationship to plasma lipoproteins and postheparin lipolytic activities. J Lipid Res 33:209–214

Goldberg IJ, Soprano DR, Wyatt ML, Vanni TM, Kirchgessner TG, Schotz MC (1989) Localization of lipoprotein lipase mRNA in selected rat tissues. J Lipid Res 30:1569–1577

Groot PHE, vanStiphout WAHJ, Krauss XH, Jansen H, Van Tol A, Van Ramhorst E, Chin-On S, Hofman A, Cresswell SR, Havekes LM (1991) Postprandial lipoprotein metabolism in normolipidemic men with and without coronary artery disease. Arterioscler Thromb 11:653–662

Hamilton JA, Small DM (1981) Solubilization and localization of triolein in phosphatidylcholine bilayers: a carbon-13 NMR study. Proc Natl Acad Sci USA 78:6878–6882

Hide WA, Chan L, Li W-H (1992) Structure and evolution of the lipase superfamily. J Lipid Res 33:167–178

Hixenbaugh EA, Sullivan TRJ, Strauss JF, Laposata EA, Komaromy MC, Paavola LG (1989) Hepatic lipase in rat ovary. Ovaries cannot synthesize hepatic lipase but accumulate it from the circulation. J Biol Chem 264:4222–4230

Hoogewerf AJ, Cisar LA, Evans DC, Bensadoun A (1991) Effect of chlorate on the sulfation of lipoprotein lipase and heparan sulfate proteoglycans. Sulfation of heparan sulfate proteoglycans affects lipoprotein lipase degradation. J Biol Chem 266:16564–16571

Hultin M, Bengtsson-Olivecrona, G, Olivecrona T (1992) Release of lipoprotein lipase to plasma by triacylglycerol emulsions. Comparison to the effect of heparin. Biochim Biophys Acta Lipids Lipid Metab 1125:97–103

Hurtley SM, Helenius A (1989) Protein oligomerization in the endoplasmic reticulum. Annu Rev Cell Biol 5:277–307

Ikeda Y, Takagi A, Yamamoto A (1989) Purification and characterization of lipoprotein lipase and hepatic triglyceride lipase from human postheparin plasma: production of monospecific antibody to the individual lipase. Biochim Biophys Acta 1003:254–269

Jackson RL, Balasubramaniam A, Murphy RF, Demel RA (1986a) Interaction of synthetic peptides of apolipoprotein C-II and lipoprotein lipase at monomolecular films. Biochim Biophys Acta 875:203–210

Jackson RL, Tajima S, Yamamura T, Yokoyama S, Yamamoto A (1986b) Comparison of apolipoprotein C-II-deficient triacylglycerol-rich lipoproteins and trioleoylglycerol/phosphatidylcholine-stabilized particles as substrates for lipoprotein lipase. Biochim Biophys Acta 875:211–219

Jansen H, Hülsmann WC (1985) Enzymology and physiological role of hepatic lipase. Biochem Soc Trans 13:24–26

Jensen DR, Bessesen DH, Etienne J, Eckel RH, Neville MC (1991) Distribution and source of lipoprotein lipase in mouse mammary gland. J Lipid Res 32:733–742

Karpe F, Olivecrona T, Walldius G, Hamsten A (1992a) Lipoprotein lipase in plasma after an oral fat load: relation to free fatty acids. J Lipid Res 33:975–984

Karpe F, Tornvall P, Olivecrona T, Steiner G, Carlson LA, Hamsten A (1992b) Composition of human low density lipoproteins. Effects of postprandial triglyceride-rich lipoproteins, lipoprotein lipase, hepatic lipase and cholesteryl ester transfer protein. Atherosclerosis (in press)

Kern PA (1991) Lipoprotein lipase and hepatic lipase. Curr Opin Lipidol 2:162–169

Khoo JC, Mahoney EM, Witztum JL (1981) Secretion of lipoprotein lipase by macrophages in culture. J Biol Chem 256:7105–7108

Kirchgessner TG, Chaut JC, Heinzmann C, Etienne J, Guilhot S, Svenson K, Ameis D, Pilon C, D'Auriol L, Andalibi A, Schotz MC, Galibert F, Lusis AJ (1989a) Organization of the human lipoprotein lipase gene and evolution of the lipase gene family. Proc Natl Acad Sci USA 86:9647–9651

Kirchgessner TG, LeBoeuf RC, Langner CA, Zollman S, Chang CH, Taylor BA, Schotz MC, Gordon JI, Lusis AJ (1989b) Genetic and developmental regulation of the lipoprotein lipase gene: loci both distal and proximal to the lipoprotein lipase structural gene control enzyme expression. J Biol Chem 264:1473–1482

Kubo M, Matsuzawa Y, Yokoyama S, Tajima S, Ishikawa K, Yamamoto A, Tarui S (1982) Mechanism of inhibition of hepatic triglyceride lipase from human postheparin plasma by apolipoproteins A-l and A-ll. J Biochem (Tokyo) 92:865–870

Lawson DM, Brzozowski AM, Dodson GG (1992) Protein structures: lifting the lid off lipases. Curr Biol 2:473–475

Liu G, Olivecrona T (1992) Synthesis and transport of lipoprotein lipase in perfused guinea pig hearts. Am J Physiol (Heart Circ Physiol) 263:H438–H446

Liu G, Bengtsson-Olivecrona G, Ostergaard P, Olivecrona T (1991b) Low-M_r heparin is as potent as conventional heparin in releasing lipoprotein lipase, but is less effective in preventing hepatic clearence of the enzyme. Biochem J 273:747–752

Liu G, Bengtsson-Olivecrona G, Olivecrona T (1992a) Assembly of lipoprotein lipase in perfused guinea pig hearts. Biochem J 292:277–282

Liu G, Hultin M, Ostergaard P, Olivecrona T (1992b) Interaction of size-fractionated heparins with lipoprotein lipase and hepatic lipase in the rat. Biochem J 285:731–736

Lookene A, Bengtsson-Olivecrona G (1993) Chymotryptic cleavage of lipoprotein lipase. Identification of cleavage site and functional studies of the truncated molecule. Eur J Biochem 213:185–194

Ma Y, Bruin T, Tuzgol S, Wilson BI, Roederer G, Liu M-S, Davignon J, Kastelein JJP, Brunzell JD, Hayden MR (1992) Two naturally occurring mutations at the first and second bases of codon aspartic acid 156 in the proposed catalytic triad of human lipoprotein lipase. In vivo evidence that aspartic acid 156 is essential for catalysis. J Biol Chem 267:1918–1923

Masuno H, Blanchette-Mackie EJ, Schultz CJ, Spaeth AE, Scow RO, Okuda H (1992) Retention of glucose by N-linked oligosaccharide chains impedes expression of lipoprotein lipase activity: effect of castanospermine. J Lipid Res 33:1343–1349

McConathy WJ, Wang CS (1989) Inhibition of lipoprotein lipase by the receptor-binding domain of apolipoprotein E. FEBS Lett 251:250–252

McConathy WJ, Gesquiere JC, Bass H, Tartar A, Fruchart J-C, Wang CS (1992) Inhibition of lipoprotein lipase activity by synthetic peptides of apolipoprotein C-lll. J Lipid Res 33:995–1003

Melin T, Qi C, Bengtsson-Olivecrona G, Åkesson B, Nilsson Å (1991) Hydrolysis of chylomicron polyenoic fatty acid esters with lipoprotein lipase and hepatic lipase. Biochim Biophys Acta Gen Subj 1075:259–266

Mickel FS, Weidenbach F, Swarovsky B, LaForge KS, Scheele GA (1989) Structure of the canine pancreatic lipase gene. J Biol Chem 264:12895–12901

Miesenböck G, Patsch JR (1992) Postprandial hyperlipemia: the search for the atherogenic lipoprotein. Curr Opin Lipidol 3:196–201

Miesenböck G, Hölzl B, Föger B, Brandstätter E, Paulweber B, Sandhofer F, Patsch JR (1993) Heterozygous lipoprotein lipase deficiency due to a missense mutation as the cause of triglyceride intolerance with multiple lipoprotein lipoprotein abnormalities. J Clin Invest 91:448–455

Mitchell JRD, Jacobsson A, Kirchgessner TG, Schotz MC, Cannon B, Nedergaard J (1992) Regulation of expression of the lipoprotein lipase gene in brown adipose tissue. Am J Physiol (Endocrinol Metab) 263:E500–E506

Mulder M, Lombardi P, Jansen H, Van Berkel TJC, Frants RR, Havekes LM (1992) Heparan sulphate proteoglycans are involved in the lipoprotein lipase-mediated enhancement of the cellular binding of very low density and low density lipoproteins. Biochem Biophys Res Commun 185:582–587

Nikkilä EA (1987) Role of lipoprotein lipase in metabolic adaptation to exercise and training. In: Borensztajn J (ed) Lipoprotein lipase. Evener, Chicago, p 187

Nilsson Å, Landin B, Schotz MC (1987) Hydrolysis of chylomicron arachidonate and linoleate ester bonds by lipoprotein lipase and hepatic lipase. J Lipid Res 28:510–517

Nilsson-Ehle P, Garfinkel AS, Schotz MC (1976) Intra- and extracellular forms of lipoprotein lipase in adipose tissue. Biochim Biophys Acta 431:147–156

Olivecrona T (1990) Characterization and quantitation of lipases involved in lipoprotein metabolism. In: Lenfant C, Albertini A, Paoletti R, Catapano AL (eds) Biotechnology of dyslipoproteinemias. Application in diagnosis and control. Raven, New York, p 199

Olivecrona T, Bengtsson-Olivecrona G (1987) Lipoprotein lipase from milk – the model enzyme in lipoprotein lipase research. In: Borensztajn J (ed) Lipoprotein lipase. Evener, Chicago, p 15

Olivecrona T, Price SR, Pekala PH, Scow RO, Chernick SS, Semb H, Vilaró S, Bengtsson-Olivecrona G (1987) Regulation of lipoprotein lipase activity. Its role in lipid-lowering therapies. In: Paoletti R, Kritchevsky D, Holmes WL (eds) Drugs affecting lipid metabolism X. Springer, Berlin Heidelberg New York, p 88

Olivecrona T, Bengtsson-Olivecrona G, Hultin M (1991a) Lipases in lipoprotein metabolism. In: Shepherd J (ed) Lipoproteins and the pathogenesis of atherosclerosis. Elsevier, Amsterdam, p 51

Olivecrona T, Hopferweiser T, Patsch J, Bengtsson-Olivecrona G (1991b) Transport of lipoprotein lipase in plasma during alimentary lipemia. In: Winkler E, Greten H (eds) Hepatic endocytosis of lipids and proteins, Zuckschwerdt, München, p 288

Osborne JC Jr, Bengtsson-Olivecrona G, Lee N, Olivecrona T (1985) Studies on inactivation of lipoprotein lipase. Role of dimer to monomer dissociation. Biochemistry 24:5606–5611

Parkin SM, Walker K, Ashby P, Robinson DS (1980) Effect of glucose and insulin on the activation of lipoprotein lipase and on protein synthesis in rat adipose tissue. Biochem J 188:193–199

Patsch JR, Prasad S, Gotto AM Jr, Patsch W (1987) High density lipoprotein: 2. Relationship of the plasma levels of this lipoprotein species to its composition, to the magnitude of postprandial lipemia, and to the activities of lipoprotein lipase and hepatic lipase. J Clin Invest 80:341–347

Pelham HRB (1989) Control of protein exit from the endoplasmic reticulum. Annu Rev Cell Biol 5:1–23

Persson B, Bengtsson-Olivecrona G, Enerbäck S, Olivecrona T, Jörnvall H (1989) Structural features of lipoprotein lipase. Lipase family relationships, binding interactions, non-equivalence of lipase cofactors, vitellogenin similarities, and functional subdivisions of lipoprotein lipase. Eur J Biochem 179:39–45

Persson B, Jörnvall H, Olivecrona T, Bengtsson-Olivecrona G (1991) Lipoprotein lipases and vitellogenins in relation to the known three-dimensional structure of pancreatic lipase. FEBS Lett 288:33–36

Peterson J, Bengtsson-Olivecrona G, Olivecrona T (1986) Mouse preheparin plasma contains high levels of hepatic lipase with low affinity for heparin. Biochim Biophys Acta 878:65–70

Peterson J, Bihain BE, Bengtsson-Olivecrona G, Deckelbaum RJ, Carpentier YA, Olivecrona T (1990) Fatty acid control of lipoprotein lipase: a link between energy metabolism and lipid transport. Proc Natl Acad Sci USA 87:909–913

Pradines-Figuères A, Vannier C, Ailhaud G (1988) Short-term stimulation by insulin of lipoprotein lipase secretion in adipose cells. Biochem Biophys Res Commun 154:982–990

Pradines-Figuères A, Vannier C, Ailhaud G (1990) Lipoprotein lipase stored in adipocytes and muscle cells is a cryptic enzyme. J Lipid Res 31:1467–1476

Rojas C, Olivecrona T, Bengtsson-Olivecrona G (1991) Comparison of the action of lipoprotein lipase on triacylglycerols and phospholipids when presented in mixed liposomes or in emulson particles. Eur J Biochem 197:315–321

Saxena U, Goldberg IJ (1990) Interaction of lipoprotein lipase with glycosaminoglycans and apolipoprotein C-II: effects of free-fatty-acids. Biochim Biophys Acta 1043:161–168

Saxena U, Witte LD, Goldberg IJ (1989) Release of endothelial cell lipoprotein lipase by plasma lipoproteins and free fatty acids. J Biol Chem 264:4349–4355

Saxena U, Klein MG, Goldberg IJ (1990) Metabolism of endothelial cell-bound lipoprotein lipase. Evidence for heparan sulfate proteoglycan-mediated internalization and recycling. J Biol Chem 265:12880–12886

Saxena U, Klein MG, Goldberg IJ (1991) Transport of lipoprotein lipase across endothelial cells. Proc Natl Acad Sci USA 88:2254–2258

Schoonderwoerd K, Hülsmann WC, Jansen H (1981) Stabilization of liver lipase in vitro by heparin or by binding to non-parenchymal liver cells. Biochim Biophys Acta 665:317–321

Schrag JD, Li Y, Wu S, Cygler M (1991) Ser–His–Glu triad forms the catalytic site of the lipase from Geotrichum candidum. Nature 351:761–764

Scow RO, Chernick SS (1987) Role of lipoprotein lipase during lactation. In: Borensztajn J (ed) Lipoprotein lipase. Evener, Chicago, p 149

Scow RO, Egelrud T (1976) Hydrolysis of chylomicron phosphatidylcholine in vitro by lipoprotein lipase, phospholipase A2 and phospholipase C. Biochim Biophys Acta 431:538–549

Semb H, Olivecrona T (1987) Mechanisms for turnover of lipoprotein lipase in guinea-pig adipocytes. Biochim Biophys Acta 921:104–115

Semb H, Olivecrona T (1989) The relation between glycosylation and activity of guinea pig lipoprotein lipase. J Biol Chem 264:4195–4200

Semb H, Peterson J, Tavernier J, Olivecrona T (1987) Multiple effects of tumor necrosis factor on lipoprotein lipase in vivo. J Biol Chem 262:8390–8394

Semenkovich CF, Chen SH, Wims M, Luo CC, Li W-H, Chan L (1991) Lipoprotein lipase and hepatic lipase mRNA tissue specific expression, developmental regulation, and evolution. J Lipid Res 30:423–431

Severson DL, Lee M, Carroll R (1988) Secretion of lipoprotein lipase from myocardial cells isolated from adult rat hearts. Mol Cell Biochem 79:17–24

Sivaram P, Klein MG, Goldberg IJ (1992) Identification of a heparin-releasable lipoprotein lipase binding protein from endothelial cells. J Biol Chem 267:16517–16522

Smith LC, Pownall HJ (1984) Lipoprotein lipase. In: Borgström B, Brockman H (eds) Lipases. Elsevier, Amsterdam, p 263

Sniderman A, Baldo A, Cianflone K (1992) The potential role of acylation stimulating protein as a determinant of plasma triglyceride clearance and intracellular triglyceride synthesis. Curr Opin Lipidol 3:202–207

Spooner PM, Chernick SS, Garrison MM, Scow RO (1979) Insulin regulation of lipoprotein lipase activity and release in 3T3-L1 adipocytes. J Biol Chem 254:10021–10029

Staels B, Auwerx J (1992) Perturbation of developmental gene expression in rat liver by fibric acid derivatives: lipoprotein lipase and α-fetoprotein as models. Development 115:1035–1043

Stein O, Halperin G, Leitersdorf E, Olivecrona T, Stein Y (1984) Lipoprotein lipase mediated uptake of non-degradable ether analogues of phosphatidylcholine and cholesteryl ester by cultured cells. Biochim Biophys Acta 795:47–59

Stewart JE, Schotz MC (1974) Release of lipoprotein lipase activity from isolated fat cells. J Biol Chem 249:904–907

Tan MH, Sata T, Havel RJ (1977) The significance of lipoprotein lipase in rat skeletal muscles. J Lipid Res 18:363–370

Taskinen MR (1987) Lipoprotein lipase in hypertriglyceridemias. In: Borensztajn J (ed) Lipoprotein lipase. Evener, Chicago, p 201

Taskinen MR, Kuusi T (1987) Enzymes involved in triglyceride hydrolysis. In: Shepherd J (ed) Clinical endocrinology and metabolism. Lipoprotein metabolism. Saunders, London, pp 639–666

Traber MG, Olivecrona T, Kayden HJ (1985) Bovine milk lipoprotein lipase transfers tocopherol to human fibroblasts during triglyceride hydrolysis in vitro. J Clin Invest 75:1729–1734

van Tilbeurgh H, Sarda L, Verger R, Cambillau C (1992) Structure of the pancreatic lipase-colipase complex. Nature 359:159–162

Vannier C, Ailhaud G (1989) Biosynthesis of lipoprotein lipase in cultured mouse adipocytes. II. Processing, subunit assembly, and intracellular transport. J Biol Chem 264:13206–13216

Vannier C, Amri E-Z, Etienne J, Négrel R, Ailhaud G (1985) Maturation and secretion of lipoprotein lipase in cultured adipose cells: I. Intracellular activation of the enzyme. J Biol Chem 260:4424–4431

Vannier C, Etienne J, Ailhaud G (1986) Intracellular localization of lipoprotein lipase in adipose cells. Biochim Biophys Acta 875:344–354

Vannier C, Deslex S, Pradines-Figuères A, Ailhaud G (1989) Biosynthesis of lipoprotein lipase in cultured mouse adipocytes: I. Characterization of a specific antibody and relationships between the intracellular and secreted pools of the enzyme. J Biol Chem 264:13199–13205

Verger R, deHaas GH (1976) Interfacial enzyme kinetics of lipolysis. Annu Rev Biophys Bioeng 5:77–117

Vilaró S, Llobera M, Bengtsson-Olivecrona G, Olivecrona T (1987) Synthesis of lipoprotein lipase in the liver of newborn rats and localization of the enzyme by immunofluorescence. Biochem J 249:549–556

Vilaró S, Llobera M, Bengtsson-Olivecrona G, Olivecrona T (1988) Lipoprotein lipase uptake by liver: localization, turnover and metabolic role. Am J Physiol 254:G711–G722

Vilaró S, Camps L, Reina M, Perez-Clausell J, Llobera M, Olivecrona T (1990) Localization of lipoprotein lipase to discrete areas of the guinea pig brain. Brain Res 506:249–253

Wallinder L, Peterson J, Olivecrona T, Bengtsson-Olivecrona G (1984) Hepatic and extrahepatic uptake of intravenously injected lipoprotein lipase. Biochim Biophys Acta 795:513–524

Wang C-S, Hartsuck J, McConathy WJ (1992) Structure and functional properties of lipoprotein lipase. Biochim Biophys Acta Lipids Lipid Metab 1123:1–17

Weintraub MS, Eisenberg S, Breslow JL (1987) Different patterns of postprandial lipoprotein metabolism in normal, type IIa, type III, and type IV hyperlipoproteinemic individuals. J Clin Invest 79:1110–1119

Williams KJ, Petrie KA, Brocia RW, Swenson TL (1991) Lipoprotein lipase modulates net secretory output of apolipoprotein B in vitro. A possible pathophysiologic explanation for familial combined hyperlipidemia. J Clin Invest 88:1300–1306

Williams KJ, Fless GM, Petrie KA, Snyder ML, Brocia RW, Swenson TL (1992) Mechanisms by which lipoprotein lipase alters cellular metabolism of lipoprotein(a), low density lipoprotein, and nascent lipoproteins. Roles for low density lipoprotein receptors and heparan sulfate proteoglycans. J Biol Chem 267:13284–13292

Wilson DE, Emi M, Iverius P-H, Hata A, Wu LL, Hillas E, Williams RR, Lalouel J-M (1990) Phenotypic expression of heterozygous lipoprotein lipase deficiency in the extended pedigree of a proband homozygous for a missense mutation. J Clin Invest 86:735–750

Winkler FK, D'Arcy A, Hunziker W (1990) Structure of human pancreatic lipase. Nature 343:771–774

Wong H, Davis RC, Nikazy J, Seebart KE, Schotz MC (1991) Domain exchange: characterization of a chimeric lipase of hepatic lipase and lipoprotein lipase. Proc Natl Acad Sci USA 88:11290–11294

Yacoub LK, Vanni TM, Goldberg IJ (1990) Lipoprotein lipase mRNA in neonatal and adult mouse tissues: comparison of normal and combined lipase deficiency (cld) mice assessed by in situ hybridization. J Lipid Res 31:1845–1852

Young E, Prins M, Levine MN, Hirsh J (1992) Heparin binding to plasma proteins, an important mechanism for heparin resistance. Thromb Haemost 67:639–643

CHAPTER 7

Animal Models of Lipoprotein Metabolism

D. KRITCHEVSKY

A. Introduction

The first study involving purely nutritional influences on experimental atherosclerosis was carried out by IGNATOWSKI (1908) over 80 years ago. He was examining the effects of animal protein on atherosclerosis in rabbits. A few years later, ANITSCHKOW (1913) demonstrated the atherogenicity for rabbits of dietary cholesterol and established the role of that sterol in atherogenesis. Despite the fact that cholesterol was implicated in cardiovascular heart disease and the relative ease (if not absolute accuracy) of cholesterol determination and measurement of levels in plasma, cholesterol did not provide a totally satisfactory diagnosis nor a clue as to possible mechanism(s) of establishment of atherosclerosis. GOFMAN et al. (1950a) were able to demonstrate separation of a number of lipid–protein complexes by means of their flotation when plasma was subjected to ultracentrifugation in a medium of known density. The lipoproteins are water-soluble complexes which appear as spherical particles with the more hydrophobic lipids (triglycerides, cholesteryl esters) in the central core and cholesterol and phospholipids in surface monolayers. Over the 40 years since Gofman's discovery, the basic methodology has undergone some changes and the nomenclature has evolved from the Sf (Svedberg units of flotation) to the currently established chylomicrons, very low density (VLDL), intermediate-density (IDL), low-density (LDL), and high-density (HDL) lipoproteins. As the techniques of separation become more sophisticated, we find that the basic complexes (as defined by physical means) have different composition and may even be subfractionated further. THOMPSON (1989) has recently written an excellent short treatise on lipoproteinemias which discusses the chemistry as well as the metabolism of the lipoproteins. Tables 1 and 2 are constructed from data summarized by THOMPSON (1989) and SCANU (1991). The apolipoproteins are not equally distributed among the lipoprotein classes. Thus, the apoAs are found principally in HDL, whereas apoB is found in chylomicrons, VLDL, and LDL, but not in HDL. ApoC-I and II are present in chylomicrons and VLDL, but ApoC-III may be found in all the lipoprotein classes. ApoE is also found in all the lipoprotein classes and is the specific protein of apo (a). The apoproteins have distinct physiological functions including activation of lecithin cholesterol acyltransferase (LCAT) (ApoA-I, A-IV,

Table 1. Characteristics of plasma lipoprotein classes

Class	Density (g/ml)	Flotation Rate		Electrophoretic mobility	Diameter (nm)	Molecular weight × 10^6
		Sf 1.063	Sf 1.20			
Chylomicron	0.93	>400	–	Origin	75–1200	5–1000
VLDL	0.93–1.006	20–400	–	Pre-β	30–80	10–80
IDL	1.006–1.019	12–20	–	Pre-β	25–35	5–10
LDL	1.019–1.063	0–12	–	β	18–25	2.3
HDL_2	1.063–1.125	–	3.5–9.0	α1	9–12	0.36
HDL_3	1.125–1.210	–	0–3.5	α1	5–9	0.18

A variant of LDL called Lp(a) occurs in plasma at levels as high as 100 mg/dl. Its hydrated density is between 1.051–1.082 g/ml; its diameter is about 26 nm and it carries an extra protein, apo(a).
VLDL, very low density lipoprotein; IDL, intermediate-density lipoprotein; LDL, low-density lipoprotein; HDL, high-density protein.

Table 2. Chemical composition of lipoprotein classes

Class	Composition (%)		% of lipid		
	Protein	Lipid	Triglyceride	Cholesterol[a]	Phospholipid
Chylomicron	2	98	88	3 (46)	9
VLDL	10	90	56	17 (57)	19
IDL	18	82	32	41 (66)	27
LDL	25	75	7	59 (70)	28
HDL_2	40	60	6	43 (74)	42
HDL_3	55	45	7	38 (81)	41

VLDL, very low density lipoprotein; IDL, intermediate-density lipoprotein; LDL, low-density lipoprotein; HDL, high-density lipoprotein
[a] Numbers in parentheses represent percentage of cholesterol present as ester.

and C-I); ligand for LDL receptor (ApoB-100, ApoE); and activator of lipoprotein lipase (C-II), among others. The chromosomal location of each apoprotein has been identified and they are: chromosome 1 (apo-A-II); chromosome 2 (apoB); chromosome 11 (apoA-I, apoA-IV, apoC-III); and chromosome 19 (apoC-I, apoC-II, and apoE). The specific protein for apo (a) is located on chromosome 6.

B. Animal Models

Over the years attempts have been made to establish atherosclerosis in a large number of experimental animal species. This area was reviewed in 1975 (KRITCHEVSKY 1975), but there are several more recent reviews of the field (MUELLER and KIESSIG 1983; JOKINEN et al. 1985; VESSELINOVITCH 1988;

Table 3. Requirements for animal models used to study atherosclerosis (after VESSELINOVITCH 1988)

Readily available and inexpensive to buy
Inexpensive to house
Appropriate size
Reproduce in captivity
Develop lesion with minimal manipulation and in relatively short time
Experimental lesions not confounded by spontaneous disease
Resemble man in physiology and biochemistry
Lesions resemble those seen in man

WHITE 1989; ARMSTRONG and HIESTAD 1990). While it has been possible to establish atherosclerosis in a number of species, the means of establishing the lesion may be extreme and the applicability to man questionable. Studying spontaneous disease is often difficult because of the time involved, hence the investigator must weigh the evidence in deciding which species to use.

Among the practical factors, one must consider life span (ranging from 2 to 4 years in the rat to 20+ years in monkeys and swine); handling and housing, which are easy in the case of smaller species such as rat, rabbit, pigeons, and chickens and difficult for swine and monkeys. The initial cost and cost of maintenance are high for minipigs, pigs, and monkeys. Sampling (bleeding) is only easy in the rabbit and rat. VESSELINOVITCH (1988) has summarized the optimal requirements for animals used in atherosclerosis research (Table 3).

Historically, the rabbit is the animal of choice for atherosclerosis studies. It is very susceptible to dietary cholesterol and the lesion has been well characterized. Unfortunately, the rabbit is *too* susceptible and a cholesterol regimen will lead to very high plasma cholesterol levels and cholesterol storage in many organs. WILSON et al. (1982) have produced lesions resembling those of humans in rabbits by feeding them a diet similar to the American diet for 5 years. It is also possible to produce atherosclerosis in rabbits by feeding a semipurified diet containing saturated fat, animal protein, and cellulose (KRITCHEVSKY and TEPPER 1965; 1968). This process leads to reasonable elevations of cholesterol and requires 8–12 months. The Watanabe rabbit (WATANABE 1980) is an inbred strain with elevated LDL. It serves as a model for familial hypercholesterolemia (BUJA et al. 1983). The St. Thomas strain of rabbit exhibits elevated levels of VLDL, IDL, and LDL and develops lesions when fed a normal diet (SEDDON et al. 1987).

Avian models have been used for several decades. Chickens are susceptible to cholesterol-rich diets and their aortic lesions contain foam cells (DAUBER and KATZ 1943). Atherosclerosis can also be induced in chickens by infecting them with the herpes virus of Marek's disease (FABRICANT et al. 1983). Japanese quail are susceptible to cholesterol feeding and they have been used for drug screening (CHAPMAN et al. 1976). The Japanese quail lesion resembles the human lesion more closely than does that seen in chickens (VESSELINOVITCH 1988).

The White Carneau pigeon develops atherosclerosis naturally (CLARKSON et al. 1965). The lesions are very similar to those of man (PRICHARD et al. 1964) and can be exacerbated by feeding cholesterol. Other breeds of pigeon, the Show Racer, for instance, are more resistant to experimentally induced atherosclerosis and can be studied for comparison (LOFLAND and CLARKSON 1960). The use of pigeons in atherosclerosis research was pioneered by the Prichard–Lofland–Clarkson group in Winston-Salem, North Carolina, and has been studied extensively by them. They have succeeded in breeding birds that are hyper- or hyporesponsive to dietary cholesterol and show varying severities of atherosclerosis (WAGNER et al. 1973; WAGNER and CLARKSON 1974). The difficulty with using avian models of atherosclerosis is that they are not mammals and thus it is difficult to relate results obtained in birds with the condition in man.

The pig is an excellent model for the study of atherosclerosis (RATCLIFFE and LUGINBUHL 1971; CEVALLOS et al. 1979; FRITZ et al. 1980). This animal exhibits naturally occurring lesions which can be exacerbated by feeding diets high in fat and cholesterol (REITMAN et al. 1982). Pigs have lipoprotein levels similar to those of man, and cerebral and myocardial infarction occur in this species. However, the pig is difficult to handle and expensive to keep.

Nonhuman primates are phylogenetically close to man and thus may be the most appropriate model for experimental atherosclerosis. The squirrel monkey (*Saimiri sciureus*) has been studied extensively (CLARKSON et al. 1976). It exhibits spontaneous atherosclerosis and is susceptible to diet-induced arterial disease. There are hyper- and hyporesponders to dietary cholesterol (CLARKSON et al. 1971). The drawbacks to using squirrel monkeys include endogenous arteritis and a tendency to develop chronic renal disease. This species was studied extensively about 20 years ago, but has not been used much recently. The rhesus monkey (*Macaca mulatta*) has been used in atherosclerosis research since the seminal publications of TAYLOR et al. (1962, 1963a,b), who reported that cholesterol feeding led to atherosclerosis of the aorta, carotid, iliac, and coronary arteries. Myocardial infarctions have been found to occur in cholesterol-fed rhesus monkeys (BOND et al. 1980). Studies in this species have waned since export from India has been banned.

JOKINEN et al. (1985) believe that the cynomolgus monkey (*Macaca fascicularis*) is one of the best primate models for atherosclerosis. They develop lesions whose morphology resembles that seen in humans, show male–female differences, and exhibit myocardial infarction. They are difficult to handle and maintain, however. Stumptail macaques (*Macaca arctoides*) and pigtail macaques (*Macaca nemestrina*) are also susceptible to cholesterol-induced aortic lesions, but they have not been studied to the same extent as rhesus or cynomolgus monkeys.

Table 4. Influence of dietary cholesterol on atherosclerosis in various monkey strains

Species	*n*	Serum cholesterol (mg/dl)	Fatty streaks (%)	Raised lesions (%)	Aortic F/E cholesterol[a]
Control	–	–	8 ± 2	<1	3.40
Squirrel	8	474 ± 54	64 ± 8	25 ± 9	2.50
Cebus	4	461	45	0	–
Stumptail	4	698 ± 33	8 ± 4	79 ± 13	1.00
Pigtail	6	810 ± 55	0	82 ± 8	0.80
Rhesus	20	627 ± 29	44 ± 9	32 ± 7	1.11
Cynomolgus	7	701 ± 47	13 ± 6	80 ± 10	1.19
Vervet	4	467 ± 27	38 ± 3	26 ± 9	2.94

F/E, Ratio of free to esterified cholesterol.
[a] After WILLIAMS and CLARKSON (1989).

The African green monkey (*Cercopithecus aethiops*) is becoming a popular model for atherosclerosis research. The morphology and distribution of its lesions resemble those of man (BULLOCK et al. 1975; WAGNER and CLARKSON 1975). This monkey is susceptible to dietary-induced atherosclerosis (KRITCHEVSKY et al. 1977; FINCHAM et al. 1987a,b). The baboon would appear to be a good candidate for studies of atherosclerosis. It is abundant and of good size, but difficult to handle. The baboon exhibits naturally occurring arterial lesions (MCGILL et al. 1960), but is resistant to dietary manipulation, usually showing few plaques after long-term administration of high-fat, high-cholesterol diets (STRONG and MCGILL 1967; STRONG et al. 1976). Baboons fed a semipurified diet containing 0.1% cholesterol for 1 year were found to develop severe atherosclerosis (KRITCHEVSKY et al. 1980).

WILLIAMS and CLARKSON (1989) have summarized the effects of cholesterol feeding on cholesterolemia and atherosclerosis in a number of monkey strains (Table 4). The table illustrates the ranges of response to cholesterol diet which one can expect.

C. Lipoproteins

I. Rabbit

Shortly after their introduction of the ultracentrifugal methodology, which opened up the lipoprotein field, GOFMAN et al. (1950b) recognized the presence of subpopulations of lipoproteins and suggested that some of those populations might possess special atherogenic potential. This heterogeneity has been recognized by other, later, investigators, and those who have the analytical capabilities may capitalize on these observations. The thrust of many of the studies is aimed at elucidating differences in LDL size caused by differences in the core lipid triglyceride and cholesteryl ester content. Unlike man, the diabetic rabbit exhibits a reduced incidence of atherosclerosis (DUFF and PAYNE 1950; McGILL and HOLMAN 1949). This protection is reversed when the alloxan-diabetic rabbit is treated with insulin (DUFF et al. 1954). It has been postulated that the large lipoprotein molecules seen in the diabetic rabbit are incapable of entering the aortic wall (NORDESTGAARD et al. 1987). Zilversmit and colleagues (NORDESTGAARD and ZILVERSMIT 1988; MINNICH and ZILVERSMIT 1989; MINNICH et al. 1989; NORDESTGAARD and ZILVERSMIT 1989) investigated this phenomenon. Their work suggests that the mechanisms for removal of triglyceride molecules are impaired in hypertriglyceridemic, diabetic, cholesterol-fed rabbits, resulting in fewer LDL molecules and the presence of large, cholesterol-rich lipoproteins which are unable to cross the endothelial barrier. Previously, STENDER and ZILVERSMIT (1981) had demonstrated that arterial incorporation of lipoprotein molecules is inversely proportional to their diameter. KANAZAWA et al. (1991) injected LDL from normocholesterolemic rabbits or saline into cholesterol-fed recipient rabbits and found the latter treatment to be more atherogenic. In rabbits fed 0.5% cholesterol, the VLDL levels of the saline-injected rabbits were almost double those of the rabbits given LDL; the differences were not evident when the diet contained 1% cholesterol. The authors do not attribute their observations to the altered LDL size. DAUGHERTY et al. (1988) found that cholesterol-fed rabbits elaborated intestinally derived and hepatically derived cholesteryl ester rich VLDL.

II. Monkeys

PRONCZUK et al. (1991) examined long-term (8–12 years) effects of diets rich in corn oil or coconut oil on plasma lipids in rhesus, cebus, and squirrel monkeys. The hypercholesterolemic response to coconut oil was similar in the three groups, but the ratio of LDL cholesterol to HDL cholesterol was highest in the rhesus monkeys and lowest in the cebus monkeys. Further investigation of the influences of dietary fat on lipid metabolism was carried out in these three monkey species (HAYES et al. 1991). In these studies, the investigators used fat blends instead of single fats. This is a welcome change in the conduct of this type of dietary experiment, since the fat blends are more representative of the fat content of the normal diet. Results obtained by exchanging all of one fat for all of another do not reflect normal nutritional practices. In this study rhesus, cebus, and squirrel monkeys were fed diets containing 31% of calories as fat with the fats blended to have different polyene to saturated fatty acid (P:S) ratios and different ratios of palmitic to myristic plus lauric acid. The fats consisted of blends of coconut, palm, soybean, and high-oleic safflower oils. The results are summarized in Table 5. The responses of the three species (which were rotated through all five

Table 5. Influence of mixed fats on plasma lipids in monkeys (after HAYES et al. 1991)

	1	2	3	4	5
Iodine Value	26	66	60	64	93
C12:0 + C:14 (%)	66.6	33.4	19.2	1.2	1.1
C16:0 (%)	10.7	8.6	25.1	40.3	23.4
Cholesterol (mg/dl)					
Total	232 ± 10	205 ± 11	203 ± 10	183 ± 9	173 ± 9
LDL	110 ± 8	92 ± 8	87 ± 7	79 ± 6	68 ± 5
HDL	102 ± 6	99 ± 4	96 ± 6	86 ± 6	89 ± 4
LDL/HDL	47.6 ± 4.3	36.8 ± 3.1	37.9 ± 3.4	34.4 ± 2.7	27.8 ± 1.9
ApoB/ApoAI	0.69 ± 0.05	0.61 ± 0.06	0.67 ± 0.05	0.58 ± 0.06	0.56 ± 0.06

Combined data from 21 monkeys (eight rhesus, eight cebus, five squirrel). Animals rotated through all five diets in 12-week periods.
LDL, low-density lipoprotein; HDL, high-density lipoprotein.

diets in 12-week periods) were roughly equivalent, so they have been combined. The data show that replacement of the lauric and myristic acids by palmitic acid led to a decrease in plasma cholesterol and in LDL/HDL cholesterol and apoB/apo-A-I ratios. The data support Hegsted's suggestion that myristic acid was responsible for two-thirds of the hypercholesterolemic effect of saturated fats (HEGSTED et al. 1965). Later studies in rhesus monkeys (KHOSLA and HAYES 1991) have suggested that the difference in cholesterolemic effect between fats rich in lauric plus myristic acids and those rich in palmitic acid is that the former doubles direct production of LDL apoB and increases the LDL apoB pool.

In monkeys, plasma levels of large cholesteryl ester rich LDL are correlated positively with severity of atherosclerosis (RUDEL et al. 1985). MARZETTA and RUDEL (1986) found that LDL of cynomolgus monkeys fed atherogenic diets were larger and less dense than those of vervet monkeys fed the same diet. The molecular weight of the cynomolgus LDL was 4.10 ± 0.45 g/μmol compared to 3.46 ± 0.44 g/μmol for African green monkey LDL. The average total plasma cholesterol for the two groups of monkeys was practically the same, but the cynomolgus monkeys carried 6% more cholesterol in their LDL and 39% less in their HDL than did the African green monkeys. This difference might account for the greater susceptibility to atherogenic diets of the cynomolgus monkey. TALL et al. (1978) reported that the difference between large and small LDL lies in the physical state of the cholesteryl ester core. When monkeys are fed saturated fat, the increase in saturated and monounsaturated fatty acid esters of cholesterol parallels the increase in LDL molecular weight. Similar changes appear in LDL of monkeys fed unsaturated fat but the presence of poly-unsaturated fatty acid esters of cholesterol keep the transition temperatures of LDL below body temperature. There are also data which suggest that less dense, cholesteryl ester rich LDL bind preferentially to the proteoglyans of the arterial wall (CAMEJO 1982; CAMEJO et al. 1976).

When cebus and squirrel monkeys were fed saturated (coconut oil) or unsaturated (corn oil) fat, they developed comparable levels of plasma cholesterol, but there were vast differences in the mode of cholesterol transport. The squirrel monkey expanded its LDL pool whereas in the cebus monkey the HDL pool was increased. The cebus monkeys did not accumulate stainable aortic lipid on either diet (NICOLOSI et al. 1977).

There is little argument that LDL is the atherogenic lipoprotein class in experimental animals as well as in man. But just as earlier studies of serum cholesterol levels failed to provide totally satisfactory evidence of its atherogenic potential, we now find that mere analytical identification of LDL levels is insufficient to explain all of its atherogenic effects. While LDL is usually derived from VLDL, some is secreted directly into the circulation. This mode of LDL production has been observed in man (SOUTAR et al. 1977), cynomolgus monkeys (GOLDBERG et al. 1983), and squirrel monkeys (ILLINGWORTH 1975). Thus, the direction of future research regarding lipoprotein metabolism in atherosclerosis aimed at elucidating the steps involved in atherogenesis will involve further investigation of LDL size and composition. An excellent update review on lipoproteins and atherosclerosis was published by BABIAK and RUDEL (1987).

One further aspect of lipid nutriture and its role in lipoprotein metabolism is exemplified by the recent studies of LINDSEY et al. (1990) and HAYES and KHOSLA (1992). These authors have shown how specific fatty acids may influence LDL and HDL production. These findings will eventually be integrated with the data on lipoprotein size and composition to clarify the precise roles of the lipoproteins in atherosclerosis.

D. Conclusion

The field of experimental atherosclerosis which stemmed from the studies of Ignatowski and Anitschkow has shown a steady, healthy development. We have established a number of animal models which have each served to advance knowledge. There are several published in-depth reviews and a brief review in this essay detailing the strengths and weaknesses of each model. One area of risk for the development of atherosclerosis, namely cholesterol, has commanded our attention for decades. From examination of levels of blood cholesterol, we advanced to plasma lipoproteins and are now examining further the lipid and protein composition of those macromolecules. We have seen that aortic uptake of LDL depends not only on its physical properties (i.e., hydrated density), but also on its content of triglycerides and cholesteryl esters and possibly on the fatty acid moieties of the esters. The other side of the uptake equation, the reactions at the aortic surface, is also being studied. Those studies are more difficult to carry out, but they, too, are progressing. Eventually, we will learn precisely the combination of conditions of aortic metabolism and blood lipid composition that initiate atherosclerosis.

References

Anitschkow N (1913) Über die Veränderungen der Kaninchenaorta bei experimentelle Cholesterinsteatose. Beitr Pathol Anat 56:379–404

Armstrong ML, Heistad DD (1990) Animal models of atherosclerosis. Atherosclerosis 85:15–23

Babiak J, Rudel LL (1987) Lipoproteins and atherosclerosis. Ballieres Clin Endocrinol Metab 1:515–550

Bond MG, Bullock BC, Bellinger DA, Hamm TE (1980) Myocardial infarction in a large colony of nonhuman primates with coronary artery atherosclerosis. Am J Pathol 101:675–692

Buja LM, Kita T, Goldstein JL, Watanabe Y, Brown MS (1983) Cellular pathology of progressive atherosclerosis in the WHHL rabbit. An animal model of familial hypercholesterolemia. Arteriosclerosis 3:87–101

Bullock BC, Lehner NDM, Clarkson TB, Feldner MA, Wagner WD, Lofland HB (1975) Comparative primate atherosclerosis: I. Tissue cholesterol concentrations and pathologic anatomy. Exp Mol Pathol 22:151–175

Camejo G (1982) The interaction of lipids and lipoproteins with the intracellular matrix of arterial tissue: its possible role in atherogenesis. Adv Lipid Res 19:1–53

Camejo G, Mateu L, Lalaguna F, Padron R, Waich S, Acquatella H, Vega H (1976) Structural individuality of human serum LDL associated with differential affinity for a macromolecular component of the arterial wall. Artery 2:79–97

Cevallos WA, Holmes WL, Myers RN, Smink DD (1979) Swine in atherosclerosis research: development of an experimental animal model and study of the effect of dietary fats on cholesterol metabolism. Atherosclerosis 34:303–317

Chapman KP, Stafford WW, Day CE (1976) Animal model for experimental atherosclerosis produced by selective breeding of Japanese quail. In: Day CE (ed) Atherosclerosis drug discovery. Plenum, New York, pp 347–356

Clarkson TB, Middleton CC, Prichard RW, Lofland HB Jr (1965) Naturally occuring atherosclerosis in Birds. Ann NY Acad Sci 127:685–693

Clarkson TB, Lofland HB Jr, Bullock BC, Goodman HO (1971) Genetic control of plasma cholesterol. Studies on squirrel monkeys. Arch Pathol 92:37–45

Clarkson TB, Lehner NDM, Bullock BC, Lofland HB, Wagner WD (1976) Atherosclerosis in New World monkeys. Primates Med 9:90–144

Dauber DV, Katz LN (1943) Experimental atherosclerosis in the chick. Arch Pathol 36:473–492

Daugherty A, Oida K, Sobel BE, Schonfeld G (1988) Dependence of metabolic and structural heterogeneity of cholesterol ester-rich very low density lipoproteins on the duration of cholesterol feeding in rabbits. J Clin Invest 82:562–570

Duff GL, Payne TPB (1950) The effect of alloxan diabetes on experimental cholesterol atherosclerosis in the rabbit: III. The mechanism of the inhibition of experimental cholesterol atherosclerosis in alloxan-diabetic rabbits. J Exp Med 92:299–371

Duff GL, Brechin DJH, Finkelstein WE (1954) The effect of alloxan diabetes on experimental cholesterol atherosclerosis in the rabbit: IV. The effect of insulin therapy on the inhibition of atherosclerosis in the alloxan-diabetic rabbit. J Exp Med 100:371–380

Fabricant CG, Fabricant J, Minick CR, Litrenta MM (1983) Herpes virus-induced atherosclerosis in chickens. Fed Proc 42:2476–2479

Fincham JE, Faber M, Weight MJ, Labadarios D, Taljaard JJF, Steyher JG, Jacobs P, Kritchevsky D (1987a) Diets realistic for westernised people significantly affect lipoproteins, calcium, zinc, vitamins C, E, B_6 and hematology in Vervet monkeys. Atherosclerosis 66:191–203

Fincham JE, Woodruff CW, van Wyk MJ, Capatos D, Weight MJ, Kritchevsky D, Rossouw JE (1987b) Promotion and regression of atherosclerosis in Vervet monkeys by diets realistic for westernised people. Atherosclerosis 66:205–213

Fritz KE, Daoud AS, Augustyn JM, Jarmolych J (1980) Morphological and biochemical differences among grossly defined types of swine aortic atherosclerotic lesions induced by a combination of injury and atherogenic diet. Exp Mol Pathol 32:61–72

Gofman JW, Lindgren FT, Elliott HA, Mantz W, Hewitt J, Strisower B, Herring V, Lyon TP (1950a) The role of lipids and lipoproteins in atherosclerosis. Science 111:166–186

Gofman JW, Jones HB, Lindgren FT, Lyon TP, Elliott HA, Strisower BA (1950b) Blood lipids in human atherosclerosis. Circulation 11:161–178

Goldberg IJ, Le NA, Ginsberg HN, Paterniti JR Jr, Brown WV (1983) Metabolism of apoprotein B in cynomolgus monkeys. Evidence for independent production of low density lipoprotein apoprotein B. Am J Physiol 244:E196–E201

Hayes KC, Khosla P (1992) Dietary fatty acid thresholds and cholesterolemia. FASEB J 6:2600–2607

Hayes KC, Pronczuk A, Lindsey S, Diersen-Schade D (1991) Dietary saturated fatty acids (12:0, 14:0, 16:0) differ in their impact on plasma cholesterol and lipoproteins in non-human primates. Am J Clin Nutr 53:491–498

Hegsted DM, McGandy RB, Myers ML, Stare FJ (1965) Quantitative effects of dietary fat on serum cholesterol in man. Am J Clin Nutr 17:281–295

Ignatowski A (1908) Influence de la nourriture animale sur l'organisme des lapins. Arch Med Exp Anat Pathol 20:1–20

Illingworth DR (1975) Metabolism of lipoproteins in nonhuman primates. Studies on the origin of low density lipoprotein apoprotein in the plasma of the squirrel monkey. Biochim Biophys Acta 388:38–51

Jokinen MP, Clarkson TB, Prichard RW (1985) Animal models in atherosclerosis research. Exp Mol Pathol 42:1–28

Kanazawa T, Tanaka M, Fukushi Y, Onodera K, Lee KT, Metoki H (1991) Effects of low-density lipoprotein from normal rabbits on hypercholesterolemia and the development of atherosclerosis. Pathobiology 50:85–91

Khosla P, Hayes KC (1991) Dietary fat saturation in rhesus monkeys affects LDL concentrations by modulating the independent production of LDL apoliprotein B. Biochim Biophys Acta 1083:46–56

Kritchevsky D (1975) Animal models for atherosclerosis research. In: Kritchevsky D (ed) Hypolipidemic agents. Springer, Berlin Heidelberg New York, pp 216–227

Kritchevsky D, Tepper SA (1965) Factors affecting atherosclerosis in rabbits fed cholesterol-free diets. Life Sci 4:1467–1471

Kritchevsky D, Tepper SA (1968) Experimental atherosclerosis in rabbits fed cholesterol-free diets: influence of chow components. J Atheroscler Res 8:357–369

Kritchevsky D, Davidson LM, Kim HK, Krendel DA, Malhotra SM, Vander Watt JJ, du Plessis JP, Winter PAD, Ipp T, Mendelsohn D, Bersohn I (1977) Influence of semipurified diets on atherosclerosis in African Green monkeys. Exp Mol Pathol 26:28–51

Kritchevsky D, Davidson LM, Kim HK, Krendel DA, Malhotra S, Mendelsohn D, Vander Watt JJ, du Plessis JP, Winter PAD (1980) Influence of type of carbohydrate on atherosclerosis in baboons fed semipurified diets plus 0.1% cholesterol. Am J Clin Nutr 33:1869–1887

Lindsey S, Benattar J, Pronczuk A, Hayes KC (1990) Dietary palmitic acid (16:0) enhances high density lipoprotein cholesterol and low density lipoprotein receptor mRNA abundance in hamsters. Proc Soc Exp Biol Med 195:261–269

Lofland HB, Clarkson TB (1960) Serum lipoproteins in atherosclerosis-susceptible and resistant pigeons. Proc Soc Exp Biol Med 103:236–241

Marzetta CA, Rudel LL (1986) A species comparison of low density lipoprotein in nonhuman primates fed atherogenic diets. J Lipid Res 27:753–762

McGill HC Jr, Holman RL (1949) The influence of alloxan diabetes on cholesterol atheromatosis in the rabbit. Proc Soc Exp Biol Med 72:72–75

McGill HC Jr, Strong JP, Holman RL, Werthessen NT (1960) Arterial lesions in the Kenya baboon. Circ Res 8:670–679

Minnich A, Zilversmit DB (1989) Impaired triglycerol catabolism in hypertriglyceridemia of the diabetic, cholesterol-fed rabbit: a possible mechanism for protection from atherosclerosis. Biochim Biophys Acta 1002:324–332

Minnich A, Nordestgaard BG, Zilversmit DB (1989) A novel explanation for the reduced LDL cholesterol in severe hypertriglyceridemia. J Lipid Res 30:347–355

Müller G, Kiessig R (1983) Tiermodelle in der atheroskleroseforschung. Dtsch Gesmndheitswesen 38:601–609, 646–651

Nicolosi RJ, Hojnacki JL, Llansa N, Hayes KC (1977) Diet and lipoprotein influence on primate atherosclerosis. Proc Soc Exp Biol Med 156:1–7

Nordestgaard BG, Zilversmit DB (1988) Hyperglycemia in normotriglyceridemic, hypercholesterolemic insulin-treated diabetic rabbits does not accelerate atherogenesis. Atherosclerosis 72:37–47

Nordestgaard BG, Zilversmit DB (1989) Comparison of arterial intimal clearances of LDL from diabetic and non-diabetic cholesterol-fed rabbits. Differences in intimal clearance explained by size difference. Arteriosclerosis 9:176–183

Nordestgaard BG, Stender S, Kjeldsen K (1987) Severe hypertriglyceridemia, large lipoproteins and protection against atherosclerosis. Scand J Clin Lab Invest 47 [Suppl 186]:7–12

Prichard RW, Clarkson TB, Goodman HO, Lofland HB Jr (1964) Aortic atherosclerosis in pigeons and its complications. Arch Pathol 77:244–257

Pronczuk A, Patton GM, Stephan ZF, Hayes KC (1991) Species variation in the atherogenic profile of monkeys. Relationship between dietary fats, lipoproteins and platelet aggregation. Lipids 26:213–222

Ratcliffe HL, Luginbuhl H (1971) The domestic pig: a model for experimental atherosclerosis. Atherosclerosis 13:133–136

Reitman JS, Mahley RW, Fry DL (1982) Yucatan miniature swine as a model for diet-induced atherosclerosis. Atherosclerosis 43:119–132

Rudel LL, Bond MG, Bullock BC (1985) LDL heterogeneity and atherosclerosis in nonhuman primates. Ann NY Acad Sci 454:248–253

Scanu AM (1991) Physiopathology of plasma lipoprotein metabolism. Kidney Int 39 [Suppl 31]:S3–S7

Seddon AM, Woolf N, La Ville A, Pittilo RM, Rowles PM, Turner PR, Lewis B (1987) Hereditary hyperlipidemia and atherosclerosis in the rabbit due to overproduction of lipoproteins: II. Preliminary report of arterial pathology. Arteriosclerosis 7:113–124

Soutar AK, Myant NB, Thompson GR (1977) Simultaneous measurement of apolipoprotein B turnover in very low density and low density lipoproteins in familial hypercholesterolemia. Atherosclerosis 28:247–256

Stender S, Zilversmit DB (1981) Transfer of plasma lipoprotein components and of plasma proteins into aortas of cholesterol-fed rabbits. Molecular size as a determinant of plasma lipoprotein influx. Arteriosclerosis 1:38–49

Strong JP, McGill HC Jr (1967) Diet and experimental atherosclerosis in baboons. Am J Pathol 50:669–690

Strong JP, Eggen DA, Jirge SK (1976) Atherosclerotic lesions produced in baboons by feeding an atherogenic diet for four years. Exp Mol Pathol 24:320–332

Tall AR, Small DM, Atkinson D, Rudel LL (1978) Studies on the structure of low density lipoproteins isolated from Macaca fascicularis fed an atherogenic diet. J Clin Invest 62:1354–1363

Taylor CB, Cox GE, Manalo-Estrella P, Southworth J (1962) Atherosclerosis in rhesus monkeys: II. Arterial lesions associated with hypercholesterolemia induced by dietary fat and cholesterol. Arch Pathol 74:16–34
Taylor CB, Manalo-Estrella P, Cox GE (1963a) Atherosclerosis in rhesus monkeys: V Marked diet induced hypercholesterolemia with xanthamatosis and severe atherosclerosis. Arch Pathol 76:239–249
Taylor CB, Patton DE, Cox GE (1963b) Atherosclerosis in rhesus monkeys: VI. Fatal myocardial infarction in a monkey fed fat and cholesterol. Arch Pathol 76:404–412
Thompson GR (1989) A handbook of hyperlipidaemia. Current Science, London
Vesselinovitch D (1988) Animal models and the study of atherosclerosis. Arch Pathol Lab Med 112:1011–1017
Wagner WD, Clarkson TB (1974) Mechanisms of the genetic control of plasma cholesterol in selected lines of Show Racer pigeons. Proc Soc Exp Biol Med 145:1050–1057
Wagner WD, Clarkson TB (1975) Comparative primate atherosclerosis: II. A biochemical study of lipids, calcium and collagen in atherosclerotic arteries. Exp Mol Pathol 23:96–121
Wagner WD, Clarkson TB, Feldner MA, Prichard RW (1973) The development of pigeon strains with selected atherosclerosis characteristics. Exp Mol Pathol 19:304–319
Watanabe Y (1980) Serial inbreeding of rabbits with hereditary hyperlipidemia (WHHL-rabbit): incidence and development of atherosclerosis and xanthoma. Atherosclerosis 36:261–268
White RA (ed) (1989) Atherosclerosis and arteriosclerosis: human pathology and experimental animal methods and models. CRC Press, Boca Raton
Williams JK, Clarkson TB (1989) Nonhuman primate atherosclerotic models. In: White RA (ed) Atherosclerosis and arteriosclerosis: human pathology and experimental animal methods and models. CRC Press, Boca Raton, pp 261–285
Wilson RB, Miller RA, Middleton CC, Kinden D (1982) Atherosclerosis in rabbits fed a low cholesterol diet for five years. Arteriosclerosis 2:228–241

Section B
Lipid Lowering Therapy

CHAPTER 8

Rationale to Treat

D.H. BLANKENHORN and H.N. HODIS

A. Introduction

The weight of evidence from primary and secondary prevention clinical trials indicates that reduction of low-density lipoprotein cholesterol (LDL-C) levels by diets, drugs, or other means can decrease the incidence of fatal and nonfatal myocardial infarction. A series of angiographic trials have demonstrated that reduced mortality and morbidity from LDL-C reduction are attributable, at least in part, to stabilization and regression of coronary atherosclerosis in both native vascular beds and venous bypass grafts. These studies provide the rationale for treatment of hyperlipoproteinemia with the goal of preventing coronary heart disease. Evidence from human clinical trials is supported by an extensive series of experiments in atherosclerotic animals of many species, including nonhuman primates. This review of treatment rationale summarizes evidence from human mortality and morbidity-based trials, human angiographic trials, and experimental animal models.

B. Mortality- and Morbidity-Based Drug Trials (Table 1)

The World Health Organization Cooperative Trial (OLIVER et al. 1978) was a randomized, double-blind trial of clofibrate versus placebo in 15745 males, aged 30–59, with an average 5.3 years of follow-up. The clofibrate-treated group of 5331 subjects had an average serum cholesterol of 6.44 mmol/l (249 mg/dl) and was compared to two control groups; the first group of 5296 subjects had an average entry serum cholesterol of 6.39 mmol/l (247 mg/dl) for comparison with clofibrate-treated subjects in the upper third of the cholesterol distribution. The second control group of 5118 subjects had an average serum cholesterol of 4.68 mmol/l (181 mg/dl) for comparison with the lower third of the cholesterol distribution in the clofibrate group. Clofibrate produced a 9% total cholesterol reduction and a 20% ($p < 0.05$) lower incidence of coronary artery disease as compared to controls in the upper third cholesterol distribution. Nonfatal myocardial infarction was reduced by 25% ($p < 0.05$). Reduction of incidence was greatest in subjects who had the highest initial cholesterol levels, smokers, and subjects with above-average blood pressure. Fatal myocardial infarction was not signi-

Table 1. Randomized controlled drug intervention trials

Trial/drug	Years of follow-up	Treatment group			Control group			Reduction in TC (%)	Reduction in CHD incidence (%)
		n	Mean TC	No. of CHD cases*	n	Mean TC	No. of CHD cases*		
Primary prevention									
WHO Oliver et al. (1978)									
Clofibrate	5	5331	224	167	5296	244	208	9.0	20.1**
Lipid Research Clinics Coronary Primary Prevention Trial (1984)									
Cholestyramine	7	1906	251	155	1900	276	187	8.5	18.9**
Helsinki Frick et al. (1987)									
Gemfibrozil	5	2051	247	56	2030	272	84	8.0	34.0**
Secondary prevention									
Newcastle Study (1971)									
Clofibrate	5	244	227	57	253	253	94	9.8	39.0**
Scottish Society Study (1971)									
Clofibrate	5	350	233	59	367	266	79	10.0	26.1
Coronary Drug Project (1975)									
Clofibrate	5	1103	235	309	2789	251	839	6.5	9.5
Nicotinic Acid	5	1119	226	287	2789	251	839	9.9	14.0**

TC, total cholesterol; CHD, coronary heart disease.
* Coronary heart disease death and/or definite nonfatal myocardial infarction.
** Significant at $p < 0.05$.

ficantly reduced by clofibrate therapy and all-cause mortality was greater in the clofibrate-treated group ($p < 0.01$). Follow-up for an average of 13.2 years indicated that the excess all-cause death rate was confined to the period of clofibrate exposure.

The Lipid Research Clinics Coronary Primary Prevention Trial (1984) was a multicenter, randomized, double-blind, placebo-controlled, primary prevention study in which 3806 men aged 35–59 years with type II hyperlipoproteinemia and total cholesterol greater than 6.85 mmol/l (265 mg/dl) received dietary therapy plus cholestyramine or placebo for an average of 7.4 years. In the cholestyramine-treated group, the average total cholesterol level decreased by 13.4%, and LDL-C by 20.3%. The average difference in levels between drug and placebo of total cholesterol was 8.5% and of LDL-C, 12.6%. Definite coronary death was reduced by 24% and nonfatal myocardial infarction by 19% in the cholestyramine-treated group. Additionally, the incidences of angina pectoris, development of a positive exercise stress test, and need for coronary artery bypass surgery were significantly reduced by 20%, 25%, and 21%, respectively, in the drug-treated group. Reduction in coronary heart disease events was directly related to reduction in total cholesterol and LDL-C, with the greatest benefit occurring in subjects with the greatest reduction in these levels. High-density lipoprotein cholesterol (HDL-C), which rose slightly in the cholestyramine-treated group, accounted for a 2% independent reduction in coronary heart disease incidence. Overall mortality was reduced by 7% in the cholestyramine-treated group, a nonsignificant difference from control.

The Helsinki Heart Study (Frick et al. 1987) was a multicenter, double-blind, placebo-controlled, 5-year, primary prevention trial testing gemfibrozil in 4081 men aged 40–55. All subjects had non-HDL-C (LDL-C + very low density lipoprotein cholesterol, VLDL-C) levels greater than 5.17 mmol/l (200 mg/dl) upon entrance into the study. The gemfibrozil-treated group demonstrated an HDL-C increase of 10% while total cholesterol, LDL-C, non-HDL-C, and triglyceride levels decreased by 8%, 8%, 12%, and 35%, respectively. These lipid changes were associated with a 34% ($p < 0.02$) reduction in overall incidence of coronary heart disease. There was a 37% reduction in fatal myocardial infarction and a 26% decrease in coronary heart disease deaths. All-cause mortality was not different between the groups. Changes in both HDL-C and LDL-C levels within the gemfibrozil-treated group were significantly associated with the reduction in incidence of coronary events.

The Coronary Drug Project (1975) was a double-blind, randomized, placebo-controlled trial in which 8341 men aged 30–64 with a previous myocardial infarction were randomly allocated to five drug treatment groups or control. Three of the treatments, 2.5 mg/day conjugated estrogens, 5.0 mg/day conjugated estrogens, and 6.0 mg/day dextrothyroxine sodium, were discontinued because of increased morbidity and mortality. Subjects randomized to the clofibrate and nicotinic acid treatment groups completed

5 years of treatment. In the clofibrate group, total cholesterol levels were reduced by 6.5% and triglyceride levels by 22%, but cardiac end points were not significantly reduced. The nicotinic-acid-treated group experienced a 10% decrease in total cholesterol and a 26% decrease in triglyceride levels, which resulted in a statistically significant 27% decrease in the incidence of nonfatal myocardial infarction. Reduction in combined coronary death and nonfatal myocardial infarction was also significant, with a 14% lower incidence in the nicotinic acid-treated group, principally due to fewer myocardial infarctions. All-cause mortality was not significantly different between placebo and nicotinic-acid-treated groups at the end of the initial 6 years. A subsequent 15-year follow-up, 9 years after termination of the initial study, revealed an 11% ($p = 0.0004$) lower overall mortality in the nicotinic-acid-treated group compared to the placebo group (Canner et al. 1986).

C. Mortality- and Morbidity-Based Diet Trials

Six randomized controlled dietary studies have been conducted in 2467 subjects. These studies are difficult to perform under the rigorous conditions applicable to drug testing. Producing sustained major differences in lipid levels in large subject groups is a formidable task, particularly if blinding is an added condition. Studies are listed in chronological order in Table 2. Significant benefit from cholesterol lowering ($p < 0.05$) was found in the Los Angeles Veterans Administration Trial (Dayton et al. 1969) and the Oslo Diet Heart Study (Leren 1970).

D. Atherosclerotic Regression Trials – Uncontrolled Case Series and Case Reports

Uncontrolled studies and case reports of atherosclerotic lesion regression and stabilization in the femoral, popliteal, carotid, renal, and coronary arteries are summarized in Table 3. The interventions included dietary changes, drugs, exercise, cessation of smoking, lowering of blood pressure, and plasmapheresis.

I. Femoral Artery

The earliest human angiographic studies were of the femoral arteries, reported by Ost and Stenson (1967) who treated 31 hyperlipidemic men for intermittent claudication with 3–6 g nicotinic acid per day. After a mean treatment period of 42 months, three patients showed regression and 11 patients arrest of progression of femoral lesions.

In a study conducted on 25 hyperlipidemic patients, Barndt et al. (1977) reported nine patients with regression and three with arrest in pro-

Table 2. Randomized controlled dietary intervention trials

Trial	Years of follow-up	Treatment group			Control group			Reduction in TC (%)	Reduction in CHD incidence* (%)
		n	Mean TC	No. of CHD cases*	n	Mean TC	No. of CHD cases*		
Primary prevention									
Los Angeles VA Diet DAYTON et al. (1969)	8	424	195	52	422	226	65	13	23**
Secondary prevention									
ROSE (1965)									
Corn oil	2	28	208	12	26	260	6	20	0
Olive oil	2	26	258	9	26	260	6	1	0
London MRC Low Fat diet (RESEARCH COMMITTEE 1965)	4	123	216	46	129	241	48	17	0.8
London MRC Soya Bean oil (RESEARCH COMMITTEE to the MEDICAL RESEARCH COUNCIL 1968)	2–6	199	224	45	194	258	51	17	18
Oslo Diet Heart Study LEREN (1970)	5	206	243	61	206	258	81	18	35**
WOODHILL et al. (1978)	2–7	221	281	38	237	282	52	11	27

TC, total cholesterol; CHD, coronary heart disease.
* CHD death or definite nonfatal myocardial infarction.
** Significant at $p < 0.05$.

Table 3. Angiographic studies: uncontrolled regression studies and case reports

Study	Arterial bed	Number of subjects				Mean study interval	% Lipid changes	Therapy
		Total	–	0	+			
OST and STENSON (1967)	Femoral	31	3	22	17	42	NR	Nicotinic acid
DEPALMA et al. (1970)	Popliteal	1	1			9	TC ↓ 42	Diet, smoking cessation, exercise
BASTA et al. (1976)	Renal	1	1			41	TC ↓ 56 TG ↓ 23	Diet
BARNDT et al. (1977)	Femoral	25	9	3	13	13	Regr: TC ↓ 21 TG ↓ 60 Prog: TC ↓ 6 TG ↓ 17	Diet, clofibrate + neomycin
KUO et al. (1979)	Coronary, brachiocephalic, peripheral	25		21		36–48	TC ↓ 35 LDL ↓ 43 HDL ↑ 7	Diet, colestipol
CRAWFORD et al. (1979)	Femoral	2	2			13; 30	NR	Exercise, weight loss
RAFFLENBEUL et al. (1979)	Coronary	25	5	9	11	12	NR	Low saturated fat diet, smoking cessation
THOMPSON et al. (1980)	Aorta	3		3	(homo)	12–24	TC ↓ 40	Plasma exchange + medications
	Coronary	2	1	1	(hetero)		TC ↓ 27	

ROTH and KOSTUK (1980)	Coronary	1	1			9	TC $\downarrow$ 25 TG $\downarrow$ 55	Diet, exercise
NASH et al. (1982)	Coronary	25 17		22 9	3 8	24 24	TC $\downarrow$ 20 TG $\downarrow$ 3 TC $\downarrow$ 2 TG $\uparrow$ 5	Colestipol placebo non-responders
NIKKILA et al. (1984)	Coronary	28		9	19	24	TC $\downarrow$ 18 TG $\downarrow$ 38 LDL $\downarrow$ 19 HDL $\uparrow$ 10	Diet + nicotinic acid + clofibrate
STEIN et al. (1986)	Coronary CABG	1		1		31	TC $\downarrow$ 43 TG $\downarrow$ 10 LDL $\downarrow$ 49	Plasma exchange and medications
YOKOYAMA et al. (1987)	Renal	1	1			72	TC $\downarrow$ 58	LDL apheresis
KELLER and SPENGEL (1988)	Carotid	3	1	1	1	16–72	TC $\downarrow$ 50	Plasma exchange, plasmapheresis
ARNTZENIUS et al. (1985)	Coronary	39		18	21	24	TC to HDL <6.9 = no progression	P/S fat >2.0, cholesterol <100 mg/day

NR, not reported; TC, total cholesterol; LDL, low-density lipoprotein cholesterol; HDL, high-density lipoprotein cholesterol; TG, triglyceride; P/S, polyunsatured/saturated; CABG, coronary artery bypass graft.

gression after an average of 13 months of treatment with lipid-lowering drugs, principally clofibrate and neomycin. Regression of disease was significantly correlated with reductions in total serum cholesterol levels, triglyceride levels, and systolic and diastolic blood pressures.

II. Popliteal Artery

In one of the first angiographic case studies, DEPALMA et al. (1970) showed plaque regression by serial arteriograms taken 9 months apart in the popliteal artery of a 57-year-old diabetic man who had reduced his total cholesterol from 8.46 to 4.91 mmol/l (327 to 190 mg/dl) after initiating dietary therapy, exercise, and cessation of cigarette smoking.

III. Renal Artery

BASTA et al. (1976) reported a case of a 49-year-old hypertensive woman with renovascular hypertension due to a 90% occlusion of the right renal artery and a 75% narrowing of the left renal artery. After 3 years of therapy with cholestyramine and clofibrate, total cholesterol was reduced from 8.79 to 3.90 mmol/l (340 to 151 mg/dl) and total triglycerides from 1.77 to 1.37 mmol/l (157 to 121 mg/dl). A repeat arteriogram showed almost complete resolution of the right renal artery occlusion and some regression of the left renal artery lesion with concomitant normalization of blood pressure and the peripheral vein plasma renin level.

YOKOYAMA et al. (1987) demonstrated complete disappearance of a 90% stenotic renal artery lesion in a homozygous familial hypercholesterolemic patient who had LDL apheresis for 6 years beginning at age 5. Prior to therapy, the patient's total cholesterol was 18.62 mmol/l (720 mg/dl); while receiving plasmapheresis it averaged 7.76 mmol/l (300 mg/dl). Atherosclerotic lesions also regressed in the supravalvular region of the ascending aorta and stabilized in the coronary arteries.

IV. Carotid Artery

Among three brothers with severe familial hypercholesterolemia treated with plasmapheresis and followed with duplex ultrasound imaging, KELLER and SPENGEL (1988) observed regression and stabilization of carotid atherosclerotic lesions in two, who had continuous plasmapheresis, and progression in one, who discontinued plasmapheresis.

V. Coronary Artery and Aorta

Long-term (1–2 years) plasma exchange for treatment of familial hypercholesterolemia was evaluated with serial aortagrams and coronary angiograms 1–2 years apart by THOMPSON et al. (1980). Plasma exchange plus nicotinic acid in three homozygous patients produced a reduction in mean total cholesterol from 15.83 to 9.52 mmol/l (612 to 368 mg/dl) and angiographic evidence of aortic supravalvular lesion stabilization. In two heterozygous patients, plasma exchange plus cholestyramine reduced mean total plasma cholesterol from 6.36 to 4.65 mmol/l (246 to 180 mg/dl) with coronary artery lesion regression in one patient and stabilization in the other. Stabilization of coronary atherosclerosis for 3 years with plasma exchange and lipid-lowering medication has also been demonstrated in homozygous familial hypercholesterolemia by STEIN et al. (1986). RAFFLENBEUL et al. (1979), in an angiographic study of coronary anatomy of 25 patients with unstable angina, noted regression in five patients treated after an average interval of 1 year with a low-saturated-fat diet and counseling for cigarette smoking cessation. Lipid-lowering medications were not used.

ROTH and KOSTUK (1980) reported a symptomatic patient with a positive exercise thallium stress test and a 90% stenosis of the left anterior descending artery. Dietary modification and increased exercise reduced the total cholesterol level from 6.96 to 5.20 mmol/l (269 to 201 mg/dl) and triglyceride level from 2.31 to 1.04 mmol/l (205 to 93 mg/dl). Repeat angiography 1 year later showed decreased stenosis in the left anterior descending artery and disappearance of retrograde collateral filling from the right coronary artery seen on a first angiogram. The patient was asymptomatic and repeat exercise thallium stress tests were normal.

KUO et al. (1979) examined the effect of lipid lowering induced by dietary and colestipol therapy on coronary artery, brachiocephalic, and peripheral arterial lesions in familial type II hyperlipoproteinemia with a 3- to 4-year interval between angiograms. Average total cholesterol level decreased from 10.68 to 6.98 mmol/l (413 to 270 mg/dl) and the LDL-C from 8.56 to 4.86 mmol/l (331 to 188 mg/dl). Stabilization of lesions was demonstrated in all three vascular beds in 21 of 25 subjects.

NASH et al. (1982) reported 25 patients who received colestipol hydrochloride and 17 nonresponders to colestipol who were given placebo. All patients were on a low-cholesterol, low-fat diet and all had baseline serum cholesterol levels greater than 6.46 mmol/l (250 mg/dl). Angiograms were repeated after 2 years, at which time the total cholesterol had decreased by 21%, 7.24 to 5.74 mmol/l (280 to 222 mg/dl) with colestipol treatment, and by 2%, 6.77 to 6.64 mmol/l (262 to 257 mg/dl) in the placebo group. In the drug-treated group, 22 of 25 patients had stable coronary artery lesions; in the placebo group, nine of 17 were stable ($p = 0.011$). NIKKILA et al. (1984) compared 28 patients on lipid-lowering diet and clofibrate and nicotinic acid therapy to 13 nonrandomized controls drawn from another study who were not on lipid-lowering diet or medication. In the diet–drug-treated group, total cholesterol, triglyceride, and LDL-C decreased by an average of 18%, 38%, and 19%, respectively, and HDL-C increased by 10%. More patients showed coronary artery lesion stabilization upon repeat angiography at 2 years in the treated group than in the control group – nine out of 28 (32%) versus one out of 13 (8%).

E. Controlled Coronary Angiographic Studies of Drug Therapy (Table 4)

COHN et al. (1975) conducted the earliest controlled coronary angiographic trial. Twenty-four patients were randomized to clofibrate therapy and 16 patients to placebo. Total cholesterol and triglyceride levels were reduced by 3% and 14%, respectively, in the clofibrate group. No significant reduction in the progression of atherosclerosis was noted after 1 year.

The National Heart, Lung, and Blood Institute Type II Coronary Intervention Study (BRENSIKE et al. (1984); LEVY et al. (1984)) was a double-blind, randomized, placebo-controlled trial test of cholestyramine. Average baseline entry serum levels for total cholesterol, triglyceride, LDL-C, and HDL-C were 8.35, 1.85, 6.52, and 1.01 mmol/l (323, 164, 252, and 39 mg/dl), respectively. A prerandomization low-cholesterol, low-fat diet reduced LDL-C by 6% in both the treatment and control groups. After randomization, this level decreased by another 5% in the control group and by another 26% in the cholestyramine-treated group. At the end of the 5-year study period, there was a significant difference in total cholesterol (7.47 versus 6.62 mmol/l; 289 versus 256 mg/dl) and LDL-C (5.66 versus 4.60 mmol/l; 219 versus 178 mg/dl) between the cholestyramine and placebo groups. Total triglyceride (1.82 versus 2.18 mmol/l; 161 versus 193 mg/dl) and HDL-C (1.01 versus 1.06 mmol/l; 39 versus 41 mg/dl) levels were not significantly different between the two groups.

Matched angiograms were evaluated in 116 patients, 57 placebo treated and 59 cholestyramine treated, after 5 years. Repeat angiograms were originally scheduled after 2 years, but only nine out of 31 patients showed any lesion change and the interval was lengthened to 5 years. Definite progression of disease occurred in 35% (20 of 57) of the placebo-treated group compared to 25% (15 of 59) of the cholestyramine-treated group. Probable progression was found in 14% (eight of 57) of the placebo patients and in 7% (four of 59) of the drug-treated patients. Definite and probable progression combined occurred in 49% (28 of 57) of the placebo-treated patients compared to 32% (19 of 59) of the cholestyramine-treated patients ($p < 0.05$). Definite regression was found in 2% (one of 57) of the placebo-treated group compared to 3% (two of 59) of the drug-treated group. After baseline demographic inequalities and lesion severity were taken into account, it was found that lesions with a 50% or more stenosis at baseline progressed more slowly in the cholestyramine-treated group than in the placebo-treated group (12% versus 33%, respectively; $p < 0.05$). Increases in HDL-C to total cholesterol and HDL-C to LDL-C ratios were found to be the best predictors of lesion stabilization (LEVY et al. 1984).

The Cholesterol-Lowering Atherosclerosis Study (CLAS; BLANKENHORN et al. 1987) was the first angiographic trial to conclusively demonstrate lesion regression in humans. CLAS was a selectively blinded, randomized, placebo-controlled trial. Subjects were 162 nonsmoking, nondiabetic, normotensive men, aged 40–59, who all had coronary bypass surgery and entrance fasting total cholesterol levels between 4.78 and 9.05 mmol/l (185 and 350 mg/dl). Coronary angiograms were obtained at baseline and repeated 2 years after therapy. The 82 control subjects were prescribed a diet that provided 26% of energy as fat (10% as polyunsaturated fat and 5% as saturated fat) and contained 250 mg of cholesterol per day. The 80 drug treatment subjects received maximal dosage of colestipol (30 g/day) and nicotinic acid (3–12 g/day, average 4.3 g/day) and a diet prescription which provided 22% energy as fat (10% polyunsaturated fat and 4% saturated fat) and contained 125 mg of cholesterol per day. After 2 years of intervention, the treatment group showed large decreases in total cholesterol (27%), from an average of 6.36 mmol/l (246 mg/dl) at baseline to 4.65 mmol/l (180 mg/dl), total triglycerides (27%), 1.70 to 1.24 mmol/l (151 to 110 mg/dl), LDL-C (43%), 4.42 to 2.51 mmol/l (171 to 97 mg/dl), and the LDL-C to HDL-C ratio (58%), 3.8 to 1.6. HDL-C increased by 36% from 1.15 to 1.57 mmol/l (45 to 61 mg/dl). In the placebo group, smaller, statistically significant decreases also occurred: total cholesterol (5%), 6.28 to 6.00 mmol/l (243 to 232 mg/dl); total triglycerides (8%), 1.74 to 1.59 mmol/l (154 to 141 mg/dl); LDL-C (5%), 4.37 to 4.14 mmol/l (169 to 160 mg/dl); and LDL-C/HDL-C (8%), 3.9 to 3.6 mmol/l (4.0 to 3.7 mg/dl). HDL-C also increased by 2%, 1.13 to 1.15 mmol/l (43.7 to 44.4 mg/dl).

Compared to the control group, the drug-treated group showed a significant reduction in the average number of lesions per patient that progressed in native arteries (1.4 versus 1.0; $p < 0.03$), as well as a reduction in the number of patients with new lesion formation in native coronary arteries (18/82 versus 8/80; $p < 0.03$). Also compared to the control group, the drug-treated group showed a significant reduction in the number of patients with new lesion formation (25/82 versus 14/80; $p < 0.04$) and any adverse changes (32/82 versus 19/80; $p < 0.03$) in the venous bypass grafts. Coronary arterial atherosclerosis regression was found in 16.2% of the drug-treated group compared to 3.6% of the placebo-treated group ($p = 0.007$). The evidence for benefit retained strong statistical significance after the cohort was divided in half and the coronary artery changes were compared to baseline total cholesterol levels from 4.78 to 6.20 mmol/l (185 to 240 mg/dl) and from 6.23 to 9.05 mmol/l (241 to 350 mg/dl). This implicates the entire range of total cholesterol from 4.78 to 9.05 mmol/l (185 to 350 mg/dl) as a risk factor for arterial lesion progression.

In a 2-year extension of the CLAS trial (CLAS II; CASHIN-HEMPHILL et al. 1990), repeat coronary artery angiograms were obtained in 103 patients who continued colestipol/nicotinic acid therapy (n = 56) and placebo (n = 47). Changes in lipid levels obtained in the first 2 years (CLAS I) were

Table 4. Angiographically controlled regression clinical trials*

Study	Arterial bed	Number of study (S) and control (C) subjects: Total	–	0	+	Mean study interval (mo)	% Lipid changes	Therapy
TERRY et al. (1976)	Femoral	S 22		19	3	24	NR	Pyridinol carbamaate
		C 16		1	15			
ERIKSON[¶] et al. (1983)	Femoral	S 8	6				TC ↓ 39 LDL ↓ 45 TG ↓ 64 HDL ↑ 30	Finofibrate and niacin
DUFFIELD et al. (1983)	Femoral	S 12	% seg progress 6.9%			19	TC ↓ 25 LDL ↓ 28 TG ↓ 45 HDL ↑ 26	Cholestyramine; nicotinic acid or clofibrate
		C 12	% segs progress 17.3%				TC ↓ 3 LDL ↓ 1 TG ↓ 7 HDL ↓ 8	
NHLBI Type II, BRENSIKE et al. (1984)	Coronary	S 59	4**	36[¶¶]	19[†]	60	TC ↓ 17 LDL ↓ 26 TG ↑ 28 HDL ↑ 8	Cholestyramine
		C 57	4**	25[¶¶]	28[†]		TC ↓ 1 LDL ↓ 5 TG ↑ 26 HDL ↑ 2	
CLAS[▲] BLANKENHORN et al. (1987)	Coronary	S 82	13	37	32	24	TC ↓ 27 LDL ↓ 43 TG ↓ 27 HDL ↑ 36 LDL/HDL ↓ 58	Colestipol and niacin
		C 80	2	29	49		TC ↓ 2 LDL ↓ 5 TG ↓ 8 HDL ↑ 2 LDL/HDL ↓ 8	
FATS BROWN et al. (1990)	Coronary	S 32*	12		8	30	LDL ↓ 32 HDL ↑ 43	Colestipol and niacin
		S 34*	11		7		LDL ↓ 46 HDL ↑ 15	Colestipol and lovastatin
		C 37*	4		17		LDL ↓ 7 HDL ↑ 5	

Lifestyle Heart Trial, ORNISH et al. (1990)	Coronary	S 22	18		4	12	TC ↓ 24 TG ↑ 22	LDL ↓ 38 HDL ↓ 2	7% fat calories 12 mg/d cholesterol
		C 19	8	1	10		TC ↓ 5 TG ↓ 8	LDL ↓ 6 HDL ↓ 2	Stop smoking, exercise, stress management
POSCH BUCHWALD et al. (1990)	Coronary	S 421	6	37	52**	120	TC ↓ 22 TG ↑ 34	LDL ↓ 39 HDL ↑ 3	Partial ileal bypass
		C 417	3	9	68#		TC ↓ 4 TG ↑ 6	LDL ↓ 6 HDL ↓ 5	
UC San Francisco SCOR, ▲▲ KANE et al. (1990)	Coronary	S 540	21	8	11	26	TC ↓ 31 TG ↑ 22	LDL ↓ 39 HDL ↑ 20	Binary/ternary drug combinations of niacin, colestipol, lovastatin
		C 32	12	3	17		TC ↓ 9 TG ↑ 4	LDL ↓ 12 HDL ↑ 1	

NR, not reported; TC, total cholesterol; TG, triglyceride; LDL, low-density lipoprotein; HDL, high-density lipoprotein.
* Study (S) versus control (C) groups, all significantly different to at least the $p < 0.05$ level.
¶ Ongoing trial.
¶¶ No change and mixed change.
* Regression and progression only.
# $N = 80$ pairs of angiograms at 10 years.
** Probable and definite regression.
† Probable and definite progression.
** $n = 95$ pairs of angiograms at 10 years.
▲ Treatment benefit in coronary artery bypass grafts.
▲▲ Treatment benefit in women.
↓ Indicates decrease in lipid changes; ↑ an increase in lipid changes.

maintained in CLAS II. Angiograms after 4 years of colestipol/nicotinic acid therapy, when compared to the control group, showed a significant reduction in the average number of lesions that progressed per patient (0.9 versus 2.0) and the percent of patients with new lesions (12.5% versus 40.4%) in native arteries. The drug-treated group also showed a significant reduction in the percent of patients with new lesions in bypass grafts as compared to the control group (17.9% versus 38.3%). Overall, the coronary artery lesion status of the drug-treated and placebo groups grew more divergent over the additional 2 years, with the drug-treated group showing stability of lesions and the placebo group showing marked progression. The power of the CLAS to detect lesion change was greater than previous angiographic trials because more subjects completed sequential angiograms and maintained larger changes in LDL-C and HDL-C levels than in any previous study.

An unexpected finding in CLAS was the importance of triglyceride-rich lipoproteins (chylomicron remnants and VLDL) in atherogenesis. This was discovered by analysis of the interrelations between apolipoprotein levels and angiographic change (BLANKENHORN et al. 1990). CLAS patients who progressed in the placebo group had significantly higher apolipoprotein B levels as well as apolipoprotein C-III levels in the LDL, VLDL, and HDL lipoproteins compared to placebo patients who did not progress. In the drug-treated group, a lower apolipoprotein C-III level found in the HDL was associated with lesion progression. CLAS apolipoprotein findings in both drug and placebo groups are interpreted as indicating a significant atherogenic effect of triglyceride-rich lipoproteins.

The Familial Atherosclerosis Treatment Study (FATS; BROWN et al. 1990) was a double-blind, randomized, placebo-controlled trial which recruited 103 subjects with elevated apolipoprotein B levels greater than 125 mg/dl and documented coronary artery disease. There were two treatment arms (colestipol/nicotinic acid and colestipol/lovastatin) and one arm treated with conventional therapy. After 2.5 years of therapy, the colestipol/nicotinic acid treated group demonstrated decreases in total cholesterol from 6.99 mmol/l (270 mg/dl) at baseline to 5.41 mmol/l (209 mg/dl), in LDL-C, 4.92 mmol/l (190 mg/dl) to 3.34 mmol/l (129 mg/dl), and in total triglycerides, 2.19 mmol/l (194 mg/dl) to 1.55 mmol/l (137 mg/dl). HDL-C increased from 1.01 mmol/l (39 mg/dl) to 1.42 mmol/l (55 mg/dl). In the colestipol/lovastatin treatment group, total cholesterol decreased from 7.12 mmol/l (275 mg/dl) to 4.71 mmol/l (183 mg/dl), LDL-C from 5.08 mmol/l (196 mg/dl) to 2.77 mmol/l (107 mg/dl), and triglycerides from 2.27 mmol/l (201 mg/dl) to 2.07 mmol/l (183 mg/dl). HDL-C increased from 0.91 mmol/l (35 mg/dl) to 1.06 mmol/l (41 mg/dl). The placebo group also showed smaller, but statistically significant changes in LDL-C, from 4.53 mmol/l (175 mg/dl) to 4.20 mmol/l (162 mg/dl) and in HDL-C, from 0.98 mmol/l (38 mg/dl) to 1.04 mmol/l (40 mg/dl). Total cholesterol decreased from 6.79 mmol/l (263 mg/dl) to 6.55 mmol/l (253 mg/dl) and total triglycerides increased from 2.59 mmol/l (229 mg/dl) to 2.98 mmol/l (264 mg/dl).

Baseline and end-of-trial angiograms, separated by 2.5 years, were analyzed by quantitative angiography. Definite regression occurred in 12 of 32 colestipol/nicotinic-acid-treated subjects and 11 of 34 colestipol/lovastatin-treated subjects as compared to four of 37 control subjects. Definite progression occurred in eight colestipol/nicotinic-acid-treated subjects, seven of the colestipol/lovastatin-treated subjects, and 17 of the control subjects. Cardiovascular events (death, myocardial infarction, or revascularization) occurred in 19% of the patients in the control group as compared with 4% in the colestipol/nicotinic-acid-treated group and 7% in the colestipol/lovastatin-treated group, a statistically significant, 73% reduction in clinical events. Multivariate analysis revealed that regression of coronary lesions was independently correlated with a reduction in apolipoprotein B or LDL-C and systolic blood pressure and an increase in HDL-C.

The Program on Surgical Control of the Hyperlipidemias (POSCH; Buchwald et al. 1990) was a multicenter, angiographic, secondary prevention trial initiated in 1975 which randomized 421 postmyocardial infarction patients to partial ileal bypass (PIB) and 417 to diet control. Entry criteria required total cholesterol levels greater than 5.66 mmol/l (219 mg/dl) or LDL-C greater than 6.18 mmol/l (239 mg/dl). Mean follow-up was 9.7 years and PIB patients had a 23.3% lower total cholesterol ($p < 0.001$), 37.7% lower LDL-C ($p < 0.0001$), and 4.3% higher HDL-C ($p = 0.02$) when compared with control. PIB produced a 21.7% reduction in total mortality and a 28% reduction in coronary heart disease mortality, differences not statistically significant. However, total mortality in a subgroup of patients with a 50% or greater left ventricular ejection fraction was significantly reduced by 36.1% (39 of 292 control and 24 of 281 surgery patients; $p = 0.02$). When atherosclerotic coronary heart disease mortality was combined with proven nonfatal myocardial infarction, there were 138 events in control and 91 events in surgery patients ($p < 0.001$). For the combined end point of atherosclerotic coronary heart disease mortality, proven or suspected myocardial infarction, and unstable angina there were 222 events in the control and 160 events in the surgery group ($p < 0.0001$). Sequential coronary arteriograms demonstrated a greater progression of coronary artery disease at each follow-up interval, but a consistently greater one in control patients ($p < 0.001$). The difference was 41.4% versus 28.1% at 3 years, 65.4% versus 37.5% at 5 years, 77.4% versus 46.5% at 7 years, and 85.7% versus 53.8% at 10 years. The need for coronary artery bypass grafting, angioplasty, or heart transplantation was significantly reduced ($p < 0.0001$), and there was a significant reduction in the development of clinical peripheral vascular events ($p = 0.038$) in the surgery group. At 5 years, 33.6% of control and 19% of surgery patients developed significant reduction in ankle–brachial index by Doppler assessment ($p < 0.01$).

The University of California, San Francisco, Specialized Center of Research (UCSF SCOR) Intervention Trial (Kane et al. 1990) recruited patients between the ages of 19 and 72 years who had clinical features compatible with heterozygous familial hypercholesterolemia. To be included, patients had to have tendon xanthomas, mean LDL-C levels above 5.17 mmol/l (200 mg/dl), and total triglyceride levels below 3.10 mmol/l (275 mg/dl) while they were consuming a diet restricted in saturated fats and cholesterol. Patients without tendon xanthomas were included if they had an LDL-C level above 6.46 mmol/l (250 mg/dl) or an LDL-C level greater than 5.16 mmol/l (200 mg/dl) and a first-degree relative with tendon xanthomas. Patients with previous angioplasty, coronary artery bypass surgery, or multiple infarcts were excluded, as were patients with systemic diseases other than atherosclerosis or hypertension. Patients with disorders known to produce secondary hyperlipidemia and those homozygous for apolipoprotein E-II were also excluded.

After demonstration of angiographically quantifiable coronary lesions, subjects were randomly assigned by sex and age to treatment and control groups. Both groups received dietary counseling. The treatment group received aggressive drug therapy to reduce LDL-C levels with combinations of lipid-lowering agents. These patients were initially given up to 30 g colestipol and up to 7.5 g niacin daily, as tolerated. Thirty-six of the 40 patients in the treatment group took niacin, 25 of whom consistently took more than 1.5 g daily. Twenty-eight patients took 30 g and four took 15 g colestipol daily throughout the study. When lovastatin became available as an investigational drug it was given also. Sixteen patients took 40–60 mg lovastatin daily in binary or ternary drug combinations.

At the outset, the control group was treated with diet alone. Following publication of the Lipid Research Clinics Trial (1984) results, patients in this group were offered 15 gm per day of bile-acid-binding resin. Fourteen patients elected to take colestipol. Seven men took it for an average of 19 months and seven women for an average of 20 months. The control group included those treated with diet alone and those treated with diet plus resin.

Coronary angiography was repeated after 2 treatment years. The primary outcome variable was change in cross-sectional percent area stenosis. The patient rather than the lesions was selected as the analytical unit and the change in percent area stenosis was averaged for all lesions in each subject. The mean change in percent area stenosis among controls was +0.80, indicating progression, while the mean change for the treatment group was −1.53, indicating regression ($p = 0.039$ by two-tailed t test for the difference between groups). Regression among women, analyzed separately, was also significant. The change in percent area stenosis was correlated with LDL-C levels on trial. The authors concluded that reduction of LDL-C levels induced regression of atherosclerotic lesions of the coronary arteries in both men and women with familial hypercholesterolemia. This angiographic trial is unique in demonstrating significant therapy benefits for women; other trials have excluded women, included too few for adequate evaluation of a treatment effect, or have not specifically analyzed the outcome for an effect in women.

F. Coronary Angiographic Studies of Diet Therapy

The Leiden Intervention Trial (Table 3; ARNTZENIUS et al. 1985) was a 2-year test of vegetarian diet with a polyunsaturated to saturated fat ratio greater than 2.0 and cholesterol intake less than 100 mg/day which included 39 patients with angiograms separated by 2 years. There was no control group. Eighteen patients showed no progression in coronary angiograms analyzed both by visual assessment and computerized image processing. Lesion progression was strongly related to the total cholesterol to HDL-C ratio, with no coronary lesion progression occurring in patients with a total cholesterol to HDL-C ratio less than 6.9 throughout the study or in those who had reduced a high ratio at baseline to less than 6.9. An important finding of the Leiden study was the indication that lesion progression can be arrested by dietary modification without weight loss. Atherosclerotic lesions have long been known to decrease with war-related severe dietary deprivation (SCHETTLER 1979) and wasting diseases (WILENS 1947).

The Lifestyle Heart Trial (Table 4; ORNISH et al. 1990) was a 1-year randomized, controlled, angiographic study testing lifestyle and dietary change on coronary atherosclerosis. Angiograms were analyzed on 22 subjects in the experimental group and on 19 in the usual-care control group by quantitative coronary angiography. The experimental group followed a low-fat vegetarian diet (mean 6.8% calories from fat, 12.4 mg cholesterol/day), stopped smoking, exercised, and underwent stress management training, while the usual-care group made minimal dietary and lifestyle changes. At the end of 1 year, total cholesterol fell from 5.88 mmol/l (227 mg/dl) to 4.45 mmol/l (172 mg/dl), LDL-C from 3.92 mmol/l (152 mg/dl) to 2.46 mmol/l (95 mg/dl), and HDL-C from 1.00 mmol/l (39 mg/dl) to 0.97 mmol/l (38 mg/dl) in the experimental group. The usual-care group showed a decrease in total cholesterol from 6.34 mmol/l (245 mg/dl) to 6.00 mmol/l (232 mg/dl), LDL-C from 4.32 mmol/l (167 mg/dl) to 4.07 mmol/l (157 mg/dl), and HDL-C from 1.35 mmol/l (52 mg/dl) to 1.31 mmol/l (51 mg/dl). In the experimental group, 18 subjects showed regression and four progression, while in the usual-care group, eight subjects showed regression, ten progression, and one no change. The experimental group also had a 91%, 42%, and 28% reduction in frequency, duration, and severity of angina, respectively. Further analysis of the data showed that patients who made the greatest dietary and lifestyle changes had the greatest angiographic improvement and that the degree of change in stenosis was correlated with the amount of these changes. Furthermore, the most severely stenosed lesions showed the greatest improvement.

G. Femoral Angiographic Drug Studies (Table 4)

In 1976, TERRY et al. (1976) reported atherosclerotic lesion stabilization in a randomized, double-blind test of pyridinol carbamate versus placebo. After 2 years of treatment, 19 of 22 drug-treated patients showed no progression in femoral angiograms, as compared with 15 of 16 placebo patients showing progression. This study, which was conducted in patients with claudication, provided early information on an appropriate treatment interval for angiographic trials, because angiograms were obtained yearly and showed a difference in progression at 2 years.

In a study of advanced femoral artery disease, DUFFIELD et al. (1983) described quantitative evidence of disease stabilization and regression. Twenty-four patients with stable intermittent claudication and baseline total cholesterol levels greater than 6.52 mmol/l (252 mg/dl) and/or triglyceride levels exceeding 1.78 mmol/l (158 mg/dl) were randomly assigned to a usual-care group or a drug-treatment group which included dietary advice and cholestyramine, nicotinic acid, or clofibrate therapy. All patients received antismoking advice and a weight-reducing diet if indicated. Patients in the usual-care group showed no significant lipid changes. The treatment group exhibited mean plasma reductions in total cholesterol, total triglycerides, LDL-C, and VLDL-C of 25%, 45%, 28%, and 57%, respectively, and a 26% increase in HDL-C. Matched femoral angiograms obtained at an average of 19 months apart were analyzed visually and by computerized image processing. The treatment group had 60% fewer arterial segments showing progression compared to the control group ($p < 0.01$). The mean increment in the arterial surface area covered by plaque (square millimeter per segment per year) among the treated patients was 33% less than that observed among the nontreated patients ($p < 0.01$). Regression was observed in 15 of 46 segments in the treatment group compared to seven of 46 in the control group, as determined by an edge irregularity index ($p < 0.05$).

In an ongoing femoral artery study in which angiography is repeated every 6 months and analyzed by computerized image processing, ERICKSON et al. (1983) have reported findings in eight patients. Asymptomatic patients with elevated blood lipid levels were selected, and the study is designed to observe early atherosclerotic lesions. Fenofibrate and nicotinic acid treatment produced a 39% reduction in total cholesterol, a 64% reduction in total triglyceride level, a 82% decrease in VLDL triglyceride, a 45% decrease in LDL-C, and a 30% increase in HDL-C. Six of eight patients had regression of the femoral atherosclerotic lesions, as evidenced by reduction in plaque area.

H. Atherosclerosis Regression in Experimental Animal Models

Reversal of atherosclerosis has been observed to occur after discontinuance of cholesterol feeding in all major species of animals used in atherosclerosis research: dogs (DePalma et al. 1977), rats (Morin et al. 1964), pigs (Fritz et al. 1976), chickens (Horlick and Katz 1949), pigeons (Clarkson et al. 1973), and nonhuman primates (Malinow 1980). Further, regression of atherosclerosis in the presence of continued cholesterol feeding has been observed to occur in nonhuman primates treated with cholestyramine (Vesselinovitch et al. 1978), exercise (Kramsch et al. 1981a), high polyunsaturated fat diets (Mendelsonh and Mendelsohn 1989), alfalfa meal (Malinow et al. 1978a), clofibrate (Fritz et al. 1975), ileal bypass (Subbiah et al. 1978), and, most recently, administration of calcium channel blockers (Kramsch et al. 1981b).

Nonhuman primates, principally rhesus (*macaca mulatta*) and cynomolgus (*macaca fascicularis*) monkeys are preferred animals for regression studies because of anatomic, physiological, and metabolic similarities to human beings. Hypercholesterolemia and atherosclerosis have been most commonly induced in monkeys by feeding 0.37%–2% cholesterol diets with variable amounts of saturated fat (usually in the form of butter), for 2–38 months. Total cholesterol levels are typically increased from 6.72 to 25.86 mmol/l (260 to 1000 mg/dl). Regression phases have ranged from 2 weeks to 48 months with cholesterol levels between 2.46 and 7.76 mmol/l (95 and 300 mg/dl), but mostly below 5.17 mmol/l (200 mg/dl). Typical observations of regression are based upon group comparisons of animals sacrificed at different times for morphologic and histochemical exploration of atherosclerosis regression at a cellular level. It has been observed that after a period of atherosclerosis induction, the regression phase leads to normalization of arterial wall morphology and histology, paralleling the angiographic changes. Physiologic changes induced by regression have also been studied by pulse wave velocity (Farrar et al. 1980), arterial wall reactivity (Heistad et al. 1987), and relaxation (Harrison et al. 1987). Absolute establishment of atherosclerotic lesion regression requires that evolution of individual atherosclerotic plaques be studied sequentially over time in the same subject. This has been accomplished in vivo by angiographic examination (DePalma et al. 1980), as well as by serial arteriotomy for direct visualiztion of change in individual lesions (DePalma et al. 1977). Portman et al. (1967) reported evidence of disease stabilization in squirrel monkeys achieved by returning the animals to regular chow following 3 months on a high-cholesterol diet. Animals returned to the nonatherogenic diet for 3–5 months showed significantly fewer aortic lesions and less arterial wall lipid content than did those who remained on the atherogenic diet. In the early 1970s, Armstrong et al. (1970) demonstrated unequivocal reduction in the size of advancing, stenosing coronary artery lesions in the rhesus monkey. A 17-month induction phase of a high-fat, high-cholesterol diet was followed by 40 months of a cholesterol-free diet, which allowed total cholesterol to return to baseline levels (3.62 mmol/l; 140 mg/dl). The lesions produced during dietary induction were found to average 60% obstruction in all main epicardial coronary arteries. After 40 months on regression diets, average coronary lumen obstructions had decreased to approximately 21%.

STARY (1972) observed normalization of cell kinetics in arterial lesions during regression after returning monkeys to a basal diet for 3–10 months. Reparative changes included a measurable decrease in intracellular lipid and a return to normal cellular proliferative patterns. Initial signs of disease reversibility were seen as early as 4 weeks after the total cholesterol had returned to baseline levels. Foam cell production stopped and normalization of smooth muscle cell morphology was observed. TUCKER et al. (1971) returned rhesus monkeys fed a high-cholesterol, high-fat diet for 2 months to a basal diet for 4 months and found a resulting decrease in the quantity of foam cells and intracellular lipid and an increase in extracellular lipid. Biochemical verification of lipid depletion has been demonstrated by SRINIVASAN et al. (1980) in cynomolgus monkeys and by ARMSTRONG and MEGAN (1972) in rhesus monkeys. ARMSTRONG and MEGAN (1972) observed that in arteries that had undergone regression, arterial wall cholesterol content decreased from 51 mg/g dry weight to 18 mg/g dry weight and that cholesteryl ester and free cholesterol decreased by 69% and 53%, respectively. Regression of the atherosclerotic lesions was also accompanied by measurable reductions in the arterial wall collagen and elastin content (ARMSTRONG and MEGAN 1975).

Wissler's group has shown that experimental lesions in rhesus monkeys can be reversed by treatment with cholestyramine (WISSLER et al. 1975). Male rhesus monkeys were kept on an atherogenic diet for 1 year and then switched to a low-cholesterol, low-fat diet, with or without the addition of cholestyramine. A third group of animals received cholestyramine while remaining on the atherogenic diet. The amount of aortic surface area affected with grossly visible lesions was significantly less at 1 and 2 years in all three treatment groups than among control animals who remained on the atherogenic diet (10%–31% versus 62%–84%). Low-cholesterol, low-fat diet with added cholestyramine produced the most favorable results, particularly in the coronary arteries where there was 1%–5% luminal narrowing versus 12%–36% narrowing among animals remaining on the atherogenic diet. Using female cynomolgus monkeys, MALINOW et al. (1978b) have also shown coronary artery regression in monkeys fed an atherogenic diet for 6 months and then treated with cholestyramine for 18 months. Cholestyramine decreased the plasma cholesterol levels to normal (4.24 mmol/l; 164 mg/dl) and normalized lipoprotein patterns.

KRAMSCH et al. (1981a) studied the effects of exercise during induction of atherosclerosis. Regular moderate exercise in cynomolgus monkeys comparable to jogging in humans increased coronary vessel caliber and decreased lesion size. A control group fed regular chow for 36 months was compared to a group of monkeys on an atherogenic diet for 36 and 42 months. Half of the latter group received regular treadmill exercising three times per week while the other half was sedentary. In the sedentary monkeys on the atherogenic diet, coronary artery mean involvement was 46%–60% compared to 15%–20% in the exercising monkeys on the atherogenic diet.

The threshold for regression levels of total cholesterol tested in monkeys are well within ranges attainable in human beings. CLARKSON et al. (1984) induced plasma cholesterol levels of approximately 15.51 mmol/l (600 mg/dl) in rhesus monkeys and then compared regression regimens with levels of 5.17 and 7.75 mmol/l (200 and 300 mg/dl). After 24 months, 13 of 19 animals in the 5.17 mmol/l group and nine of 18 animals in the 7.75 mmol/l group showed coronary artery lesion regression, a nonsignificant difference between the two groups. At 48 months, 5.17 mmol/l prevented progression in practically all of the animals. The authors also reported an inverse relationship between HDL-C and coronary artery lesion progression. Only regression and no progression occurred at a total cholesterol to HDL-C ratio of 2.5. No regression was seen in animals with a total cholesterol to HDL-C ratio greater than 3.5. Changes seen in the abdominal aorta were comparable to those observed in the coronary arteries.

ARMSTRONG et al. (1970) found that a serum cholesterol level of 3.62 mmol/l (140 mg/dl) resulted in regression of severe atherosclerotic lesions in the coronary artery and aorta of rhesus monkeys. VESSELINOVITCH et al. (1976) observed that returning rhesus monkeys to a low-cholesterol, low-fat diet for 18 months after 18 months of a high-cholesterol, high-fat diet leads to regression of advanced coronary artery and aortic atherosclerotic lesions when plasma cholesterol levels are maintained at or below 4.52 mmol/l (175 mg/dl). DEPALMA et al. (1980) found that regression of fatty-fibrous plaques in rhesus monkeys occurred at plasma cholesterol levels below 5.17 mmol/l (200 mg/dl).

I. Summary

Three major morbidity- and mortality-based cholesterol intervention studies have provided clear evidence for benefit from drug-induced reduction of blood lipid levels. One mechanism underlying this benefit, shown in controlled human angiographic trials, is stabilization and regression of coronary and femoral lesions. Other vascular beds, including the carotid and renal arteries, are also reported to benefit in noncontrolled trials and case series. Evidence for benefit from diet treatment alone in morbidity- and mortality-based trials is encouraging, but less conclusive. Two angiographic studies, one a controlled trial and the other uncontrolled, indicate benefits from diet counseling and lifestyle change. Reports in man are substantiated by histologic evidence for lesion regression in experimental animal models. These include controlled, quantitative regression studies in nonhuman primates.

The sum of evidence for success in prevention and reversibility of atherosclerosis is of sufficient strength that lowering of blood lipid level should be initiated soon after diagnosis of coronary heart disease. Further, coronary risk factor reduction, including appropriate diet counseling, should be initiated early in adult life to prevent disease development.

References

Armstrong ML, Megan MB (1972) Lipid depletion in atheromatous coronary arteries in rhesus monkeys after regression diets. Circ Res 30:675–680

Armstrong ML, Megan MB (1975) Arterial fibrous proteins in cynomolgus monkeys after atherogenic and regression diets. Circ Res 36:256–261

Armstrong ML, Warner ED et al. (1970) Regression of coronary atheromatosis in rhesus monkeys. Circ Res 27:59–67

Arntzenius AC, Kromhout D et al. (1985) Diet, lipoproteins, and the progression of coronary atherosclerosis. The Leiden Intervention Trial. N Engl J Med 312:805–811

Barndt R JR, Blankenhorn DH et al. (1977) Regression and progression of early femoral atherosclerosis in treated hyperlipoproteinemic patients. Ann Intern Med 86:139–146

Basta LL, Williams C et al. (1976) Regression of atherosclerotic stenosing lesions of the renal arteries and spontaneous cure of systemic hypertension through control of hyperlipidemia. Am J Med 61:420–423

Blankenhorn DH, Nessim SA el al. (1987) Beneficial effects of combined colestipol–niacin therapy on coronary atherosclerosis and coronary venous bypass grafts. JAMA 257:3233–3240

Blankenhorn DH, Alaupovic P et al. (1990) Prediction of angiographic change in native human coronary arteries and aortocoronary bypass grafts: lipid and non-lipid factors. Circulation 81:470–476

Brensike JF, Levy RI et al. (1984) Effects of therapy with cholestyramine on progression of coronary arteriosclerosis: results of the NHLBI Type II Coronary Intervention Study. Circulation 69:313–324

Brown G, Albers JJ et al. (1990) Regression of coronary artery disease as a result of intensive lipid-lowering therapy in men with high levels of apolipoprotein B. N Engl J Med 323:1289–1298

Buchwald H, Varco RL et al. (1990) Effect of partial ileal bypass surgery on mortality and morbidity from coronary heart disease in patients with hyper-cholesterolemia. Report of the Program on the Surgical Control of the Hyper-lipidemias (POSCH). N Engl J Med 323:946–955

Canner PL, Berge KG et al. (1986) Fifteen year mortality in Coronary Drug Project patients: long-term benefit with niacin. J Am Coll Card 8:1245–1255

Cashin-Hemphill L, Mack WJ et al. (1990) Beneficial effects of colestipol–niacin on coronary atherosclerosis. A 4-year follow-up. JAMA 264:3013–3017

Clarkson TB, King JS et al. (1973) Pathologic characteristics and composition of diet-aggravated atherosclerotic plaques during "regression". Exp Mol Pathol 19:267–283

Clarkson TB, Bond MG et al. (1984) A study of atherosclerosis regression in Macaca mulatta: V. Changes in abdominal aorta and carotid and coronary arteries from animals with atherosclerosis induced for 38 months and then regressed for 24 or

48 months at plasma cholesterol concentrations of 300 or 200 mg/dl. Exp Mol Pathol 41:96–118
Cohn K, Sakai FJ et al. (1975) Effect of clofibrate on progression of coronary disease: a prospective angiographic study in man. Am Heart J 89:591–598
Coronary Drug Project Research Group (1975) The Coronary Drug Project. Clofibrate and niacin in coronary heart disease. JAMA 231:360–381
Crawford DW, Sanmarco ME et al. (1979) Spatial reconstruction of human femoral atheromata showing regression. Am J Med 66:784–789
Dayton S, Pearce ML et al. (1969) A controlled clinical trial of a diet high in unsaturated fat in preventing complications of atherosclerosis. Circulation 40 [Suppl II]:1-63
DePalma RG, Hubay CA et al. (1970) Progression and regression of experimental atherosclerosis. Surg Gynecol Obstet 131:633–647
DePalma GR, Bellon EM et al. (1977) Approaches to evaluating regression of experimental atherosclerosis. Adv Exp Med Biol 82:459–470
DePalma RG, Klein L et al. (1980) Regression of atherosclerotic plaques in rhesus monkeys. Angiographic, morphologic, and angiochemical changes. Arch Surg 115:1268–1278
Duffield RGM, Miller NE et al. (1983) Treatment of hyperlipidemia retards progression of symptomatic femoral atherosclerosis. A randomized controlled trial. Lancet 1:639–642
Erikson U, Helmius G et al. (1983) Measurement of atherosclerosis by arteriography and microdensitometry. Model and clinical investigations. In: Schettler G, Gotto AM et al. (eds) Atherosclerosis VI. Springer, Berlin Heidelberg New York, p 197
Farrar DJ, Green HD et al. (1980) Reduction in pulse wave velocity and improvement of aortic distensibility accompanying regression of atherosclerosis in the rhesus monkey. Circ Res 47:425–432
Frick MH, Elo O et al. (1987) Helsinki Heart Study: primary-prevention trial with gemfibrozil in middle-aged men with dyslipidemia. Safety of treatment, changes in risk factors, and incidence of coronary heart disease. N Engl J Med 317: 1237–1245
Fritz KE, Augustyn JM et al. (1975) Effect of moderate diet and clofibrate on regression of swine atherosclerosis. Circulation 52:II–16
Fritz KE, Augustyn JM et al. (1976) Regression of advanced atherosclerosis in swine. Arch Pathol Lab Med 100:380–385
Harrison DG, Armstrong ML et al. (1987) Restoration of endothelium-dependent relaxation by dietary treatment of atherosclerosis. J Clin Invest 80:1808–1811
Heistad DD, Breese K et al. (1987) Cerebral vasconstrictor responses to serotonin after dietary treatment of atherosclerosis: implications for transient ischemic attacks. Stroke 18:1068–1073
Hjermann I, Enger SC et al. (1970) The effect of dietary changes on high density lipoprotein cholesterol. The Oslo Study. Am J Med 66:105–109
Horlick K, Katz LN (1949) Retrogression of atherosclerotic lesions on cessation of cholesterol feeding in the chick. J Lab Clin Med 34:1427
Kane JP, Malloy MJ et al. (1990) Regression of coronary atherosclerosis during treatment of familial hypercholesterolemia with combined drug regimens. JAMA 264:3007–3012
Keller C, Spengel FA (1988) Changes of atherosclerosis of the carotid arteries due to severe familial hypercholesterolemia following long-term plasmapheresis, assessed by duplex scan. Klin Wochenschr 66:149–152
Kramsch DM, Aspen AJ et al. (1981a) Reduction of coronary atherosclerosis by moderate conditioning exercise in monkeys on an atherogenic diet. N Engl J Med 305:1483–1489
Kramsch DM, Aspen AJ et al. (1981b) Atherosclerosis: prevention by agents not affecting abnormal levels of blood lipids. Science 213:1511–1512

Kuo PT, Hayase K et al. (1979) Use of combined diet and colestipol in long-term (7½ years) treatment of patients with type II hyperlipoproteinemia. Circulation 59:199–211
Leren P (1970) The Oslo diet-heart study: eleven year report, Circulation 42:935–942
Levy RI, Brensike JF et al. (1984) The influence of changes in lipid values induced by cholestyramine and diet on progression of coronary artery disease: results of NHLBI Type II Coronary Intervention Study. Circulation 69:325–337
Lipid Research Clinics Coronary Primary Prevention Trial (1984) Results. II. The relationship of reduction in incidence of coronary heart disease to cholesterol lowering. JAMA 251:365–374
Malinow MR (1980) Atherosclerosis. Regression in nonhuman primates. Circulation 46:311–320
Malinow MR, McLaughlin P et al. (1978a) Effect of alfalfa meal on shrinkage (regression) of atherosclerotic plaques during cholesterol feeding in monkeys. Atherosclerosis 30:27–43
Malinow MR, McLaughlin P et al. (1978b) Treatment of established atherosclerosis during cholesterol feeding in monkeys. Atherosclerosis 31:185–193
Mendelsohn D, Mendelsohn L (1989) Effect of polyunsaturated fat on regression of atheroma in the non-human primate. S Afr Med J 76:371–373
Morin RJ, Bernick L et al. (1964) Effects of essential fatty acid deficiency and supplementation of atheroma formation and regression. J Atherosler Res 4:387
Nash DT, Gensini G et al. (1982) Effect of lipid-lowering therapy on the progression of coronary atherosclerosis assessed by scheduled repetitive coronary arteriography. Int J Cardiol 2:43–55
Newcastle Study (1971) Trial of clofibrate in the treatment of ischaemic heart disease: five-year study by a group of physicians of the Newcastle upon Tyne region. BMJ 4:767–775
Nikkila EA, Viikinkoski P et al. (1984) Prevention of progression of coronary atherosclerosis by treatment of hyperlipidaemia: a seven year prospective angiographic study. Br Med J [Clin Res] 289(6439):220–223
Oliver MF, Heady JA et al. (1978) A cooperative trial in the primary prevention of ischemic heart disease using clofibrate. Br Heart J 40:1069–1118
Ornish D, Brown SE et al. (1990) Can lifestyle changes reverse coronary heart disease? The Lifestyle Heart Trial. Lancet 336:129–133
Ost RC, Stenson S (1967) Regression of peripheral atherosclerosis during therapy with high doses of nicotinic acid. Scand J Clin Lab Invest [Suppl] 99:241–245
Portman OW, Alexander M et al. (1967) Nutritional control of arterial lipid composition on squirrel monkeys. Major ester classes and types of phospholipids. J Nutr 91:35
Rafflenbeul W, Smith LR et al. (1979) Quantitative coronary arteriography. Coronary anatomy of patients with unstable angina pectoris reexamined 1 year after optimal medical therapy. Am J Cardiol 43:699–707
Research Committee (1965) Low-fat diet in myocardial infarction: a controlled trial. Lancet I:501–504
Research Committee to the Medical Research Council (1968) Controlled trial of soya-bean oil in myocardial infarction. Lancet I:693–700
Rose GA, Thomson WB et al. (1965) Corn oil in the treatment of ischaemic heart disease. BMJ 1:1531–1533
Roth D, Kostuk WJ (1980) Noninvasive and invasive demonstration of spontaneous regression of coronary artery disease. Circulation 62:888–896
Schettler G (1979) Cardiovascular diseases during and after World War II: a comparison of the Federal Republic of Germany with other European countries. Prev Med 8:581
Scottish Society Study (1971) Ischaemic heart disease: a secondary prevention trial using clofibrate. BMJ 4:775–784
Srinivasan SR, Patton D et al. (1980) Lipid changes in atherosclerotic aortas of Macaca fascicularis after various regression regimens. Atherosclerosis 37:591–601

Stary HC (1972) Progression and regression of experimental atherosclerosis in rhesus monkeys. In: Goldsmith EF, Morr-Hankowsky J (eds) Medical primatology. Karger, Basel, p 356
Stein EA, Adolph R et al. (1986) Nonprogression of coronary artery atherosclerosis in homozygous familial hypercholesterolemia after 31 months of repetitive plasma exchange. Clin Cardiol 9:115–119
Subbiah MT, Dicke BA et al. (1978) Regression of naturally occurring atherosclerotic lesions in pigeon aorta by intestinal bypass surgery. Atherosclerosis 31:117
Terry EN, Rouen LR et al. (1976) Attempts to delay progression in occlusive atherosclerosis. Ann NY Acad Sci 275:379–385
Thompson GR, Myant NB et al. (1980) Assessment of long-term plasma exchange for familial hypercholesterolaemia. Br Heart J 43:680–688
Tucker CF, Catsulis C et al. (1971) Regression of early cholesterol-induced aortic lesions in rhesus monkeys. Am J Pathol 65:494–502
Vesselinovitch D, Wissler RW et al. (1976) Reversal of advanced atherosclerosis in rhesus monkeys: I. Light-microscopic Studies. Atherosclerosis 23:155–176
Vesselinovitch D, Wissler RW et al. (1978) The effect of diets with or without cholestryamine on the lesion components of atherosclerotic plaques. Fed Proc 37:835
Wilens SL (1947) The resorption of arterial atheromatous deposits in wasting disease. Am J Pathol 23:793–804
Wissler RW, Vesselinovitch D et al. (1975) Regression of severe atherosclerosis in cholestryamine treated rhesus monkeys with or without a low fat, low-cholesterol diet. Circulation 52:II-16
Woodhill JM, Palmer AJ et al. (1978) Low fat, low cholesterol diet in secondary prevention of coronary heart disease. Adv Exp Med Biol 109:317–330
Yokoyama S, Yamamoto A et al. (1987) LDL-apheresis; potential procedure for prevention and regression of atheromatous vascular lesion. Jpn Circ J 51: 1116–1122

CHAPTER 9

The Dietary Therapy of Hyperlipidemia: Its Important Role in the Prevention of Coronary Heart Disease

W.E. CONNOR and S.L. CONNOR

A. Introduction

Dietary therapy is the mainstay for the treatment of hyperlipidemia. This is true regardless of the cause, genetic or nutritional. Even in hyperlipidemia that persists after the correction of metabolic errors (e.g., diabetes mellitus) or endocrine abnormalities (e.g., hypothyroidism), diet should play a predominant role. The advantages of dietary therapy, which will be amplified in the subsequent discussion, are categorized in Table 1.

The use of hypolipidemic drugs, which may be necessary in some patients, should not be regarded as an alternative form of therapy, but as an additive therapy. Dietary factors can assist or interfere with the actions of all hypolipidemic drugs. The mechanisms by which dietary factors and drugs affect the plasma lipoproteins are reasonably well known, and their interactions can be deduced. A good example is the interaction between the 3-hydroxy-3-methylglutaryl coenzyme A (HMG-CoA) reductase inhibitor, lovastatin, and dietary cholesterol and saturated fat. Lovastatin increases the hepatic low-density lipoprotein (LDL) receptor activity, promoting the removal of LDL from the plasma and the lowering of plasma cholesterol levels (BROWN and GOLDSTEIN 1986). Dietary cholesterol and saturated fat do the opposite; they decrease the hepatic LDL receptor activity and, ultimately, raise the plasma total and LDL cholesterol levels (SPADY and DIETSCHY 1988). The full benefit of lovastatin will not be achieved unless the diet also contains low amounts of cholesterol and saturated fat.

The dietary factors of the typical American diet cause hyperlipidemia in susceptible individuals by affecting both the synthesis and secretion of lipids into the plasma and the removal of lipids from the plasma (Table 2). The liver is the organ responsible both for the synthesis and, most importantly, for the catabolism of lipoproteins. The LDL receptor plays a crucial role in the removal process (BROWN and GOLDSTEIN 1983).

Historically and to the present time, a vast amount of evidence in both humans and animals has pointed to certain dietary factors which have hyperlipidemic effects. Some have a hypolipidemic action. Many other factors have no or minimal effects. These nutritional factors will be discussed briefly to provide a theoretical and practical basis for dietary prescription in the treatment of hyperlipidemia. It must also be appreciated that certain dietary factors can affect thrombosis, an event which adds greatly to the organ ischemia produced by atherosclerotic blockage of blood flow.

The major dietary factors to be considered include the following:

Hyperlipidemic dietary factors

- Dietary cholesterol
- Saturated fat
- *Trans* fatty acids
- Total fat
- Total calories with adiposity
- Alcohol (in some individuals)

Hypolipidemic dietary factors

- Polyunsaturated fat: omega-6-rich vegetable oils[1] and omega-3-rich fish and fish oils
- Monounsaturated fat[1]
- Soluble fibers (pectin, guar gum)
- Carbohydrate as starches replacing fat
- Possibly vegetable protein or other substances from vegetables

Dietary factors with no discrete long-term effects

- protein generally
- Vitamins and minerals
- Lecithin

Table 1. The advantages of dietary therapy

1. Inexpensive
2. Can become habitual and can be used by the entire family
3. Safe over the lifetime
4. Will achieve therapeutic goals without drugs in many patients
5. Will augment the action of all lipid lowering drugs
6. Is antithrombotic as well

Table 2. Causes of hyperlipidemia

1. Overproduction of triglyceride, VLDL, cholesterol, LDL, and apoB and subsequent secretion into the plasma.
2. Defective removal by the liver: chylomicron and VLDL remnants by the apoE receptor LDL by the apoB-100 receptor

VLDL, very low density lipoprotein; LDL, low-density lipoprotein

[1] Very high amounts of polyunsaturated or monounsaturated vegetable fat, while resulting in a hypocholesterolemic effect overall, would produce postprandial hypertriglyceridemia and increased remnant formation.

B. Dietary Cholesterol

Dietary cholesterol enters the body via the chylomicron pathway and is removed from the plasma by the liver as a component of chylomicron remnants. Only about 40% of ingested cholesterol is absorbed. The remaining 60% is excreted in the stool. Since the feedback inhibition of cholesterol biosynthesis in the body is only partial, even with a large dietary cholesterol intake, dietary cholesterol is added to the cholesterol synthesized by the body (CONNOR and CONNOR 1982). The sterol nucleus cannot be broken down like fat, protein and carbohydrate. As a consequence, cholesterol must be either excreted or stored. Cholesterol is excreted in the bile and, ultimately, in the stool. It is excreted intact or as bile acids synthesized in the liver. There, an efficient enterohepatic reabsorption and circulation returns much of what is excreted into the bile back into the body. Thus, it is easy to see how a particular tissue, e.g., a coronary artery, can become overloaded with cholesterol if the influx of cholesterol exceeds its efflux. Such is the case when the plasma LDL cholesterol concentration is elevated.

Dietary cholesterol is removed by the liver as a component of chylomicron remnants. It does not directly enter into the formation of the lipoproteins synthesized in the liver, very low density lipoproteins (VLDL) and LDL. It does, however, profoundly affect the catabolism of LDL. This effect is mediated through the LDL receptor (BROWN and GOLDSTEIN 1983). Since dietary cholesterol ultimately contributes to the total amount of hepatic cell cholesterol, it can affect both the biosynthesis of cholesterol and the activity of the LDL receptor in the liver (SPADY and DIETSCHY 1988). In particular, an increase in hepatic cell cholesterol, usually not compensated for by a reduction in cholesterol biosynthesis, will decrease synthesis of messenger RNA for the LDL receptor, which decreases the LDL receptor activity and, subsequently, causes an increase in the level of LDL cholesterol in the plasma (SORCI-THOMAS et al. 1989; FOX et al. 1987). Conversely, a drastic decrease in dietary cholesterol will increase the LDL receptor activity in the liver, enhance LDL removal, and, hence, lower plasma LDL cholesterol levels.

Twenty-six separate metabolic experiments over the past 30 years involving 196 human subjects have shown profound effects of dietary cholesterol upon plasma total and LDL cholesterol levels (CONNOR and CONNOR 1989; HOPKINS 1992). These data document the importance of dietary factors in hyperlipidemia associated with any phenotype or genotype. At one extreme are the Tarahumara Indians, who have an average plasma cholesterol level of 120 mg/dl. They consume a low-cholesterol, low-fat diet and respond to an increase in dietary cholesterol with a 20% increase in the plasma cholesterol and a 24% increase in LDL cholesterol concentrations (MCMURRY et al. 1982). This is similar to an increase of 17% in the mean plasma cholesterol that occurred when 1 g of dietary cholesterol was added to a cholesterol-free diet in 25 subjects (11 normal and seven with type II-a mild,

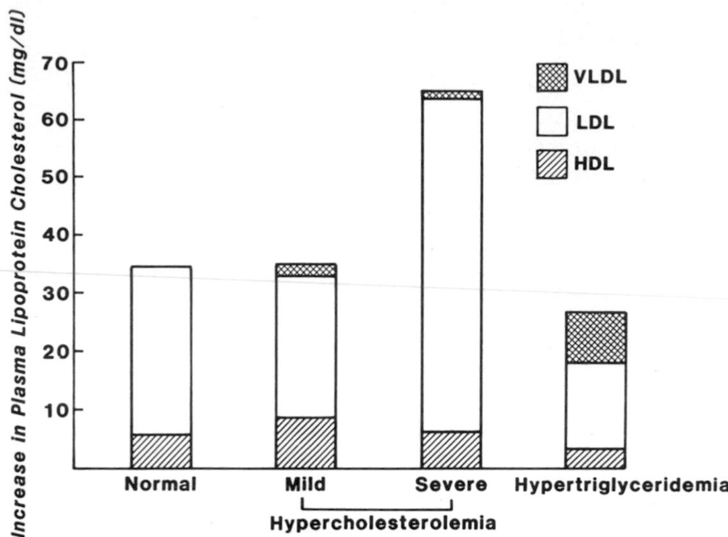

Fig. 1. The effect of a 1 g cholesterol diet upon the plasma lipoproteins in 25 subjects. The control diet was cholesterol free. *VLDL*, very low density lipoprotein; *LDL*, low-density lipoprotein; *HDL*, high-density lipoprotein

five with II-a severe, and nine with type IV hyperlipidemia; Fig. 1). LDL cholesterol increased very significantly in all groups, again showing indirectly the effects of dietary cholesterol upon the LDL receptor. At the other extreme are patients with familial hypercholesterolemia. These patients respond well to the reduction of dietary cholesterol. The plasma cholesterol level even decreased by 18% and 21% in two homozygote patients in response to the removal of cholesterol from the diet.

Large changes in the amounts of dietary cholesterol will not necessarily increase the plasma cholesterol if dietary cholesterol intake is already substantial. For example, an increase in dietary cholesterol from 475 to 950 mg per day will not elevate the plasma cholesterol level. This phenomenon has been restudied from time to time and mistakenly interpreted as showing that dietary cholesterol has no effect on the plasma cholesterol levels. These dietary cholesterol feeding studies have been exhaustively reviewed for those who wish to explore the subject more fully (Hopkins 1992; Roberts et al. 1981).

The effect of gradually increasing amounts of dietary cholesterol upon the plasma cholesterol level is shown in Fig. 2. These findings are supported by both animal and human experiments. In the context of a cholesterol-free diet, the amount of dietary cholesterol necessary to produce a measurable increase in plasma cholesterol is termed "the threshold amount." As the dietary cholesterol is increased, the plasma cholesterol increases and then plateaus. The dietary cholesterol at this inflection point is termed "the ceiling amount." Increasing the dietary cholesterol further does not lead

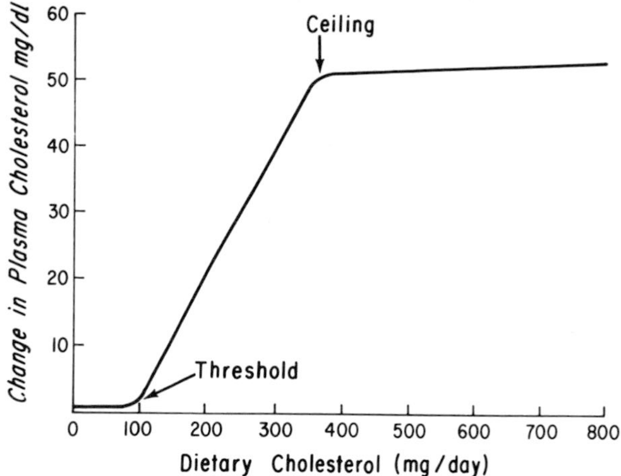

Fig. 2. The effects upon the plasma cholesterol level of gradually increasing the amount of dietary cholesterol in human subjects whose background diet is very low in cholesterol content. See the text for discussion of the threshold and ceiling concepts

to higher levels of plasma cholesterol, even though phenomenally large amounts may be fed. Each animal or human being probably has its own unique threshold and ceiling amount. Generally speaking, the average threshold amount for human beings is about 100 mg/day. The average ceiling is in the neighborhood of 300–400 mg/day. Thus, a baseline dietary cholesterol intake of 500 mg/day from two eggs would, for most individuals, exceed the ceiling. The addition of two more egg yolks for a total of 1000 mg/day, however, would not then further increase the plasma cholesterol concentration. Beginning with a low cholesterol diet under 100 mg/day and adding the equivalent of two egg yolks, or 500 mg, would produce a striking change in plasma cholesterol concentration. This change can be as much as 60 mg/dl.

Recent surveys indicate that the average American dietary cholesterol intake is about 400 mg/day for women and 500 mg/day for men (GORDON et al. 1982). Decreasing these amounts of dietary cholesterol to 100 mg/day, the objective of the therapeutic and preventive diets to be amplified below, would have a profound effect on plasma cholesterol concentrations. Operationally, one would be on the descending limb of the curve illustrated in Fig. 2.

C. Effects of Dietary Fats upon the Plasma Lipids and Lipoproteins

The amount and kind of fat in the diet affect plasma lipid concentrations. The total amount of dietary fat is important in that the formation of chylomicrons in the intestinal mucosa and their subsequent circulation in the blood is directly proportional to the amount of fat which has been consumed in the diet. A fatty meal will result in the production of large numbers of chylomicrons and will impart the characteristic lactescent appearance to postprandial plasma 3–5 h after eating. A typical American diet with 110 g of fat produces 110 g of chylomicron triglyceride per day. "Remnant" production from chylomicrons is proportional to the number of chylomicrons synthesized. Chylomicron remnants resulting from the action of lipoprotein lipase are cholesterol rich and are atherogenic particles (ZILVERSMIT 1979). Postprandial lipemia is, of course, intense after the usual American diet and may be present for many hours before being cleared. Not only is this lipemia (the composite of chylomicrons and remnants) atherogenic, but it may also promote thrombosis. Postprandial lipemia is lessened by physical activity and by a diet low in fat and/or a diet containing omega-3 fatty acids from fish. It is worse and very prolonged in patients with fasting hypertriglyceridemia whose clearance mechanisms are already impaired.

Different types of fat have different effects on the plasma cholesterol levels. Fats may be divided into three major classes based on the degree of saturation of the fatty acid. Long-chain, saturated fatty acids have no double bonds, are not essential nutrients, and are readily synthesized in the body from acetate. Saturated fatty acids in the diet have a powerful hypercholesterolemic effect, increase the concentrations of LDL, and are thrombogenic as well. All animal fats are highly saturated (30% or more of the fat is saturated) and contain little polyunsaturated fatty acid, except for the fat of fish and shellfish, which is, in contrast, highly polyunsaturated. The molecular basis for the effects of dietary saturated fat on the plasma cholesterol level is now well understood. It rests upon its influence on the LDL receptor activity of liver cells, as described by BROWN and GOLDSTEIN (1983, 1986). Dietary saturated fat suppresses messenger RNA synthesis for the LDL receptor, which decreases hepatic LDL receptor activity, decreases the removal of LDL from the blood, and thus increases the concentration of LDL cholesterol in the blood (SPADY and DIETSCHY 1988; FOX et al. 1987). Cholesterol augments the effect of saturated fat by further suppressing hepatic LDL receptor activity and raising the plasma LDL cholesterol level. Conversely, a decrease in dietary cholesterol and saturated fat increases the LDL receptor activity of the liver cells, enhances the hepatic pickup of LDL cholesterol, and lowers the concentration of LDL cholesterol in the blood (SPADY and DIETSCHY 1988). Some saturated fats, such as coconut oil, increase the synthesis of cholesterol and LDL in the liver. Metabolic studies suggest that one can expect an average plasma cholesterol decrease of 20% by maximally decreasing dietary cholesterol and saturated fat.

Besides natural sources of saturated fats, the hydrogenation of liquid vegetable oils can saturate some or all of the unsaturated fatty acids. The soft margarines and shortenings are lightly hydrogenated. The softer a margarine is at room temperature the less hydrogenated it is. Peanut butter is so lightly hydrogenated that its fatty acid composition is little affected.

Table 3. Effects of saturated fatty acids on plasma cholesterol levels

Fatty Acid	Action
C8, C10: medium chain	Neutral
C12: lauric	Increase
C14, C16: myristic, palmitic	Increase
C18: stearic	Neutral

Large quantities of highly hydrogenated fat should be avoided in order to keep the total saturated fat low. The daily use of small quantities of the soft margarines is quite acceptable in the context of a low-fat diet. Monounsaturated *trans* fatty acids, isomers of the *cis* oleic acid, are important byproducts of the hydrogenation process. *Trans* fatty acids are oxidized for energy, as are other fatty acids. In a recent study in which large amounts of *trans* fatty acids were consumed, the plasma LDL cholesterol level was significantly elevated (Memsink and Katan 1990). However, the presence of small amounts of *trans* fatty acids in lightly hydrogenated margarines (those in which a liquid vegetable oil is listed as the first ingredient) does not constitute a problem in the diet for the treatment of hyperlipidemia.

Attention has been called to the fact that some saturated fats do not seem to cause hypercholesterolemia (see Table 3). Medium chain triglycerides (C8 and C10 saturated fatty acids) are water soluble and are handled metabolically more like carbohydrate than fat. They are transported to the liver via the portal vein blood rather than as chylomicrons. These fatty acids do not elevate the plasma cholesterol concentration.

Stearic acid, an 18-carbon saturated fatty acid, also has a limited effect upon the plasma cholesterol concentration. Excessive stearic acid from the diet is converted into oleic acid, a monounsaturated fatty acid, by a desaturase enzyme. Feeding animals large quantities of stearic acid, such as cocoa butter containing a considerable percentage of its total fatty acids as stearic acid (33%), does not result in the deposition of stearic acid in the adipose tissue, as would occur with mono- and polyunsaturated fat feeding. This, again, is because of the action of the desaturase enzyme. The practical importance of these observations about stearic acid is limited because it is present in foods that also contain appreciable amounts of other fatty acids that cause hypercholesterolemia. Palmitic acid is the most common saturated fat found in our food supply. It has 16 carbons and is intensely hypercholesterolemic. Myristic acid and lauric acid with 14 and 12 carbons, respectively, are also intensely hypercholesterolemic. It is these fatty acids present in "saturated" dietary fats that cause their untoward effects. Amounts of stearic acid in the American diet are not great compared with palmitic acid (Table 4). Also, the equations developed for the prediction of plasma cholesterol change have been based upon the changes produced by a given

Table 4. Sources of specific saturated fatty acids in foods and fats

Food	Fatty acids
MCT	C8, C10
Coconut oil	C12, C14
Butter	C14, C16, C18
Beef tallow, lard	C16, C18
Palm oil	C16
Cocoa butter (chocolate)	C16, C18

MCT, medium chain triglyceride.

fat including its concentration of stearic acid. Thus, all of the information which have accumulated about the hypercholesterolemic and atherogenic properties of a given fat such as beef fat, butterfat, lard, palm oil, cocoa butter, and coconut oil remain completely valid.

D. The Cholesterol-Saturated Fat Index of Foods

The major plasma cholesterol elevating effects of a given food reside in its cholesterol and saturated fat content. To help understand the contribution of these two factors in a single food item and to compare one food with another, we have computed a cholesterol-saturated fat index (CSI) for selected foods (Connor et al. 1986; see Fig. 3). The formula for the CSI is: CSI = (1.01 × g saturated fat) + (0.05 × mg cholesterol), where the amounts of saturated fat and cholesterol in a given amount of a food item

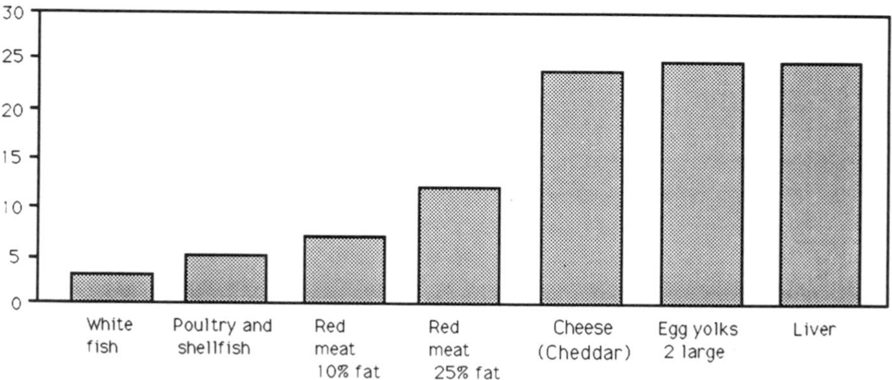

Fig. 3. The cholesterol-saturated fat index (CSI) of 3 oz fish, poultry, shellfish, meat, cheese, egg yolk, and liver. The CSI for poultry is the average CSI for cooked light and dark chicken without skin. The CSI for shellfish is the average CSI of cooked crab, lobster, shrimp, clams, oysters, and scallops. The CSI for cheese is the average CSI of cheddar, Swiss, and processed cheese

are entered into this equation. The higher the CSI of a food, the greater the hypercholesterolemic and atherogenic effect. This cholesterol-saturated fat index is a representation of how much a given food will decrease the activity of the LDL receptor and, hence, will raise the level of LDL cholesterol in plasma.

In this context it is particularly instructive to compare the CSI of fish versus that of moderately fat beef. An 85 g portion of cooked fish contains 51 mg cholesterol and 0.14 g saturated fat. This contrasts to 80 mg cholesterol and 11.3 g saturated fat in 85 g 30%-fat beef. The CSI for the fish is 3 and 15 for the beef, five times greater. The caloric value of these two portions also differs greatly (71 for fish and 323 kcal for beef). The CSI of cooked chicken and turkey (without the skin) is also lower than that of beef and other red meats. The total fat content is considerably lower. The saturated fat in an 85 g serving is 1.6 g and the cholesterol is 71 mg. The CSI of poultry is 5. Table 5 lists the CSI for various foods. Shellfish have low

Table 5. The cholesterol-saturated fat index and kilocalorie content of selected foods

	CSI[a]	kcalories[a]
Fish, poultry, red meat (3 oz or 85 g cooked)		
Shellfish (oysters, scallops, clams)	3	100
Whitefish (snapper, perch, sole, cod, halibut, etc.)	3	100
Salmon	5	168
Shellfish (shrimp, crab, lobster)	5	85
Poultry, no skin	5	154
Beef, pork and lamb:		
10% fat (ground sirloin, flank steak)	7	170
15% fat (ground extra lean, round steak, pork chop)	9	215
25% fat (ground lean, rump roasts)	12	278
30% fat (typical ground beef, ground lamb, steaks, ribs, lamb chops, pork sausage, roasts)	15	323
Cheese (1 oz or 28 g)		
Low-fat ricotta, tofu (bean curd), Dorman's Light[b], Alpine Lace Free N Lean, Kraft Free Singles	<1	53
Imitation mozzarella[b], Hickory Frams Lyte[b], Heart Beat Sharp Cheddar[b]	1	74
Lite-line, part-skim ricotta, Reduced Calories Laughing Cow, lite part-skim mozzarella, Mini Chol (Swedish low fat)[b]	2	62
Lite n'Lively, Olympia's Low Fat, Green River Part-Skim, Kraft Light Naturals, Light cream cheese, Heidi Ann Low Fat Ched-Style, Lappi, String, part-skim mozzarella	4	73
Cheese spreads (jars), Neufchâtel (lower-fat cream cheese), Velveeta, Brie, Swiss, Gruyère	6	93
Cheddar, Roquefort, Jack, American, cream cheese, Havarti, feta	8	111
Eggs		
Whites (two)	0	32
Egg substitute (equivalent to two eggs)	0	98
Whole (two)	24	158

Table 5. *Continued*

	CSI[a]	kcalories[a]
Fats (1/4 cup or 4 tablespoons)		
Peanut butter	6	380
Mayonnaise, fat free, cholesterol free	trace	48
Mayonnaise, light	3	200
Mayonnaise	8	404
Most vegetable oils	7	491
Olive oil	8	486
Soft vegetable margarines	8	405
Soft shortenings	13	464
Bacon grease	20	464
Butter	36	432
Coconut oil, palm oil	37	491
Frozen desserts (1 cup)		
Fruit ices, sorbets	0	255
Nonfat frozen yogurt	trace	224
Sherbet or frozen yogurt (low-fat)	3	249
Ice milk	4	224
Ice cream, 10% fat	15	329
Rich ice cream, 12% fat	19	480
Specialty ice cream, 18% fat	30	580
Milk products (1 cup)		
Skim milk or powdered nonfat	1	85
1% milk, buttermilk	2	100
2% milk	4	122
Whole milk (3.5% fat)	7	149
Cottage cheese, low-fat (2%)	4	204
Cottage cheese, regular	8	222
Liquid nondairy creamers: soybean oil	4	326
Liquid nondairy creamers: coconut oil	23	326
Sour cream	36	468
Imitation sour cream (IMO)	40	480

CSI, cholesterol-saturated fat index.
[a] Averages.
[b] Cheeses made with skim milk and vegetable oils.

CSIs because their saturated fat content is extremely low, despite the fact that their cholesterol or total sterol content is 2.5 to three times higher than fish, poultry, or red meat. Shellfish have an average CSI of 4 per 3 oz. This means that, when considering both cholesterol and saturated fat, shellfish, like poultry, is a better choice than even the leanest red meats. Salmon also has a low CSI and is preferred to meat. We have now calculated the CSI for a thousand foods and incorporated this concept in a new book along with low-fat, low-cholesterol recipes (Connor et al. 1989; Connor and Connor 1991).

The second class of dietary fats consists of the characteristic *monounsaturated fatty acids* present in all animal and vegetable fats. For practical

FATTY ACID NOMENCLATURE | DIETARY SOURCES

FAMILY	FATTY ACID	STRUCTURE	DIETARY SOURCES
ω3	Eicosapentaenoic Acid (C20:5 ω3)	H_3C–3–RCOOH	Marine Oils, Fish
ω6	Linoleic Acid (C18:2 ω6)	H_3C–6–R′COOH	Vegetable Oils
ω9	Oleic Acid (C18:1 ω9)	H_3C–9–R″COOH	Vegetable Oils; Animal Fats

Fig. 4. Nomenclature for fatty acids. Fatty acids can be organized into families according to the position of the first double-bond from the terminal methyl group. Typical fatty acids from three common families are shown. Omega-3 fatty acids all have three carbons between the methyl end and the first double bond. Besides eicosapentaenoic acid (C20:5), other common omega-3 fatty acids are linolenic acid (C18:3) and docosahexaenoic acid (C22:6). Linoleic acid (C18:2) and arachidonic acid (20:4) are the most important omega-6 fatty acids. Oleic acid (C18:1) is the most common fatty acid in the omega-9 family

purposes, oleic acid, having one double bond at the omega-9 position, is the only significant dietary monounsaturated fatty acid (Fig. 4). In general, the effects of dietary monounsaturated fatty acids have been "neutral" in terms of their effects on the plasma cholesterol concentrations, neither raising nor lowering them. They are, however, cholesterol lowering when compared to saturated fat. Reports that Mediterranean basin populations that consume olive oil in large quantities have fewer heart attacks than people in America, however, have led to further investigations of the atherogenic properties of the monounsaturates. Recent studies have shown that large amounts of monounsaturated fat, like polyunsaturated oils, lower plasma total and LDL cholesterol levels when compared with saturated fat (MATTSON and GRUNDY 1985). Furthermore, unlike polyunsaturated oils, monounsaturated fat did not lower the plasma HDL cholesterol level. However, distinct from omega-3 fatty acids from fish oil, monounsaturated fat does not decrease the plasma triglyceride concentrations. Furthermore, monounsaturated fat has no known effect upon prostaglandin metabolism or upon platelet function. As will be indicated later, omega-3 fatty acids are antithrombotic; monounsaturated fat has no such action.

There are several additional points to be made in regards to these recent studies: (a) the "Mediterranean diet" is also rich in fish, beans, fruit, and vegetables and is low in both saturated fat and cholesterol; these could be the decisive factors which influence the lessened incidence of coronary disease and lower plasma cholesterol levels; (b) olive oil is low in saturated fatty acids (which raise plasma cholesterol levels); this is why the recent metabolic experiments have shown some cholesterol lowering from large amounts of monounsaturated fat in the diet; and (c) large amounts of any kind of fat should be avoided to lower the risk of other diseases such as colon or breast cancer and obesity. In addition, all fats, after absorption as chylomicrons, are acted upon by lipoprotein lipase to form remnant particles which circulate in the blood. These are atherogenic. One translation of the latest research on monounsaturated fats is to recommend that patients include them as part of a general lower-fat eating style, but not to consider them as particularly antiatherogenic or hypolipidemic.

E. Polyunsaturated Fatty Acids

The third class of fatty acids are vital constituents of cellular membranes and serve as prostaglandin precursors. Because they cannot be synthesized by the body and are only obtainable from dietary sources, they are "essential" fatty acids (NEURINGER and CONNOR 1986). The two classes of polyunsaturated fatty acids are the omega-6 and omega-3 fatty acids (Fig. 4). The most common examples of omega-6 fatty acids are arachadonic acid, 20 carbons in length with four double bonds, and its dietary precursor, linoleic acid. Linoleic acid is converted to arachidonic acid in the liver. Since the basic structure of omega-6 fatty acids cannot be synthesized by the body, 2%–3% of the total energy in the diet must consist of linoleic acid to meet the metabolic requirements of the body for the omega-6 structure.

Omega-3 fatty acids differ in the position of the first double bond. Counting from the methyl end of the molecule, this double bond is at the third rather than the sixth carbon. Omega-3 fatty acids are also "essential." They constitute important membrane components of the brain, retina, and sperm. Omega-6 and Omega-3 fatty acids are not interconvertible. The dietary sources of omega-3 fatty acids are plant foods, some, but not all, vegetable oils, and leafy vegetables. Fish and shellfish are especially rich in omega-3 fatty acids. Linolenic acid, C18:3, is obtained from vegetable products. Eicosapentaenoic acid (EPA), C20:5, and docosahexaenoic acid (DHA), C22:6, are derived from fish, shellfish, and phytoplankton (the plants of the ocean). These fatty acids are highly concentrated in fish oils. Like omega-6 fatty acids, omega-3 fatty acids are viewed as essential nutrients and a safe intake would be 0.4%–0.6% of total calories. Once either the omega-3 or omega-6 structure comes into the body as the 18-carbon linoleic or linolenic acid, the body can synthesize the longer-chain and more highly polyunsaturated omega-6 or omega-3 fatty acids (20 and 22 carbons).

EPA and DHA are present in the diet in two forms: as triglyceride in the adipose tissue of fish, usually present between muscle fibers, and as membrane phospholipids of the muscle of fish. These highly polyunsaturated fatty acids occupy the middle position of the glycerol skeleton for both triglycerides and phospholipids. In either of these dietary forms, EPA and

DHA are efficiently absorbed from the intestinal tract. After absorption, EPA and DHA readily associate with membranes and are found in the four lipid classes: triglyceride, free fatty acids, cholesterol ester, and phospholipids. Ultimately, they are stored in the adipose tissue and, in experimental animals, even reach the brain and the retina.

There are distinctly different functions in the body for omega-3 and omega-6 fatty acids (GOODNIGHT et al. 1982). Both serve as substrates for the formation of different prostaglandins and leukotrienes and are abundant in phospholipid membranes. Both omega-3 and omega-6 fatty acids are concentrated in nervous tissue. Omega-3 fatty acids are rich in the retina, brain, spermatozoa, the gonads, and many other organs. Omega-6 fatty acids are concentrated in the different plasma lipid classes (cholesterol esters, phospholipids, etc.) and, in addition, play a role in lipid transport.

Polyunsaturated fatty acids in large amounts, either the omega-6 or omega-3 structure, reduce plasma total and LDL cholesterol concentrations in normal and hypercholesterolemic individuals (CONNOR and CONNOR 1989). The situation is somewhat different in hypertriglyceridemic individuals or in those with combined hyperlipidemia. Only the omega-3 fatty acids from fish and fish oil have a decided hypotriglyceridemic effect. VLDL, in particular, is decreased by the omega-3 fatty acids (PHILLIPSON et al. 1985). Isotopic studies have shown that the hypotriglyceridemic effect of omega-3 fatty acids occurs as a result of the depression of triglyceride and VLDL synthesis in the liver and also accelerated catabolism of VLDL from the plasma (HARRIS et al. 1990). In some patients with type IV hypertriglyceridemia and in some with combined hyperlipidemia, there have been reports of an increase in LDL as the plasma triglyceride falls. This also occurs when gemfibrozil is given to these patients. In the severely hypertriglyceridemic type V patients, however, large doses of fish oil (10–15 g/day) produce a dramatic clearing of chylomicronemia and lower both triglyceride and cholesterol concentrations. Fish oil also seems to promote the clearing of chylomicrons after the administration of a fatty meal. This would be of particular benefit in the type V patient, in whom there is great difficulty in clearing chylomicrons. Most experiments have not shown plasma triglyceride lowering from the omega-6 vegetable oils.

In the early days of dietary therapy for hyperlipidemia, it was suggested that large amounts of a vegetable oil such as corn oil could be used to treat hypercholesterolemia. Its large content of the omega-6 fatty acid, linoleic acid, was probably not beneficial. It promoted obesity, gallstone formation, and, more importantly, may have been carcinogenic, as suggested by animal studies. In current diets for the treatment of hyperlipidemia, omega-6 polyunsaturated fatty acids are not increased from the current amount in the American diet, which averages about 6%–8% of the total calories. When saturated and monounsaturated fatty acids are decreased, they should not be replaced by equivalent amounts of polyunsaturated fatty acids from liquid vegetable oil, margarines, or shortenings.

It is not known exactly how much the intake of omega-3 fatty acids should be increased to achieve optimum effect. One study from the Netherlands indicated that men who included fish in their diet twice a week were less likely to die of heart disease. A similar protective effect from eating fish occurred in Japan. In a prospective trial in Wales, the prescription of two fish meals per week led to both a reduction in total mortality and in coronary events (BURR et al. 1989). Even very low fat seafood contains an appreciable amount of omega-3 fatty acids, up to 40% or more of total fatty acids. Eating a total of 12 oz of a variety of fish and shellfish each week would provide 3–5 g omega-3 fatty acids as well as protein, vitamins, and minerals. The fish could be fresh, frozen, or canned without affecting the quantity of omega-3 fatty acids. Patients with hyperlipidemia can only expect to have beneficial effects if they follow this dietary advice, especially if the fish replaced meat in the diet, meat being a major source of saturated fat. Also to be considered are the antithrombotic actions of fish oil mediated through the inhibition of the tromboxane A2 in platelets (GOODNIGHT et al. 1982) and the enhanced clearance of chylomicrons (BURR et al. 1989; HARRIS et al. 1988; WEINTRAUB et al. 1988). Other effects of fish oil include inhibition of platelet-derived growth factor, alteration of certain functions of leukocytes, reduction of blood viscosity, greater fibrinolysis, and also inhibition of intimal hyperplasia in vein grafts used for arterial bypass. Fish oil serves as a therapeutic agent in certain hyperlipidemic states, especially the chylomicronemia of type V hyperlipidemia. Thus, fish oils have not only discrete effects upon the plasma lipids and lipoproteins, but also upon the atherosclerotic and thrombotic process.

F. Carbohydrate

If the total fat content of a hypolipidemic diet is reduced from the current American fat intake of 40% of the total calories to only 20% and if the protein of the diet is kept constant, then the difference in caloric intake between a high-fat diet and a low-fat diet must be made up by increasing the carbohydrate content of the diet. Epidemiological evidence buttresses this basic concept, since populations ingesting a high-carbohydrate diet from complex carbohydrates have low plasma cholesterol levels and a low incidence of coronary disease.

Are there harmful effects from a high-carbohydrate diet? It was demonstrated more than 25 years ago that a sudden increase in the amount of dietary carbohydrate in Americans accustomed to a high-fat diet would dramatically increase the plasma triglyceride concentration. After many weeks of the new diet, however, adaptation occurs and the hypertriglyceridemia subsides. If the dietary carbohydrate is gradually increased as fat is reduced, the hypertriglyceridemia does not occur. Thus, high dietary carbohydrate should not be regarded as a problem, but rather as a caloric replacement for the saturated fat removed from the diet. For example, we recently increased the dietary carbohydrate intake gradually from 45% kcal to 65% kcal over a 28-day period in eight mildly hypertriglyceridemic subjects. There was a significant fall in the mean plasma cholesterol level, from 232 to 198 mg/dl. The mean plasma triglyceride level remained constant, 213–230 mg/dl (ULLMANN et al. 1991).

In a low-fat, high-carbohydrate diet, the majority of the carbohydrate is in the form of cereals and legumes, and not as sucrose or other simple sugars. Americans commonly consume about 20% of the total calories as sucrose or other simple sugars or about half of their carbohydrate intake. In the dietary changes being suggested, sugars would fall to about 10% of the total calories. Sucrose, in large quantities, is mildly hypertriglyceridemic. The primary point is that all simple sugars, including sucrose, are potent promoters of obesity; they are extremely low in bulk and have a high caloric density.

G. Fiber, Saponins, and Antioxidants

Dietary fiber is a broad, nondescript term that includes several carbohydrates thought to be indigestible by the human gut. These include cellulose, hemicellulose, lignin, pectin, and β-glucans. Dietary fiber is only found in plants and is commonly present in unprocessed cereals, legumes, vegetables, and fruits. In ruminant animals, dietary fiber is completely digested by the microbial flora of the rumen. In these animals, fiber provides a major source of energy. In humans, however, dietary fiber contributes little to the caloric content of the diet, but promotes satiety through its bulk. Bulk also greatly affects colonic function. A high-fiber diet produces larger stools and a more rapid intestinal transit, factors which may prevent certain diseases of the colon (i.e., diverticulitis, colon cancer).

Interest in the hypolipidemic effects of fiber dates back at least 30 years. Fiber added to semisynthetic diets fed to rats has usually had a plasma cholesterol lowering effect. Feeding fiber, predominantly in the insoluble form such as wheat bran, has probably not had a hypocholesterolemic effect in humans. Soluble fiber may be different. Large amounts of soluble fiber (17 g/2000 kcal), such as that contained in oat bran or beans, produced a 20% reduction of the plasma total and LDL cholesterol levels (ANDERSON et al. 1984). These decreases were, at least in part, the result of concurrent weight loss. Rich sources of soluble fiber include fruits, oats, and some other cereals, legumes, and vegetables. One way that soluble fiber acts to lower cholesterol levels is to bind bile acids in the gut and prevent their reabsorption. This is the same mechanism as the bile acid binding resins such as cholestyramine.

A high-fiber diet is certainly integral to the dietary concepts for the treatment of hyperlipidemia. The consumption of more foods from vegetable sources will automatically mean a higher consumption of both total and soluble fiber. Such foods are bulky and have a low caloric density, which is helpful for the overweight patient. A feeling of satiety occurs from the consumption of high-fiber foods that also are low in calories. It is likely that foods high in fiber have a mild hypolipidemic effect. If oat bran is consumed in place of bacon and eggs for breakfast, this high-fiber food would also have a beneficial effect indirectly, since it replaced foods high in cholesterol and saturated fat. Finally, plant foods rich in fiber may also contain other hypocholesterolemic substances. One of these is a group of compounds called saponins, which, in monkeys, had a hypocholesterolemic action (MALINOW et al. 1981). Saponins bind cholesterol present in the gut lumen and prevent its absorption.

Other dietary factors which may be important in prevention of the atherogenic process are contained in fruits, vegetables, grains, and beans. Besides the soluble fiber and saponins already mentioned, we should also consider the role of the antioxidants contained in these foods in blocking the oxidation of LDL, since oxidized LDL is believed to be particularly pathogenic in the atherosclerotic process (WITZAM and STEINBERG 1985; PRINCEN et al. 1992). The antioxidant factors in plant foods include β-carotene, ascorbic acid or vitamin C, and α-tocopherol or vitamin E. At this time, there are few data to suggest that consuming these vitamins in the form of vitamin supplements would be helpful. The important point is that when dietary fat is reduced, the consumption of fruits, vegetables, and beans is enhanced, which would increase the intake of the desirable antioxidants in a more natural and physiological form.

H. Protein

The dietary treatment of hyperlipidemia involves, in general, a shift from the consumption of protein derived from animal sources, such as meat and dairy products, to the consumption of more protein from plants. The nutritional adequacy of such protein shifts is assured because mixtures of vegetable proteins, plus the provision of ample low-fat animal protein sources, provide for essential amino acid requirements. Ranges of protein intake from 25 g/day to 150 g/day have been tested for effects upon blood lipids and have been found to have no effect within amounts commonly consumed by Americans. Experiments in animals, however, suggest that an animal protein, such as casein, is definitely hypercholesterolemic and a vegetable protein, such as soy protein, is hypocholesterolemic. There are suggestions, moreover, that the consumption of vegetable protein may have some hypocholesterolemic action in humans. Thus, it may be postulated that a shift in protein intake to include more vegetable protein carries no risk and may confer some benefit to the hyperlipidemic individual.

I. Calories

Excessive caloric intake and adiposity can contribute to both hypertriglyceridemia and hypercholesterolemia by stimulating the liver to overproduce triglyceride and VLDL. This is especially true for abdominal or visceral obesity, in which mobilized free fatty acids from the adipose tissue are carried directly to the liver via the portal circulation. The plasma triglyceride and VLDL concentrations of hypertriglyceridemic patients are greatly reduced by weight reduction and are increased by weight gain. There is little direct evidence, however, that the LDL receptor and plasma cholesterol and LDL concentrations are directly affected by caloric excess. Nonetheless, it is known that obese individuals have a total body cholesterol production higher than that of normal weight individuals. Weight reduction and fasting, which involve a decrease in the consumption of cholesterol and saturated fat, would be expected to increase LDL receptor activity and improve LDL levels in hyperlipidemic patients. It is, therefore, reasonable to advise caloric control and the avoidance of obesity in the dietary management of hyperlipidemia. The role of increased physical activity is most important in weight control.

When excessive dietary cholesterol and fat are combined with excessive calories, a particularly potent hyperlipidemic effect occurs. Such a "holiday" diet occurs particularly in the United States between Thanksgiving and New Year. The effects of such dietary excesses were actually studied in the Tarahumara Indians of Mexico, a people habitually accustomed to a very low-fat diet (McMurry et al. 1991). After a suitable baseline period in which they consumed their typical diet, Tarahumara Indians of both sexes were given a diet high in cholesterol, fat, and calories. Within 2 weeks, their plasma lipids had increased tremendously: total cholesterol 31%, LDL 39%, HDL 31%, and a plasma triglyceride of 18%. Body weight gain averaged 7–8 pounds for each individual. The composite of all of these potent dietary hyperlipidemic factors was certainly greater than any one of them alone.

J. Alcohol

In the past few years, there has been confusion over the relationship of alcohol consumption and factors related to coronary heart disease. Alcohol consumption correlates well with higher values of HDL in the blood and reduced mortality from coronary heart disease (Marmot and Brunner 1991; Hennekens et al. 1978; Hegsted and Ausman 1988; Suh et al. 1992). Since HDL plays a role in "reverse" cholesterol transport, this finding has encouraged some to drink more with the idea that alcohol is "good for the heart." These findings related to alcohol, HDL, and coronary disease, however, are too scant, contradictory, and complex to lend themselves to any such definite conclusions. From a clinical point of view, the typical patient attending a lipid clinic who is overweight and consuming two or more drinks per day could have some problems that are directly related to alcohol consumption itself. Such patients are usually hypertriglyceridemic and have low HDL concentrations (FRY et al. 1973; KUDZMA and SCHONFELD 1970). One aspect of the treatment for hypertriglyceridemia is to reduce alcohol consumption or even stop it completely, since alcohol ingestion increases plasma triglyceride levels. Alcohol also has a high caloric content of 140 kcal per drink. Two to three drinks a day can contribute 300–400 kcal to an already hypercaloric state, thus stimulating further the synthesis of triglyceride and VLDL by the liver (GINSBERG et al. 1974). Even more importantly, alcohol has so many adverse social, psychological, and physical effects that the astute physician would not recommend alcohol even if it does increase HDL levels. On the contrary, the best evidence indicates that those who drink are least likely to enjoy good health. In our Lipid Clinic, we suggest that people restrict alcohol consumption to no more than two to four drinks per week in order to avoid adverse effects upon the plasma lipids and lipoproteins.

K. Coffee and tea

Because of the enjoyment people derive from coffee, almost a national pastime, there is great interest in the ever-conflicting reports on the relationship between coffee and coronary heart disease. One study from Europe does make some sense in that boiled coffee, which would extract most of the ingredients that might have any effect upon health, is positively associated with a higher incidence of coronary heart disease ZOCK et al. 1990). The methods of coffee preparation in the United States are generally different (instant coffee, percolated coffee, drip coffee). In a recent U.S.

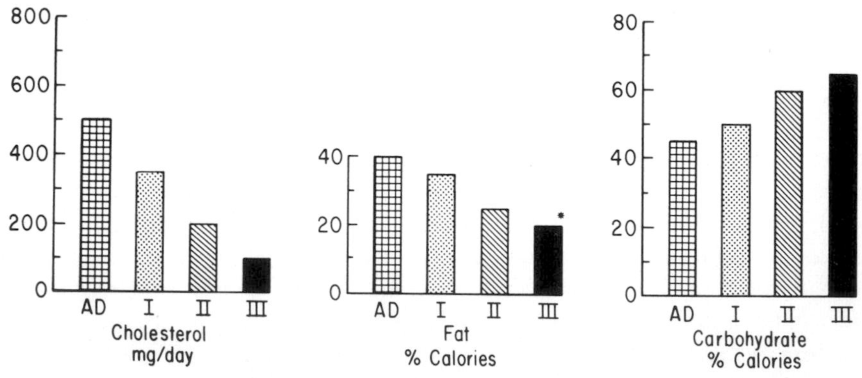

Fig. 5. The cholesterol, fat, and cabohydrate content of the typical American diet (*AD*) and the three phases of the low-fat, high complex carbohydrate diet (*I*, *II*, *III*)

study, only the consumption of 720 ml filtered caffeinated coffee raised the plasma cholesterol slightly (both LDL and HDL alike) in contrast to no effect from the same amount of decaffeinated coffee (FRIED et al. 1992). A theoretical reason why large quantities of coffee might be harmful relates to its caffeine content. Caffeine stimulates the release of epinephrine, wich increases free fatty acid concentrations. The production of VLDL and triglyceride would be expected to increase as a consequence. No one has defined precisely what an excessive coffee intake is. When patients ask us, we generally suggest no more than four cups per day. Tea, on the other hand, has not been associated with any disease. In fact, tea drinkers, if anything, seem to have less coronary heart disease. Whether a patient chooses to drink decaffeinated, regular, or no coffee or strong tea at all really becomes a matter of personal choice.

L. The Dietary Design to Achieve Optimal Plasma Lipid-Lipoprotein Levels

In view of the evidence about the role of dietary factors in the causation of hyperlipidemia and subsequent coronary heart disease, it should be possible to indicate the features of an effective diet to prevent and treat hyperlipidemia (CONNOR and CONNOR 1985). The major features of this "maximal" diet is that it is low in cholesterol, low in fat and saturated fat, and, reciprocally, high in complex carbohydrate and fiber (Fig. 5). The first objective is to reduce cholesterol consumption from 500 mg/day to less than 100 mg/day. Note that only foods of animal origin contain cholesterol. This requires keeping egg yolk consumption to a minimum, since much of the dietary cholesterol comes from egg yolk. Half of this is from visible eggs and half from eggs incorporated into processed foods. Meat, poultry, and fish also are limited. Nonfat dairy products are recommended. The second objective is to reduce fat intake by one-half from 40% to 20% of calories. This can be done by avoiding fried foods, reducing the fat used in baked goods by one-third, and using low-fat dairy products. Added fat should be limited to 3 teaspoons per day for women and children and 5 teaspoons per day for teenagers and men. Peanut butter should be used as part of a meal and not as a snack, and nuts used sparingly as condiments.

The third objective is to decrease the current saturated fat intake by two-thirds, from 14% to 5% of calories. This requires eating limited amounts of red meat or cheese no more than twice a week, using lower-fat cheeses (20% fat or less), avoiding products containing coconut and palm oil, limiting ice cream and chocolate to once a month and using soft margarines and oils sparingly.

When people are advised to decrease the amount of fat in their diets, they usually think only of visible fat and are surprised to learn that fat added at the table represents only 22% of their total fat intake. Decreasing dietary fat would be very difficult without knowing that 78% is invisible, with the majority coming from red meat, cheese, ice cream and other dairy products, baked products, and fat used in food preparation.

The fourth major dietary objective is to increase the carbohydrate content. If dietary fat is reduced from 40% to 20% of calories and protein kept constant at 15% of calories, then carbohydrate intake must be increased from 45% to 65% of total calories. In practical terms, this means that at least two complex-carbohydrate-containing foods should be eaten at each meal. For example, eating toast and cereal for breakfast, a sandwich (two slices of bread) or bean soup and nonfat crackers at lunch, and one to two cups of rice, pasta, potatoes, corn, etc. with bread at dinner. Snacks should be of a complex carbohydrate type, such as baked chips, popcorn, low-fat crackers, or low-fat cookies. This is a significant change as most Americans currently limit carbohydrate foods to no more than one per meal. To reach the increased carbohydrate objective, the patient must also eat two to four cups of legumes per week and two to four cups of vegetables per day. While research increasingly supports the value of a high-carbohydrate diet, many people are reluctant to adopt it because "starchy" foods are falsely associated with weight gain and are viewed as the food of the poor. The carbohydrate objective, however, is essential if the fat intake is reduced. Another objective is to eat three to five pieces of fruit per day with a concomitant decrease in refined sugar intake from 20% of calories to 10%. This means that sweets (soda pop, candy, or desserts) must be limited to no more than one serving per day.

M. A Phased Approach to the Dietary Treatment of Hyperlipidemia

A realistic view is that even well-motivated patients have difficulty making abrupt changes in their dietary habits. It may take many months and even years to change patterns of food consumption. Therefore, the changes recommended to the current American diet of most hyperlipidemic patients should be approached in a gradual manner, with each of three phases introducing more changes toward the eating pattern ultimately required for maximal therapy.

We call this changing from phase I (making substitutions) to phase II (eating meatless, cheeseless lunches and trying new recipes) to phase III (doing every day what you used to do once a week see Table 6). People need to change slowly and gradually by trying new recipes (CONNOR and CONNOR 1986, 1991) and new food products one at a time and incorporating those they like into their lifestyles. How far and how fast one progresses depends on the energy for trying new recipes and new products. There needs to be continual exposure to new foods and recipes if change is to become permanent and the eating style is to keep up with current tastes. For

Table 6. Summary of the suggested dietary changes in the different phases of the diet

Phase I
Avoid foods very high in cholesterol and saturated fat:
delete egg yolk, butterfat, lard, and organ meats
Substitute soft margarine for butter, vegetable oils and shortening for lard, skim milk for whole milk, egg whites for whole eggs
Phase II
Gradually use less meat and more fish, chicken, and turkey:
no more than 6–8 oz a day
Use less fat and cheese
Acquire new recipes
Phase III
Eat mainly cereals, legumes, fruit, and vegetables:
Use meat as a condiment
Use low-cholesterol cheeses
Save these foods for use only on special occasions: extra meats, regular cheese, chocolate, candy, and coconut

example, 5 years ago people had not eaten fahitas, stir-fried vegetables, and chicken in a flour tortilla. Now there are recipes for fajitas and many restaurants serve fajitas. Eating styles are continually being remodeled, so people need continual inspiration.

The goal of phase I is to modify the customary consumption of foods very high in cholesterol and saturated fat (Table 6). This can be accomplished by deleting egg yolk, butterfat, lard, and organ meats from the diet and by using substitute products when possible: soft margarine for butter, vegetable oils and shortenings for lard, skim milk for whole milk, and egg whites for whole eggs.

In phase II, a reduction in meat consumption is the goal, with a gradual transition from the presumed American ideal of up to a pound of a meat a day to no more than 6–8 oz per day (Table 6). Meat can no longer be the center of the meal, particularly for two or three meals a day. Some ideas for lunch with and without sandwiches have been detailed in sample menus and recipes (Connor and Connor 1986, 1991). In addition, in phase II we propose the use of less fat and cheese.

Substitute recipes have been developed to replace recipes which are centered on meat or high-fat dairy products (cream cheese, butter, sour cream, cheese) as the principal ingredients. Since these foods are to be eaten in smaller amounts or even omitted (butterfat), the patient needs to find recipes which use larger amounts of grains, legumes, vegetables, and fruits. Examples of such recipes are included in Connor and Connor (1986, 1991).

In phase III, the maximal diet for the treatment of hyperlipidemia is attained (Table 6). The cholesterol content of the diet is reduced to 100 mg per day, and saturated fat is lowered to 5%–6% of total calories. Since cholesterol is contained only in foods of animal origin, these changes mean that meat consumption in particular must be further reduced. Meat, fish, and poultry should be used as "condiments" rather than "aliments." With this philosophy, the meat dish will no longer occupy the center of the table. Instead, meat in smaller quantities will spice up dishes based on vegetables, rice, cereal, and legumes much as Asian, Indian, and Mediterranean cookery has been doing for eons. The total of meat and poultry should average 3–4 oz per day, but the use of poultry should be stressed because of its lower content of saturated fat. The recommendation is to eat fish twice a week. Note the low CSI of fish in Table 5, shared also by shellfish. Because of a low CSI, up to 6 oz of fish can be used in place of meat during phase III of the diet. All fish contain omega-3 fatty acids.

Shellfish are divided into two groups: high-cholesterol shellfish (shrimp, crab, and lobster) and low-cholesterol shellfish (oysters, clams, and scallops). Both contain omega-3 fatty acids and have a low-fat content. Because of these differences, high-cholesterol shellfish are more restricted in the daily diet than are the low-cholesterol shellfish, i.e., 3 oz vs. 6 oz per day. The low-cholesterol shellfish contain other sterols (e.g., brassicasterol) that are more analogous to plant sterols. These are poorly absorbed by humans.

The use of special low-cholesterol cheeses is an important component of phase III (see Table 5). The sample menus in CONNOR and CONNOR (1986, 1991) give some idea about the eating pattern in phase III.

A new eating habit questionnaire, the Diet Habit Survey, can be used to classify the diet into the typical U.S. diet and into phases I, II, and III of the low-fat, high complex carbohydrate diet (CONNOR et al. 1992). The Diet Habit Survey is related to plasma cholesterol change, measures eating habits directly, reflects nutrient composition indirectly, is quick to administer, and is inexpensive to analyze. Thus, it is useful as a research questionnaire and helpful in the dietary management of patients with hyperlipidemia and coronary heart disease.

N. Predicted Plasma Cholesterol Lowering from the Three Phases of the Low-Fat, High Complex Carbohydrate Diet

As has been emphasized, both dietary cholesterol and saturated fat elevate plasma cholesterol levels, whereas monounsaturated and polyunsaturated fats have a mildly depressing effect. In stepwise fashion, the cholesterol and saturated fat of each phase of the diet are successively reduced, with phase III providing for the lowest intakes. According to calculations derived from HEGSTED et al. (1965), phase III of the diet would provide for maximal plasma cholesterol lowering, an estimated average change of 20%. Phase II would produce a lowering of 14% and phase I, 7%. These plasma cholesterol changes for all phases offer the possibility of improved plasma levels, depending on the amount of dietary modifications, with phase III as

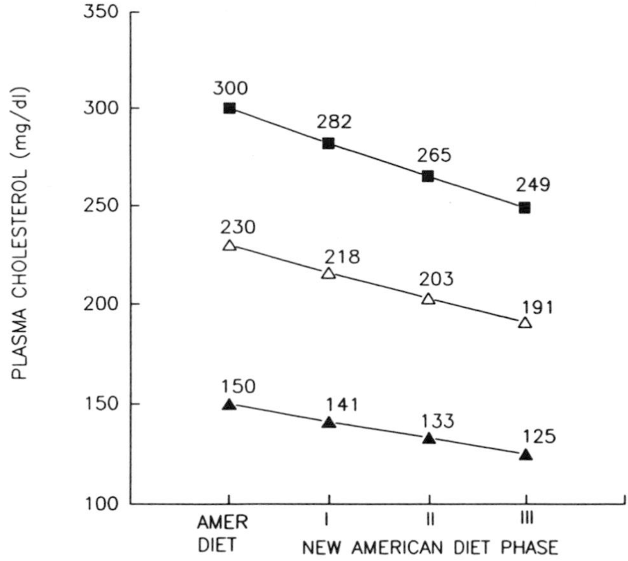

Fig. 6. Theoretical plasma cholesterol changes for three people changing from the American diet to a low-fat, high complex carbohydrate diet. *The New American Diet* phases I, II, and III

the ultimate goal. Figure 6 illustrates the theoretical plasma cholesterol lowering for three people who have very different initial plasma cholesterol levels based on genetic differences, while consuming the current American diet. Regardless of the inital plasma cholesterol level, everyone's level decreases as dietary changes are made, by approximately 5%–7% per phase.

O. The Applicability of the Low-Cholesterol, Low-Fat, High-Carbohydrate Diet in the Treatment of the Various Phenotypes and Genotypes of Hyperlipidemia

We emphasize a single diet approach for the treatment of all forms of hyperlipidemia (Connor and Connor 1982, 1989; Fried et al. 1992). There need not be a different diet for each phenotype of this clinical problem. Instead, as indicated and further illustrated in Table 7, this particular dietary approach deals effectively with the disturbed pathophysiology in all of these phenotypes, regardless of whether the abnormal lipoprotein is chylomicrons, VLDL, intermediate-density lipoproteins (IDL), or LDL (for a thorough discussion of the phenotypes and genotypes of hyperlipidemia, please see Connor and Connor 1982 and Illingworth and Connor 1987). The reduction in dietary cholesterol and saturated fat will increase LDL receptor

Table 7. How the low-fat, low-cholesterol diet will affect the plasma lipids and lipoproteins

Dietary factor	Plasma lipids-lipoproteins					Phenotype affected
	Cholesterol	LDL	Triglyceride	VLDL	Chylomicrons & remnants	
Cholesterol 100 mg/day	↓↓↓	↓↓↓				II-a, II-b
Saturated fat 5% of calories	↓↓↓	↓↓↓				II-a, II-b
Total fat to 20% of calories	↓	↓	↓		↓↓↓	All phenotypes esp. I and V
Complex CHO, fiber, and plant foods	↓	↓	↓			All phenotypes
Protein, more vegetable, less animal	↓	↓				No information
Caloric reduction for the overweight	↓	↓	↓↓	↓↓	↓↓	II-b, III, IV, V

LDL, low-density lipoprotein; VLDL, very low protein lipoprotein; CHO, carbohydrate activity. The plasma cholesterol and LDL will decrease. Lower dietary fat intake means fewer chylomicrons.

In types I and V hyperlipidemia, it may be necessary to restrict the fat below the suggested 20% of total calories. In these conditions, the amount of dietary fat is a trial and error proposition. Fat may need to be restricted to less than 10% with the diet composed chiefly of fruits, vegetables, grains, beans, and fish, with a little chicken. In chylomicronemia, the cholesterol intake is of lesser importance, so shellfish can be consumed safely. Fish and shellfish also contain omega-3 fatty acids, which would tend to depress the synthesis of VLDL by the liver. Should caloric restriction be necessary in order to attain ideal body weight, the use of the very low fat diet plus increased physical activity would be the approach to follow. Even in familial hypercholesterolemia, we have found the influence of diet to be crucial in attaining the lowest possible LDL concentrations. In fact, we suspect that the LDL receptor of these patients is even more susceptible to the effects of dietary cholesterol and saturated fat, since the number of LDL receptors is already reduced by about 50%.

Of particular importance is the increase in the HDL cholesterol level that occurs with a high-fat, high-cholesterol diet and the decrease in HDL from a low-fat diet (LIN and CONNOR 1981). However, populations such as the Tarahumara Indians are known to have low HDL cholesterol levels, indeed pathologically low by U.S. standards. Both Tarahumara men and women have average HDL cholesterol levels of 28–29 mg/dl. Other such populations that eat low-fat diets have also been observed to have low HDL cholesterol levels (KNUIMAN et al. 1980; MILLER and MILLER 1982). Since these populations have low LDL cholesterol levels and little coronary heart disease, the low HDL cholesterol level is not a risk factor. The physiological response to a low-fat diet is a 10%–20% decrease in the HDL cholesterol level (ULLMANN et al. 1991). Since LDL cholesterol lowering is even greater, the LDL to HDL ratio is little changed or even improved. Also, a recent study indicated that a lower HDL cholesterol level resulting from a low-fat, high complex carbohydrate diet is different from a genetically low HDL cholesterol level and that the coronary risk is not enhanced (BRINTON et al. 1990).

P. The Use of the Low-Fat, High Complex Carbohydrate Diet in Diabetic Patients, Pregnant Patients, Children, and Hypertensive Patients

The approach to diabetic patients who are also hyperlipidemic involves the same dietary considerations for the treatment of the hyperlipidemia. Phase III of the diet has been used successfully in both juvenile-onset and maturity-onset diabetic patients (insulin-dependent and non-insulin-dependent). Clearly involved in this treatment is the appropriate control of carbohydrate as well as lipid metabolism by means of adequate amounts of insulin and weight reduction when the patient is overweight. The great propensity of diabetic patients to atherosclerotic vascular disease makes control of their hyperlipidemia of particular importance. The principles we have outlined can be utilized fully with benefit to these patients.

Pregnancy constitutes a particularly difficult situation, because in most pregnant women there will be a 40%–50% increase in plasma lipids and lipoproteins (chiefly LDL) in physiological circumstances. A hyperlipidemic patient who becomes pregnant should continue on the same diet advised previously for the treatment of her hyperlipidemia, supplemented by vitamins and minerals as is usual in pregnancy. In patients with familial hypercholesterolemia, the phase III diet is utilized as before, with some increase in calories to permit the desired weight gain. In the type I or V hypertriglyceridemic pregnant patient, there is apt to be a profound augmentation of the usual hyperchylomicronemia, and strict adherence to the 5%–10% fat diet is often necessary to avoid acute pancreatitis.

The single-diet concept for the treatment of hyperlipidemic children can be applied as for adults, with the exception that a child up to 4 years of age is allowed less cholesterol than an older child. Above the age of 4 years, no more than 3 oz of meat per day is the goal. Egg yolk, organ meat, and butterfat are eliminated from the diet, even for infants. Dietary iron is supplemented from fortified cereals. Human breast milk or whole cow's milk is recommended prior to weaning or until the infant eats sufficient table food to provide adequate calories for growth. This is usually from 1 to 2 years. From that time on, the child should drink skim milk. The greatest thing parents can do is be a role model. Know that by serving low-fat, high complex carbohydrate foods they will be diluting the fat and cholesterol they are consuming elsewhere. What children see happening at home in food preparation will ultimately help them develop correct food habits useful throughout their lifetime.

For the rare infant with type I hyperlipidemia, the fat content of human breast or infant formulas is much too high. A basic skim milk formula will have to be used to avoid the abdominal pain and episodes of pancreatitis to which these patients are so prone. However, sufficient amounts of essential fatty acids must be provided. These can be prescribed separately as canola or soybean oil, to yield at least 2%–4% of the total calories. Such oils will provide both omega-6 and omega-3 essential fatty acids. In this way, much of the fat intake will be from linoleic and linolenic acids and very little other fat will be taken in. Success in several infants with the type I disorder has been achieved using this dietary approach and has resulted in the abolition of episodes of abdominal pain.

Special attention must be given to hypertensive, hyperlipidemic patients for several reasons. First, coronary heart disease and atherosclerotic brain disease are common causes of death in hypertensive patients. Second, some diuretic agents (thiazides) and beta blockers are hyperlipidemic in themselves, so that the usual hypertensive patient will have an increase in plasma lipids from the use of these agents alone. Thus, hypertensive individuals given thiazides should have additional dietary therapy in the form of the single-diet approach. Furthermore, the genetic syndrome of both hypertension and hypertriglyceridemia (hypertensive dyslipidemia) requires simultaneous treatment of both conditions. Finally, a salt-restriction program can readily be incorporated to provide for the additional treatment of the essential hypertension as this is desired by the physician. The correction of obesity and lessening of alcohol intake may also have blood pressure lowering effects. In order to combine the dietary treatment of hypertension and that of hyperlipidemia in a single diet, a stepwise reduction in salt use is also advocated and incorporated into the different phases (CONNOR and CONNOR 1986).

Q. Interrelationships Between Dietary and Pharmaceutical Therapy of Hyperlipidemic States

Although dietary therapy remains the cornerstone in the treatment of hyperlipidemia, many patients will require additional pharmaceutical treatment in order to obtain the therapeutic objectives. In this connection, some patients may believe that once they are started on drug therapy they no longer need to pay as much attention to their diet as was the case when they were not receiving drug treatment. Nothing could be further from the truth. The two modes of therapy are not mutually exclusive, but are instead complementary. The maximum of dietary treatment may obtain a lowering of the plasma LDL and total cholesterol which averages 10%, but which may be as much as 15%–20%. This will continue to be the case when pharmaceutical treatment is added. Then the patient will receive plasma lipid lowering from diet plus the additional lowering from medication.

The complementary nature of diet and drugs together is easily noted during the holiday period from Thanksgiving to New Year when weight gain and additional intakes of cholesterol and saturated fat from holiday foods occur. Patients who are well controlled on diet and drug therapy continue their medication, but their levels of plasma cholesterol and LDL as well as VLDL and triglyceride often increase during this luxurious holiday period. This particular premiss was put to the test in a metabolic feeding study in patients with familial hypercholesterolemia who were well controlled on diet and then were given 40 mg lovastatin per day. When they were given a diet high in cholesterol and saturated fat, while still maintaining the usual dose of lovastatin, the plasma cholesterol level increased from 246 to 289 mg/dl, a 17.5% increase. LDL cholesterol increased even more, by 20%. When it is considered that the dietary factors of cholesterol and saturated fat have a primary effect in decreasing LDL receptor activity (SPADY and DIETSCHY 1988) and that lovastatin has a primary action in increasing LDL receptor activity (BROWN and GOLDSTEIN 1986), it then becomes apparent that a diet high in cholesterol and saturated fat may nullify some of the beneficial effects of lovastatin.

Similar responses may occur in patients with types IV or V hyperlipidemia. Weight gain and a high-fat diet can nullify the effects of gemfibrozil in maintaining lower plasma triglyceride concentrations. When a patient whose lipid values have previously been under good control, with good adherence to diet and medication, has a deterioration in lipid values, it is reasonable to question whether the previous dietary lifestyle changes have been disregarded. In diabetic hypertriglyceridemic patients, the dietary deviations will upset glycemic control as well. Many times when this matter is explained to the patient, he/she is then able to return to their previous excellent dietary lifestyle with a concomitant decrease in the plasma triglyceride and cholesterol levels and, if diabetic, better glucose control.

R. Summary

The principal goal of dietary prevention of atherosclerotic coronary heart disease is the achievement of physiological levels of the plasma total and LDL cholesterol, triglyceride, VLDL, and chylomicrons. These goals have been well delineated by the National Cholesterol Education Program of the National Heart, Lung, and Blood Institute and the American Heart Association. Dietary treatment is first accomplished by enhancing LDL receptor activity and at the same time depressing liver synthesis of cholesterol and triglyceride. Both dietary cholesterol and saturated fat decrease LDL receptor activity and inhibit the removal of LDL from the plasma by the liver. Saturated fat decreases LDL receptor activity, especially when cholesterol is concurrently present in the diet. The total amount of dietary fat is also of importance. The greater the flux of chylomicron remnants into the liver, the greater is the influx of cholesterol ester. In addition, factors which affect VLDL and LDL synthesis could be important. Of most importance is obesity. Excessive calories enhance triglyceride and VLDL synthesis and thereby LDL synthesis. Weight loss and omega-3 fatty acids from fish oil inhibit synthesis of both VLDL and triglyceride by the liver.

The optimal diet for the treatment of children and adults is similar and has the following characteristics: cholesterol (100 mg/day or less), total fat (20% of kcal, 5% of kcal as saturated fat with the balance from omega-3 and omega-6 polyunsaturated and monounsaturated fat), carbohydrate (65% kcal, at least two-thirds from starch including 11–15 g soluble fiber), and protein (15% kcal). This low-fat, high-carbohydrate diet can lower the plasma cholesterol by 18%–21%. This diet is also an antithrombotic diet, thrombosis being another major consideration in preventing coronary heart disease. Dietary therapy is the mainstay of the prevention and treatment of coronary heart disease through the control of plasma lipid and lipoprotein levels.

Acknowledgement. Special thanks is accorded to Lois Wolfe for the meticulous preparation of this manuscript. This work was supported by researth grants from the National Heart, lung, and Bolld Institute (HL 25, 687), the Clinical Nutrition Research Unit (DK40, 566) of the National Institute of Diabetia, Digestive, and Kidney Diseases, and the Clinical Research Center Program (RR334) of the Division of Research Resources of the National Institutes of Health.

References

Anderson JW, Story L, Sieling B, Chen WJL, Petro MS, Story J (1984) Hypocholesterolemic effects of oat-bran or bean intake for hypercholesterolemic men. Am J Clin Nutr 40:1146–1155

Brinton EA, Eisenberg S, Breslow JL (1990) A low-fat diet decreases high density lipoprotein (HDL) cholesterol levels by decreasing HDL apolipoprotein transport rates. J Clin Invest 85:144–151

Brown MS, Goldstein JL (1983) Lipoprotein receptors in the liver: control signals of plasma cholesterol traffic. J Clin Invest 72:743–747

Brown MS, Goldstein JL (1986) A receptor-mediated pathway for cholesterol homeostasis. Science 232:34–47

Burr ML, Gilbert JF, Holliday RM, Elwood PC, Fehly AM, Rogers S, Sweetnam PM, Deadman NM (1989) Effects of changes in fat, fish, and fibre intakes on death and myocardial reinfarction: diet and reinfarction trial (Dart). Lancet 2:756–761

Connor SL, Connor WE (1986) The New American Diet. Simon and Schuster, New York

Connor SL, Artaud-Wild SM, Classick-Kohn CJ, Gustafson JR, Flavell DP, Hatcher LF, Connor WE (1986) The cholesterol-saturated fat index: an indication of the hypercholesterolaemic and atherogenic potential of food. Lancet 1:1229–1232

Connor SL, Gustafson JR, Artaud-Wild SM, Classick-Kohn CJ, Connor WE (1989) The cholesterol-saturated fat index for coronary prevention: background use and a comprehensive table of foods. J Am Diet Assoc 89:807–816

Connor SL, Gustafson JR, Sexton G, Becker N, Artaud-Wild SR, Connor WE (1992) The diet habit survey: a new method of dietary assessment that relates to plasma cholesterol changes. J Am Diet Assoc 92:41–47

Connor WE, Connor SL (1982) The dietary treatment of hyperlipidemia: rationale, technique and efficacy. Med Clin North Am 66:485–518

Connor WE, Connor SL (1985) The dietary prevention and treatment of coronary heart disease. In: Connor WE, Bristow JD (eds) Coronary heart disease: prevention, complications, and treatment. Lippincott, Philadelphia, pp 43–64

Connor WE, Connor SL (1989) Diet, atherosclerosis and fish oil. Adv Intern Med 35:139–172

Connor SL, Connor WE (1991) The New American Diet System. Simon and Schuster, New York

Fox JC, McGill HC Jr, Carey KD, Getz GS (1987) In vivo regulation of hepatic LDL receptor mRNA in the baboon – differential effects of saturated and unsaturated fat. J Biol Chem 262:7014

Fried RE, Levine DM, Kwiterovich PO, Diamond EL, Wilder LB, Moy TF, Pearson TA (1992) The effect of filtered-coffee consumption on plasma lipid levels. JAMA 267:811–815

Fry MM, Spector AA, Connor SL et al. (1973) Intensification of hypertriglyceridemia by either alcohol or carbohydrate. Am J Clin Nutr 26:798–802

Ginsberg H, Olefsky J, Farquhar JW et al. (1974) Moderate ethanol ingestion and plasma triglyceride levels – a study in normal and hypertriglyceridemic persons. Ann Intern Med 80:143–149

Goodnight SH Jr, Harris WS, Connor WE, Illingworth DR (1982) Polyunsaturated fatty acids, hyperlipidemia and thrombosis. Arteriosclerosis 2:87–113

Gordon T, Fisher M, Ernst M et al. (1982) Relation of diet to LDL cholesterol, VLDL cholesterol and plasma total cholesterol and triglycerides in white adults. Atherosclerosis 2:502

Harris WS, Connor WE, Alam N, Illingworth DR (1988) The reduction of postprandial triglyceridemia in humans by dietary n-3 fatty acids. J Lipid Res 29:1451–1460

Harris WS, Connor WE, Illingworth DR, Rothrock DW, Foster DM (1990) Effect of fish oil on VLDL triglyceride kinetics in humans. J Lipid Res 31:1549–1558

Hegsted DM, Ausman LM (1988) Diet, alcohol and coronary heart disease in men. J Nutr 118:1184–1189

Hegsted DM, McGandy RB, Myers ML, Stare FJ (1965) Quantitative effects of dietary fat on serum cholesterol in man. Am J Clin Nutr 17:281–295

Hennekens CH, Rosner B, Cole DS (1978) Daily alcohol consumption and coronary heart disease. Am J Epidemiol 107:196–200

Hopkins PN (1992) Effects of dietary cholesterol on serum cholesterol: a meta-analysis and review. Am J Clin Nutr 55:1060–1070

Illingworth DR, Connor WE (1987) Disorders of lipid metabolism. In: Felig P, Baxter JD, Broadus AE, Frohman LA (eds) Endocrinology and metabolism, 2nd edn. McGraw-Hill, New York, pp 1244–1314

Knuiman JT, Hermus RJJ, Hautvast JGAJ (1980) Serum total and high density lipoprotein (HDL) cholesterol concentrations in rural and urban boys from 16 countries. Atherosclerosis 36:529–537

Kudzma DJ, Schonfeld G (1970) Alcoholic hyperlipidemia: induction by alcohol but not by carbohydrate. J Lab Clin Med 77:384–395

Lin DS, Connor WE (1981) The long-term effects of dietary cholesterol upon the plasma lipids, lipoproteins, cholesterol absorption and the sterol balance in man: the demonstration of feedback inhibition of cholesterol biosynthesis and increased bile acid excretion. J Lipid Res 21:1042–1052

Malinow MR, Connor WE, McLaughlin P et al. (1981) Sterol balance in Macaca fascicularis: effects of alfalfa saponins. J Clin Invest 67:156–162

Marmot M, Brunner E (1991) Alcohol and cardiovascular disease – the status of the U shaped curve. Br Med J 303:565–568

Mattson FH, Grundy SM (1985) Comparison of effects of dietary saturated, monounsaturated and polyunsaturated fatty acids on plasma lipids and lipoproteins in men. J Lipid Res 26:194–202

McMurry MP, Connor WE, Cerqueira MT (1982) Dietary cholesterol and the plasma lipids and lipoproteins in the Tarahumara Indians: a people habituated to a low-cholesterol diet after weaning. Am J Clin Nutr 35:741–744

McMurry MP, Cerqueira MT, Connor SL, Connor WE (1991) Changes in lipid and lipoprotein levels and body weight in Tarahumara Indians after consumption of an affluent diet. N Engl J Med 325:1704–1708

Mensink RP, Katan MB (1990) Effect of dietary trans fatty acids on high-density and low-density lipoprotein cholesterol levels in healthy subjects. N Engl J Med 323:439–445

Miller GJ, Miller NE (1982) Dietary fat, HDL cholesterol, and coronary disease: one interpretation. Lancet 2:1270–1271

Neuringer MD, Connor WE (1986) Omega-3 fatty acids in the brain and retina: evidence for their essentiality. Nutr Rev 44:285–294

Phillipson BE, Rothrock DW, Connor WE, Harris WS, Illingworth DR (1985) The reduction of plasma lipids, lipoproteins and apoproteins in hypertriglyceridemic patients by dietary fish oils. N Engl J Med 312:1210–1216

Princen HMG, van Poppel G, Vogelezang C, Buytenhek R, Kok FJ (1992) Supplementation with vitamin E but not β-carotene in vivo protects low density lipoprotein from lipid peroxidation in vitro: effect of cigarette smoking. Arterioscler Thromb 12:548–553

Roberts SL, McMurry M, Connor WE (1981) Does egg feeding (i.e., dietary cholesterol) affect plasma cholesterol levels in humans? The results of a double-blind study. Am J Clin Nutr 34:2092–2099

Sorci-Thomas M, Wilson MD, Johnson FL, Williams DL, Rudel LL (1989) Studies on the expression of genes encoding apolipoproteins B100 and B48 and the low density lipoprotein receptor in nonhuman primates. J Biol Chem 264:9039

Spady DK, Dietschy JM (1988) Interaction of dietary cholesterol and triglycerides in the regulation of hepatic low density lipoprotein transport in the hamster. J Clin Invest 81:300–309

Suh II, Shaten BJ, Cutler JA, Kuller LH (1992) Alcohol use and mortality from coronary heart disease: the role of high-density lipoprotein cholesterol. Ann Intern Med 116:881–887

Ullmann D, Connor WE, Hatcher LF, Connor SL, Flavell DP (1991) Will a high carbohydrate, low-fat diet lower plasma lipids and lipoproteins without producing hypertriglyceridemia? Arterioscler Thromb 11:1059–1067

Weintraub MS, Zechner R, Brown, Eisenberg S, Breslow JL (1988) Dietary polyunsaturated fats of the w-6 and w-3 series reduce postprandial lipoprotein levels: chronic and acute effects of fat saturation on postprandial lipoprotein metabolism. J Clin Invest 82:1884–1893

Witzam JL, Steinberg D (1985) Role of oxidized low density lipoprotein in atherogenesis. J Lipid Res 26:194–202

Zilversmit DB (1979) Atherogenesis: a postprandial phenomenon. Circulation 60:473–485

Zock PL, Katan MB, Merkus MP, van Dusseldorp M, Harryvan JL (1990) Effect of a lipid-rich fraction from boiled coffee on serum cholesterol. Lancet 335:1235–1237

CHAPTER 10
Lipid Apheresis

D. SEIDEL

A. Introduction

There is ample evidence that diseases resulting from premature atherosclerosis, of which the leading event is myocardial infarction (MI), have great impact on health care in all industrialized countries (CASTELLI et al. 1990; CREMER and MUCHE 1990). These illnesses are the most common cause not only of death, but also of early retirement. No doubt this has contributed to the fact that atherosclerosis, together with cancer, is now the primary consideration in experimental, clinical, and preventive medicine. Results of many studies have shown that there are a large number of factors involved in atherogenesis. While disturbances in lipid metabolism (hyperbetalipoproteinemia, hypoalphalipoproteinemia, accumulation of chylomicron and/or very low density lipoprotein, VLDL, remnants), structural abnormalities of lipoproteins (Lp) and family history of MI are the most important risk factors. Hypertension, smoking, elevated blood glucose, and overweight also have an impact on early cardiovascular events. More recently Lp(a), fibrinogen, and the biological modification of lipoproteins have been added to the list of risk factors for atherosclerosis (SEIDEL et al. 1991; KOENIG and ERNST 1992; SMITH 1986, 1990; KIENAST et al. 1990).

In many patients suffering from coronary heart disease (CHD), LDL (low-density lipoprotein), Lp(a), and fibrinogen are elevated at the same time and may potentiate the cardiovascular risk derived from each factor alone.

Most forms of hypercholesterolemia result from a defect in the removal of LDL from plasma by the liver, and the LDL receptor is now recognized as the crucial element in the control of LDL cholesterol homeostasis (BROWN and GOLDSTEIN 1986; SEIDEL et al. 1985). If the physiological clearing mechanisms for LDL are insufficient, diet and drug therapy alone are often ineffective. This holds true also for Lp(a) and fibrinogen, either of which at present can hardly be lowered by diet or drugs.

In humans, plasma LDL cholesterol levels below 110 mg/dl seem to be necessary to inhibit the development of atherosclerosis or to induce regression of the vessel wall lesions (CREMER et al. 1991). This has been impressively demonstrated in six different secondary intervention studies: Life-Style Heart Study (ORNISH et al. 1990); Program on the Surgical Control of the Hyperlipidemias (POSCH) Study (BUCHWALD et al. 1990); Familial Atherosclerosis Treatment Study (FATS) (BROWN et al. 1990); Cholesterol-Lowering Atherosclerosis Study (CLAS) (BLANKENHORN et al. 1987); Monitored Atherosclerosis Regression Study (MARS) Study (BLANKENHORN et al. 1992, personal communication); and the Heparin-Induced Extracorporeal Low-Density Lipoprotein Plasmapheresis (H.E.L.P.) Multicenter Study (SCHUTT-WERNER et al. 1993; HENNERICI et al. 1991). Although the therapeutic approach and strategy were different in these studies, the promising outcome in each was the same. Lowering of LDL cholesterol by 35%–50% is followed by a twofold increase in the progression to regression ratio when controls were compared with treated patients.

On the basis of these and other studies, cholesterol-lowering strategies to reduce the risk of CHD in the population have now been proposed (STUDY GROUP OF THE EUROPEAN ATHEROSCLEROSIS SOCIETY 1987; THE EXPERT PANEL 1988). The corner-stone of such strategies is undoubtedly diet and, when necessary, drug therapy. With the advent of the 3-hydroxy-3-methylglutaryl coenzyme A (HMG-CoA) reductase inhibitors, a new class of powerful lipid-lowering drugs has been introduced with great potential in the treatment of hypercholesterolemia. The use of these drugs is now increasing, and long-term safety will be of profound importance for their general application in the treatment of atherosclerotic disease. It is still an open question as to whether this family of compounds should be used in primary or only in secondary prevention of CHD.

More radical measures such as partial ileal bypass (BUCHWALD 1964), portocaval shunt (STARZL et al. 1983), liver transplantation (STARZL et al. 1984), plasma exchange (DE GENNES et al. 1976; THOMPSON et al. 1975), and LDL apheresis (LUPIEN et al. 1976) have also been introduced for the treatment of severe hypercholesterolemia.

Plasma exchange has proven to be particularly successful in the management of severe hypercholesterolemia such as homozygous familial hypercholesterolemia (FH; DE GENNES et al. 1976; THOMPSON et al. 1975). Since this therapy requires substitution of plasma fractions with its inherent danger, several LDL apheresis procedures with varying degrees of selectivity and efficiency have subsequently been developed, some of which are at present being evaluated in clinical trials.

While the use of such therapies for the primary prevention of CHD will largely be restricted to the most severe forms of hypercholesterolemia, in secondary prevention the combination of diet and drugs together with plasmapheresis seems to be an attractive therapeutic possibility if diet and drug therapy alone is not sufficient to achieve the therapeutic goal of LDL cholesterol levels less than 110 mg/dl. The combination of plasma LDL apheresis together with diet and drugs now allows a maximal lowering of LDL cholesterol of up to 80%. This also holds true for patients who, only a few years ago, were classified as resistant to treatment of hypercholesterolemia. Besides LDL, some apheresis procedures also eliminate other risk factors for CHD, such as Lp(a) and or fibrinogen, and are thus of clinical importance, being able to greatly improve the hemorheological status of the patients.

B. Low-Density Lipoprotein Apheresis Procedures

In the last decade, several systems have been developed for the extracorporeal elimination of LDL cholesterol from plasma. These procedures are collectively referred to as LDL apheresis. Today, LDL apheresis has largely replaced plasma exchange therapy as introduced by De Gennes et al. (1976) and later also by Thompson et al. (1975).

Now, with the experience gathered in the course of several years of clinical application, the efficiency, specificity, and safety of the different LDL apheresis methods can be compared. Besides the marked reduction in LDL cholesterol concentrations by all techniques, it has become apparent that at least one of the procedures (the H.E.L.P. system) results in an equally significant change in hemorheology. The long-term clinical benefit of LDL apheresis treatment, in particular for secondary prevention of CHD, is still under investigation. One section of this review will cover and compare the issue of safety, efficiency and specificity as well as clinical results of various LDL apheresis techniques based on recent publications.

Three methods of LDL apheresis have been clinically established and are now used for the treatment of severe hypercholesterolemia: a) various LDL immunoadsorption techniques, using immobilized mono or polyclonal antibodies of apoB-100 (Stoffel and Demant 1981; Riesen et al. 1986); b) LDL binding by dextrane sulfate attached to cellulose (Yokoyama et al. 1985); and c) heparin-induced extracorporeal LDL precipitation (Seidel and Wieland 1982; Eisenhauer et al. 1987).

Plasma membrane filtration has also been proposed, but this retains other macromolecules apart from LDL such as high-density lipoproteins (HDL), immunglobulins, and albumin and therefore cannot be considered as being specific. This technique closely resembles plasma exchange, with its disadvantages for long-term therapy.

The H.E.L.P. system has been widely used. Its efficiency and safety, alone and in combination with HMG-CoA reductase inhibitors, has been investigated in great detail. Therefore, this review will focus on the H.E.L.P. system as a new therapeutic tool to simultaneously lower LDL, Lp(a), and fibrinogen, to improve hemorheology, and to achieve regression of coronary sclerosis in patients who were otherwise refractory to the treatment of severe hypercholesterolemia.

C. The Heparin-Induced Extracorporeal Low-Density Lipoprotein Plasmapheresis System

This technique operates by an increase in the positive charges on LDL and Lp(a) particles at low pH, allowing them to specifically form a network with heparin and fibrinogen in the absence of divalent cations (SEIDEL and WIELAND 1982; ARMSTRONG 1987; SEIDEL 1990). Only a limited number of other heparin-binding plasma proteins are coprecipitated by heparin at low pH. Other proteins, such as apoA_1, apoA_2, albumin, or immunoglobulins, do not significantly bind to heparin at low pH and are not precipitated in the system (EISENHAUER et al. 1987; ARMSTRONG 1987). Complement activation takes place in all extracorporeal therapy systems. However, as a specific feature of the H.E.L.P.-system, activated complement C3 and C4 as well as the terminal complement complex are largely adsorbed to the precipitation filter, resulting in plasma concentrations which are actually below those measured before apheresis (WÜRZNER et al. 1991). Leukocytopenia, a hallmark of complement activation, has not been observed under H.E.L.P. therapy. The H.E.L.P. system (manufactured by B. Braun Melsungen, Germany) has unique features:

1. It removes LDL, Lp(a), and fibrinogen with high efficiency.
2. It does not remove HDL.
3. It does not alter or modify plasma lipoproteins.
4. It does not change plasma concentrations of cell mediators.
5. It avoids the use of compounds with immunogenic or immunostimulatory activity.
6. It uses only disposable material and avoids regeneration of any of the used elements.
7. It is a technically safe and well-standardized procedure.
8. In short and long-term treatment, tolerance and benefit are excellent.

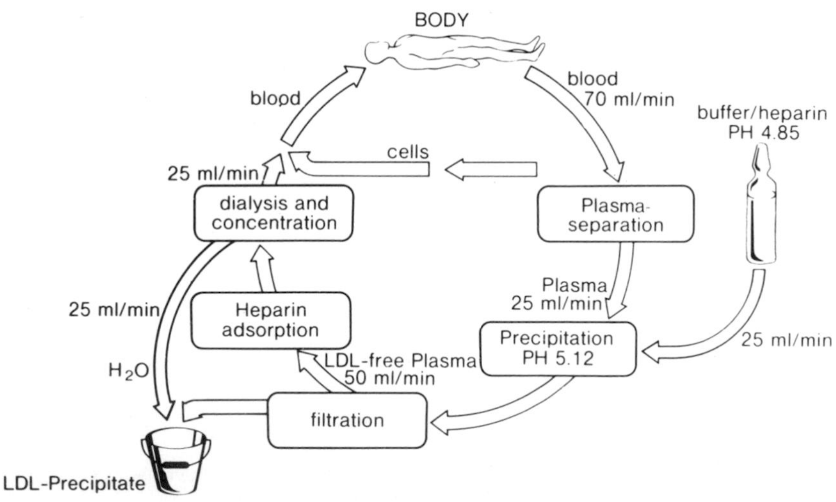

Fig. 1. Flow sheet of the heparin-induced extracorporeal low-density lipoprotein plasmapheresis procedure. *LDL*, low-density lipoprotein

The major steps of the H.E.L.P. system to remove the atherogenic compounds are illustrated in the flow sheet (see Fig. 1). In the first step, plasma is obtained by filtration of whole blood through a plasma separator. This is then mixed continuously with a 0.3-*M* acetate buffer of pH 4.85 containing 100 IU heparin/ml. The sudden precipitation occurs at a pH of 5.12, and the suspension is circulated through a 0.4-*M* polycarbonate filter to remove the precipitated LDL, Lp(a), and fibrinogen. Excess heparin is absorbed by passage through an anion-exchange column, which binds only heparin at the given pH. The plasma buffer mixture is finally subjected to bicarbonate dialysis and ultrafiltration to remove excess fluid and to restore the physiological pH, before the plasma is mixed with the blood cells and returned to the patient. All filters and tubings required for the treatment are sterile, disposable, and intended for one use only. This makes it easy and reliable to work with the system and guarantees a steady quality for each treatment, independent of the clinic performing the procedure. Safety is assured by a visual display and two microprocessors operating in parallel (see Fig. 2). Due to the excellent tolerance of the procedure, the patients leave the hospital shortly after the end of the treatment session.

D. Clinical Experience with the Heparin-Induced Extracorporeal Low-Density Lipoprotein Plasmapheresis System

The clinical experience with the H.E.L.P. system goes back to 1985. Between then and 1992, approximately 300 patients were treated in over 30 000 single treatments. Some patients were treated for more than 5 years. Currently, the system operates in approximately 100 centers in Germany, Italy, USA, Austria, and Ireland.

The efficiency of the system is 100% for the elimination of LDL, Lp(a), and fibrinogen. Per single treatment (lasting 1.5–2 h), 2.8–3 l plasma is treated, causing a reduction of approximately 50% of these three compounds in plasma of the treated patients.

The rates of return to preapheresis concentrations for LDL differ between normocholesterolemics and heterozygous as well as homozygous FH patients, while they are almost identical for Lp(a) (Armstrong et al. 1989; Thiery et al. 1990b). Normocholesterolemics return rather quickly towards the steady-state pretreatment levels. Heterozygous FH patients display a rate of return intermediate between normocholesterolemics and a homozygous FH patient, the latter being slowest in their rate of return to pretreatment LDL concentrations. In biweekly treatment intervals, the pretreatment values usually reach a new steady state after 4–8 treatments.

Long-term effects of the H.E.L.P. treatment based on interval values between two treatments (LDL-cholesterol concentration after H.E.L.P. +

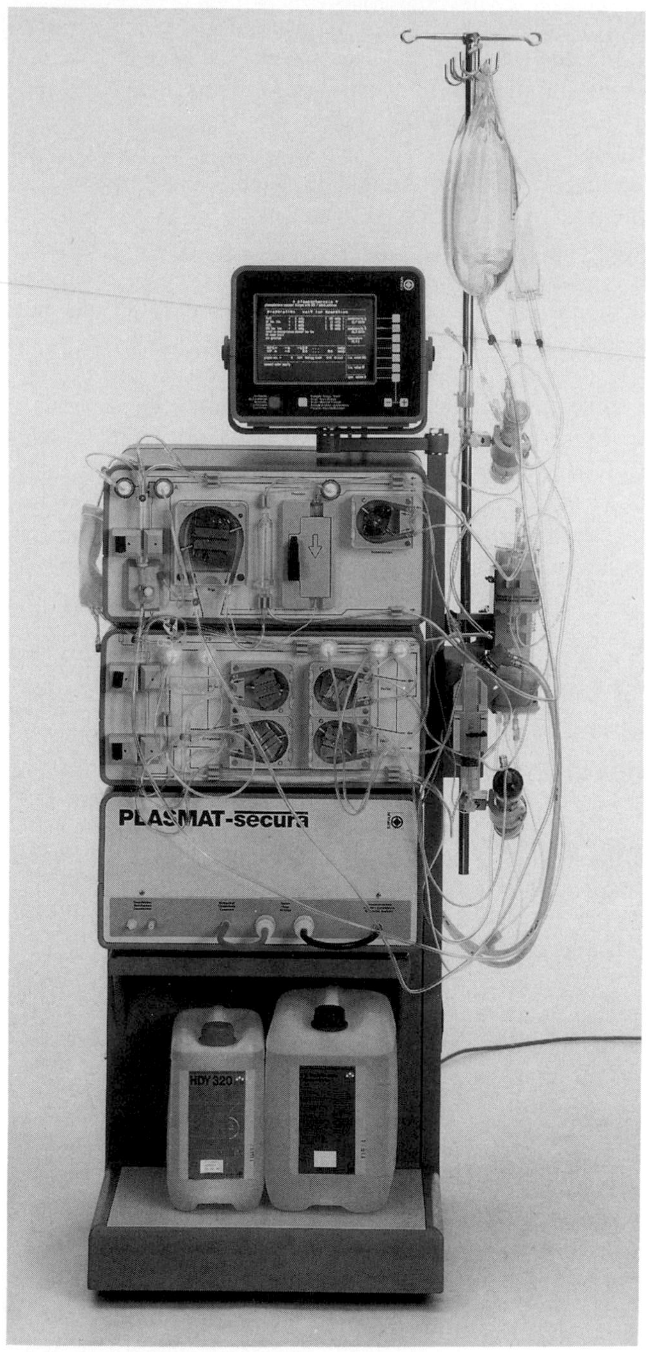

a

Fig. 2a,b. Hemodialysis Secura system (manufactured by B. Braun Melsungen AG, Germany)

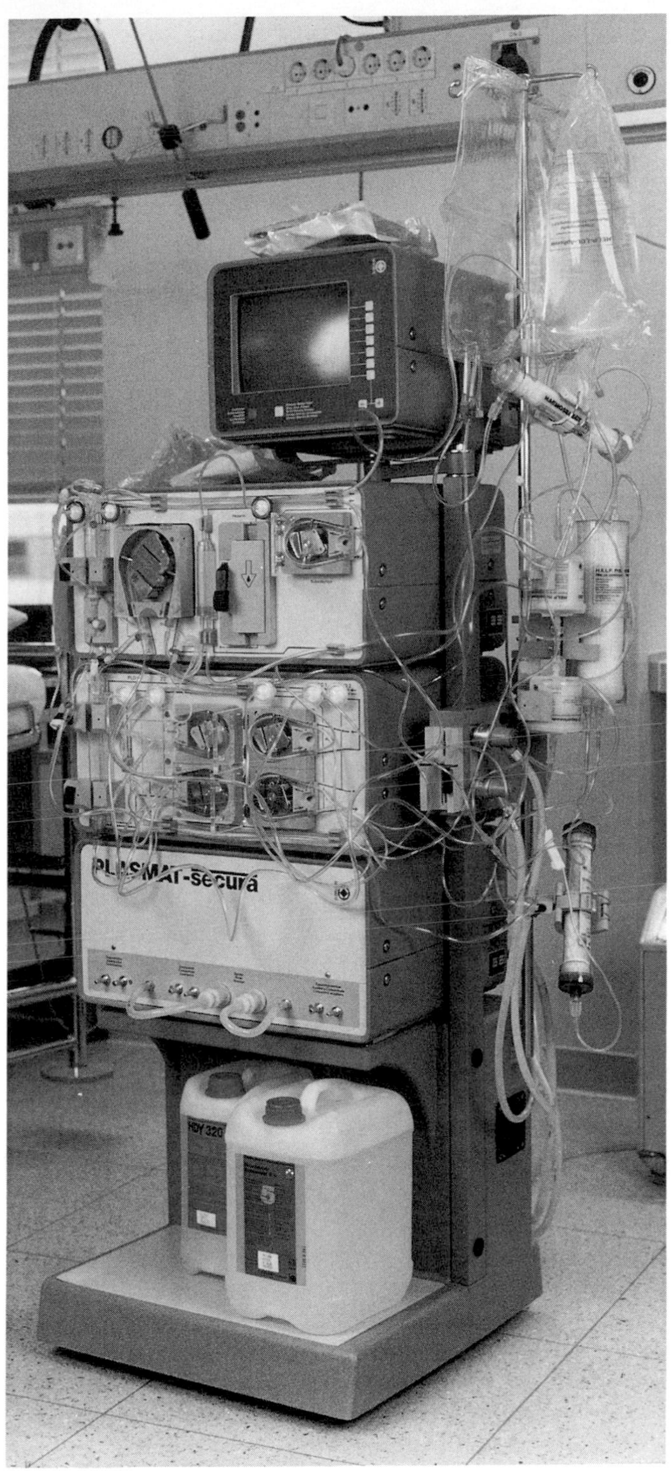

b

Table 1. Long-term effects of the heparin-induced extracorporeal low-density lipoprotein plasmapheresis treatment

	Mean interval values of approximately 6000 treatments
LDL cholesterol	−51% ± 14
Lp(a)	−45% ± 5
Fibrinogen	−46% ± 15
Apoprotein B-100	−45% ± 10
HDL cholesterol	+12% ± 2
Apoprotein A-1	+9% ± 2

LDL, low-density lipoprotein; HDL, high-density lipoprotein; Lp(a); lipoprotein(a).

LDL-cholesterol concentration before the next H.E.L.P. treatment: 2) and expressed as percentage of concentrations at the start are shown in Table 1.

The H.E.L.P. treatment also significantly improves plasma viscosity (−15%), erythrocyte aggregation (−50%), and erythrocyte filtration (+15%), which is followed by an acute (20%–30%) increase in the oxygen tension in muscle tissue (SCHUFF-WERNER et al. 1989; KLEOPHAS et al. 1990). The changes in plasma viscosity are due to the reduction in both LDL and fibrinogen. The change in erythrocyte aggregation is primarily due to fibrinogen reduction. Changes in erythrocyte filterability correlate with an improvement in the cholesterol to phospholipid ratio of cell membranes

Table 2. Evaluation of ten patients with angiographically documented coronary heart disease who have been treated by heparin-induced extracorporeal low-density lipoprotein plasmapheresis for more than 1 year

Patient	Sex (m/f)	Age	Angina		Effort dyspnea		Exercise electrocardiogram	
			0	1 year	0	1 year	0	1 year
1	f	47	+++	−	+	−	75 W	125 W
2	m	34	−	−	−	−	225 W	225 W
3	m	49	+++	−	++	−	50 W	100 W
4	m	62	+	−	+	−	75 W	100 W
5	f	30	+++	−	+++	−	75 W	100 W
6	m	36	+	−	−	−	175 W	225 W
7	m	39	+	−	−	−	175 W	200 W
8	m	52	++	−	++	−	100 W	100 W
9	m	44	+++	+	+++	−	50 W	125 W
10	m	50	++	−	++	−	125 W	125 W

+++ Symptoms at rest.
++ Symptoms with minor effort.
+ Symptoms with major effort.
− No symptoms.

(Schuff-Werner et al. 1989). It is conceivable to associate the rheological findings with the impressive relief from angina, together with the improvement in exercise electrocardiogram (ECG) and in physical condition that we observe in most (over 90%) of the patients shortly (2–3 months) after the start of therapy (Seidel 1990; Schuff-Werner et al. 1989; Kleophas et al. 1990) (Table 2).

The first coronary angiographies 2 years after H.E.L.P. treatment in over 50 patients (to be reported elsewhere by the H.E.L.P. Multicenter Study Group; Schuff-Werner et al. 1993) lend support to the hope that regression of CHD is possible in humans (Hennerici et al. 1991). First evaluation of the 2-year follow-up by blinded angiograms revealed a regression of the coronary artery disease twofold more than progression.

HMG-CoA reductase inhibitors were not available when the H.E.L.P. Multicenter Study was started. In the meantime, since these compounds are now on the market, we and others have investigated the effect of a combined therapy, using lovastatin, simvastatin, or pravastatin together with the H.E.L.P. apheresis system.

E. Experience with Combined Heparin-Induced Extracorporeal Low-Density Lipoprotein Plasmapheresis and 3-Hydroxy-3-methylglutaryl Coenzyme A Reductase Inhibitor Therapy

In cases with plasma cholesterol levels exceeding 300 mg/dl, the use of specific diets and drugs may not be sufficient if LDL concentrations less than 110 mg/dl and/or regression of CHD is aimed at as a means of secondary intervention.

We have, therefore, investigated the efficacy of a combined therapy, using HMG-CoA reductase inhibitors (lovastatin, simvastatin, pravastatin) together with H.E.L.P. apheresis, and treated approximately 20 patients with severe FH on a long-term basis.

These compounds significantly decrease the rate of return after H.E.L.P. apheresis in both heterozygous and homozygous FH patients by 20%–30% (Seidel 1990; Armstrong et al. 1989; Thiery et al. 1990a,b). When the two treatments are combined, a reduction of the interval LDL cholesterol (LDL-C) level of 70% may be achieved, while Lp(a) and fibrinogen are not further affected (over the effect by the H.E.L.P. treatment alone; see Table 3). In the combined form, therapy intervals between two H.E.L.P. treatments may, in many cases, be stretched from 7 to 14 days, depending on the synthetic rates for LDL or the severity of CHD.

Table 3. Long-term effects of heparin-induced extracorporeal low-density lipoprotein plasmapheresis and 3-hydroxy-3-methylglutaryl coenzyme A reductase inhibitor treatment

	Mean interval values of approximately 1400 treatments	
HMG-CoA reductase inhibitor		
LDL Cholesterol	−38%	±12
HDL Cholesterol	+10%	±9
ApoB	−30%	±9
ApoA-1	+13%	±4
Lp(a)		no change
Fibrinogen		no change
H.E.L.P. + HMG-CoA reductase inhibitor		
LDL Cholesterol	−69%	±12
HDL Cholesterol	+14%	±6
ApoB	−53%	±8
ApoA-1	+12%	±9
Lp(a)	−43%	±7
Fibrinogen	−44%	±10

HMG-CoA, 3-hydroxy-3-methylglutaryl coenzyme A; H.E.L.P., heparin-induced extracorporeal low-density lipoprotein plasmapheresis; LDL, low-density lipoprotein; HDL; high-density lipoprotein; Lp(a), lipoprotein(a).

F. Treatment Tolerance and Safety of Heparin-Induced Extracorporeal Low-Density Lipoprotein Plasmapheresis

Overall treatment tolerance has been very good, and no major complications have been observed after 30000 treatments in approximately 300 patients. The treatment effects have been maintained on long-term treatment for over 6 years. At the end of the H.E.L.P. therapy, plasma concentrations of proteins that are not selectively precipitated by heparin at low pH were generally in the range of 80%–90% of the initial values and returned to their original level no later than 24h after the end of the treatment (EISENHAUER et al. 1987; SEIDEL 1990; SEIDEL et al. 1991). Substitution of any kind has not been necessary in 6 years of clinical experience. In contrast to some other LDL apheresis systems, the H.E.L.P. procedure does not alter the physicochemical characteristics of LDL, nor does it alter the ligand quality of LDL for lipoprotein receptors (SCHULTIS et al. 1990). Particular attention has been focused on the effect of H.E.L.P. on hemostasis. All posttreatment controls were typical for extracorporeal procedures, and no critical bleeding complications have been observed. Complement activation is found in all extracorporeal procedures. However, as a specific feature of the H.E.L.P. system, activated complement C3, C4, and the terminal complement complex are largely adsorbed to the filter system of H.E.L.P., resulting in plasma concentrations which are actually below those measured before LDL apheresis. C5a is not retained in the filter system, but plasma levels at the end of the treatment were within the normal range and leukocytopenia, a hallmark of complement activation, was never observed under H.E.L.P. treatment (WÜRZNER et al. 1991). Plasma electrolytes, hormones, vitamins, enzymes, and immunoglobulin concentrations as well as hematological parameters remained virtually unchanged at the end of each treatment and on long-term application (SEIDEL 1990; THIERY et al. 1990a,b; SEIDEL et al. 1991; see Table 4).

Table 4. Heparin-induced extracorporeal low-density lipoprotein plasmapheresis therapy in combination with 3-hydroxy-3-methylglutaryl coenzyme A reductase inhibitor – laboratory data

Parameters	Baseline		24 months Simvastatin +HELP treatment	
	$\bar{X}$	±SEM n.s.	$\bar{X}$	±SEM
Substrates				
Sodium	140.0	0.7	141.0	0.3
Potassium	3.9	0.12	4.0	0.05
Calcium	9.2	0.11	8.9	0.1
Phosphate	3.7	0.16	3.3	0.03
Iron	88.2	9.3	95.5	3.9
Creatinine	0.85	0.04	0.9	0.02
BUN	15.2	1.7	14.5	0.4
Uric acid	5.3	0.4	5.3	0.4
Glucose	94.0	0.9	100.0	6.4
Total bilirubin	0.43	0.03	0.56	0.4
Total protein	7.0	0.1	6.9	0.1
Albumin %	61.6	1.73	61.4	0.42
α_1-Protein %	3.6	0.3	3.6	0.1
α-Protein %	8.0	0.42	8.3	0.14
β_2-Protein %	13.0	0.56	12.0	0.03
γ-Protein %	13.7	0.99	14.8	0.14
Enzymes				
ALAT (GOT)	10.0	0.4	13.5	0.4
ASAT (GPT)	11.0	2.0	19.0	1.0
γ-GT	21.0	5.8	25.0	2.1
CK	45.0	7.0	45.0	2.0
LDH	143.0	10.8	151.0	4.6
Amylase	16.0	2.7	16.0	0.3
CHS	5151.0	525.0	5455.0	530.0
ALP	101.0	6.6	110.0	2.8
Hematological indices				
Hemoglobin	14.0	0.44	14.3	0.07
Hematocrit	41.8	1.1	42.0	0.73
Erythrocytes	4.4	0.14	4.6	0.1
Thrombocytes	226.0	10.2	220.0	9.5
Leukocytes	5.18	0.39	5.22	0.48

Table 4. *Continued*

Parameters	Baseline $\bar{X}$	±SEM n.s.	24 months Simvastatin +HELP treatment $\bar{X}$	±SEM
Lymphocytes	37.4	2.76	33.3	2.1
Monocytes	7.2	1.02	6.2	2.45
Neutrophils	51.3	3.16	57.4	3.12
Eosinophils	2.6	0.43	1.8	0.61
Basophils	0.7	0.18	0.7	0.1
Hemostasis				
Quick test (PT) %	98.0	1.25	99.0	0.91
TT (s)	14.0	0.12	14.0	0.21
Endocrinological indices				
Cortisol	12.6	1.05	13.3	1.15
Testosterone	6.7	1.07	6.4	0.26
ACTH	40.3	3.78	40.4	6.18
LH[a]	15.9	8.36	11.1	5.91
FSH[a]	16.0	0.22	28.0	10.3
T3	133.5	7.35	123.5	12.4
T4	7.0	0.61	7.3	0.1
FT4	7.5	0.62	7.5	0.6
FT3	142.5	7.64	137.5	6.01

[a] In men and premenopausal women.

G. Typical Case Reports

I. Case 1

A typical follow-up kinetic for LDL and lipoprotein(a) under H.E.L.P. treatment of a patient with severe progressive coronary heart disease is shown in Fig. 3.

At the start of our therapy, the 33-year-old MI patient was coronary bypassed and treated with percutaneous transluminal coronary angioplasty (PTCA), with LDL-C levels of 350 mg/dl and a marked Lp(a) elevation of 165 mg/dl.

LDL-C could be lowered with an HMG-CoA reductase inhibitor (simvastatin) by about 48% to 170 mg/dl, but no effect on lipoprotein(a) levels was observed. In combination with regular H.E.L.P. treatment, we were able to maintain LDL concentration at an interval value of 110 mg/dl. In addition, H.E.L.P. treatment resulted in a marked decrease (70%) of postapheresis lipoprotein(a) concentrations. The interval Lp(a) levels remained around 90 mg/dl. Fibrinogen was lowered from a baseline value of 317 mg/dl to a H.E.L.P. interval value of 177 mg/dl, which is a 44% reduction.

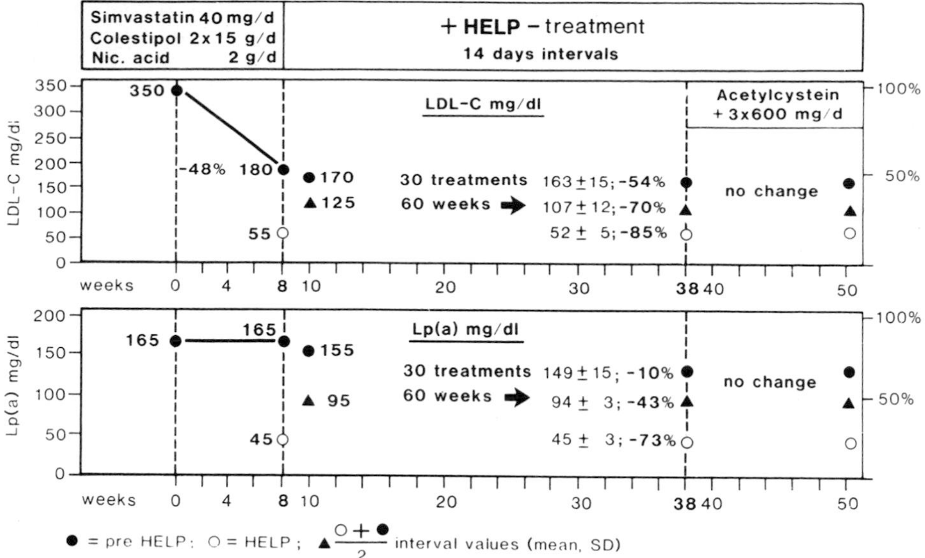

Fig. 3. Maximal treatment of familial hypercholesterolemia (FH) and elevated plasma lipoprotein(a) (*Lp(a)*) concentrations. Patient, N.J., 33 years, male. Baseline low-density lipoprotein cholesterol (*LDL-C*) 350 mg/dl, Lp(a) 165 mg/dl, aortocoronary venous bypass, percutaneous transluminal coronary angioplasty (PTCA). Well-maintained PTCA results after 1 year of treatment and no further progression of coronary heart disease. *Nic. acid*, nicotinic acid

A control angiography after 2 years revealed that the combined treatment was able to stop the very progressive CHD which was developing in the patient previous to the treatment. The clinical situation has also considerably improved.

II. Case 2

Experience with the H.E.L.P. treatment in a homozygous form of FH (see Fig. 4).

Early death from cardiac consequences of premature coronary sclerosis and aortic stenoses is the usual outcome of homozygous FH (Goldstein and Brown 1989). Inherited as an autosomal dominant defect of the LDL receptor gene, this disease is characterized by very high plasma LDL-C concentrations (between 200 and 1000 mg/dl) and the development of severe cutaneous and tendon xanthomata in childhood. All conventional lipid-lowering treatments with diet and medication are completely insufficient.

Since 1985 we have been following and treating an FH patient, born in 1979, with the H.E.L.P. apheresis procedure (Thiery et al. 1990b). LDL-C concentrations before the start of treatment exceeded 800 mg/dl. The follow-up of LDL concentrations under the H.E.L.P. treatment alone and in

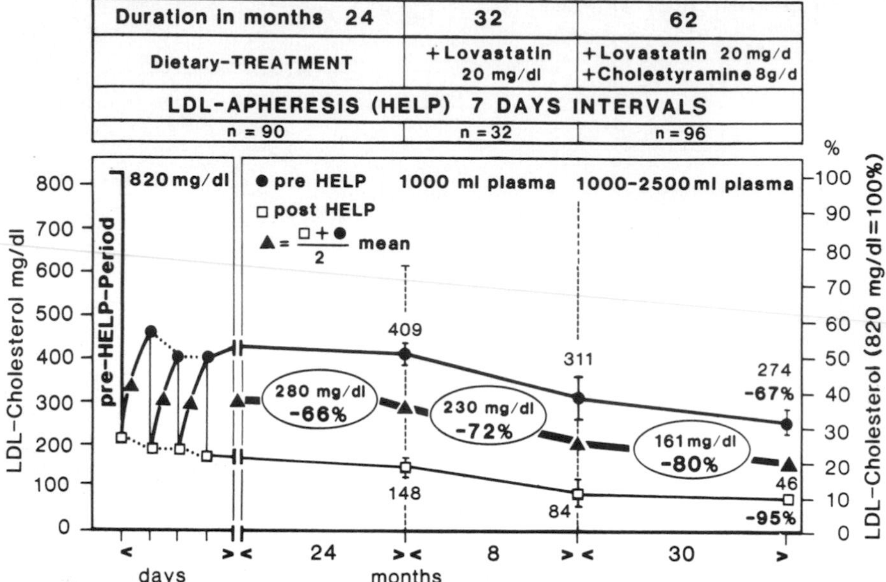

Fig. 4. Maximal treatment of familial hypercholesterolemia (FH): follow-up of low-density lipoprotein cholesterol (*LDL-C*). Patient, Ch.J., female, 7 years, homozygous FH. Baseline LDL-C 820 mg/dl

combination with lovastatin and regular cholestyramine is shown in Fig. 4. The girl was treated for 2 years with a weekly H.E.L.P. apheresis. Under this procedure, the LDL-C interval levels were maintained below 280 mg/dl. At this time a rapid regression of multiple xanthomata could be observed. With additional medication of lovastatin and cholestyramine, a further LDL decrease to 180 mg/dl could be achieved. The treated plasma volume could be recently enhanced from 1.5 to 2.5 l. This has now resulted in a mean LDL-C level of 160 mg/dl, which is equivalent to a decrease of 80% as compared to pretreatment values.

The therapy is excellently tolerated. The girl is well and shows normal growth and development. No signs of cardiovascular symptoms have been noted.

H. The Heparin-Induced Extracorporeal Low-Density Lipoprotein Plasmapheresis Treatment in Heart Transplant Patients with Severe Hypercholesterolemia: Report of an Ongoing Study

The goal of this ongoing trial is to decrease recurrent CHD in patients with heart grafts. In this study the patients will be followed for 4 years. LDL-C concentrations in all patients are maintained at a level below 120 mg/dl. Treatment was started from a baseline LDL-C concentration greater than 280 mg/dl with simvastatin, which resulted in a reduction of LDL-C by 40%, but exceeded 170 mg/dl LDL. In the combination of simvastatin and H.E.L.P. treatment, an LDL-C level less than 120 mg/dl was achieved. As in other H.E.L.P. patients, overall treatment tolerance in this group ($n = 5$) has been very good and no major complications have been observed in the first 6 months of therapy.

Special attention has been given to the tolerance and pharmacokinetics of both simvastatin and cyclosporine A. No signs of myopathy were observed. No change of cell mediators such as interleukin II receptor, interleukin VI receptor, interferon gamma, and tumor necrosis factor before and after H.E.L.P. treatment were observed.

From our first clinical experience in heart-transplanted patients with severe hypercholesterolemia, the additional therapy with H.E.L.P. LDL apheresis may not only be useful, but necessary to achieve long-lasting benefit from the transplantation. Annual examination by angiograms of the patients should provide a rationale for the drastic LDL-lowering therapy in the prevention of graft atherosclerosis in heart transplant patients.

I. The Heparin-Induced Extracorporeal Low-Density Lipoprotein Plasmapheresis U System for a Simultaneous Hemodialysis: Low-Density Lipoprotein Apheresis

Disturbances in plasma lipoprotein metabolism and progressive atherosclerosis are well-known clinical complications in patients on long-term hemodialysis treatment. The therapy with lipid-lowering drugs in many of these patients has proved to be difficult and/or unsatisfactory.

Therefore, we took advantage of the dialysis block, which is part of the H.E.L.P. system, to combine both hemodialysis with H.E.L.P. LDL apheresis at the same time. For this purpose, only minor modifications of the system were required. The system used for hemodialysis in combination with LDL apheresis is manufactured by B. Braun Melsungen, Germany, and has been named Plasmat Secura-U^{R}. LDL apheresis begins at the start of the hemodialysis and continues for about $2–2\frac{1}{2}$ h only. Blood flow is kept at 250 ml/min. Two hundred milliliters of whole blood is dialyzed immediately. Fifty milliliters/min is directed over a plasma filter to enter the H.E.L.P. cascade system.

Weekly treatment of hypercholesterolemia in renal insufficient patients with the basic H.E.L.P. LDL apheresis system or with the simultaneous H.E.L.P. hemodialysis procedure (the U model) for up to 1 year was well tolerated and no major complications have been observed in more than 120 treatments.

The H.E.L.P. U system in its present configuration reduces the LDL-C by 25%–50%, triglycerides by 25%, and Lp(a) and fibrinogen by 45%–50%, depending on the ultrafiltration rate of the patient. HDL-C is increased from 8%–30%. Shifting the LDL apheresis towards the end of the hemodialysis, in particular for patients with high ultrafiltration rates, may improve the efficiency.

The H.E.L.P. U LDL apheresis has thus proved to be very efficient in normalizing elevated LDL, Lp(a), triglycerides, and plasma fibrinogen concentrations in patients on hemodialysis. As well as the reduction of these cardiovascular risk factors, the patients treated with the H.E.L.P. U model benefit greatly from the improvement of plasma and blood viscosity, which is a well-known and inherent problem in all hemodialysis treatments.

I. Case Report (Table 5)

The effect of a H.E.L.P. U treatment in a patient with low ultrafiltration rate (UFR) (A) reveals an approximately 50% reduction in total cholesterol, apoB, and fibrinogen.

The marked increase in total cholesterol, apoB, and fibrinogen in a case with high UFR (B) on hemodialysis is not only compensated by simultaneous H.E.L.P. U treatment, but significantly reduced below predialysis values.

Table 5. Effects of simultaneous heparin-induced extracorporeal low-density lipoprotein plasmapheresis and hemodialysis treatment in a patient with low and with high ultrafiltration rate (H.E.L.P.-U system)

	Hct %	TC	ApoB	Fib (mg/dl)	Alb
Low UFR (<2000 ml)					
Pre-HD	31	176	92	375	4.6
Post-HD	32	188	89	394	4.7
Difference (%)	+3	+7	−3	+5	+2
Pre-HELP-U	33.2	185	116	385	4.5
Post-HELP-U	35.6	109	55	173	4.3
Difference (%)	+7	−51	−52	−54	−4
High UFR (>3000 ml)					
Pre-HD	30	226	137	380	4.0
Post-HD	38.3	317	184	475	4.8
Difference (%)	+28	+40	+34	+25	+20
Pre-HELP-U	29.6	221	134	377	3.4
Post-HELP-U	36.8	196	112	271	3.9
Difference (%)	+25	−11	−16	−28	+16

HD, hemodialysis; UFR, ultrafiltration rate; HELP-U, heparin-induced extracorporeal low density lipoprotein plasmapheresis with simultaneous hemodialysis; Hct, hematocrit; TC, total cholesterol; ApoB, apolipoprotein B; Fib, fibrinogen; Alb, albumin.

J. Comparison of Techniques to Lower Low-Density Lipoprotein Levels by Apheresis

Three methods for the selective removal of LDL from plasma have been established and are now used for treatment of severely hypercholesterolemic patients: (a) LDL immunoadsorption, using immobilized anti-apolipoprotein B antibodies for LDL binding (STOFFEL and DEMANT 1981; RIESEN et al. 1986); (b) LDL binding to dextran sulfate cellulose (DSC; YOKOYAMA et al. 1985); and (c) H.E.L.P. (SEIDEL and WIELAND 1982; EISENHAUER et al. 1987).

In two recent reviews (KELLER 1991; DEMANT and SEIDEL 1992) the different procedures were compared with regard to their efficiency in lowering LDL concentrations and safety.

Immunoabsorption, heparin precipitation, and dextran sulphate binding all achieved an approximately 60% decrease of LDL plasma concentrations in the course of a single LDL apheresis session. The reduction in HDL levels and immunoglobulins is usually less than 20%, with no significant difference between the three LDL apheresis methods. These apparent losses may to some extent be due to nonspecific plasma dilution by the saline priming solution from the extracorporeal plasma circuit.

Double plasma filtration, although also effective in reducing LDL, is not selective, Total plasma protein loss (HDL, immunglobulins, and albumin) is significant and concomitant albumin substitution is regularly required. Therefore, double plasma filtration occupies a position close to plasma exchange. This technique should not be recommended for FH treatment.

Immunoadsorption and DSC apheresis are highly specific for apoB-containing lipoproteins which include VLDL, intermediate density lipoproteins (IDL), and Lp(a). It has recently been demonstrated that increased Lp(a) concentrations are significantly correlated with increased risk for CHD. Immunoadsorption, DSC LDL apheresis, and H.E.L.P. LDL apheresis all eliminate Lp(a) to about the same extent as LDL. In contrast to immunoadsorption and DSC apheresis, H.E.L.P. LDL apheresis also eliminates fibrinogen. Parallel measurements of plasma viscosity and erythrocyte aggregation before and after H.E.L.P. LDL apheresis revealed a significant reduction of 15% and 50%, respectively (SCHUFF-WERNER et al. 1989). The muscle oxygen tension was found to be significantly higher directly after treatment, when compared with pretreatment values, probably as a result of improved microcirculation under H.E.L.P. therapy (SCHUFF-WERNER et al. 1989; KLEOPHAS et al. 1990).

Results from multicenter studies using H.E.L.P. LDL apheresis and DSC LDL apheresis as a means of drastic lipid-lowering therapy are now available (SEIDEL et al. 1991; HENNERICI et al. 1991; GORDON et al. 1991; TATAMI et al. 1992).

In the H.E.L.P. Multicenter Study, 51 participants were examined by coronary angiography at the start and after 2 years of treatment. First evaluations of the angiographs with quantitative determination of stenosis revealed a regression rate twofold higher than the progression rate of the coronary arteries in the patients under H.E.L.P. therapy. Close to 40% of all patients showed a clear tendency of stenosis regression (SEIDEL et al. 1991; SCHUFF-WERNER et al. 1993). Another small study of seven patients with heterozygous FH produced evidence that H.E.L.P. LDL apheresis administered once a week for 7–24 months induced regression of carotid atherosclerotic plaques (HENNERICI et al. 1991). Plaques were evaluated by a three-dimensional reconstruction of ultrasound images. Out of 21 observed plaques only one progressed, 12 did not change, and eight regressed within 6–12 months.

Data of one multicenter study using DSC LDL apheresis (GORDON et al. 1991) are at present only available in preliminary form. Sixty-four patients with FH (54 heterozygous and ten homozygous) were treated at 7- to 14-day intervals for 18 weeks. Baseline LDL-C concentrations were 243 mg/dl and 447 mg/dl, respectively. Time-averaged LDL-C levels on treatment were 139 mg/dl in heterozygotes and 210 mg/dl in homozygotes. HDL-C increased slightly, but changes were not significant. Lp(a) levels were reduced markedly, but long-term concentrations are not given.

The second DSC LDL Apheresis Multicenter Study (TATAMI et al. 1992) uses LDL apheresis combined with cholesterol-lowering drugs to treat homozygous or heterozygous FH. As a result, by visual judgement or computer analysis the coronary angiograms revealed a regression rate in approximately 38%, no change in 48%, and progression in 14%, indicating an encouraging result of aggressive cholesterol-lowering therapy in coronary atherosclerosis of FH patients.

Another multicenter study, using LDL immunoadsorption, has been started, but at present no data from this study are available.

LDL immunoadsorption, DSC LDL apheresis, and H.E.L.P. LDL apheresis are all safe and equally potent methods of extracorporeal LDL elimination. Lp(a) can also specifically be removed from plasma by these procedures. In addition, H.E.L.P. LDL apheresis selectively reduces plasma fibrinogen, which seems to have a beneficial effect on the microcirculation. Long-term observations show that, besides the marked reduction in LDL-C, some increase of HDL-C occurs, which may add to the antiatherogenic effect of LDL apheresis treatment.

K. Indication for Heparin-Induced Extracorporeal Low-Density Lipoprotein Apheresis

Based on the experience of many centers, a German consensus panel has recently published differentiated guidelines as to when H.E.L.P. LDL apheresis should be used (GRETEN et al. 1992). These are as follows:

- In the presence of homozygous FH;
- In the primary prevention of CHD in young patients with severe hypercholesterolemia, mild CHD (stage I to II), and a family history of CHD, provided LDL-C cannot be decreased below 200 mg/dl by a hyperlipidemic diet and maximal drug therapy;
- In the secondary prevention of CHD in patients with severe CHD (stage III–IV) and marked hypercholesterolemia, provided LDL-C cannot be decreased below 135 mg/dl by maximal dietary and pharmacological regimen. In any case, diet and drug therapy should be continued while patients are on H.E.L.P. LDL apheresis.
- Further indications for primary prevention of CHD in patients with renal insufficiency and after heart transplantation will have to be reconsidered in the light of the results from the ongoing trials.

L. Conclusion

LDL apheresis is the most potent technique to eliminate LDL and Lp(a) if the physiological clearing mechanisms are insufficient. The H.E.L.P. LDL apheresis system in addition can also very efficiently remove fibrinogen and by this improve plasma viscosity and microcirculation.

Long-term observations show that as well as the marked reduction of LDL-C, a remarkable increase of HDL occurs, which may add to the antiatherogenic effect of the extracorporeal procedures. For the future the availability of safe and efficient apheresis techniques will not only provide a new dimension in the treatment of severe hypercholesterolemia in patients with CHD, but will also offer possibilities to answer key questions of atherogenesis in man.

References

Armstrong VW (1987) Die säureinduzierte Präzipitation von Low-Density-Lipoproteinen mit Heparin, Grundlagen zum H.E.L.P.-Verfahren. *Habilitationsschrift*, University of Göttingen

Armstrong VW, Schleef J, Thiery J, Muche R, Schuff-Werner P, Eisenhauer T, Seidel D (1989) Effect of H.E.L.P.-LDL apheresis on serum concentrations of human lipoprotein(a): kinetic analysis of the post-treatment return to baseline levels. Eur J Clin Invest 19:235–240

Blankenhorn D, Nessim S, Johnson R, Sanmarco ME, Azen SP et al. (1987) Beneficial effects of combined colestipol-niacin therapy on coronary atherosclerosis and coronary venous bypass grafts. JAMA 257:3233–3240

Brown BG, Albers JJ, Fisher LD, Schaeffer SM, Lin JT, Kaplan C, Zhao XQ, Bisson BD, Fitzpatrick VF, Dodge HT (1990) Regression of coronary artery disease as a result of intensive lipid lowering therapy in men with high levels of apolipoprotein B. N Engl J Med 323:1289–1298

Brown MS, Goldstein JL (1986) A receptor mediated pathway for cholesterol homeostasis. Science 232:34–37

Buchwald H (1964) Lowering of cholesterol adsorption and blood levels by ileal exclusion. Circulation 29:713–720

Buchwald H, Varco RL, Matts JP, Long JM, Fitch LL, Campbell GS, Pearce MB, Yellin AE, Edmiston WA, Smink RD Jr, Sawin HS Jr, Campos CT, Hansen BJ, Tuna N, Karnegis JN, Sanmarco ME, Amplatz K, Castaneda-Zuniga WR, Hunter DW, Bissett JK, Weber FJ, Stevenson JW, Leon AS, Chalmers TC, the POSCH Group (1990) Effect of partial ileal bypass surgery on mortality and morbidity from coronary heart disease in patients with hypercholesterolemia. N Engl J Med 323:946–955

Castelli WP, Wilson PWF, Levy D, Anderson K (1990) Serum lipids and risk of coronary artery disease. Atheroscler Rev 21:7–19

Cremer P, Muche R (1990) Göttinger Risiko-, Inzidenz- und Prävalenzstudie (GRIPS). Empfehlungen zur Prävention der koronaren Herzkrankheit. Ther Umsch 6:482–491

Cremer P, Nagel D, Labrot B, Muche R, Elster H, Mann H, Seidel D (1991) Göttinger Risiko-, Inzidenz- und Prävalenzstudie (GRIPS). Springer, Berlin Heidelberg New York

De Gennes J-L, Touraine R, Maunand B, Truffert J, Laudat Ph (1967) Formes homozygotes cutanéo-tendineuses de xanthomatose hypercholestérolémique dans une observation familiale exemplaire. Essai de plasmaphérèse à titre de traitement héroïque. Société Médicale des Hôpitaux de Paris 118:1377–1402

Demant T, Seidel D (1992) Recent developments in low-density lipoprotein apheresis. Curr Opin Lipidol 3:43–48

Eisenhauer T, Armstrong VW, Wieland H, Fuchs C, Scheler F, Seidel D (1987) Selective removal of low density lipoproteins (LDL) by precipitation at low pH: first clinical application of the H.E.L.P. system. Klin Wochenschr 65: 161–168

Goldstein JL, Brown MS (1989) Familial hypercholesterolemia. In: Scriver CR, Beaudet AL, Sly WS, Valle D (eds) The metabolic basis of inherited disease. McGraw-Hill, New York, pp 1215–1250

Gordon BR, Bilheimer DW, Brown DC, Dau PC, Gotto AM, Illingworth DR, Jones PH, Kelsey SF, Leitman SF, Stein EA, Stern TN, Zavoral JH, Zwiener J, for Liposorber Study Group (1991) Multicenter study of the treatment of familial hypercholesterolemia (FH) by LDL-apheresis using the liposorber[R] LA-15 system (abstr). Arterioscler Thromb 11:1409

Greten H, Bleifeld W, Beil FU, Daerr W, Strauer BE, Kleophas W, Gries FA, Schuff-Werner P, Thiery J, Seidel D (1992) LDL-Apherese. Ein therapeutisches Verfahren bei schwerer Hypercholesterinämie. Dtsches Arztebl 89:48–49

Hennerici M, Kleophas W, Gries FA (1991) Regression of carotid plaques during low density lipoprotein cholesterol elimination. Stroke 22:989–992

Keller C (1991) LDL-apheresis: results of long-term treatment and vascular outcome. Atherosclerosis 86:1–8

Kienast J, Berning B, van de Loo J (1990) Fibrinogen als Risikoindikator bei arteriosklerotischen Veränderungen und Koronararterien-Erkrankungen. Diagn Lab 40:162

Kleophas W, Leschke M, Tschöpe D, Martin J, Schauseil S, Schottenfeld Y, Strauer BE, Gries FA (1990) Akute Wirkungen der extrakorporalen LDL-Cholesterin- und Fibrinogen-Elimination auf Blutrheologie und Mikrozirkulation. Dtsch Med Wochenschr 115:7–11

Koenig W, Ernst E (1992) The possible role of hemorheology in atherothrombogenesis. Atherosclerosis 94(2/3):93–107

Lupien PJ, Moojani S, Award J (1976) A new approach to the management of familial hypercholesterolemia: removal of plasma cholesterol based on the principle of affinity chromatography. Lancet 1:1261–1265

Ornish D, Brown SE, Scherwitz LW, Billings JH, Armstrong WT, Ports TA, McLanahan SM, Kirkeeide RL, Brand RJ, Gould KL (1990) Can lifestyle changes reverse coronary heart disease? Lancet 336:129–133

Riesen WT, Imhof C, Sturzenegger E, Descoeudres C, Mordasini R, Oetliker OH (1986) Behandlung der Hypercholesterinämie durch extrakorporale Immunadsorption. Schweiz Med Wochenschr 116:8

Schuff-Werner P, Schütz E, Seyde WC, Eisenhauer T, Janning G, Armstrong VW, Seidel D (1989) Improved haemorheology associated with a reduction in plasma fibrinogen and LDL in patients being treated by heparin-induced extracorporeal LDL precipitation (H.E.L.P.). Eur J Clin Invest 19:30–37

Schultis H-W, von Bayer H, Neitzel H, Riedel E (1990) Functional characteristics of LDL particles derived from various LDL-apheresis techniques regarding LDL-drug-complex preparation. J Lipid Res 31:2277–2284

Seidel D (1990) The H.E.L.P. system: an efficient and safe method of plasmatherapy in the treatment of severe hypercholesterolemia. Ther Umsch 47:514–519

Seidel D, Wieland H (1982) Ein neues Verfahren zur selektiven Messung und extrakorporalen Elimination von Low-Density-Lipoproteinen. J Clin Chem Clin Biochem 20:684–685

Seidel D, Cremer P, Thiery J (1985) Plasmalipoproteine und Atherosklerose. Intern Welt 5:114–124, 6:159–165

Seidel D, Armstrong VW, Schuff-Werner P, for the H.E.L.P. Study Group (1991) The H.E.L.P.-LDL-apheresis multicenter study, an angiographically assessed trial on the role of LDL-apheresis in the secondary prevention of coronary heart disease: I. Evaluation of safety and cholesterol-lowering effects during the first 12 months. Eur J Clin Invest 21:375–383

Seidel D, Neumeier D, Cremer P, Nagel D (1992) Lipoprotein(a) in internal medicine. In: Stein O, Eisenberg S, Stein Y (eds) Atherosclerosis IX. Proceedings of the 9th International Symposium on Atherosclerosis. R&L Creative Communications, Tel Aviv, pp 127–130

Smith EB (1986) Fibrinogen, fibrin and fibrin degradation products in relation to atherosclerosis. In: Fidge NH, Nestel PJ (eds) Atherosclerosis VI. Elsevier, Amsterdam, pp 459–462

Smith EB (1990) Transport, interactions and retention of plasma proteins in the intima: the barrier function of the internal elastic lamina. Eur Heart J 11 Suppl E:72–81

Starzl TE, Bilheimer DW, Bahnson HT, Shaw BW, Hardesty RL, Griffith BP, Iwatsuki S, Zitelli BJ, Gartner JC, Malatack JJ, Urbach AH (1984) Heart–liver transplantation in a patient with familial hypercholesterolemia. Lancet 1:1382–1383

Starzl TE, Chase HP, Ahrens EH, McNamara DJ, Bilheimer DW, Schaefer EF, Rey J, Porter KA, Stein E, Francavilia A, Benson LN (1983) Portocaval shunt in patients with familial hypercholesterolemia. Ann Surg 198:273–283

Stoffel W, Demant T (1981) Selective removal of apolipoprotein B-containing serum lipoproteins from blood plasma. Proc Natl Acad Sci USA 78:611–615

Study Group of the European Atherosclerosis Society (1987) Strategies for the prevention of coronary heart disease: a policy statement of the European Atherosclerosis Society. Eur Heart J 8:77–88

Tatami R, Inoue N, Itoh H, Kishino B, Koga N, Nakashima Y, Nishide T, Okamura K, Saito Y, Teramoto T, Yasugi T, Yamamoto A, Goto Y, for the LARS Investigators (1992) Regression of coronary atherosclerosis by combined LDL-apheresis and lipid-lowering drug therapy in patients with familial hypercholesterolemia: a multicenter study. Atherosclerosis 95:1–13

The Expert Panel (1988) Report of the National Cholesterol Education Program Expert Panel on detection, evaluation and treatment of high blood cholesterol in adults. Arch Intern Med 148:36–39

Thiery J (1988) Maximaltherapie der Hypercholesterinämie bei koronarer Herzkrankheit. Kombination einer Plasmatherapie (H.E.L.P.) mit HMG-CoA-Reduktasehemmer. Therapiewoche 38:1–12

Thiery J, Armstrong V, Bosch T, Eisenhauer T, Schuff-Werner P, Seidel D (1990a) Maximaltherapy der Hypercholesterinämie bei koronarer Herzerkrankung. Ther Umsch 47(6):520–529

Thiery J, Walli AK, Janning G, Seidel D (1990b) Low density lipoprotein plasmapheresis with and without lovastatin in the treatment of the homozygous form of familial hypercholesterolemia. Eur J Pediatr 149:716–721

Thompson GR, Lowenthal R, Myant NB (1975) Plasma exchange in the management of homozygous familial hypercholesterolemia. Lancet 1:1208–1211

Würzner R, Schuff-Werner P, Franzke A, Nitze R, Oppermann M, Armstrong VW, Eisenhauer T, Seidel D, Götze O (1991) Complement activation and depletion during LDL-apheresis by heparin-induced extracorporeal LDL-precipitation (H.E.L.P.). Eur J Clin Invest 21:288–294

Yokoyama S, Hayashi R, Satani M, Yamamoto A (1985) Selective removal of low density lipoprotein by plasmapheresis in familial hypercholesterolemia. Atherosclerosis 5:613

Note Added in Proof

Schuff-Werner P, Gohlke H, Bartmann U, Corti MC, Dinsenbacher A, Eisenhauer T, Grützmacher P, Keller C, Kettner U, Kleophas W, Köster W, Olbricht CJ, Richter WO, Seidel D and the HELP-Study Group (1993) The Help-LDL-Apheresis Multicenter Study, an angiographically assessed trial on the role of LDL-apheresis in the secondary prevention of coronary heart disease. II. Final evaluation of the effect of regular treatment on LDL-cholesterol plasma concentrations and the course of coronary heart disease. Eur J Clin Invest (in press)

Section C
Lipid Lowering Drugs

CHAPTER 11

3-Hydroxy-3-methylglutaryl Coenzyme A Reductase Inhibitors

D.R. ILLINGWORTH and E.B. SCHMIDT

A. Introduction

Delineation of the mechanisms responsible for the maintenance of cholesterol homeostasis in humans has provided a sound scientific basis for the development of drugs which act to inhibit key enzymes in the pathway of cholesterol biosynthesis. In humans, the majority of cholesterol which is transported in plasma lipoproteins is derived from de novo synthesis rather than from dietary sources (BROWN and GOLDSTEIN 1986). Conversion of 3-hydroxy-3-methylglutaryl coenzyme A (HMG-CoA) to mevalonic acid by the enzyme HMG-CoA reductase is the rate-limiting enzyme in cholesterol biosynthesis and is subject to metabolic regulation. Other enzymes, including HMG-CoA synthase and squalene synthase, are also subject to metabolic regulation.

Depletion of the cellular pool of cholesterol leads to an increase in cholesterol biosynthesis with an increase in the mass and activity of HMG-CoA reductase, whereas cholesterol biosynthesis is repressed if the cellular cholesterol content is transiently increased. In humans, the liver is the major organ responsible for cholesterol biosynthesis and is also the site of synthesis of the major lipoproteins containing apoprotein B-100 (very low density lipoproteins, VLDL, and low-density lipoproteins, LDL), as well as contributing to the synthesis of high-density lipoproteins (HDL). The liver is also the major organ responsible for the removal of plasma lipoproteins and, in humans, approximately 70% of the plasma pool of LDL which is removed per day is taken up by this organ (BROWN and GOLDSTEIN 1983).

The plasma concentrations of LDL are determined by the rates of hepatic production of apoprotein-B-containing lipoproteins, the rate of conversion of VLDL to LDL, and by the catabolic rate for LDL; the latter is primarily determined by the number of specific, high-affinity LDL receptors present on hepatocyte and other cell membranes. In view of this central role of the cholesterol biosynthetic pathway in the regulation of the cellular pools of cholesterol, which, in turn, influence the expression of high-affinity LDL receptors and the rates of hepatic production of VLDL and LDL, the development of specific competitive inhibitors of HMG-CoA reductase has provided a targeted and attractive therapeutic approach to the treatment of patients with elevated plasma concentrations of atherogenic lipoproteins, particularly LDL and VLDL remnants. The present review focusses on the mechanisms of action and clinical efficacy of this new class of drugs in the therapy of adult patients with hyperlipidemia.

B. Structure and Mechanism of Action of 3-Hydroxy-3-methylglutaryl Coenzyme A Reductase Inhibitors

I. Structure

The first two specific competitive inhibitors of HMG-CoA reductase to be tested in human subjects were originally isolated from fungal cultures. Mevastatin (originally called compactin) was isolated from cultures of *Penicillium citrinum*, whereas lovastatin has been isolated from cultures of the soil fungus *Aspergillus terreus* (Alberts 1988). Mevastatin was used as an investigational hypocholesterolemic agent in Japan and, despite promising early clinical results (Mabuchi et al. 1981), this drug was withdrawn from clinical use because of changes in intestinal morphology in dogs. Fortunately, similar adverse effects have not been seen with other drugs in this class (Gerson et al. 1989), and lovastatin, a methylated derivative of this drug (simvastatin), and a hydroxy derivative of mevastatin (pravastatin) have all undergone extensive clinical trials in North America, Western Europe, and the Far East and have now been approved by regulatory agencies in a large number of countries. In addition, other totally synthetic HMG-CoA reductase inhibitors have recently been developed (e.g., fluvastatin, dalvastatin), and it is likely that other agents in this class will also be synthesized.

The structure of HMG-CoA and a number of the approved and investigational HMG-CoA reductase inhibitors are shown in Fig. 1; note the similarity between the open acid portions of these molecules and the structure of HMG-CoA. Lovastatin and simvastatin are administered as pro-drugs, and hydrolysis of the lactone ring to the active open acid form occurs in the liver; in contrast, pravastatin is administered as the open acid. Results of studies in which the uptakes of lovastatin and pravastatin by the liver and peripheral tissues of animals have been examined are contradictory (Alberts 1988; Tsujita et al. 1986). Further studies are necessary to assess whether or not the greater hydrophilicity of pravastatin, as compared to the more lipophilic pro-drug forms of lovastatin and simvastatin, may result in lower rates of uptake of pravastatin by peripheral tissues when administered to hypercholesterolemic patients. At doses administered in clinical practice, however, lovastatin, pravastatin, and simvastatin are all taken up primarily by the liver and exert their hypolipidemic effects by inhibiting hepatic HMG-CoA reductase, thereby inducing an increased expression of high-affinity LDL receptors on hepatocyte membranes (Ma et al. 1986; Reihner et al. 1990).

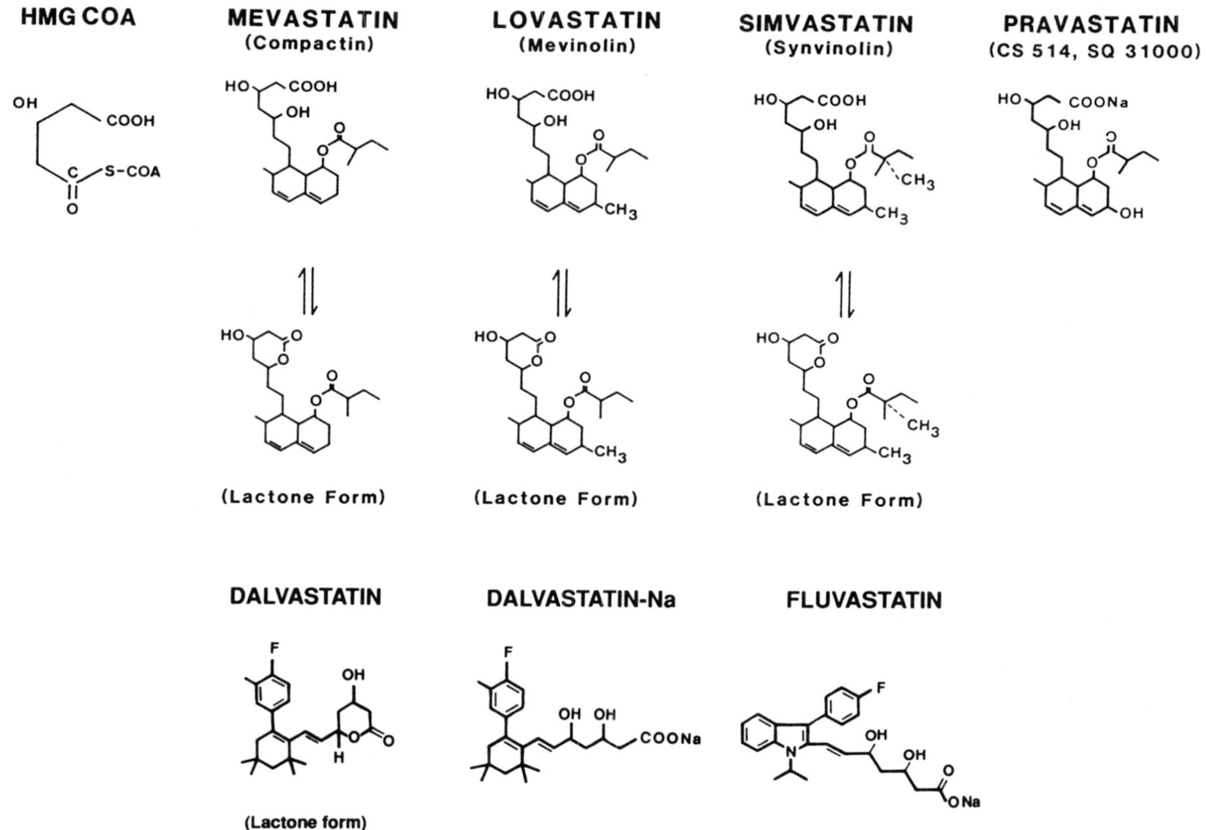

Fig. 1. Structure of 3-hydroxy-3-methylglutaryl coenzyme A (HMG-CoA) and the currently available and some of the investigational HMG-CoA reductase inhibitors

II. Pharmacokinetic Properties of Lovastatin, Pravastatin, and Simvastatin

Absorption of lovastatin, simvastatin, and pravastatin after oral administration is incomplete in humans, and peak plasma concentrations are obtained 2–4 h after oral administration in the case of lovastatin and simvastatin and 1–2 h after oral administration of pravastatin (HENWOOD and HEEL 1988; TODD and GOA 1990; MCTAVISH and SORKIN 1991). The fractional absorption of all three drugs is in the range of 30%–50% following oral administration; steady-state concentrations occur within 2–3 days, but do not rise further, indicating that drug accumulation does not occur. In the case of lovastatin, less than 5% of the active drug reaches the general circulation and the systemic availability of all three drugs is low. Concentrations of active metabolites of these drugs in plasma are reduced by one third when the drugs are administered with food; this is believed to be due to an enhanced first-pass hepatic extraction. Absorption of lovastatin and simvastatin may be better when the drugs are administered with food, whereas pravastatin is recommended to be taken at bedtime without food. The presence of a carboxyl group on pravastatin renders this drug more acidic and absorption is reduced if this drug is coadministered with cholestyramine; similar binding of cholestyramine to lovastatin and simvastatin does not appear to occur. Lovastatin and simvastatin are highly protein bound (greater than 95% in human plasma), whereas for pravastatin protein binding is only approximately 55% (MCTAVISH and SORKIN 1991). Lovastatin and simvastatin are more lipophilic than pravastatin and this may contribute to the higher protein binding of the former two drugs, and since albumin crosses the blood–brain barrier, this may also explain the ability of lovastatin to cross the blood brain barrier, where it can be detected in the cerebrospinal fluid (BOTTI et al. 1991). As previously discussed, studies on the uptakes of lovastatin and simvastatin, as compared to pravastatin, by peripheral tissues have provided conflicting data (TSUJITA et al. 1986; ALBERTS 1988; GERMERSHAUSEN et al. 1989) and any potential clinical differences in side-effect profiles based upon apparent differences in "tissue selectivity" remain to be demonstrated.

Metabolic transformation of lovastatin, simvastatin, and pravastatin to a number of metabolites occurs in both animals and humans, but none accumulate in tissues after prolonged administration. Lovastatin and simvastatin are primarily excreted via the biliary route and less than 10% of orally administered drug is recovered in the urine. In contrast, both renal and hepatic mechanisms are responsible for the excretion of pravastatin; when intravenously administered, 60% of this drug was recovered in the urine as compared to 34% present in the feces (MCTAVISH and SORKIN 1991). These data suggest that impaired excretion and, potentially, increases in the plasma concentrations of lovastatin and simvastatin may be more likely to occur in patients with cholestasis than may be the case with pravastatin, whereas increased plasma concentrations of pravastatin may be more likely to occur in patients with renal insufficiency. At present, it would seem prudent to use all three of these drugs at substantially reduced doses (if at all) in patients with cholestasis and to reduce the dose of pravastatin in patients with renal failure.

III. Mechanism of Action

The primary mechanism of action of all of the currently available HMG-CoA reductase inhibitors is to competitively inhibit the conversion of HMG-CoA to mevalonic acid by the enzyme HMG-CoA reductase. Both sterol and nonsterol products of the cholesterol biosynthetic pathway exert regulatory effects on HMG-CoA reductase, and the ability of lovastatin and related drugs to inhibit HMG-CoA reductase results in reduced concentrations of these regulatory products and a compensatory increase in the mass of HMG-CoA reductase (EDWARDS et al. 1983). However, the magnitude of induction of HMG-CoA reductase during chronic administration of lovastatin, simvastatin, or pravastatin in humans is insufficient to render the drugs ineffective during long-term therapy, but, in individual patients, differences in the magnitude of induction of HMG-CoA reductase may contribute to the variability in hypolipidemic response observed during chronic therapy.

The influence of HMG-CoA reductase inhibitors on cholesterol biosynthesis in human subjects has been evaluated in both normal subjects and patients with hypercholesterolemia using a variety of different techniques. GRUNDY and BILHEIMER (1984) examined the influence of lovastatin on sterol balance in five patients with primary hypercholesterolemia; a reduction in neutral and acidic steroids was observed in three of the five patients but, overall, cholesterol synthesis assessed by this technique was not reduced during therapy with lovastatin. A similar conclusion was reported by GOLDBERG et al. (1990) in nine patients who underwent cholesterol turnover studies on diet alone and then after 15 months of treatment with lovastatin at a dose of 40 mg/day. In this study, LDL cholesterol concentrations decreased by 21% during treatment with lovastatin, but the rates of cholesterol production did not change, nor did the authors observe any changes in the total exchangeable pool of cholesterol in the body or in the size of the rapidly and slowly miscible pools. GOLDBERG et al. (1990) concluded that lovastatin therapy reduced plasma concentrations of total and LDL cholesterol without influencing the steady-state rates of whole body cholesterol biosynthesis or the tissue pools of cholesterol. In contrast to these results, studies in which the influence of lovastatin or pravastatin on the concentrations of intermediates in the cholesterol biosynthetic pathway have been determined have shown that these drugs reduce the plasma concentrations of lathosterol (REIHNER et al. 1990) or the urinary excretion of mevalonic acid (PAPPU et al. 1989; HAGEMENAS et al. 1990). In patients with heterozygous familial hypercholesterolemia, the 24-h urinary excretion of mevalonic acid fell by 19%, 35%, and 31%, respectively, during treatment with lovastatin at doses of 20, 40, and 80 mg/day (Pappu et al. 1989), whereas in a different group of patients treated with simvastatin, the urinary excretion of mevalonic acid decreased by 17% and 31% on doses of 20 and 40 mg/day (Hagemenas et al. 1990). Lovastatin, pravastatin, and simvastatin exert their primary effects in the liver, and it is possible that plasma concentrations of sterol precursors more accurately reflect changes in hepatic cholesterol biosynthesis than do measurements of whole body cholesterol synthesis assessed by either sterol balance techniques or turnover studies of isotopically labeled plasma cholesterol.

In addition to their ability to reduce plasma or urinary concentrations of sterol intermediates in human subjects, HMG-CoA reductase inhibitors have been shown to induce HMG-CoA reductase activity in freshly isolated mononuclear leukocytes (Hagemenas and Illingworth 1989) or in human liver samples obtained at elective surgery (Reihner et al. 1990). In both of these studies, the induction of HMG-CoA reductase was accompanied by a twofold increase in the expression of high-affinity LDL receptor activity on both mononuclear leukocytes (Hagemenas and Illingworth 1989) and liver biopsy specimens (Reihner et al. 1990).

IV. Effects on Lipoprotein Metabolism

Kinetic studies of radiolabeled VLDL and/or LDL have been conducted in patients with hyperlipidemia, before and during treatment with HMG-CoA reductase inhibitors, to assess their influence on the synthesis and fractional rate of catabolism of these lipoproteins. In patients with heterozygous familial hypercholesterolemia in whom the fractional catabolic rate of LDL is reduced secondary to the inherently lower number of high-affinity LDL receptors, treatment with lovastatin at a dose of 20 mg twice daily increased the fractional catabolic rate of ^{125}I-labeled LDL by 37%, but did not cause a concurrent reduction in the rate of LDL synthesis (Bilheimer et al. 1983). In contrast to these results, studies in patients with other genetic causes of hypercholesterolemia not associated with an impaired FCR for LDL have shown that treatment with HMG-CoA reductase inhibitors reduces the production rate of LDL apoprotein B, but does not increase the FCR (Grundy and Vega 1985; Vega et al. 1988, 1990). These results are supported by kinetic studies of ^{125}I-labeled VLDL in patients with combined hyperlipidemia in which treatment with lovastatin has been shown to reduce the synthetic rate for both VLDL apoprotein B and VLDL triglycerides (Arad et al. 1990, 1992). These results suggest that in patients with primary hypercholesterolemia not attributable to mutations in the LDL receptor gene, the ability of HMG-CoA reductase inhibitors to reduce LDL concentrations may result from a direct inhibition of VLDL apoB synthesis as well as an enhanced rate of receptor-mediated clearance of VLDL and

VLDL remnant particles from plasma with a concurrent reduction in the formation of LDL.

Lovastatin has been shown to enhance the hepatic clearance of chylomicron remnant particles (CIANFLONE et al. 1990), but HMG-CoA reductase inhibitors do not appear to reduce the rates of formation of chylomicron particles in the small intestine or decrease the absorption of exogenous triglycerides. In patients with receptor-negative homozygous familial hypercholesterolemia, who lack all functional high-affinity LDL receptors, treatment with lovastatin has been shown to be ineffective in reducing LDL concentrations and exerts no effect on the metabolism of ^{125}I-labeled LDL (UAUY et al. 1988). On the basis of kinetic studies of lipoprotein metabolism in humans, the therapeutic potential of HMG-CoA reductase inhibitors would be anticipated to be greatest in patients with increased plasma concentrations of LDL or in patients with combined hyperlipidemia, but the drugs would not be expected to be effective in the treatment of patients with severe hypertriglyceridemia; they are known to be ineffective in patients with receptor-negative homozygous familial hypercholesterolemia.

V. Other Potential Effects of 3-Hydroxy-3-methylglutaryl Coenzyme A Reductase Inhibitors

On theoretical grounds, inhibition of HMG-CoA reductase, particularly in patients with defects in LDL receptor activity, could lead to impaired production of steroid hormones if the drugs exerted inhibitory effects on cholesterol biosynthesis in steroidogenic tissues such as the adrenal gland and corpus luteum of the ovary. Fortunately, studies in patients with heterozygous familial hypercholesterolemia have shown no impairment in steroid hormone production in patients treated with lovastatin, simvastatin, or pravastatin as assessed by either serum concentrations of cortisol, testosterone, progesterone, or in the adrenal response to adrenocorticotrophic hormone (ACTH) stimulation (MCTAVISH and SORKIN 1991; PRIHODA et al. 1991). HMG-CoA reductase inhibitors do not affect bile acid biosynthesis or the activity of 7-α-hydroxylase in the human liver, and studies of biliary composition have indicated that the cholesterol saturation index is reduced by 20%–30% during treatment with lovastatin, pravastatin, or simvastatin. These results indicate that long-term treatment with HMG-CoA reductase inhibitors would not be anticipated to result in an increased risk of gallstones, and in fact the reduction in the cholesterol saturation index suggests that these agents may exert a potentially beneficial effect on biliary lithogenicity. HMG-CoA reductase inhibitors do not alter the synthesis of vitamin D and despite initial concerns that these drugs may promote the development of cataracts in humans, long-term trials (3–5 years) and the experience of the first author over a period of 5–10 years with simvastatin and lovastatin have failed to demonstrate any effect of the currently available HMG-CoA reductase inhibitors on the development of lens opacities. Slit-lamp examinations of the lens are not recommended as part of the therapeutic monitoring of patients with hypercholesterolemia treated with any of the currently available HMG-CoA reductase inhibitors by regulatory authorities throughout the world.

C. Clinical Efficacy of 3-Hydroxy-3-methylglutaryl Coenzyme A Reductase Inhibitors in the Treatment of Hyperlipidemia

I. Effects in Primary Hypercholesterolemia

Lovastatin, pravastatin, and simvastatin exert their greatest effect on the plasma concentrations of LDL particles as determined by LDL cholesterol concentrations or the concentrations of apoprotein B. This class of drugs is more effective than either the bile acid sequestrants or nicotinic acid in reducing plasma concentrations of LDL cholesterol and are an appropriate first choice agent for use in the treatment of adult patients with primary hypercholesterolemia attributable to increased concentrations of LDL. With all three drugs, the dose–response curves are nonlinear and the largest reductions in LDL concentrations per milligram of drug administered are observed with the first 20 mg/day of lovastatin or pravastatin and the first 10 mg/day of simvastatin. Table 1 illustrates comparative data from different published studies in which the dose–response relationships for lovastatin, simvastatin, and pravastatin have been evaluated in Caucasian patients with heterozygous familial hypercholesterolemia. These studies indicate that, on a milligram for milligram basis, lovastatin and pravastatin are of approximately equal efficacy. whereas simvastatin is twice as potent. This view is substantiated by a recent crossover comparison of lovastatin and simvastatin in patients with heterozygous familial hypercholesterolemia (Illingworth et

Table 1. Hypolipidemic effects of 3-hydroxy-3-methylglutaryl coenzyme A reductase inhibitors in patients with heterozygous familial hypercholesterolemia

Daily dose (mg)	Percent decrease in LDL cholesterol				
	Lovastatin		Simvastatin		Pravastatin
	a	b	c	d	e
10	20 (13)	17 (20)	28 (8)	ND	ND
20	29 (13)	25 (20)	30 (4)	38 (10)	21 (40)
40	35 (13)	31 (20)	37 (7)	44 (10)	28 (40)
80	38 (13)	40 (20)	ND	ND	ND

Numbers in parentheses refer to the number of patients studied on each dosage of drug. Drugs given twice daily. LDL, low density lipoprotein; ND, not determined. a, Illingworth and Sexton (1984); b, Havel et al. (1987); c, Mol et al. (1986); d, Illingworth and Bacon (1989); e, Wiklund et al. (1990).

al. 1992a) in which the magnitude of reduction in LDL cholesterol in patients treated with 20 mg/day of simvastatin was similar to that observed during treatment with 40 mg/day of lovastatin. Although the difference was not statistically significant, LDL concentrations were reduced by 39% in patients treated with simvastatin at a dose of 40 mg/day and fell by 35% in the same patients during treatment with lovastatin at a dose of 80 mg/day. In this and other studies, the magnitude of the hypolipidemic response to lovastatin and simvastatin in individual patients with heterozygous familial hypercholesterolemia was heterogeneous, but those patients who responded well to lovastatin were also good responders to simvastatin, whereas the hyporesponders to one drug were hyporesponders to the other (ILLINGWORTH et al. 1992a). Factors which contribute to the variability in hypolipidemic response to HMG-CoA reductase inhibitors in patients with heterozygous familial hypercholesterolemia remain poorly understood, but may include differences in apolipoprotein E phenotypes (O'MALLEY and ILLINGWORTH 1990), potential differences in the LDL receptor mutation responsible for familial hypercholesterolemia (HOBBS et al. 1988), and potential differences in drug pharmacokinetics between individual patients. Studies in the author's laboratory have examined the relationship between inhibition of cholesterol biosynthesis assessed by urinary excretion of mevalonic acid and the magnitude of reduction in LDL cholesterol concentrations in adult patients with heterozygous familial hypercholesterolemia treated with lovastatin (80 mg/day). As illustrated in Fig. 2, no correlation has been found between the magnitude of decrease in the urinary excretion of mevalonic acid and the reduction in LDL cholesterol concentrations in individual patients. Although the reasons responsible for this lack of correlation are unclear, they may reflect, at least in part, the dual effects of compensatory homeostatic mechanisms which are invoked during therapy with HMG-CoA reductase inhibitors. These drugs are known to promote an increase in the synthesis and mass of HMG-CoA reductase as well as an increased expression of high-affinity LDL receptors in the liver, but the magnitude of these changes is some sixfold greater for the former enzyme than for the LDL receptor (REIHNER et al. 1990). Increases in the mass and activity of HMG-CoA reductase will tend to reduce the inhibitory effects of a given dose of lovastatin, pravastatin, or simvastatin on the formation of mevalonic acid, whereas even modest further increases in the expression of high-affinity LDL receptors on hepatocyte membranes will lead to further reductions in the plasma concentrations of this lipoprotein.

The biosynthesis of cholesterol in humans displays a diurnal rhythm with increased rates of synthesis at night. When administered in a once-daily dosage schedule, the HMG-CoA reductase inhibitors are more effective when given as a single dose in the evening as compared to the same dose given once daily in the morning (ILLINGWORTH 1986). At dosages of 40 and 80 mg/day, lovastatin appears to be more effective when given in a twice-daily dosage regimen, whereas simvastatin and pravastatin are of equal

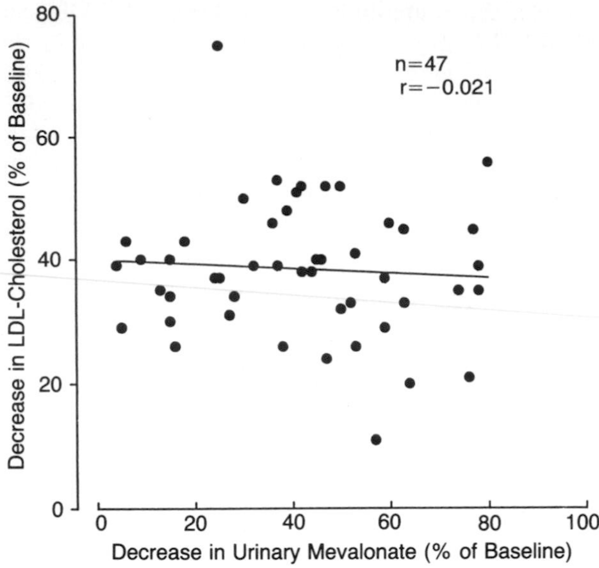

Fig. 2. Relationship between changes in the plasma concentrations of low-density lipoprotein (LDL) cholesterol and the 24-h urinary excretion of mevalonic acid in adult patients with heterozygous familial hypercholesterolemia treated with lovastatin. The data are from patients studied under steady-state conditions on diet only and then on diet plus lovastatin at a dosage of 80 mg/day. No correlation was found between the magnitude of decrease in LDL cholesterol concentrations and the reduction in cholesterol synthesis assessed by the urinary excretion of mevalonic acid

efficacy when given once daily in the evening or twice daily (MOL et al. 1988a; JONES et al. 1991).

The hypolipidemic effects of HMG-CoA reductase inhibitors in patients with different genetic causes of hypercholesterolemia appear to be similar, but considerable individual variability exists in the magnitude of reduction in LDL concentrations in response to a given dose of lovastatin, simvastatin, or pravastatin in each patient population. The percentage decrease in LDL cholesterol concentrations obtained in controlled clinical trials in which different doses of lovastatin were administered to patients with heterozygous familial hypercholesterolemia, patients with primary hypercholesterolemia (including familial combined hyperlipidemia, but excluding patients with heterozygous familial hypercholesterolemia) and patients with moderate primary hypercholesterolemia is illustrated in Table 2. Plasma concentrations of total and LDL cholesterol are higher in the patients with familial hypercholesterolemia, and the absolute decrease in LDL cholesterol observed in these patients is greater than that seen in other patient populations, although the percentage decrease is similar (HUNNINGHAKE et al. 1986; HAVEL et al. 1987; BRADFORD et al. 1991). Recent studies (ILLINGWORTH et al. 1992b) have indicated that patients with familial defective apolipoprotein B-100 also

Table 2. The influence of lovastatin on plasma low-density lipoprotein cholesterol levels in various patient groups with primary hypercholesterolemia

Lovastatin (mg/day)[a]	Percentage decrease in LDL-cholesterol		
	FH[b]	Non-FH[c]	Moderate hypercholesterolemia[d]
10	−20	−25	ND
20	−29	−24	−21
40	−33	−34	−34
80	−39	−39	−40

LDL, low-density lipoprotein; FH, familial hypercholesterolemia; ND, not determined.
[a] Lovastatin was taken in a twice daily dosage at the two higher doses.
[b] HAVEL et al. (1987). Data from 20 patients with heterozygous familial hypercholesterolemia (FH) on each dose.
[c] HUNNINGHAKE et al. (1986). Data from 20 patients with primary hypercholesterolemia (excluding FH patients) on each dose.
[d] BRADFORD et al. (1991). Data from 1642–1663 patients with moderate hypercholesterolemia on each dose.

respond to treatment with lovastatin; in nine patients treated with 40 mg/day of this drug, concentrations of LDL cholesterol were reduced by an average of 32% with a range of 19%–42%.

Monotherapy with lovastatin, simvastatin, and pravastatin in patients with primary hypercholesterolemia has been associated with a 20%–30% decrease in the concentrations of plasma triglycerides and an overall tendency for HDL cholesterol to increase by 2%–10% (ILLINGWORTH and SEXTON 1984; HUNNINGHAKE et al. 1986; HAVEL et al. 1987; BRADFORD et al. 1991; JONES et al. 1991). HMG-CoA reductase inhibitors are, however, not effective in reducing plasma concentrations of lipoprotein(a) (Lp(a); KOSTNER et al. 1989).

Clinical experience with lovastatin in the treatment of patients with primary hypercholesterolemia treated in Portland now exceeds 10 years, and in compliant patients such therapy has been associated with sustained reductions in the plasma concentrations of total and LDL cholesterol. This is consistent with earlier studies of shorter duration in which sustained efficacy has been demonstrated in patients treated with lovastatin or simvastatin for periods of 3–7 years (ILLINGWORTH et al. 1988; OJALA et al. 1990). Cessation of drug treatment, however, results in a rapid increase in the plasma concentrations of total and LDL cholesterol, which show significant increases by 1 week and have reverted back to their original values after 4 weeks. The time relationships between discontinuation of lovastatin therapy and changes in LDL cholesterol in patients with heterozygous familial

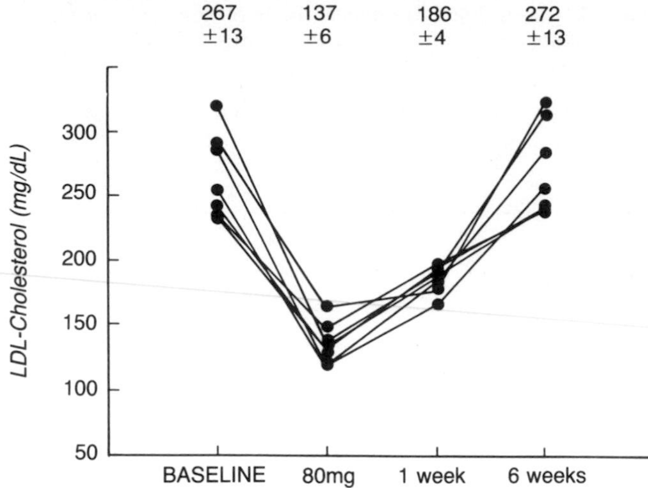

Fig. 3. Time course of changes in the plasma concentrations of low-density lipoprotein (LDL) cholesterol in patients with heterozygous familial hypercholesterolemia during treatment with lovastatin and following its discontinuation

hypercholesterolemia treated with 80 mg/day of lovastatin is illustrated in Fig. 3.

II. Effects in Combined Hyperlipidemia

The ability of HMG-CoA reductase inhibitors to reduce plasma concentrations of both LDL cholesterol and, to a lesser extent, triglycerides in patients with primary hypercholesterolemia suggests that these drugs would also be effective in the treatment of patients with combined hyperlipidemia attributable to increased plasma concentrations of both VLDL and LDL particles in plasma. The enhanced chylomicron remnant metabolism induced by lovastatin (Cianflone et al. 1990) suggests that HMG-CoA inhibitors may also be effective in the treatment of patients with type III hyperlipidemia, in whom chylomicron and VLDL remnant particles accumulate in plasma.

Tikkanen et al. (1988) treated 36 patients with primary elevations of both VLDL and LDL cholesterol (type IIB hyperlipidemia) with lovastatin at a dosage of 40 mg/day and observed a 39% reduction in the concentration of LDL cholesterol, which was paralleled by a modest increase in the concentration of HDL cholesterol and a decrease in plasma triglycerides. Similar results have been observed by Mol et al. (1988b) in eight patients with familial hypercholesterolemia in whom plasma triglyceride concentrations at baseline were above 200 mg/dl and in whom treatment with simvastatin at dosages of 20 or 40 mg per day reduced plasma cholesterol concentrations by 34% and plasma triglycerides by 39%. Comparative studies in which lovastatin or simvastatin have been compared with gemfibrozil in patients with combined hyperlipidemia have indicated that the HMG-CoA reductase inhibitors are superior in their ability to lower LDL cholesterol concentrations and that these drugs concurrently reduce plasma triglycerides by up to 40% (TIKKANEN et al. 1988, 1989). The ability of HMG-CoA reductase inhibitors to favorably affect concentrations of VLDL, LDL, and HDL makes these drugs attractive agents to use in the therapy of patients with combined hyperlipidemia and provides an excellent alternative to nicotinic acid for this patient population. In the opinion of the authors, HMG-CoA reductase inhibitors are the drugs of choice for patients with combined hyperlipidemia who have medical or metabolic contraindications to the use of nicotinic acid.

Recent studies have indicated that lovastatin and simvastatin are effective in the treatment of patients with type III hyperlipidemia (VEGA et al. 1988; STALENHOEF et al. 1989; ILLINGWORTH and O'MALLEY 1990). In a recent comparative study, the hypolipidemic effects of lovastatin and clofibrate were compared in 12 patients with type III hyperlipidemia (ILLINGWORTH and O'MALLEY 1990). Concentrations of VLDL cholesterol were reduced by 53% during treatment with lovastatin (20 mg twice daily) and by 61% during treatment with clofibrate (1 g twice daily); LDL cholesterol concentrations were reduced by 32% on lovastatin, but remained unchanged during treatment with clofibrate. These results demonstrate that lovastatin is as effective as clofibrate in reducing VLDL cholesterol concentrations in patients with type III hyperlipidemia, but appears to be superior in its ability to concurrently reduce the concentrations of VLDL remnants and LDL in these patients.

III. Effects in Hypertriglyceridemia and Hypoalphalipoproteinemia

The ability of HMG-CoA reductase inhibitors to reduce concentrations of plasma triglycerides and concurrently increase HDL concentrations in patients with primary hypercholesterolemia has stimulated research to assess the effects of these drugs on the lipoprotein profile of patients with primary hypertriglyceridemia or those with hypoalphalipoproteinemia. The criteria for the use of lipid-lowering drugs in these two patient populations remain ill defined and, with the exception of patients with established coronary artery disease, most patients with moderate hypertriglyceridemia or those with inherently low levels of HDL cholesterol, but normal levels of LDL cholesterol, are unlikely to be treated with hypolipidemic drugs. The influence of lovastatin on the lipoprotein profile of patients with moderate hypertriglyceridemia was evaluated by VEGA and GRUNDY (1990). In 13 patients, concentrations of total plasma cholesterol decreased from 237 mg/dl on diet to only 171 mg/dl during treatment with lovastatin (20 mg twice daily), whereas plasma triglycerides fell from 444 mg/dl at baseline to 313 mg/dl on drug. The decrease in triglycerides was accompanied by a reduction in the cholesterol content of VLDL plus IDL, which fell from 88 mg/dl at baseline to 55 mg/dl during treatment with lovastatin; concentrations of LDL cholesterol decreased from 120 to 81 mg/dl on drug therapy, whereas HDL cholesterol concentrations only increased from 35 mg/dl on diet to 39 mg/dl during lovastatin (VEGA and GRUNDY 1990). Although lovastatin favorably influenced the lipoprotein profile of these patients, it remains unclear whether or not treatment with HMG-CoA reductase in this patient population is either justified or beneficial.

HMG-CoA reductase inhibitors are not approved for the treatment of severe hypertriglyceridemia and available data indicates that they are ineffective. At least one of the reported cases of rhabdomyolysis associated with combined use of lovastatin plus gemfibrozil had severe hypertriglyceridemia and was initially and inappropriately treated with lovastatin (MARAIS and LARSEN 1990). The hypertriglyceridemia persisted during lovastatin therapy and the patient subsequently developed rhabdomyolysis and acute renal failure after gemfibrozil was added as a second drug.

The potential usefulness of lovastatin has been evaluated in patients with isolated hypoalphalipoproteinemia who do not have concurrent hypertriglyceridemia (VEGA and GRUNDY 1989). Lovastatin at a dosage of 20 mg twice daily was administered to a group of 22 patients in whom baseline concentrations of total cholesterol were under 200 mg/dl, but in whom HLD cholesterol concentrations were less than 35 mg/dl. In these patients, lovastatin reduced LDL cholesterol concentrations by 34%, but HDL cholesterol concentrations remained low and only increased from 29 to 32 mg/dl during treatment with lovastatin. In the opinion of the authors, drug therapy with HMG-CoA reductase inhibitors is not justified in asymptomatic patients with isolated hypoalphalipoproteinemia.

IV. Potential Utility of 3-Hydroxy-3-methylglutaryl Coenzyme A Reductase Inhibitors in Patients with Secondary Causes of Hyperlipidemia

The potential utility of HMG-CoA reductase inhibitors in certain disorders associated with secondary hyperlipidemia in which the underlying disorder is not amenable to therapy (e.g., the nephrotic syndrome) has been evaluated by several investigators.

Lovastatin has been shown to be effective in reducing LDL concentrations in patients with the nephrotic syndrome, and in one study (GOLPER et al. 1989) LDL cholesterol concentrations were reduced by 29%, 34%, and 45%, respectively, during treatment with lovastatin at dosages of 10, 20, and 40 mg twice daily. In this study lovastatin was well tolerated and no significant clinical or biochemical side effects were noted.

Abnormalities in plasma lipoproteins are commonly seen in patients with both type I and type II diabetes and include increases in the plasma concentrations of VLDL, VLDL remnants, and LDL as well as reduced concentations of HDL cholesterol. Although no clinical trials to date have assessed the potential benefit from hypolipidemic therapy in patients with either type I or type II diabetes, the dyslipidemia which occurs in these patients has been causally implicated in the accelerated rates of atherosclerosis which occur so commonly in these patients. The ability of HMG-CoA reductase inhibitors to reduce plasma concentrations of both VLDL remnants and LDL has led to their use in the treatment of the dyslipidemia seen in diabetic patients, particularly those who do not have severe hypertriglyceridemia. Lovastatin at a dosage of 20 mg twice daily was reported to reduce VLDL cholesterol concentrations by 42% with a concurrent 28% decrease in LDL cholesterol levels in 16 adult patients with type II diabetes (GARG and GRUNDY 1988). In these studies, the use of lovastatin was not associated with any deterioration in diabetic control and this has also been observed in patients with type I diabetes (ILLINGWORTH, unpublished observations). HMG-CoA reductase inhibitors may therefore prove to be useful in the therapy of selected diabetic patients in whom concentrations of VLDL cholesterol, VLDL remnants, or LDL are increased but, as noted previously, they are not likely to be beneficial in the therapy of patients with severe hypertriglyceridemia.

Severe hypercholesterolemia due to increased concentrations of both LDL cholesterol and lipoprotein X (LpX) is commonly seen in patients with intrahepatic cholestasis due to such disorders as primary biliary cirrhosis. HMG-CoA reductase inhibitors, which are primarily excreted in bile, should not be used in the treatment of this disorder unless drug levels can be monitored. Although this class of drugs is potentially attractive for use in patients with cholestasis, the potential for toxicity due to impaired hepatic excretion is high and their use is associated with a substantially increased risk of myopathy and cannot, therefore, be recommended.

Hyperlipidemia is commonly seen in patients undergoing renal or cardiac transplantation and is promoted by the use of both corticosteroids and cyclosporine. HMG-CoA reductase inhibitors are effective in the treatment of transplant patients in whom concentrations of both VLDL and LDL are increased (KASISKE et al. 1990). This class of drugs should, however, be used with extreme caution in patients receiving cyclosporine, due to the increased risk of myopathy and, potentially, rhabdomyolysis (EAST et al. 1988; NORMAN et al. 1988). Cyclosporine impairs the excretion of all of the currently available HMG-CoA reductase inhibitors (SMITH et al. 1991) and in clinical use it is recommended that the maximal dose of lovastatin, pravastatin, and simvastatin should be reduced to 25% of the approved maximal doses.

D. Safety and Side-Effect Profile of 3-Hydroxy-3-methylglutaryl Coenzyme A Reductase Inhibitors

The clinical use of lovastatin, pravastatin, and simvastatin in both investigational trials and in postmarketing surveillance has been associated with a low incidence of clinical and biochemical side effects, and these drugs facilitate a high degree of patient compliance. Clinical side effects observed to date have included headaches, transient and inconsistent changes in bowel habits, nausea, insomnia, and, less commonly, the development of myalgias and muscle tenderness with, in some patients, concurrent increases in plasma concentrations of creatine kinase (Tobert 1988). Other rare side effects have included hepatitis, cholestatic jaundice, and hypersensitivity reactions which have included arthralgias and thrombocytopenia (Mantel et al. 1990). Myopathy is an uncommon, but serious, side effect which appears to be more frequent in patients treated with maximal or near-maximal approved doses of HMG-CoA reductase inhibitors and in women over 60 years of age (Bradford et al. 1991; Pierce et al. 1990). The mechanisms responsible for the myopathy remains unknown, but the risk appears to be increased in patients with preexistent abnormalities of muscle metabolism or in patients who have untreated hypothyroidism (Illingworth 1991). An increased risk of myopathy has also been observed in patients treated with lovastatin in combination with cyclosporine, nicotinic acid, gemfibrozil, or erythromycin, (Illingworth 1991) and such drug interactions should be viewed as a class effect of all HMG-CoA reductase inhibitors and not specifically limited to lovastatin. Despite reports that combination drug therapy with an HMG-CoA reductase inhibitor and gemfibrozil is safe and efficacious (Glueck et al. 1992), this view is not shared by the authors or the United States Food and Drug Administration (Pierce et al. 1990; Illingworth 1991).

Safe use of HMG-CoA reductase inhibitors necessitates careful clinical and biochemical monitoring to assess both efficacy and safety. Biochemical changes which have been observed in patients treated with HMG-CoA reductase inhibitors have included increases in the concentrations of alkaline phosphatase, aminotransferases, and creatine kinase. When they have occurred, pronounced increases in aminotransferases to values greater than three times the upper limit of normal have usually been observed in the first year of therapy and appear to be dose dependent. In the Expanded Clinical Evaluation of Lovastatin (EXCEL) study (Bradford et al. 1991), aminotransferase elevations to greater than three times normal in patients receiving placebo or lovastatin (at dosages of 20, 40, and 80 mg/day) were observed in 0.1%, 0.1%, 0.9%, and 1.5% of patients, respectively. The increases in aminotransferases have been reversible upon drug discontinuation, but necessitate the monitoring of liver function tests at 6- to 8-week intervals during the first year of therapy and at 3- to 6-month intervals thereafter. More frequent monitoring is indicated in patients after a change in dosage or in association with concurrent use of other drugs which may themselves adversely affect liver function tests. Prolongation of the prothrombin time has been reported after initiation of lovastatin therapy in patients on warfarin anticoagulation (Ahmad 1990), and more frequent monitoring of the prothrombin time is recommended after the addition of an HMG-CoA reductase inhibitor to the medical regimen of a patient on oral anticoagulants. Initial concerns that HMG-CoA reductase inhibitors may cause cataracts in humans have not materialized, and the data available at present indicates that at clinically effective doses, these drugs do not exert any deliterous effects upon the human lens (BRADFORD et al. 1991). Regulatory authorities in North America and Europe do not currently recommend routine ophthalmologic examinations in adult patients treated with lovastatin, pravastatin, or simvastatin.

E. Contraindications and Inappropriate Uses of 3-Hydroxy-3-methylglutaryl Coenzyme A Reductase Inhibitors

HMG-CoA reductase inhibitors are contraindicated in patients with known hypersensitivity to this class of drugs and in those with abnormal liver function tests, including those with intrahepatic cholestasis, and should not be used in the treatment of hypercholesterolemia in women during the period of conception, during pregnancy (GHIDINI et al. 1992), or during lactation. The use of HMG-CoA reductase inhibitors in the treatment of premenopausal female patients of childbearing potential should be limited to those with heterozygous familial hypercholesterolemia or other severe causes of hypercholesterolemia, in whom the benefit of lipid-lowering therapy justifies the use of systemically acting hypolipidemic medications. This class of drugs is not stored in body tissues and can, in compliant patients, be used up to the time at which methods of birth control are discontinued in younger female patients with heterozygous familial hypercholesterolemia. With the exception of the rare child with homozygous familial hypercholesterolemia in whom HMG-CoA reductase inhibitor therapy is relatively ineffective (UAUY et al. 1988), the risk of atherosclerosis in children with heterozygous familial hypercholesterolemia is sufficiently low to justify a conservative approach to the use of this class of drugs in the pediatric population. Clinical trials are currently evaluating the efficacy and safety of lovastatin in adolescent boys with heterozygous familial hypercholesterolemia but, in the opinion of the authors, the use of this class of drugs in children should be regarded as investigational until better data is available concerning long-term safety. Lovastatin is effective in reducing LDL concentrations in pediatric patients with heterozygous familial hypercholesterolemia, and its use in investigational studies conducted to date has been associated with a low incidence of side effects and excellent patient compliance.

As previously discussed, HMG-CoA reductase inhibitors are ineffective in patients with severe hypertriglyceridemia (type V phenotype) and should be used very cautiously in patients with known metabolic myopathies. Responsible use of this class of drugs necessitates prior exclusion of correctable secondary causes of hypercholesterolemia and persistence of hyperlipidemia after dietary therapy.

F. Use of 3-Hydroxy-3-methylglutaryl Coenzyme A Reductase Inhibitors in Combination Drug Therapy

The efficacy and safety of lovastatin, pravastatin, and simvastatin as components of combined drug regimens has been extensively evaluated in the last several years (ILLINGWORTH and BACON 1989). For patients with severe hypercholesterolemia attributable to increased levels of LDL cholesterol, combined drug therapy with an HMG-CoA reductase inhibitor and a bile acid sequestrant has proven to be a safe and effective regimen and, in compliant patients, reduces LDL cholesterol concentrations by 45%–55% (ILLINGWORTH and BACON 1989). Other combinations, particularly those involving an HMG-CoA reductase inhibitor plus nicotinic acid or a fibrate, are associated with an increased risk of myopathy and can not be generally recommended (ILLINGWORTH 1991). For patients with moderate hypercholesterolemia, a particularly attractive approach is to use low dosages (e.g., one to two scoops per day) of a bile acid sequestrant in combination with relatively low dosages of an HMG-CoA reductase inhibitor (e.g., lovastatin or pravastatin 20 mg/daily or simvastatin 10 mg daily). This combination may be more effective than higher doses of an HMG-CoA reductase inhibitor and is more cost effective.

G. Conclusions

The development of specific inhibitors of HMG-CoA reductase represents a major advance in the therapy of lipoprotein disorders, and these drugs are effective in the therapy of patients with primary causes of hypercholesterolemia or combined hyperlipidemia. The drugs are also effective in the treatment of selected patients with secondary causes of hyperlipidemia, including patients with the nephrotic syndrome or diabetes. As a class, the HMG-CoA reductase inhibitors are more effective in reducing plasma concentrations of LDL cholesterol than are either the bile acid sequestrants or nicotinic acid, and their efficacy can be enhanced when used in combination with bile acid sequestrants. This combined drug regimen has been utilized in two angiographic trials in which such therapy was associated with an overall tendency to regression of coronary atherosclerosis (BROWN et al. 1990; KANE et al. 1990), and long-term use of lovastatin has also been associated with the regression of tendon xanthomas in patients with heterozygous familial hypercholesterolemia (ILLINGWORTH et al. 1990).

The side-effect profile of HMG-CoA reductase inhibitors has been delineated and several drug interactions which may lead to the development of myopathy have been described. No new safety concerns have emerged in the follow-up of patients treated continuously with lovastatin for periods of 7–10 years or for shorter durations of therapy with simvastatin or pravastatin. Further studies are necessary to define the potential role of this class of drugs in the treatment of patients with moderate hypertriglyceridemia and in patients with low levels of HDL cholesterol as well as in other patient populations, including postmenopausal women with moderately increased levels of LDL cholesterol. Appropriate use of HMG-CoA reductase inhibitors mandates a careful assessment of the risk to benefit ratio in each patient for whom this class of drugs is prescribed and, once therapy has been initiated, the response to treatment needs to be monitored in terms of both efficacy and safety. Studies which have been initiated and are currently in progress should, when completed, provide more definitive information concerning whether or not treatment of high-risk patients with hyperlipidemia or those with lesser degrees of hypercholesterolemia, but established atherosclerosis, will lead to a reduction in cardiovascular and total mortality.

Acknowledgements. This work was supported in part by National Institutes of Health Research Grant HL28399, the General Clinical Research Center's Program (RR334), and the Clinical Nutrition Research Unit (P30DK40566). Dr. Schmidt was supported by a grant from the Danish Heart Foundation. We are grateful to Marcia Hindman for careful preparation of this manuscript, to colleagues whose work has contributed to the topic under discussion, and to the patients, involved in many of the studies cited in this review for their ongoing interest and cooperation.

References

Ahmad S (1990) Lovastatin warfarin interaction. Arch Intern Med 150:2407

Alberts AW (1988) Discovery, biochemistry, and biology of lovastatin. Am J Cardiol 62:10J–15J

Arad Y, Ramakrishnan R, Ginsberg HN (1990) Lovastatin therapy reduces low density lipoprotein apo B levels in subjects with combined hyperlipidemia by reducing the production of apoB containing lipoproteins. Implications for the pathophysiology of apo B production. J Lipid Res 31:567–582

Arad Y, Ramakrishnan R, Ginsberg HN (1992) Effects of lovastatin therapy on very low density lipoprotein triglyceride metabolism in subjects with combined hyperlipidemia. Evidence for reduced assembly and secretion of triglyceride-rich lipoproteins. Metabolism 41:487–493

Bilheimer DW, Grundy SM, Brown MS, Goldstein JL (1983) Mevinolin and colestipol stimulate receptor-mediated clearance of low density lipoprotein from plasma in familial hypercholesterolemia heterozygotes. Proc Natl Acad Sci USA 80:4124–4128

Botti RE, Triscari J, Pan HY, Zayat J (1991) Concentrations of pravastatin and lovastatin in cerebrospinal fluid in healthy subjects. Clin Neuropharmacol 14:256–261

Bradford RH, Shear CS, Chremos AN (1991) Expanded clinical evaluation of lovastatin (EXCEL) study results: I. Efficacy in modifying plasma lipoproteins and adverse event profile in 8245 patients with moderate hypercholesterolemia. Arch Intern Med 151:43–49

Brown G, Albers JJ, Fisher LD, Schaffer SM, Lynn JT, Kaplan C, Zhao XQ, Bisson BD, Fitzpatrick VF, Dodge HT (1990) Regression of coronary artery disease as

a result of intensive lipid lowering therapy in men with high levels of apolipoprotein B. N Engl J Med 323:1289–1298
Brown MS, Goldstein JL (1983) Lipoprotein receptors in the liver: control signals of plasma cholesterol traffic. J Clin Invest 72:743–747
Brown MS, Goldstein JL (1986) A receptor mediated pathway for cholesterol homeostasis. Science 232:34–47
Cianflone KM, Bilodeau M, Davignon J, Sniderman AD (1990) Modulation of chylomicron remnant metabolism by an hepatic hydroxymethyl glutaryl CoA reductase inhibitor. Metabolism 39:274–281
East C, Alivizatos PA, Grundy SM (1988) Rhabdomyolysis in patients receiving lovastatin after cardiac transplantation. N Engl J Med 318:47
Edward PA, Lan SF, Fogelman AM (1983) Alterations in the rate of synthesis and degradation of rat liver 3-hydroxy 3-methyl glutaryl coenzyme A reductase produced by cholestyramine and mevinolin. J Biol Chem 258:10219–10222
Garg A, Grundy SM (1988) Lovastatin for lowering cholesterol levels in non-insulin dependent diabetes mellitus. N Engl J Med 318:81–86
Germershausen JI, Hunt VM, Bostedor RJ, Bailey BJ, Karkas JD (1989) Tissue selectivity of the cholesterol lowering agents lovastatin, simvastatin and pravastatin in rats in vivo. Biochem Biophys Res Commun 158:3–9
Gerson RJ, McDonald JS, Alberts AW, Carnbrust DJ, Majka JA (1989) Animal safety and toxicology of simvastatin and related hydroxymethyl glutaryl coenzyme A reductase inhibitors. Am J Med 87 Suppl 4A:23–38
Ghidini A, Sicherer S, Willner J (1992) Congenital abnormalities (VATER) in baby born to mother using lovastatin. Lancet 339:1416–1417
Glueck CJ, Oaks N, Speirs J, Tracey T, Lang J (1992) Gemfibrozil lovastatin therapy for primary hyperlipoproteinemia. Am J Cardiol 70:1–9
Goldberg IJ, Holleran S, Ramakrishnan R, Adams M, Palmer RH, Del RB, Goodman DS (1990) Lack of effect of lovastatin therapy on the parameters of whole body cholesterol metabolism. J Clin Invest 86:801–808
Golper TA, Illingworth DR, Morris CD, Bennett WM (1989) Lovastatin in the treatment of multifactorial hyperlipidemia associated with proteinuria. Am J Kidney Dis 13:312–320
Grundy SM, Bilheimer DW (1984) Inhibition of 3-hydroxy 3-methyl glutaryl coenzyme A reductase by mevinolin in familial hypercholesterolemic heterozygous: effects on cholesterol balance. Proc Natl Acad Sci USA 81:2538–2542
Grundy SM, Vega GL (1985) Influence of mevinolin on metabolism of low density lipoproteins in primary moderate hypercholesterolemia. J Lipid Res 26:1464–1475
Hagemenas FC, Illingworth DR (1989) Cholesterol homeostasis in mononuclear leukocytes from patients with familial hypercholesterolemia treated with lovastatin. Arteriosclerosis 9:355–361
Hagemenas FC, Pappu AS, Illingworth DR (1990) The effects of simvastatin on plasma lipoproteins and cholesterol homeostasis in patients with heterozygous familial hypercholesterolemia. Eur J Clin Invest 20:150–157
Havel RG, Hunninghake DB, Illingworth DR et al. (1987) Lovastatin (mevinolin) in the treatment of heterozygous familial hypercholesterolemia: a multicenter trial. Ann Intern Med 107:609–615
Henwood JM, Heel RC (1988) Lovastatin: a preliminary review of pharmacodynamic properties and therapeutic use in hyperlipidemia. Drugs 36:429–454
Hobbs HH, Leitersdorf E, Goldstein JL, Brown MS (1988) Multiple crm mutations in familial hypercholesterolemia. Evidence for thirteen alleles including four deletions. J Clin Invest 81:909–916
Hunninghake DB, Miller VT, Palmer RH (1986) Therapeutic response to lovastatin (mevinolin) in nonfamilial hypercholesterolemia. JAMA 256:2829–2835
Illingworth DR (1986) Comparative efficacy of once versus twice daily mevinolin in the therapy of familial hypercholesterolemia. Clin Pharmacol Ther 40:338–343

Illingworth DR (1991) Use and abuse of lovastatin. Endocrinologist 1:323–330

Illingworth DR, Bacon SP (1989) Treatment of heterozygous familial hypercholesterolemia with lipid lowering drugs. Arteriosclerosis Suppl1:121–134

Illingworth DR, O'Malley JP (1990) The hypolipidemic effects of lovastatin and clofibrate alone and in combination in patients with type III hyperlipoproteinemia. Metabolism 39:403–409

Illingworth DR, Sexton G (1984) Hypocholesterolemic effects on mevinolin in patients with heterozygous familial hypercholesterolemia. J Clin Invest 74:1972–1978

Illingworth DR, Bacon SP, Larsen KK (1988) Long-term experience with HMG CoA reductase inhibitors in the therapy of hypercholesterolemia. Atheroscler Rev 18:161–187

Illingworth DR, Cope R, Bacon SP (1990) Regression of tendon xanthomas in patients with familial hypercholesterolemia treated with lovastatin. South Med J 83:1053–1057

Illingworth DR, Bacon SP, Pappu AS, Sexton GJ (1992a) Comparative hypolipidemic effects of lovastatin and simvastatin in patients with heterozygous familial hypercholesterolemia. Atherosclerosis 96:53–64

Illingworth DR, Vakar F, Mahley RW, Weisgraber KH (1992b) Hypocholesterolemic effects of lovastatin in familial defective apolipoprotein B100. Lancet 339:598–600

Jones PH, Farmer JA, Crestman MD et al. (1991) Once daily pravastatin in patients with primary hypercholesterolemia. A dose response study. Clin Cardiol 14:146–151

Kane JP, Malloy MJ, Ports TA, Phillips NR, Diehl JC, Havel RJ (1990) Regression of coronary atherosclerosis during treatment of familial hypercholesterolemia with combined drug regimens. JAMA 264:3007–3012

Kasiske BL, Tortorice KL, Heim-Duthoy KL (1990) Lovastatin treatment of hypercholesterolemia in renal transplant recipients. Transplantation 49:95–100

Kostner GM, Gavish D, Leopold B, Bolzano K, Weintraub MS, Breslow JL (1989) HMG CoA reductase inhibitors lower LDL cholesterol without reducing LPa levels. Circulation 80:131–139

Ma PTS, Gil G, Sudhof TC, Bilheimer DW, Goldstein JL, Brown MS (1986) Mevinolin, an inhibitor of cholesterol synthesis, induces mRNA for low density lipoprotein receptor in livers of hamsters and rabbits. Proc Natl Acad Sci USA 83:8370–8374

Mabuchi H, Haba T, Tatami R (1981) Effects of an inhibitor of 3-hydroxy-3-methylglutaryl coenzyme A reductase on serum lipoproteins and ubiquinone-10 levels in patients with familial hypercholesterolemia. N Engl J Med 305:478–482

Mantell G, Berg T, Staggers K (1990) Extended clinical safety profile of lovastatin. Am J Cardiol 66:11B–15B

Marais G, Larsen K (1990) Rhabdomyolysis and acute renal failure induced by combination and gemfibrozil. Ann Intern Med 112:228–229

McTavish D, Sorkin EM (1991) Pravastatin. A review of its pharmacological properties and therapeutic potential in hypercholesterolemia. Drugs 42:65–89

Mol MJTM, Erkelens DW, Gevers-Leuven JA et al. (1986) The effects of synvinolin (MK733) on plasma lipids in familial hypercholesterolemia. Lancet 2:936–939

Mol MJTM, Erkelens DW, Gevers-Leuven JA, Schouten JA, Stalenhoef AFH (1988a) Simvastatin (MK733): a potent cholesterol synthesis inhibitor in heterozygous familial hypercholesterolemia. Atherosclerosis 69:131–137

Mol MJTM, Stuyt PMH, Stalenhoef AFH, Vant Laar A (1989b) Cholesterol synthesis inhibitors in hyperlipidemia. Lancet 1:597

Norman DJ, Illingworth DR, Munson J, Hosenpud J (1988) Myolysis and acute renal failure in a heart transplant recipient receiving lovastatin. N Engl J Med 318:47

Ojala JP, Helve E, Karjalainen K, Tarkkanen A, Tikkanen MJ (1990) Long-term maintenance of therapeutic response to lovastatin in patients with familial and nonfamilial hypercholesterolemia. A three year follow-up. Atherosclerosis 82: 85–95

O'Malley JP, Illingworth DR (1990) The influence of apolipoprotein E phenotype on the response to lovastatin therapy in patients with heterozygous familial hypercholesterolemia. Metabolism 39:150–154

Pappu AS, Illingworth DR, Bacon SP (1988) Reduction in plasma low density lipoprotein cholesterol and urinary mevalonic acid by lovastatin in patients with heterozygous familial hypercholesterolemia. Metabolism 38:542–549

Pierce LR, Wysowski DK, Gross TP (1990) Myopathy and rhabdomyolysis associated with lovastatin gemfibrozil combination therapy. JAMA 264:71–75

Prihoda JS, Pappu AS, Smith FE, Illingworth DR (1991) The influence of simvastatin on adrenal corticosteroid production and urinary mevalonate during adrenocorticotropin stimulation in patients with heterozygous familial hypercholesterolemia. J Clin Endocrinol Metab 72:567–574

Reihner E, Rudlang M, Stahlburg D, Bergland L, Ewerth S, Bjorkhem I, Einarsson K, Angelin B (1990) Influence of pravastatin, a specific inhibitor of HMG CoA reductase, on hepatic metabolism of cholesterol. N Engl J Med 323:224–228

Smith PF, Eydolloth RS, Grossman SJ, Stubbs RJ, Schwartz MS, Germershausen JI, Vuias KP, Kari PH, McDonald JS (1991) HMG CoA reductase inhibitors induced myopathy in the rat. Cyclosporine A interaction and mechanism studies. J Pharmacol Exp Ther 257:1225–1235

Stalenhoef AFH, Mol MJTM, Stuyt BMJ (1989) Efficacy and tolerability of simvastatin (MK733). Am J Med 87 Suppl 4A:39S–43S

Tikkanen MJ, Helve E, Jaattala A (1988) The Finish lovastatin study group comparison between lovastatin and gemfibrozil in the treatment of primary hypercholesterolemia. Am J Cardiol 62:35J–43J

Tikkanen MJ, Bocanegra TS, Walker JF, Cook T (1989) Comparison of low dose of simvastatin and gemfibrozil in the treatment of elevated plasma cholesterol. A multicenter study. Am J Med 87 Suppl 4A:47S–53S

Tobert JA (1988) Efficacy and long-term adverse effect pattern of lovastatin. Am J Cardiol 62:28J–34J

Todd PA, Goa KL (1990) Simvastatin. A review of pharmacological properties and therapeutic potential in hypercholesterolemia. Drugs 40:583–607

Tsujita Y, Kuroda M, Shimada Y (1986) CS514, a competitive inhibitor of 3-hydroxy-3-methylglutaryl coenzyme A reductase: tissue selective inhibition of sterol synthesis and hypolipidemic effect in various animal species. Biochim Biophys Acta 877:50–60

Uauy R, Vega GL, Grundy SM, Bilheimer DW (1988) Lovastatin therapy in receptor negative homozygous familial hypercholesterolemia. Lack of effect on low density lipoprotein concentrations or turnover. J Pediatr 113:387–392

Vega GL, Grundy SM (1989) Comparison of lovastatin and gemfibrozil in normal lipidemic patients with hypoalphalipoproteinemia. JAMA 262:3148–3154

Vega GL, Grundy SM (1990) Primary hypertriglyceridemia with border line high cholesterol and elevated apolipoprotein B concentrations. Comparison of gemfibrozil versus lovastatin therapy. JAMA 264:2759–2765

Vega GL, East C, Grundy SM (1988) Lovastatin therapy in familial dysbetalipoproteinemia. Effects on kinetics of apolipoprotein B. Atherosclerosis 70:131–143

Vega GL, Krauss RM, Grundy SM (1990) Pravastatin therapy in primary moderate hypercholesterolemia. Changes in metabolism of apolipoprotein B containing lipoproteins. J Intern Med 227:81–94

Wiklund O, Angelin B, Fager G (1990) Treatment of familial hypercholesterolemia. A controlled trial of the effects of pravastatin or cholestyramine therapy on lipoprotein and apoprotein levels. J Intern Med 228:241–247

CHAPTER 12
Fibrates

A. GAW, C.J. PACKARD, and J. SHEPHERD

A. Introduction

I. Clinical Use

The fibric acid derivatives, or fibrates, have, until recently, been the most widely used lipid-lowering drugs in clinical practice. In the UK, for the year ending June 1990, the fibrates available at that time (bezafibrate, gemfibrozil and clofibrate) constituted 62% of all prescribed items for hyperlipidaemia. In the same period, bezafibrate was the most widely prescribed lipid-lowering drug (O'BRIEN 1991). In the US, from 1984–1987, gemfibrozil held first place in the ranking of lipid-lowering drugs in terms of number of prescriptions written and was only overtaken in 1988, after the introduction of lovastatin (WYSOWSKI et al. 1990).

II. History and Development

The fibrates form a family of hypolipidaemic drugs whose structure and mode of action are closely related to that of the parent compound clofibrate (ethyl 2-(4-chlorophenoxy)-2-methylpropionate; Fig. 1). Clofibrate was introduced into the UK in 1962 and was granted Food and Drug Administration approval in the US in 1967. It found wide application as a lipid-lowering agent and was used in the first major primary prevention trial testing the hypothesis that lowering cholesterol would result in reduced coronary morbidity and mortality (COMMITTEE OF PRINCIPAL INVESTIGATORS 1978). This multicentre, placebo-controlled trial reported limited hypolipidemic efficacy. More importantly, however, deaths from non-cardiac causes were higher in the treatment group. Although no single cause of death was identified, there was an excess of deaths from gastrointestinal tract malignancy and associated with cholecystectomy. This poor benefit to risk ratio has led to a steady decline in the use of clofibrate since the publication of these results in 1978 (WYSOWSKI et al. 1990). The excess mortality observed in the WHO trial was further examined in a follow-up of the trial participants (COMMITTEE OF PRINCIPAL INVESTIGATORS 1984). This study failed to demonstrate any persistence of the difference in non-cardiac mortality between the treatment and placebo groups, which suggests that

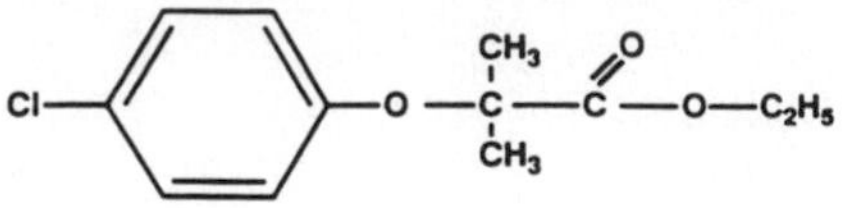

Clofibrate

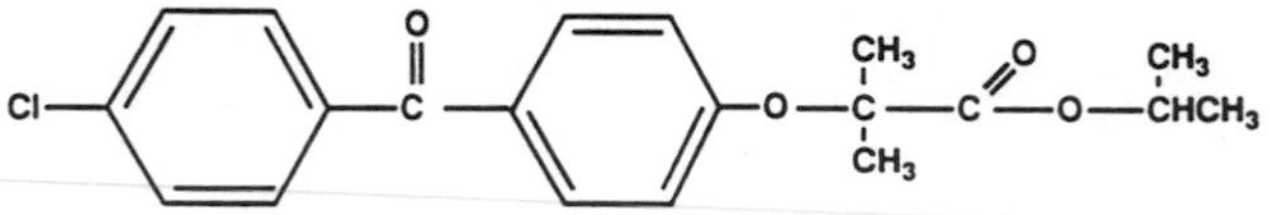

Fenofibrate

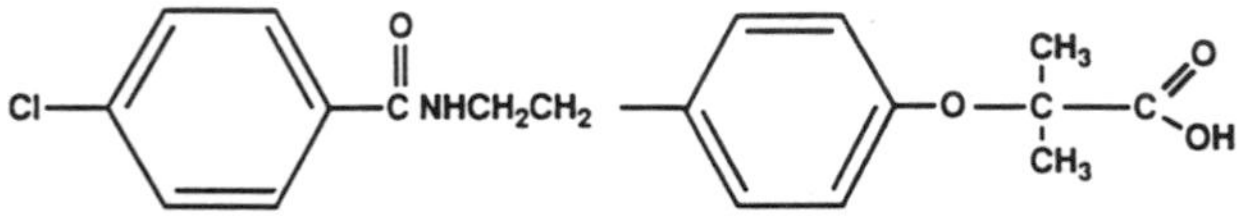

Bezafibrate

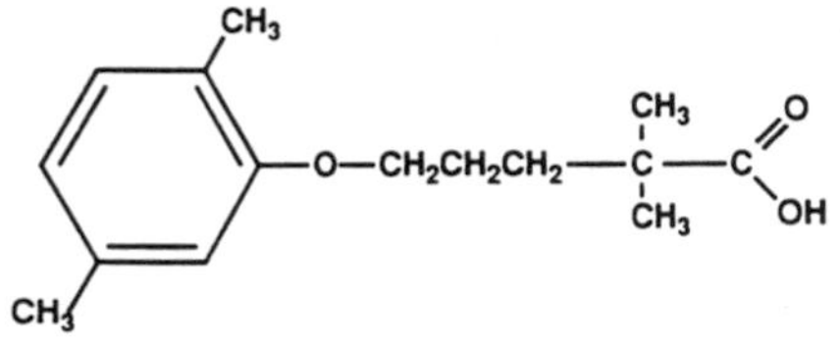

Gemfibrozil

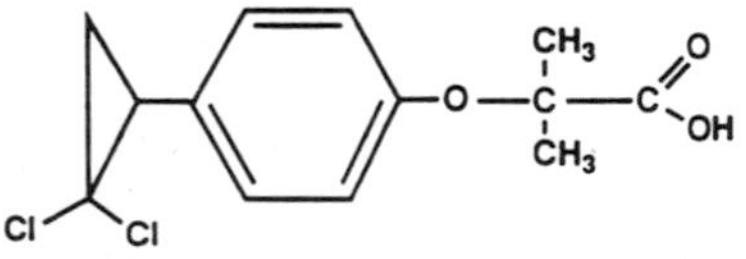

Ciprofibrate

Fig. 1. Chemical structures of the fibrates

the randomization between the groups was not inherently faulty, but that some unknown factor associated with clofibrate increased non-cardiac deaths during the study period.

These findings, coupled with the relatively poor efficacy of clofibrate (LEVY et al. 1972) and the increased bile lithogenicity reported with its use (CORONARY DRUG PROJECT RESEARCH GROUP 1977), have prompted the development of derivatives of this compound and their examination as potential lipid-lowering drugs. This exercise proved fruitful and has provided the clinician with a range of agents with improved efficacy and safety.

The second and third generation fibrates, viz., bezafibrate, gemfibrozil, fenofibrate and ciprofibrate have been developed and marketed as useful agents for the treatment of hypertriglyceridaemia, where they are first-line therapies, and hypercholesterolaemia, where they are often used as second-line treatments when bile acid sequestrant resin therapy fails. They also have an important place in the management of combined hyperlipidaemia (raised cholesterol and triglyceride) and commonly appear in combination regimens with the bile acid sequestrant resins cholestyramine and colestipol (GAW and SHEPHERD 1990).

III. Helsinki Heart Study

The clinical efficacy of one of the fibrates, gemfibrozil, was confirmed in the Helsinki Heart Study (FRICK et al. 1987), a randomized, double-blind, placebo-controlled trial designed to test the efficacy of drug in reducing the risk of coronary heart disease (CHD). A group of 4081 middle-aged men (40–55 years) with primary dyslipidaemia (non-high-density lipoprotein (HDL) cholesterol $\geqslant$5.2 mmol/l) were recruited. Half received 600 mg gemfibrozil twice daily, while the others received placebo. Gemfibrozil therapy was associated with reductions in triglyceride of 35% and in low-density lipoprotein (LDL) cholesterol of 11% and an increase in HDL cholesterol of 11%. The cumulative rate of cardiac end points (fatal and non-fatal myocardial infarctions combined) at 5 years was 27.3/1000 in the gemfibrozil-treated group and 41.4/1000 in the placebo group: a reduction in CHD of 34% ($p < 0.02$). A further analysis of the Helsinki Heart Study data (MANNINEN et al. 1988) provided more detail on the relationship between the lipid changes and CHD incidence. The success of gemfibrozil in reducing CHD events was related not only to its ability to lower LDL cholesterol but also to its HDL cholesterol-raising effect. All the modern fibrates share with gemfibrozil these beneficial effects on the lipoprotein profile, and the results of the Helsinki Heart Study have often been extrapolated to bezafibrate, fenofibrate and ciprofibrate.

B. Comparative Pharmacology of the Fibrates

The principal members of the fibrate family will be examined individually in terms of their structure, clinical pharmacology and clinical efficacy. The group as a whole will then be discussed in terms of mechanism of action, toxicology, adverse clinical events and specialized uses.

I. Clofibrate

Clofibrate is the parent compound of the fibrates and, as described above, was the first to be introduced into clinical practice. It is completely absorbed from the gut and is rapidly hydrolysed to the active acid form. The usual dose of clofibrate is shown in Table 1 and it has a plasma half-life of 13–25 h

Table 1. Usual daily doses of the major fibrates

Drug	Recommended regimen	Total daily dose (mg)
Clofibrate	1000 mg twice daily	2000
Bezafibrate[1]	200 mg thrice daily	600
Fenofibrate	100 mg thrice daily	300
Gemfibrozil	600 mg twice daily	1200
Ciprofibrate	100–200 mg daily	100–200

[1] A 400-mg once daily sustained release preparation (Bezalip Mono) is also available.

(GUGLER and HARTLAPP 1978). Clofibric acid is mainly excreted in the urine after approximately 60% is converted to the glucuronide form (BROWN and GOLDSTEIN 1985). Clofibrate is available throughout the world, although, as described above, its clinical use is now dwindling. The only remaining clinical indication for clofibrate is type III hyperlipoproteinaemia (THOMPSON 1989) and those hyperlipoproteinaemias associated with apoE phenotypes $E_{2/2}$, $E_{2/3}$ and $E_{2/4}$ (JANUS et al. 1985).

II. Bezafibrate

Bezafibrate (2-[4-[2-(chlorobenzamido)ethyl]phenoxy]-2-methylpropionic acid) is administered thrice daily or as a single daily dose of the sustained release preparation Bezalip Mono (Table 1). This drug is reported to have a good long-term safety and efficacy record after up to 4.5 years treatment (OLSSON et al. 1985), and it has been used effectively in types IIa and IIb (GAVISH et al. 1986), type III (KLOSIEWICZ-LATOSZEK et al. 1987) and types IV and V hyperlipoproteinaemias (SHEPHERD et al. 1984). It is well absorbed from the gut and has a short plasma half-life of approximately 2 h (MONK and TODD 1987). About half of the ingested dose of bezafibrate is excreted unchanged in the urine, while the other half is first metabolized (ABSHAGEN et al. 1979). Bezafibrate is widely available in Europe and in New Zealand.

III. Ciprofibrate

Ciprofibrate (2-[4-(2,2-dichlorocyclopropyl)phenoxy]-2-methylpropanoic acid) is administered as a single daily dose (Table 1). It has the longest half-life (DAVISON et al. 1975) of the fibrates (approximately 42 h). The efficacy of ciprofibrate has been demonstrated in types IIa, IIb, III and IV hyperlipoproteinaemias (OLSSON and ORO 1982; DAVIGNON et al. 1982). In these studies, laboratory safety monitoring has in general not revealed any clinically significant changes, and the drug has been well tolerated both during short-term (ILLINGWORTH et al. 1982) and long-term usage (SCHIFFERDECKER et al. 1984). Ciprofibrate is currently available in several European countries and has recently been licensed for use in the UK.

IV. Fenofibrate

Fenofibrate (isopropyl 2-[4-(4-chlorobenzoyl)-phenoxy]-2-methyl propanoate) is variably absorbed from the gut, but rapidly hydrolyzed to yield the active metabolite fenofibric acid. It has a plasma half-life of 20–27 h in healthy subjects (BALFOUR et al. 1990) and in humans is mainly excreted in the urine, although this seems to depend on the formulation and dose given (BRODIE et al. 1976; DESAGER and HARVENGT 1978). This drug has been used effectively in all forms of primary hyperlipoproteinaemia, with the exception of patients with type I hyperchylomicronaemia (CANZLER and BOJANOVSKI 1980; HARVENGT et al. 1980; LEHTONEN and VIIKARI 1981; FRUCHART et al. 1987). Fenofibrate has been available in France for over a decade and is widely used throughout Europe, Africa and Asia.

V. Gemfibrozil

Gemfibrozil (5-(2,5-dimethylphenoxy)-2,2-dimethylpentanoic acid) is well absorbed from the gut and has a plasma half-life of 7.6 h (TODD and WARD 1988). It undergoes extensive conjugation and metabolites are excreted in the urine (OKERHOLM et al. 1976). This drug has been used effectively to treat patients with types IIa, IIb and IV hyperlipoproteinaemia in the large Helsinki Heart Study (FRICK et al. 1987). It has also been used with benefit in patients with type III (HOULSTON et al. 1988) and type V hyperlipoproteinaemia (LEAF et al. 1989). Gemfibrozil is the only fibrate other than clofibrate available in North America and it is also widely available in Europe.

The comparative efficacy of the fibrates in relation to lipoprotein phenotypes are summarized in Tables 2–6. It should be noted that there is, as with any group of drugs, a degree of interindividual variation in hypolipidaemic response to the fibrates. In type IIa hyperlipoproteinaemia, the most effective LDL cholesterol lowering is produced by bezafibrate, fenofibrate and ciprofibrate. In type IIb hyperlipoproteinaemia, there is little to choose between the fibrates in terms of their triglyceride-lowering ability. In type III hyperlipoproteinaemia, however, the most effective fibrate in lowering triglyceride and raising HDL cholesterol is gemfibrozil. The hypertriglyceridaemia associated with type IV hyperlipoproteinaemia is most readily corrected by fenofibrate and ciprofibrate while in the rarer condition of type V hyperlipoproteinaemia, gemfibrozil and slow-release bezafibrate seem to be most effective in correcting the aberrant lipid profile.

C. Mechanism of Action

Clofibrate was not discovered as a result of serendipity. On the contrary, it was found after a process of systematic, pharmacological screening of branch-chain fatty acids. It was hoped that such compounds would interfere

Table 2. Comparative effects of the fibrates in type IIa hyperlipoproteinaemia

Drug	Reference	Daily dose	*n*	% Change on therapy				
				TC	TG	LDL-C	HDL-C	VLDL-C
Clofibrate	Grundy et al. (1972)	2 g	5	−19	−26	NR	NR	NR
	Levy et al. (1972)	2 g	10	−7	−2	NR	NR	NR
Bezafibrate	Stewart et al. (1982)	600 mg	4	−16	−35	−18	+2	−42
	Olsson et al. (1985)	600 mg	8	−16	−31	−22	+10	−67
	Gavish et al. (1986)	600 mg	12	−19	−25	−23	+11	NR
	Curtis et al. (1988)	600 mg	16	−16	−18	−19	−10	−25
Gemfibrozil	Manninen et al. (1988)	1200 mg	1293	−11	−38	−13	+11	NR
Fenofibrate	Canzler and Bojanowski (1980)	300 mg	10	−21	−16	NR	NR	NR
	Lehtonen and Viikari (1981)	300–600 mg	16	−20	−27	−20	+15	NR
	Rouffy et al. (1985)	300 mg	21	−20	−16	−25	+14	NR
	Rouffy et al. (1985)	400 mg	21	−28	−18	−33	+11	NR
Ciprofibrate	Olsson and Oro (1982)	100 mg	6	−25	−31	−28	+11	−49
	Illingworth et al. (1982)	50 mg	7	−11	−22	−13	+8	NR
	Illingworth et al. (1982)	100 mg	9	−20	−30	−24	+10	NR
	Rouffy et al. (1985)	100 mg	20	−21	−21	−20	+20	NR
	Dairou and Regy (1986)	100 mg	2418	−23	−21	−30	+5	NR

TC, total cholesterol; TG, triglyceride; LDL-C, low-density lipoprotein cholesterol; HDL-C, high-density lipoprotein cholesterol; VLDL-C, very low density lipoprotein cholesterol; NR, not reported.

Table 3. Comparative effects of the fibrates in type IIb hyperlipoproteinaemia

Drug	Reference	Daily dose	n	% Change on therapy				
				TC	TG	LDL-C	HDL-C	VLDL-C
Clofibrate	NESTEL et al. (1980)	2 g	3	−16	−27	NR	NR	NR
Bezafibrate	STEWART et al. (1982)	600 mg	3	−3	−25	−2	+4	−20
	GAVISH et al. (1986)	600 mg	8	−16	−54	−24	+43	NR
	KLOSIEWICZ-LATOSZEK et al. (1987)	600 mg	10	−28	−42	−32	+32	NR
Gemfibrozil	MANNINEN et al. (1988)	1200 mg	570	−10	−37	−8	+13	NR
Fenofibrate	CANZLER and BOJANOWSKI (1980)	300 mg	6	−3	−26	NR	NR	NR
	LEHTONEN et al. (1981)	300–600 mg	7	−30	−58	+30	+22	NR
Ciprofibrate	OLSSON and ORO (1982)	100 mg	9	−14	−33	−11	+5	−48
	DAIROU and REGY (1986)	100 mg	3896	−23	−46	−27	+8	NR

TC, total cholesterol; TG, triglyceride; LDL-C, low-density lipoprotein cholesterol; HDL-C, high-density lipoprotein cholesterol; NR, not reported.

Table 4. Comparative effects of the fibrates in type III hyperlipoproteinaemia

Drug	Reference	Daily dose	*n*	% Change on therapy				
				TC	TG	LDL-C	HDL-C	VLDL-C
Clofibrate	GRUNDY et al. (1972)	2 g	3	−43	−48	NR	NR	NR
	LEVY et al. (1972)	2 g	11	−32	−44	−20	+11	−54
	ILLINGWORTH (1990)	2 g	12	−40	−52	−1	+31	−61
Bezafibrate	PACKARD et al. (1986)	600 mg	6	−47	−68	−19	+28	−75
	KLOSIEWICZ-LATOSZEK et al. (1987)	600 mg	9	−20	−50	+19	−2	NR
Gemfibrozil	HOULSTON et al. (1988)	1200 mg	13	−40	−70	NR	+45	NR
Fenofibrate	CANZLER and BOJANOWSKI (1980)	300 mg	4	−48	−58	NR	NR	NR
	FRUCHART et al. (1987)	300–600 mg	9	−37	−56	−7	+25	NR
	LUSSIER-CACAN et al. (1989)	300 mg	9	−40	−60	−6	+28	NR
Ciprofibrate	DAVIGNON et al. (1982)	100 mg	1	−35	−43	NR	NR	NR

TC, total cholesterol; TG, triglyceride; LDL-C; low-density lipoprotein cholesterol; HDL-C, high-density lipoprotein cholesterol; VLDL-C, very low density lipoprotein cholesterol; NR, not reported.

Table 5. Comparative effects of the fibrates in type IV hyperlipoproteinaemia

Drug	Reference	Daily dose	*n*	% Change on therapy				
				TC	TG	LDL-C	HDL-C	VLDL-C
Clofibrate	Grundy et al. (1972)	2 g	3	−28	−45	NR	NR	NR
	Rabkin et al. (1988)	2 g	16	−8	−53	+10	+19	NR
Bezafibrate	Shepherd et al. (1984)	600 mg	8	−20	−59	+14	+14	−52
	Olsson et al. (1985)	600 mg	15	−11	−28	−9	+16	−45
Gemfibrozil	Manninen et al. (1988)	1200 mg	182	−5	−28	+2	+10	NR
	East et al. (1988)	1200 mg	8	+7	−40	+29	+26	NR
Fenofibrate	Canzler and Bojanowski (1980)	300 mg	4	−10	−60	NR	NR	NR
	Harvengt et al. (1980)	300 mg	5	−20	−60	NR	NR	NR
	Lehtonen et al. (1981)	300–600 mg	10	−17	−44	−23	+11	NR
Ciprofibrate	Olsson and Oro (1982)	100 mg	11	−10	−30	−6	+5	−79
	Davignon et al. (1982)	100 mg	14	−14	−70	NR	NR	NR
	Dairou and Regy (1986)	100 mg	497	−9	−48	−8	+7	NR

TC, total cholesterol; TG, triglyceride; LDL-c, low-density lipoprotein cholesterol; HDL-C, high-density lipoprotein cholesterol; VLDL-C, very low density lipoprotein cholesterol; NR, not reported.

Table 6. Comparative Effects of the fibrates in type V hyperlipoproteinaemia

Drug	Reference	Daily dose	*n*	% Change on therapy				
				TC	TG	LDL-C	HDL-C	VLDL-C
Clofibrate	Grundy et al. (1972)	2 g	5	−12	−40	NR	NR	NR
Bezafibrate	Shepherd et al. (1984)	600 mg	2	−39	−55	+177	+25	−70
Gemfibrozil	Saku et al. (1985)	1200 mg	6	−6	−58	+32	+34	NR
	Leaf et al. (1989)	1200 mg	13	−48	−74	+68	+38	−68
Fenofibrate	Harvengt et al. (1980)	300 mg	5	−15	−50	NR	NR	NR
	Shepherd et al. (1985)	400 mg	7	−41	−77	+41	+43	−74
Ciprofibrate	No known study							

TC, total cholesterol; TG, triglyceride; LDL-C, low-density lipoprotein cholesterol; HDL-C, high-density lipoprotein cholesterol; VLDL-C, very low density lipoprotein cholesterol; NR, not reported.

with normal fatty acid metabolism and have a useful hypolipidaemic effect (THORP 1963). From this "designer drug" approach based on the knowledge of lipoprotein metabolism as it stood in the early 1960s, the mechanism of action of the fibrates has been extensively studied and is still, some 30 years later, the subject of controversy. The influence of the fibrates on lipoprotein metabolism is represented schematically in Fig. 2.

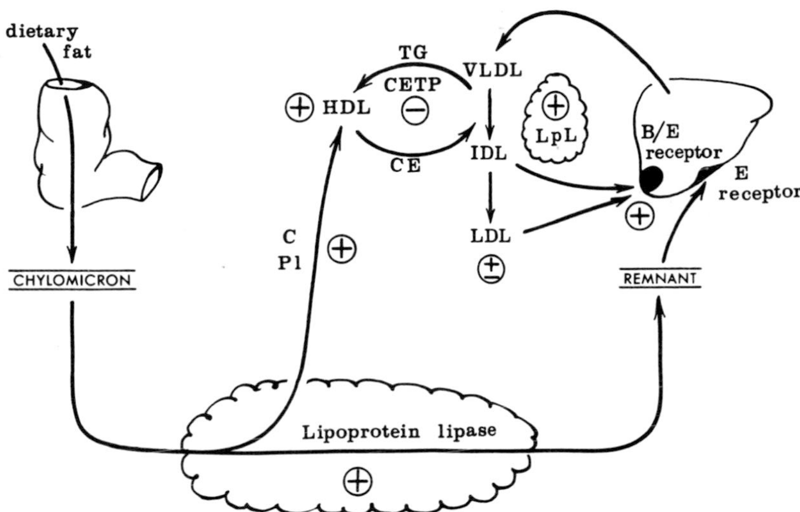

Fig. 2. Fibrates and lipoprotein metabolism. Dietary lipids are absorbed from the gut and packaged into large triglyceride (*TG*) rich particles called *chylomicrons*. These particles undergo lipolysis by the endothelium-bound enzyme *lipoprotein lipase* (*LpL*) in the capillary beds of adipose tissue and skeletal muscle. This process yields small chylomicron *remnant* particles that are thought to be cleared from the circulation via the apoE receptor on hepatocytes. Redundant surface material containing cholesterol (*C*) and phospholipid (*Pl*) is also lost into the high-density lipoprotein (*HDL*) density interval. Very low density lipoprotein (*VLDL*) is continually secreted by the liver to carry endogenously produced lipid, and these particles also undergo lipolysis by the actions of LpL. As VLDL is delipidated by this enzyme, intermediate-density lipoproteins (*IDL*) and finally low-density lipoproteins (*LDL*) are formed. Both these products are removed from the plasma by the LDL (or *B/E*) receptor. By stimulating LpL activity the fibrates increase the clearance rate of chylomicrons and the delipidation rate of apoB-containing lipoproteins (VLDL, IDL). Because of the compositional changes that take place in the LDL interval, the fibrates result in increased receptor-mediated clearance of this lipoprotein, although the final level of LDL in the plasma is largely dictated by the phenotype of the patient prior to fibrate therapy (see main text). Cholesteryl ester transfer protein (*CETP*) facilitates the intravascular remodelling of HDL, resulting in this lipoprotein becoming TG enriched and cholesteryl ester (CE) depleted. The fibrates reduce the net transfer of TG into HDL by reducing the availability of TG rich substrates. This results in increased concentrations of large CE-rich HDL particles in the circulation

I. Effects on Very Low-Density Lipoprotein Metabolism

The primary action of the fibrates is to reduce plasma levels of triglyceride rich lipoproteins. This action is thought to have two main facets: (1) decreased hepatic triglyceride synthesis, which is secondary to reduced peripheral lipolysis and diminished free fatty acid flux back to the liver and (2) enhanced VLDL catabolism, due to stimulated lipoprotein lipase activity and consequently increased intravascular VLDL triglyceride lipolysis.

This increase in postheparin lipoprotein lipase activity is documented in two recent, but differently designed studies. In 25 patients receiving continuous ambulatory peritoneal dialysis, gemfibrozil was used to correct the aberrant lipid profile (CHAN 1989). In doing so, this fibrate caused an increase in both postheparin lipoprotein lipase and hepatic lipase activity. Even long-term fibrate therapy has been shown to have a sustained effect on postheparin lipoprotein lipase activity, which fell when bezafibrate treatment of a group of 13 patients was interrupted (VESSBY and LITHELL 1990).

II. Effects on Chylomicron Metabolism

The fibrate-induced increase in lipoprotein lipase activity noted above offers an explanation for the fall in postprandial lipaemia seen in both normo- and hypercholesterolaemic patients treated with fenofibrate (SIMPSON et al. 1990). This improvement in chylomicron metabolism was associated with a highly significant increase in postheparin lipoprotein lipase activity (37%; $p < 0.001$) and only a marginal rise in that of hepatic lipase. Such an effect of the fibrates may be a very important contribution to their anti-atherogenic potential, as postprandial lipaemia is considered by many to be an additional CHD risk factor (ZILVERSMIT 1979). The presence of CHD has also been associated with prolonged and exaggerated hypertriglyceridaemia following a fat load. Such abnormal clearance of chylomicrons and their remnants seen in patients with CHD is corrected by fibrate therapy (SIMPSON et al. 1990). The fibrates, therefore, reduce levels of triglyceride-rich lipoproteins in a consistent and readily explicable fashion.

III. Effects on Low-Density Lipoprotein Metabolism

The action of the fibrates on LDL levels are variable depending on the initial LDL cholesterol value. In hypercholesterolaemia where LDL levels are high, the fibrates cause a decrease in LDL apoB, while in hypertriglyceridaemia, which is commonly associated with low plasma LDL concentrations, fibrate treatment will raise the latter (SHEPHERD et al. 1984, 1985). This differential in LDL response dependent on initial LDL and triglyceride levels has most recently been reported by MANTTARI et al. (1990). The fractional catabolic rate (FCR) of LDL apoB, in response on fibrate therapy, has previously been shown to increase in hypercholesterolaemia (STEWART et al. 1982) and decrease in hypertriglyceridaemia (Shepherd et al. 1984) while LDL apoB synthetic rates in both groups remain unchanged. This apparent paradox is resolved when these changes are regarded as normalization phenomena: in hypercholesterolaemia, an initially low FCR is increased to normal, and in hypertriglyceridaemia, an initially high FCR is reduced to normal.

IV. Effects on Low-Density Lipoprotein Composition and Subfraction Profile

The product of the lipolytic cascade, LDL, has been recognized for some years to be structurally heterogeneous in patients with hypertriglyceridaemia (Hammond et al. 1977; Fisher 1983) and more recently has been found to exist in the plasma of all individuals as a group of discrete, but overlapping, particle populations or subfractions (Shen et al. 1981; Krauss and Burke 1982). These subfractions have traditionally been separated on the basis of hydrated density and particle size, using density gradient ultracentrifugation and gradient gel electrophoresis, respectively, and are defined as LDL-I–LDL-IV (Krauss 1987). Gofman et al. (1950) were the first to demonstrate that LDL is heterogeneous in the analytical ultracentrifuge and to suggest that certain subfractions may be more closely associated with CHD than others. These early observations have now been substantiated in a series of cross-sectional studies that have revealed a constellation of abnormalities in LDL structure and concentration that lead to increased risk of CHD (Crouse et al. 1985; Musliner and Krauss 1988). A predominance of small, dense LDL in combination with a raised triglyceride level and low levels of HDL has been associated with a threefold increase in risk of myocardial infarction (Austin et al. 1988). These changes have been collectively described as an atherogenic lipoprotein phenotype (ALP; Austin et al. 1990a). There is strong evidence to suggest that ALP, and more specifically the distribution of LDL subfractions, is influenced by a major gene (Austin and Krauss 1986; Austin et al. 1990b). However, the profile is also subject to the environmental effects of diet and drugs (Griffin et al. 1992).

The effect of fibrate therapy in hypertriglyceridaemia, as described above, is to correct an initially low plasma LDL level. This change in the total LDL mass is accompanied by significant compositional changes (Shepherd et al. 1985). The particles become enriched in cholesteryl ester and relatively depleted in phospholipid and triglyceride, thereby increasing the lipid core to coat ratio. This change in lipid composition is consistent with an increase in LDL particle size, away from the small, dense LDL characteristic of hypertriglyceridaemia.

In patients who are normotriglyceridaemic, yet hypercholesterolaemic, fibrate therapy lowers the total LDL mass. This is predominantly due to a significant decrease in the LDL subfraction of mid-density (LDL-II). The changes in the LDL subfractions of lower density (LDL-I) and higher density (LDL-III) are inversely related, the most consistent finding being an increase in LDL-I and a decrease in LDL-III (Shepherd et al. 1991).

Thus, in both hyperlipidaemic states fibrate therapy induces a redistribution of LDL density and size towards lighter and larger particles: an LDL subfraction profile that has been associated with reduced coronary risk (Austin et al. 1988).

V. Effects on Cholesterol Synthesis

Besides their effects on triglyceride and apolipoprotein B metabolism, the fibrates are also reputed to have an inhibitory effect on hepatic cholesterol synthesis. This action is thought to be mediated through the inhibition of 3-hydroxy-3-methylglutaryl coenzyme A (HMG-CoA) reductase, the rate-limiting enzyme of cholesterol biosynthesis. Early work by Berndt et al. (1978) demonstrated both the inhibition of this enzyme by clofibrate and bezafibrate in rat liver microsomes in vitro and its reduced activity in liver microsomes isolated from fibrate-fed rats. Schneider et al. (1985) have also demonstrated reduced activity of HMG-CoA reductase in freshly isolated mononuclear cells from patients treated with fenofibrate, but in similar studies using bezafibrate, Stange et al. (1991) report no drug-induced effect on HMG-CoA reductase activity. McNamara et al. (1980) also observed no change in the rate of cholesterol synthesis in mononuclear cells freshly isolated from patients receiving clofibrate therapy. Other recent studies have demonstrated fibrate-induced inhibition of the regulatory enzymes in cholesterol biosynthesis. Castillo et al. (1990) showed reduced activity of the enzyme HMG-CoA reductase in vivo and in vitro in chick liver in response to clofibrate, while Cosentini et al. (1989) demonstrated in a total of 15 patients that bezafibrate reduced the incorporation of labelled acetate into non-saponifiable lipids in freshly isolated blood mononuclear cells. Again, however, there is some equivocation in the literature in that another fibrate, gemfibrozil, has been shown to increase hepatic sterol biosynthesis, albeit in rats (Maxwell et al. 1983).

Published reports to date, have been criticized on a number of counts. Firstly, just as there are differences between rodents and humans in the hepatic toxicology of the fibrates (see below), there also appear to be important species dissimilarities in the effect of the fibrates on the enzymes of cholesterol metabolism (Stahlberg et al. 1991). These workers report an approximately twofold increase in HMG-CoA reductase activity in human liver microsomes from bezafibrate-treated individuals, in contrast to the reductions seen in other studies in animals. Work done in non-human animal models may therefore not be readily extrapolated to man.

Much of the published data have also been criticized, either because of invalid methodology (Fears 1983) or the use of unrealisitically high concentrations of drug in vitro (Newton and Krause 1986). The latter workers point out that many in vitro experiments may be flawed because the concentrations of fibrates used greatly exceed potential in vivo pharmacological levels. Similar levels of HMG-CoA reductase inhibition were achieved by these workers when compactin was used as a positive control at in vitro levels 10 000-fold lower than that required of the fibrates. On this basis, NEWTON and KRAUSE (1986) suggest that the observed inhibitory effect of the fibrates on HMG-CoA reductase in vitro may represent a non-specific, rather than a physiological effect at a specific regulatory site on the enzyme. We cannot, therefore, class the fibrates with the specific HMG-CoA reductase inhibitors or "statins", even though there is good evidence that they do indeed inhibit cholesterol synthesis, for they must effect this action by another means. Indirect effects of the fibrates on cholesterol biosynthesis have been proposed by KLEINMAN et al. (1985). They have shown that the compositional changes in LDL induced by bezafibrate therapy enhances LDL receptor-mediated uptake of this lipoprotein and its ability to down-regulate LDL receptor activity and to suppress cholesterol synthesis in cultured fibroblasts. This enhanced delivery of sterol to cells may indirectly inhibit cholesterol synthesis in fibrate-treated patients, in accord with the concept of BROWN et al. (1974), who found that cholesterol itself was the feedback suppressor of HMG-CoA reductase activity in fibroblasts. Another indirect effect of the fibrates on the enzymes of cholesterol synthesis may be mediated by drug-induced changes in membrane cholesterol content and, in turn, membrane fluidity. MITROPOULOS and VENKATESAN (1985) describe possible mechanisms by which the enzyme HMG-CoA reductase, a membrane-bound protein, may be regulated by the cholesterol content of its supporting phospholipid bilayer. NEEDHAM et al. (1985) have studied the effects of fibrate therapy on membrane fluidity and conclude that the drug clofibrate decreases the molar cholesterol phospholipid ratio of plasma membranes and that this environmental change alters the functioning of integral membrane proteins. Thus, fibrate-induced changes in the membranes of the endoplasmic reticulum, where HMG-CoA reductase is mainly located, may indirectly influence this enzyme's activity. While there has been some equivocation in the literature from in vivo and in vitro experiments, a consensus is now emerging that the fibrates do indeed inhibit cholesterol synthesis in man, but that they do so by some other means than the direct inhibition of the enzyme HMG-CoA reductase. Further work in this area is clearly needed before we fully understand the nature of the mechanisms by which the fibrates exert their cholesterol-lowering effect.

VI. Effects on High-Density Lipoprotein Metabolism

In hypercholesterolaemia, fibrate-induced reductions in LDL cholesterol are commonly accompanied by increases in HDL cholesterol, leading to important changes in the calculated atherogenic risk index. Much has been made of this overall reversal of an atherogenic profile (lowering LDL and raising HDL) and indeed the strategies employed to market the modern fibrates are largely built upon it.

In hypertriglyceridaemic patients with a type V phenotype, gemfibrozil significantly increased plasma HDL cholesterol and apoA-I and A-II. When the kinetic changes associated with these alterations in plasma levels were studied, it was noted that the synthetic rates of apoA-I and A-II were increased as a result of fibrate therapy without any change in their FCR (SAKU et al. 1985). In heterozygous familial hypercholesterolaemia (FH) patients, similar studies using fenofibrate revealed increases in the synthetic rate and FCR of apoA-I (MALMENDIER and DELCROIX 1985). Because the synthetic rate increased more than the clearance, the end result was again a fibrate-induced increase in plasma apoA-I level.

This HDL cholesterol-raising action begs the question of whether individuals with reduced levels of HDL cholesterol, but with normal levels of total cholesterol and triglyceride, should be treated with fibrate therapy. VEGA and GRUNDY (1989) address this very important issue in a recent study. They examined 22 normolipidemic patients, most of whom had CHD and all of whom had low HDL cholesterol levels (<0.91 mmol/l). In a randomised crossover design they evaluated the lipid-regulating effects of gemfibrozil and lovastatin in these patients. Both drugs significantly lowered the LDL to HDL ratio, but lovastatin more so than gemfibrozil. Lovastatin was, therefore, deemed superior to the fibrate in these individuals, but the issue of whether such individuals with reduced plasma HDL levels should be treated with any form of lipid-regulating drugs is unresolved.

D. Toxicology

I. Hepatomegaly and Carcinogenicity

When given to rodents, all fibrates have been shown to cause hepatomegaly and peroxisome proliferation (BLANE 1987). Hepatocellular carcinomata have also been observed in rodents receiving fibrate therapy, but the carcinogenic dose of clofibrate, gemfibrozil and fenofibrate per kilogramme in these species appears to be at least 12 times higher than the recommended human therapeutic dose (BLANE 1987). However, this characteristic hepatic toxicity appears to be species specific. Studies in non-human primates and human liver biopsies have shown no such hepatic changes (BALFOUR et al. 1990). Large clinical studies in humans bear this out, with no excess of neoplasms in the gemfibrozil treatment arm of the Helsinki Heart Study (FRICK et al. 1987) and indeed no hepatic tumours at all.

II. Bile Lithogenicity

Treatment with clofibrate has been shown to increase the biliary secretion of cholesterol and reduce that of bile acids (GRUNDY et al. 1972). These compositional changes induce supersaturation of the bile, leading to an increased risk of gallstone formation during long-term treatment with clofibrate (Coronary Drug Project Research Group 1977; Committee of Principal Investigators 1978). Subsequently a number of studies have demonstrated similar increases in biliary cholesterol secretion with the other fibrates.

While the increased bile lithogenicity associated with the fibrates has been described as a class effect occurring in all fibrates (Blane 1987), there is evidence to suggest that ciprofibrate does not significantly increase the relative concentration of cholesterol in the bile (Angelin et al. 1984). In this study, 6–18 months therapy using 100 mg ciprofibrate once daily did not increase the lithogenicity of the bile in hyperlipidemic patients.

It is important to recognize that the level of bile saturation is not the sole index of the risk of gallstone formation, as many other factors influence this (Palmer 1987). Interestingly, there was no increase in the incidence of cholecystectomy in the gemfibrozil-treated patients in the Helsinki Heart Study (Frick et al. 1987), and no excess risk of gallstone formation has been reported to date in the recommended clinical use of fenofibrate, the fibrate with which we have longest clinical experience after clofibrate (Balfour et al. 1990).

The fibrates should, however, be used with caution in patients with pre-existing gallstones, because any increase in bile lithogenicity in such individuals may accelerate stone formation and precipitate the need for surgery. The evidence so far would suggest that the risk in patients who are initially free of gall bladder disease is too small to merit any contra-indication on these grounds.

E. Adverse Clinical Effects

I. Side Effects

The fibrates are generally very well tolerated drugs. The most common side effects that are seen in clinical practice are gastrointestinal disturbances such as nausea and diarrhoea. More uncommon side effects were noted by the Coronary Drug Project Research Group (1975) and the WHO trial (Committee of Principal Investigators 1978). These included impotence and loss of libido, skin rashes, increased appetite and weight gain. Clofibrate has also been noted to cause increases in serum creatine kinase (CK) and the transaminases associated with malaise, myalgia and myaesthenia (Langer and Levy 1968). The as yet unidentified adverse properties of clofibrate brought to light in the WHO trial (Committee of Principal Investigators 1978) and discussed above do not seem to be shared by the newer fibrates. However, the prescriber must still maintain a sense of vigilance when using these other fibrates, for it is not impossible that such adverse effects may still come to light, e.g. fenofibrate has been reported in four cases to cause drug-induced hepatitis associated with anti-smooth muscle antibodies (HOMBERG et al. 1985).

II. Drug Interactions

The fibrates bind extensively to plasma albumin and this in turn displaces other acidic drugs from these plasma protein-binding sites. The clinical consequence of this is that the co-administration of a fibrate with other drugs such as the oral anti-coagulants may potentiate the effect of the latter. This potentially hazardous drug interaction is common to all the fibrates.

F. Special Uses

I. Combination Therapy

Hypolipidemic drugs have been used in combination with the hope that two or more drugs may act synergistically to correct aberrant lipoprotein metabolism or that they may serve to offset each other's unwanted effects on the lipid profile. The use of fibrates in combined regimens with bile acid sequestrant resins was being examined in studies in the early 1970s (GOODMAN et al. 1973). This form of combination is now in wide use and raises the question as to whether the joint administration of an anion exchange resin affects the absorption of a fibrate. To answer this question, the combination of gemfibrozil and colestipol was examined by FORLAND et al. (1990), who concluded that simultaneous administration of the two drugs did indeed reduce the bioavailability of the fibrate, but added that separating the administration of the two drugs by at least 2 h would obviate this interaction. In designing new combined regimens, a more exciting combination was initially thought to be that of fibrate plus HMG-CoA reductase inhibitor. Enthusiasm for such a regimen was, however, curbed when reports of myopathy associated with lovastatin plus gemfibrozil therapy were published (PIERCE et al. 1990). Despite this, and because the combination had shown such early promise, GLUECK et al. (1990) examined 25 patients for an average duration of 12.5 months, during which patients received both drugs. While mandating careful follow-up with serial CK and liver function tests, they report the combination to be safe and effective, with only a 2.4% incidence of raised CK and no symptomatic myositis. This is an important study, but should not be viewed as a blanket approval of such therapy. Caution with fibrate plus HMG-CoA reductase inhibitor is also urged by ILLINGWORTH and O'MALLEY (1990), who recently evaluated the combination of clofibrate and lovastatin in six type III patients and monotherapy with one or other drug in 12 patients. They showed that addition of lovastatin to gemfibrozil therapy or vice versa in patients resistant to monotherapy was of benefit in further reducing concentrations of total, VLDL, and LDL cholesterol. These findings offer clinicians an alternative strategy for the management of type III hyperlipidaemia.

II. Paediatric Use of the Fibrates

As with all hypolipidemic drugs, the data available to support the use of fibrates in childhood is severely limited. Several studies have been performed, however, which indicate the efficacy and short-term safety of these drugs when given to hypercholesterolaemic children. Two small studies (Steinmetz et al. 1981; Chicaud et al. 1984) report the beneficial use of fenofibrate in hypercholesterolaemic children over 3 months and 18 months, respectively. Wheeler et al. (1985) have examined the effects of bezafibrate in 14 children with familial hypercholesterolaemia. In a 6-month, double-blind, crossover, placebo-controlled trial, bezafibrate at a dosage of 10–20 mg/kg per day resulted in a 22% fall in total cholesterol and 15% increase in HDL cholesterol. In spite of these beneficial changes, these authors conclude that because the long term safety of the fibrates has not been evaluated, their use in children with FH should be restricted to those children who cannot tolerate the bile acid sequestrant resins. Furthermore, they advocate careful clinical and biochemical monitoring when these drugs are used in children. However, in the opinion of Illingworth (1991), the benefit to risk ratio does not justify the use of the fibrates in children with heterozygous FH, and he suggests that they should be further restricted to the rare paediatric patient presenting with severe primary hypertriglyceridaemia or type III hyperlipoproteinaemia.

G. Conclusions

The fibrates are a powerful group of drugs that have found an important place in our lipid-regulating formulary. While their clinical efficacy is generally accepted, their mechanism of action has yet to be fully defined and considerable efforts are now being made to examine this issue. Workers are also directing their efforts towards the continued development of new fibric acid derivatives and towards the novel application of established fibrates in new combinations and new clinical situations. The current degree of research activity in the field of the fibric acid derivatives stands as a testament to the clinical value of this group of drugs.

References

Abshagen U, Bablok W, Koch K, Lang PD, Schmidt HAE, Senn M, Stork H (1979) Disposition pharmacokinetics of bezafibrate in man. Eur J Clin Pharmacol 16:31–38

Angelin B, Einarsson K, Leijd B (1984) Effect of ciprofibrate treatment on biliary lipids in patients with hyperlipoproteinaemia. Eur J Clin Invest 14:73–78

Austin MA, Krauss RM (1986) Genetic control of low density lipoprotein subclasses. Lancet 2:592–595

Austin MA, Breslow JL, Hennekens CH, Buring JE, Willett WC, Krauss RM (1988) Low-density lipoprotein subclass patterns and risk of myocardial infarction. JAMA 260:1917–1921

Austin MA, King MC, Vranizan KM, Krauss RM (1990a) Atherogenic lipoprotein phenotype. A proposed genetic marker for coronary heart disease risk. Circulation 82:495–506

Austin MA, Brunzell JD, Fitch WL, Krauss RM (1990b) Inheritance of low density lipoprotein subclass pattern in familial combined hyperlipoproteinaemia. Arteriosclerosis 10:520–530

Balfour JA, McTavish D, Heel RC (1990) Fenofibrate. A review of its pharmacodynamic and pharmacokinetic properties and therapeutic use in dyslipidaemia. Drugs 40:260–290

Berndt J, Gaumert R, Still J (1978) Mode of action of lipid lowering agents clofibrate and BM15075 on cholesterol biosynthesis in rat liver. Atherosclerosis 30:147–152

Blane GF (1987) Comparative toxicity and safety profile of fenofibrate and other fibric acid derivatives. Am J Med 83 Suppl 5B:26–36

Brodie RR, Chasseaud LF, Elsom FF, Franklin ER, Taylor T (1976) Antilipidemic drugs: the metabolic fate of the hypolipidemic agent isopropy-[4′-(p-chlorobenzoyl)-2-phenoxy-2-methyl]-propionate (LE 178) in rats, dogs and man. Arzneimitteeforschung 26:896–901

Brown MS, Goldstein JL (1985) Drugs used in the treatment of hyperlipoproteinemia. In: Gilman AG, Goodman LS, Rall TW, Murad F (eds) The pharmacological basis of therapeutics. Macmillan, New York, pp 827–845

Brown MS, Dana SE, Goldstein JL (1974) Regulation of 3-hydroxy-3-methylglutaryl coenzyme A reductase activity in cultured human fibroblasts. Comparison of cells from a normal subject and from a patient with homozygous familial hypercholesterolemia. J Biol Chem 249:789–796

Canzler H, Bojanovski D (1980) Lowering effect of fenofibrate (procetofene) on lipoproteins in different types of hyperlipoproteinemias. Artery 8:171–178

Castillo M, Burgos C, Rodriguez-Vico F, Zafra MF, Garcia-Peregrin E (1990) Effects of clofibrate on the main regulatory enzymes of cholesterogenesis. Life Sci 46(6):397–403

Chan MK (1989) Gemfibrozil improves abnormalities of lipid metabolism in patients on continuous ambulatory peritoneal dialysis: the role of postheparin lipases in the metabolism of high-density lipoprotein subfractions. Metabolism 38:939–945

Chicaud P, Demange J, Drouin P, Debry G (1984) Long term (18 months) effects of fenofibrate in hypercholesterolaemic subjects. Action du fenofibrate chez des enfants hypercholesterolemiques: recul de 18 mois. Presse Med 13:417–419

Committee of Principal Investigators (1978) A cooperative trial in the primary prevention of ischaemic heart disease using clofibrate. Br Heart 10:1069–1118

Committee of Principal Investigators (1984) WHO cooperative trial on primary prevention of ischaemic heart disease with clofibrate to lower serum cholesterol. Lancet 2:600–604

Coronary Drug Project Research Group (1977) Gall bladder disease as a side effect of drugs influencing lipid metabolism. N Engl J Med 296:1188–1190

Cosentini R, Blasi F, Trinchera M, Sommariva D, Fasoli A (1989) Inhibition of cholesterol biosynthesis in freshly isolated blood mononuclear cells from normolipidemic subjects and hypercholesterolemic patients treated with bezafibrate. Atherosclerosis 79:253–255

Crouse JR, Parks JS, Schey HM, Kahl FR (1985) Studies of low density lipoprotein molecular weight in human beings with coronary artery disease. J Lipid Res 26:566–574

Curtis LD, Dickson AC, Ling KLE, Betteridge J (1988) Combination treatment with cholestyramine and bezafibrate for heterozygous familial hypercholesterolaemia. Br Med J 297:173–175

Dairou F, Regy C (1986) Ciprofibrate multicentric study in 6812 hyperlipidemic patients. (Abstract) In: Proceedings of the 9th International Symposium on Drugs Affecting Lipid Metabolism, p 62

Davignon J, Gascon B, Brossard D, Quidoz S, Leboeuf N, Lelorier J (1982) The use of ciprofibrate in the treatment of familial hyperlipidemias. In: Noseda G, Fragiacomo C, Fumagalli R, Paoletti R (eds) Lipoproteins and coronary atherosclerosis. Elsevier, Amsterdam, pp 213–221

Davison C, Benziger D, Fritz A, Edelson J (1975) Absorption and disposition of 2-[4-(2,2-dichlorocyclopropyl)phenoxy]-2-methylpropanoic acid, WIN 35833, in rats, monkeys, and men. Drug Metab Dispos 3:520–524

Desager JP, Harvengt C (1978) Clinical pharmacokinetic study of procetofene, a new hypolipidemic drug, in volunteers. Int J Clin Pharmacol Biopharm 16:570–574

East C, Bilheimer DW, Grundy SM (1988) Combination drug therapy for familial combined hyperlipidemia. Ann Intern Med 109:25–32

Fears R (1983) Pharmacological control of 3-hydroxy-3-methylglutaryl coenzyme A reductase activity. In: Sabine JR (ed) 3-Hydroxy-3-methylglutaryl coenzyme A reductase. CRC, Boca Raton

Fisher WR (1983) Heterogeneity of plasma low density lipoproteins. Manifestations of the physiologic phenomenon in man, Metabolism 32:283–291

Forland SC, Feng Y, Cutler RE (1990) Apparent reduced absorption of gemfibrozil when given with colestipol. J Clin Pharmacol 30:29–32

Frick MH, Elo O, Haapa K, Heinonen OP, Heinsalmi P, Helo P, Huttunen JK, Kaitaniemi P, Koskinen P, Manninen V, Maenpaa H, Malkonen M, Manttari M, Norola S, Pasternack A, Pikkarainen J, Romo M, Sjoblom T, Nikkila EA (1987) Helsinki Heart Study: primary-prevention trial with gemfibrozil in middle-aged men with dyslipidemia. N Engl J Med 317:1237–1245

Fruchart JC, Davignon J, Bard JM, Grothe AM, Richard A, Fievet C (1987) Effect of fenofibrate treatment on type III hyperlipoproteinemia. Am J Med 83 Suppl 5B:71–74

Gavish D, Oschry Y, Fainaru M, Eisenberg S (1986) Change in very low-, low-, and high-density lipoproteins during lipid lowering (bezafibrate) therapy: studies in type IIA and type IIb hyperlipoproteinaemia. Eur J Clin Invest 16:61–68

Gaw A, Shepherd J (1990) Combination drug therapy for hyperlipidaemia. J Drug Dev 3 Suppl 1:227–231

Glueck CJ, Speirs J, Tracy T (1990) Safety and Efficacy of combined gemfibrozil-lovastatin therapy for primary dyslipoproteinemias. J Lab Clin Med 115:603–609

Gofman JW, Lindgren FT, Elliott HM, Mantz W, Hewitt J, Strisower B, Herring B (1950) The role of lipids and lipoproteins in atherosclerosis. Science 111: 166–171

Goodman DS, Noble RP, Dell RB (1973) The effects of colestipol resin and of colestipol plus clofibrate on the turnover of plasma cholesterol in man. J Clin Invest 52:2646–2655

Griffin BA, Caslake MJ, Gaw A, Yip B, Packard CJ, Shepherd J (1992) Effects of cholestyramine and acipimox on subfractions of plasma low density lipoprotein. Studies in normolipidaemic and hypercholesterolaemic subjects. Eur J Clin Invest 22:383–390

Grundy SM, Ahrens EH, Salen G (1972) Mechanisms of action of clofibrate on cholesterol metabolism in patients with hyperlipidemia. J Lipid Res 13:531–551

Gugler R, Hartlapp J (1978) Clofibrate kinetics after single and mulitiple doses. Clin Pharmacol Ther 24:432–438

Hammond MG, Mengel MC, Warmke GL, Fisher WR (1977) Macromolecular dispersion of human plasma low density lipoprotein in hypertriglyceridemia. Metabolism 26:231–242

Harvengt C, Heller F, Desager JP (1980) Hypolipidemic and hypouricemic action of fenofibrate in various types of hyperlipoproteinemias. Artery 7:73–82

Homberg JC, Abuaf N, Helmy-Khalil S, Biour M, Poupon R, Islam S, Darnis F, Levy VG, Opolon P, Beaugrand M, Toulet J, Danan G, Benhamou JP (1985) Drug-induced hepatitis associated with anticytoplasmic organelle autoantibodies. Hepatology 5:722–727

Houlston R, Quiney J, Watts GF, Lewis B (1988) Gemfibrozil in the treatment of resistant familial hypercholesterolaemia and type III hyperlipoproteinaemia. J R Soc Med 81:274–276

Illingworth DR (1991) Fibric acid derivatives. In: Rifkind BM (ed) Drug treatment of hyperlipidemia. Dekker, New York

Illingworth DR, O'Malley JP (1990) The hypolipidemic effects of lovastatin and clofibrate alone and in combination in patients with type III hyperlipidemia. Metabolism 39:403–409

Illingworth DR, Olsen GD, Cook SF, Sexton GJ, Wendel HA, Connor WE (1982) Ciprofibrate in the therapy of type II hypercholesterolaemia. A double blind trial. Atherosclerosis 44:211–221

Janus ED, Grant S, Lintott CJ, Wardell R (1985) Apolipoprotein E phenotypes in hyperlipidaemic patients and their implications for treatment. Atherosclerosis 57:249–266

Kleinman Y, Eisenberg S, Oschry Y, Gavish D, Stein O, Stein Y (1985) Defective metabolism of hypertriglyceridemic low density lipoprotein in cultured human skin fibroblasts. Normalization with bezafibrate therapy. J Clin Invest 75: 1796–1803

Klosiewicz-Latoszek L, Nowicka G, Szostak WB, Naruszewicz M (1987) Influence of bezafibrate and colestipol on LDL-cholesterol, LDL-apolipoprotein B and HDL-cholesterol in hyperlipoproteinaemia. Atherosclerosis 63:203–209

Krauss RM (1987) Physical heterogeneity of apolipoprotein B-containing lipoproteins. In: Lippel K (ed) Proceedings of the workshop on lipoprotein heterogeneity. Government Printing Office Washington, pp 15–21 (NIH publication no 87-2646)

Krauss RM, Burke DJ (1982) Identification of multiple subclasses of plasma low density lipoproteins in normal humans. J Lipid Res 23:97–104

Langer T, Levy RI (1968) Acute muscular syndrome associated with administration of clofibrate. N Engl J Med 279:856–858

Leaf DA, Connor WE, Illingworth DR, Bacon SP, Sexton G (1989) The hypolipidemic effects of gemfibrozil in type V hyperlipidemia. A double blind crossover study. JAMA 262:3154–3160

Lehtonen A, Viikari J (1981) Effect of procetofen on serum total cholesterol, triglyceride, and high density lipoprotein-cholesterol concentrations in hyperlipoproteinemia. Int J Clin Pharmacol Ther Toxicol 19:534–538

Levy RI, Fredrickson DS, Shulman R, Bilheimer DW, Breslow JL, Stone NJ, Lux SE, Sloan HR, Krauss RM, Herbert PN (1972) Dietary and drug treatment of primary hyperlipoproteinemia. Ann Intern Med 77:267–294

Lussier-Cacan S, Bard JM, Boulet L, Nestruck AC, Grothe AM, Fruchart JC, Davignon J (1989) Lipoprotein composition changes induced by fenofibrate in dysbetalipoproteinemia type III. Atherosclerosis 78:167–182

Malmendier CL, Delcroix C (1985) Effects of fenofibrate on high and low density lipoprotein metabolism in heterozygous familial hypercholesterolemia. Atherosclerosis 55:161–169

Manninen V, Elo O, Frick MH, Haapa K, Heinonen OP, Heinsalmi P, Helo P, Huttunen JK, Kaitaniemi P, Koskinen P, Maenpaa H, Malkonen M, Manttari

M, Norola S, Pasternack A, Pikkarainen J, Romo M, Sjoblom T, Nikkila EA (1988) Lipid alterations and decline in the incidence of coronary heart disease in the Helsinki Heart Study. JAMA 260:641–651

Manttari M, Koskinen P, Manninen V, Huttunen JK, Frick MH, Nikkila EA (1990) Effect of gemfibrozil on the concentration and composition of serum lipoproteins. Atherosclerosis 81:11–17

Maxwell RE, Nawrocki JW, Uhlendorf PD (1983) Some comparative effects of gemfibrozil, clofibrate, bezafibrate, cholestyramine and compactin on sterol metabolism in rats. Atherosclersis 48:195–203

McNamara DJ, Davidson NO, Fernandez S (1980) In vitro cholesterol synthesis in freshly isolated mononuclear cells of human blood: effect of in vivo administration of clofibrate and/or cholestyramine. J Lipid Res 21:65–71

Mitropoulos KA, Venkatesan S (1985) Membrane mediated control of reductase activity. In: Preiss B (ed) Regulation of HMG-CoA reductase. Academic, Orlando, pp 1–48

Monk JP, Todd PA (1987) Bezafibrate. A review of its pharmacodynamic and pharmacokinetic properties and therapeutic use in hyperlipidaemia. Drugs 33: 539–576

Musliner TA, Krauss RM (1988) Lipoprotein subspecies and risk of coronary disease. Clin Chem 34(8B):B78–B83

Needham L, Finnegan I, Housley MD (1985) Adenylate cyclase and a fatty acid spin probe detect changes in plasma membrane lipid phase separations induced by dietary manipulation of the cholesterol: phospholipid ratio. FEBS Lett 183: 81–86

Nestel PJ, Hunt D, Wahlqvist ML (1980) Clofibrate raises plasma apoprotein A-I and HDL-cholesterol concentrations. Atherosclerosis 37:625–629

Newton RS, Krause BR (1986) Mechanisms of action of gemfibrozil: comparison of studies in the rat to clinical efficacy. In: Fears R, Levy RI, Shepherd J, Packard CJ, Miller NE (eds) Pharmacological control of hyperlipidaemia. Prous, Barcelona, pp 171–186

O'Brien BJ (1991) Cholesterol and coronary heart disease: consensus or controversy? Office of Health Economics, London (Studies of Current Health Problems No 98)

Okerholm RA, Keeley FJ, Peterson FE, Glazko AJ (1976) The metabolism of gemfibrozil. Proc R Soc Med 69:11–14

Olsson AG, Oro L (1982) Dose–response study of the effect of ciprofibrate on serum lipoprotein concentrations in hyperlipoproteinaemia. Atherosclerosis 42: 229–243

Olsson AG, Lang PD, Vollmar J (1985) Effect of bezafibrate during 4.5 years of treatment of hyperlipoproteinaemia. Atherosclerosis 55:195–203

Packard CJ, Clegg RJ, Dominiczak MH, Lorimer AR, Shepherd J (1986) Effects of bezafibrate on apolipoprotein B metabolism in type III hyperlipoproteinemic subjects. J Lipid Res 27:930–938

Palmer RH (1987) Effects of fibric acid derivatives on biliary lipid composition. Am J Med 83 Suppl 5B:37–43

Pierce LR, Wysowski DK, Gross TP (1990) Myopathy and rhabdomyolysis associated with lovastatin–gemfibrozil combination therapy. JAMA 264:71–75

Rabkin SW, Hayden M, Frohlich J (1988) Comparison of gemfibrozil and clofibrate on serum lipids in familial combined hyperlipidemia. A randomized placebo-controlled, double-blind, crossover clinical trial. Atherosclerosis 73:233–240

Rouffy J, Chanu B, Bakir F, Djian F, Goy-Loeper J (1985) Comparative evaluation of the effects of cipofibrate and fenofibrate on lipids, lipoproteins and apoproteins A and B. Atherosclerosis 54:273–281

Saku K, Gartside PS, Hynd BA (1985) Mechanism of gemfibrozil action on lipoprotein metabolism. J Clin Invest 75:1702–1712

Schifferdecker E, Rosak C, Schoffling K (1984) Long term treatment with the lipid lowering agent ciprofibrate. Inn Med 11:107–112

Schneider A, Stange EF, Ditschuneit HH, Ditschuneit H (1985) Fenofibrate treatment inhibits HMG CoA reductase activity in mononuclear cells from hyperlipoproteinemic patients. Atherosclerosis 56:257–262
Shen MS, Krauss RM, Lindgren FT, Forte TM (1981) Heterogeneity of serum low density lipoproteins in normal human subjects. J Lipid Res 22:236–244
Shepherd J, Packard CJ, Stewart JM, Atmeh RF, Clark RS, Boag DE, Carr K, Lorimer AR, Ballantyne D, Morgan HG, Lawrite TDV (1984) Apolipoprotein A and B (S_f 100–400) metabolism during bezafibrate therapy in hypertriglyceridemic subjects. J Clin Invest 74:2164–2177
Shepherd J, Caslake MJ, Lorimer AR, Vallance BD, Packard CJ (1985) Fenofibrate reduces low density lipoprotein catabolism in hypertriglyceridemic subjects. Arteriosclerosis 5:162–165
Shepherd J, Griffin B, Caslake M, Gaw A, Packard C (1991) The influence of fibrates on lipoprotein metabolism. Atheroscler Rev 22:163–169
Simpson HS, Williamson CM, Olivecrona T, Pringle S, Maclean J, Lorimer AR, Bonnefous F, Bogaievsky Y, Packard CJ, Shepherd J (1990) Postprandial lipemia, fenofibrate and coronary artery disease. Atherosclerosis 85:193–202
Stahlberg D, Reihner E, Ewerth S, Einarsson K, Angelin B (1991) Effects of bezafibrate on hepatic cholesterol metabolism. Eur J Clin Pharmacol 40 Suppl 1:S33–S36
Stange EF, Fruhholz M, Osenbrugge M, Reimann F, Ditschuneit H (1991) Bezafibrate fails to directly modulate HMG-CoA reductase or LDL catabolism in human mononuclear cells. Eur J Clin Pharmacol 40 Suppl 1:S37–S40
Steinmetz J, Morin C, Panek E, Siest G, Drouin P (1981) Biological variations in hyperlipidemic children and adolescents treated with fenofibrate. Clin Chim Acta 112:43–53
Stewart JM, Packard CJ, Lorimer AR, Boag DE, Shepherd J (1982) Effects of bezafibrate on receptor mediated and receptor independent low density catabolism in type II hyperlipoproteinemic subjects. Atherosclerosis 44:355–364
Thompson GR (1989) Drug treatment of hyperlipidaemia. In: A handbook of hyperlipidaemia. Current Science, London, pp 177–194
Thorp JM (1963) An experimental approach to the problem of disordered lipid metabolism. J Atheroscler Res 3:351–360
Todd PA, Ward A (1988) Gemfibrozil. A review of its pharmacodynamic and pharmacokinetic properties and therapeutic use in dyslipidaemia. Drugs 36: 314–339
Vega GL, Grundy SM (1989) Comparison of lovastatin and gemfibrozil in normolipidemic patients with hypoalphalipoproteinemia. JAMA 262:3148–3153
Vessby B, Lithell H (1990) Interruption of long-term lipid-lowering treatment with bezafibrate in hypertriglyceridaemic patients. Atherosclerosis 82:137–143
Wheeler KAH, West RJ, Lloyd JK, Barley J (1985) Double blind trial of bezafibrate in familial hypercholesterolaemia. Arch Dis Child 60:34–37
Wysowski DK, Kennedy DL, Gross TP (1990) Prescribed use of cholesterol lowering drugs in the United States, 1978 through 1988. JAMA 263:2185–2188
Zilversmit DB (1979) Atherogenesis: a postprandial phenomenon. Circulation 60: 473–485

CHAPTER 13
Nicotinic Acid and Derivatives

A.G. Olsson

A. Introduction

Each year about 50 papers on nicotinic acid (NA) or niacin (pyridine-3-carboxylic acid; Fig. 1) are published in scientific journals. About half of them concern nicotinic acid as a vitamin and half concern it as a drug. NA, therefore, is still a Janus figure (Gey 1971), i.e., like the Roman god Janus, it has two faces: It plays the role both of an essential element in food as a vitamin and as a drug against disease. In this review, the term niacin will be used when dealing with its role as a vitamin, and NA when dealing with its role as a drug.

The pyridine nucleus with a carboxyl group in the 3-position is a chemical structure in many calcium channel-blocking agents such as felodipine, isradipine, nicardipine, nifedipine, and nimodipine. These substances, however, lie outside the scope of this chapter.

Even if NA was identified chemically as long ago as in 1867, it was not until 1937 that Elvehjem (1952) demonstrated that niacin could cure pellagra in the dog. Eighteen years later, Altschul et al. (1955) reported that plasma cholesterol decreased upon the administration of NA. This was confirmed in a number of studies that soon followed, in which a decrease in the concentration of plasma triglycerides was also reported. In 1962, Carlson and Orö reported that NA inhibits lipolysis in adipose tissue, thereby providing a tentative explanation for the plasma lipid-lowering effect.

Since then, and in spite of the introduction of several other highly effective plasma lipid-lowering drugs, NA and its derivatives play an important role in the efficient regulation of plasma lipid. It has been referred to as the "broad spectrum" plasma lipid-lowering drug (Carlson 1984). The recommendation of NA as a first-line drug for treatment of hyperlipidemia in many recommendations has recently drawn more attention to it. It is still the only drug so far to have been shown to have an effect on total mortality in long-term treatment against coronary heart disease (CHD).

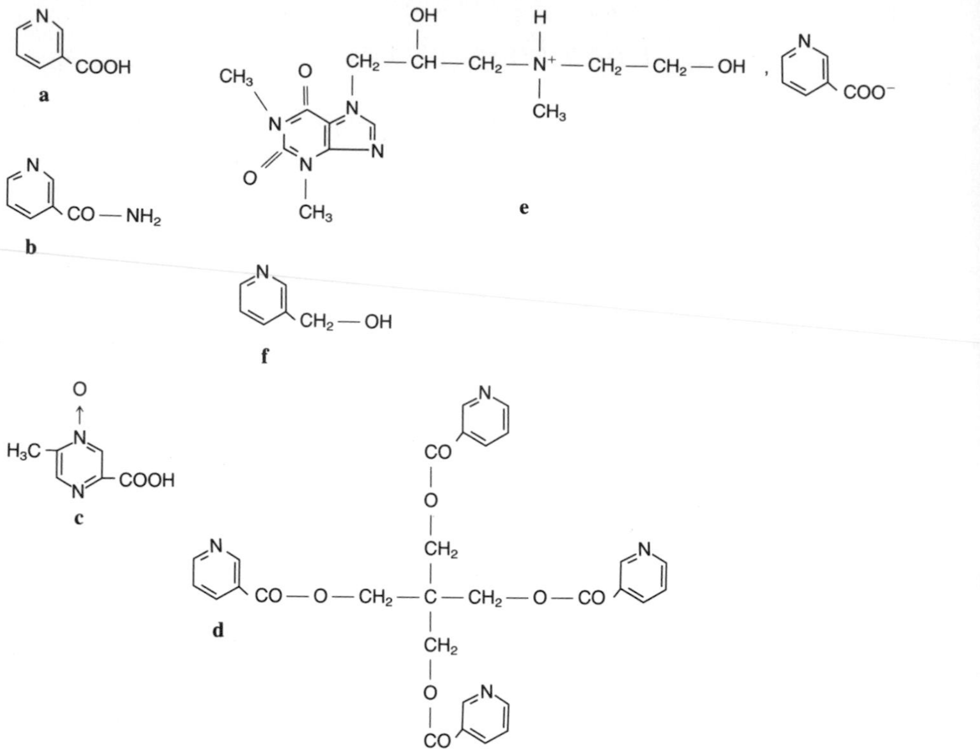

Fig. 1a–f. Chemical structure of nicotinic acid and some of its dervatives, mentioned in this chapter. **a** Nicotinic acid; **b** nicotinamide; **c** acipimox; **d** niceritrol; **e** xantinol nicotinate; **f** β-pyridyl carbinol

B. The Vitamin

I. Biochemistry and Requirements

The term niacin includes in practice two vitamers, NA and nicotinamide (Fig. 1). Much of the mammalian metabolism of niacin has been brought together in a review by HENDERSON (1983). In one sense it is not a true vitamin, since it can be synthesized by the body from the essential amino acid tryptophan. In assessing the requirement of niacin, therefore, the intake of tryptophan must be taken into account. A dietary intake of 60 mg tryptophan is equivalent to 1 mg niacin. Thus, the requirement of this vitamin is often expressed in equivalent milligrams of niacin, taking into consideration the simultaneous tryptophan intake. The recommended daily allowances for niacin is the equivalent of 15–20 mg niacin for men and 13–15 mg for women (ROSENBERG 1991).

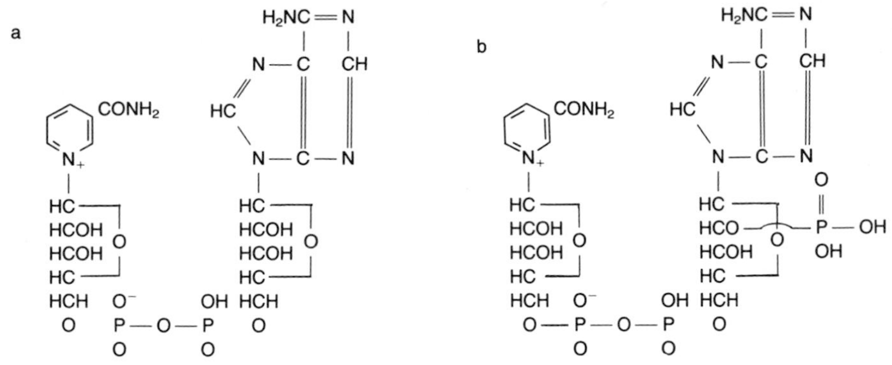

Fig. 2a,b. Structures of niacin-containing coenzymes. **a** NAD; **b** NADP

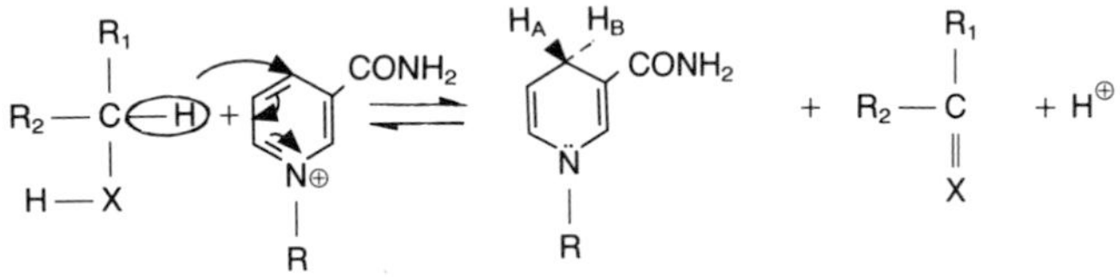

Fig. 3. Mechanism of action of niacin as coenzyme

II. Mechanism of Action

Nicotinamide participates in cellular oxidation systems as a hydrogen-transport agent, i.e., in fundamentally important intracellular enzyme systems.

Niacin is incorporated into the coenzymes nicotinamide adenine dinucleotide (NAD) and nicotinamide dinucleotide phosphate (NADP) (Fig. 2). It is niacin itself that is reactive agent in its function both as a coenzyme and as a drug, rather than a metabolite. These coenzymes are essential in many oxidation-reduction reactions, reactions which are ubiquitous in the body. Many of these oxidoreductases function as dehydrogenases and catalyze such diverse reactions as the conversion of alcohols to aldehydes or ketones, hemiacetals to lactones, aldehydes to acids, and certain amino acids to keto acids (McCormick 1988). The common operation mechanism of niacin is shown in Fig. 3. NAD, however, also functions as a non-redox coenzyme (McCormick 1988). This activity is found in mitochondria and is bound to ribosomes. It is also bound in nuclei, where it affects the operation of DNA (Ogata et al. 1981). This non-redox function of NAD probably accounts for the rapid turnover of NAD in human cells.

III. Deficiency

The symptomatology of pellagra, recognized in 1908, comprises a chronic wasting disease typically associated with dermatitis, dementia, and diarrhea. It was mostly seen in populations with a high and monotonous intake of maize. The cutaneous symptoms are due to photosensitivity and are therefore present in areas exposed to sunlight. Mental changes are lassitude, fatigue, insomnia, and apathy, eventually leading to psychosis. Diarrhea results from mucus membranes being affected (Wilson 1991). Endemic pellagra disappeared coincidentally with nutritional education and with the widespread supplementation of grain cereals with niacin. Pellagra can also manifest itself as secondary symptoms of two diseases that affect tryptophan metabolism: the carcinoid syndrome and Hartnup disease. In the former condition, up to 60% of tryptophan is catabolized by an ordinarily minor pathway of metabolism (Kaplan 1991), therefore leaving inadequate substrate for niacin synthesis. Hartnup disease is an inherited disorder in which tryptophan is poorly absorbed from the diet (Coe and Kathpalia 1991).

Small amounts of dietary supplementation of niacin (10 mg/day) in addition to adequate amounts of tryptophan suffices to cure pellagra. Larger amounts of the vitamin (40–200 mg/day) may be required in the carcinoid syndrome and in Hartnup disease (Wilson 1991).

C. The Drug

I. Pharmacology

The pharmacology and pharmacokinetics of NA were already extensively studied in animals and man during the 1960s (Fumagalli 1971). The absorption, distribution, metabolism, and excretion of the drug is readily documented in several handbooks (e.g., Gey and Carlson 1971). Briefly, NA, when given orally, is readily absorbed from the intestine, as documented by its sharp rise in blood concentrations and output of urinary metabolites after administration. Of particular importance for the understanding of the different effects of various NA derivatives is the length of time during which the drug circulates in blood and is able to exert its pharmacological effects. For 1 g plain NA, peak concentration is already reached after 1 h; 2 h after intake, the concentration is only half the peak concentration, and after 4 h it has almost returned to initial (close to zero) concentrations (Carlson et al. 1968). This is in sharp contrast to many NA derivatives such as β-pyridyl carbinol (Ronicol Retard) (Carlson and Orö 1969; Fig. 1) and acipimox (5-methyl-pyrazin-carboxylic acid-4-oxide, Olbetam; Fig. 1). In the latter case, when 400 mg was given to man, it also showed a rapid increase in blood concentrations, reaching a peak 2 h after administration. Elimination showed a two-phase pattern, with a first, rapid phase with an average half-life of 2 h and a second, slower phase with a half-life of 12–24 h (Musatti et al. 1981). A slower intravascular metabolism of NA seems to be of importance to avoid an overshoot of free fatty acid (FFA) concentrations. This in turn probably influences other metabolic consequences such as insulin resistance, which may be of importance for the cardiovascular risk (see below).

II. Effects in Lipid Metabolism

1. Free Fatty Acid Metabolism

a) Extracellular Metabolism

α) Inhibition of Fat-Mobilizing Lipolysis. The effect of NA on FFA metabolism was elucidated about 30 years ago by Carlson and his group. In a series of publications it was demonstrated that NA inhibits fat-mobilizing lipolysis in adipose tissue. In brief, 200 mg oral NA was shown to promptly lower the arterial FFA levels in man (Carlson and Orö 1962). This was followed by an increased FFA concentration occurring about 60 min after ingestion (overshoot). Also, NA completely blocked the stimulation of FFA mobilization induced by norepinephrine (Carlson and Orö 1962). FFA levels, however, never reached zero. Turnover studies demonstrated that the production of FFA was diminished. Therefore, it was anticipated that the effect was an inhibition of lipolysis in adipose tissue by hormone-sensitive triglyceride lipase, which was also confirmed in in vitro studies of adipose tissue (Carlson 1963).

A difference of great importance in the pattern of lipolysis inhibition between NA and acipimox was demonstrated by Fuccella et al. in 1980. NA and acipimox were given to healthy volunteers in three different dosages: 100, 200, and 400 mg NA and 20, 40, and 80 mg acipimox. For NA, minimum FFA levels were reached between 40 and 90 min, followed by a FFA overshoot that lasted for the entire 5 h of the experiment (Fig. 4). After acipimox, the FFA levels were most markedly depressed between 90 and 120 min, but in contrast to NA, they remained below the basal level throughout the 5-hour observation period without any overshoot. The antilipolytic activity of acipimox was 20 times that of NA and the effect of acipimox lasted longer. The study points to an important difference between NA and acipimox in metabolic behaviour. A slower elimination of acipimox from plasma than of NA is probably the reason for the lack of a rebound effect with this drug. This lack of rebound, which lasts for many hours, probably prevents the high FFA levels over 24 h that are seen with NA treatment. As FFA and glucose metabolism are closely related in the glucose–fatty acid cycle (Randle et al. 1963), this property of acipimox may be the mechanism by which this drug does not hinder glucose metabolism or induce insulin resistance (Tornvall and Walldius 1991).

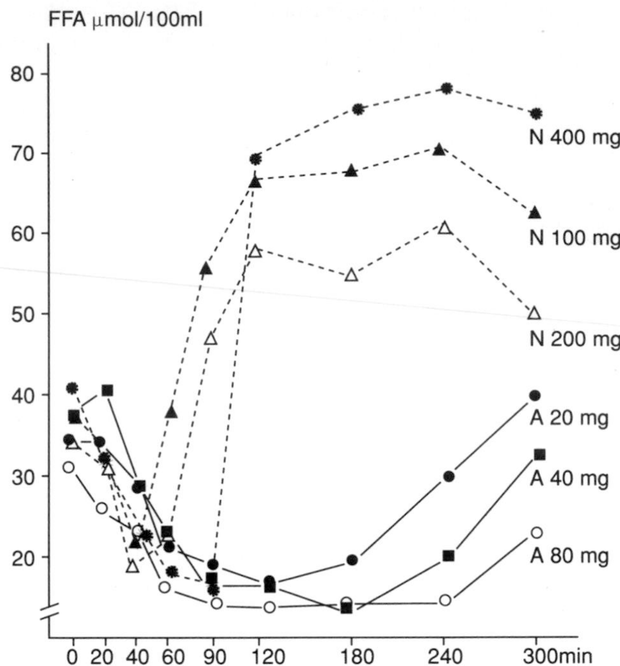

Fig. 4. Mean plasma free fatty acid (*FFA*) levels after three oral doses of acipimox (*A*) and sodium nicotinate (*N*) (FUCCELLA et al. 1980)

β) Fatty Acid Incorporation into Adipose Tissue. NA treatment also augments fatty acid incorporation into adipose tissue (FIAT) in patients with hypertriglyceridemia (CARLSON et al. 1973). These patients have a decreased FIAT (CARLSON and WALLDIUS 1976), which is normalized by NA treatment. Both in vivo and in vitro, an inverse relationship was demonstrated between lipolysis and the FIAT process (WALLDIUS 1976). It was suggested that the rate of lipolysis moderates adipose tissue capacity to incorporate exogenous fatty acids. It was also proposed that in addition to its known antilipolytic effects, NA also promotes FIAT, which may account in part for the plasma triglyceride-lowering effect of this drug (WALLDIUS 1976). Recent evidence indicates that the increased FIAT activity during long-term NA treatment continues, despite an increased rate of lipolysis in the fasting state (WALLDIUS, personal communication).

b) Intracellular Metabolism

Lipolysis in adipose tissue is regulated by the action of hormone-sensitive lipase (HSL; Fig. 5). HSL is present in an active, phosphorylated form and an inactive form. Phosphorylation occurs through the activity of the enzyme

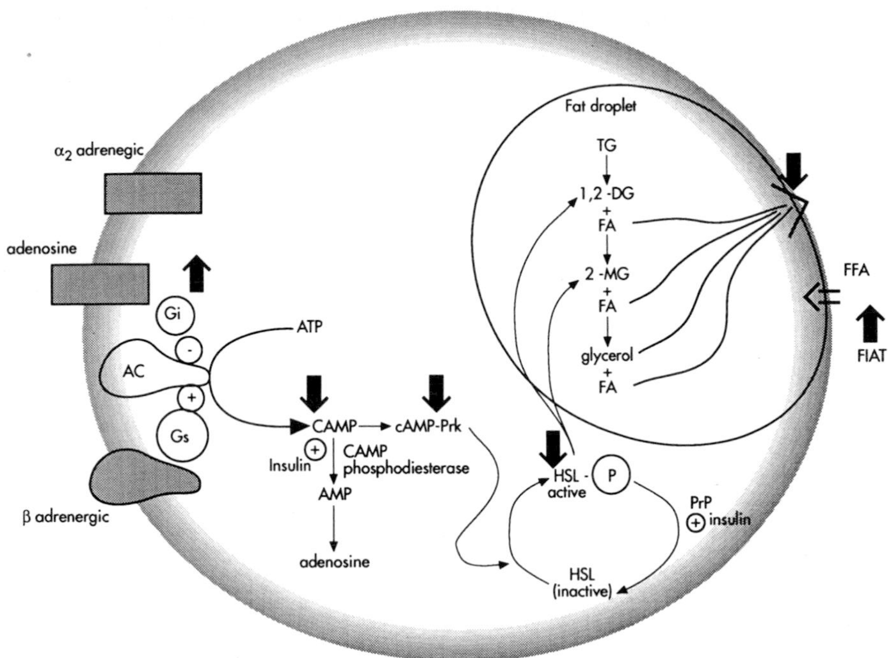

Fig. 5. Schematic representation of the mechanism by which nicotinic acid (NA) exerts its antilipolytic effect in an adipocyte. *Broad arrows* indicate NA effects. *AC*, adenylate cyclase; *cAMP*, cyclic adenosine monophosphate; *TG*, triglyceride; *DG* diglyceride; *MG*, monoglyceride; *(F)FA*, (free) fatty acid(s); *FIAT*, FA incorporation into adipose tissue; *HSL*, hormone-sensitive lipase; *Prk*, protein kinase; *PrP*, protein phosphatase; *TAG*, triacyl glycerol. A specific "NA receptor" has not been identified

protein kinase, which in turn is activated through cyclic adenosine monophosphate (cAMP). cAMP is formed by the action of adenylate cyclase and its synthesis is mediated from cell surface receptors (such as the A_1 adenosine receptor) through a protein (G) of which stimulatory (G_s) and inhibitory (Gd2i) forms exist. Three subtypes of G_i have been identified in adipose tissue, G_{i1}, G_{i2}, and G_{i3}.

The antilipolytic effect of NA was demonstrated by Butcher et al. in 1968 to be an inhibition of the accumulation of cAMP in adipose tissue by the inhibition of adenylate cyclase (Aktories and Jacobs 1982; Fredholm 1981; Fig. 5). The exact intracellular events finally leading to this effect are not fully elucidated. Recent evidence suggests that G_{i2} is responsible for the inhibition of adenylate cyclase (McKenzie and Milligan 1990; Senogles et al. 1990). NA inhibition of adenylate cyclase seems to work through a G_i-coupled receptor (Green et al. 1991), but this does not seem to be the A_1 adenosine receptor. Furthermore, at least in rat adipocytes, NA does not downregulate G_i, in contrast to other inhibitors of adenylate cyclase such as prostaglandin E1 or phenylisopropyl adenosin (PIA), the latter of which exerts its action through the A_1 adenosine receptor (Fig. 5).

2. Plasma Lipoprotein Metabolism

When NA is given in large doses such as 4 g daily to patients with hyperlipidaemia, all plasma lipoprotein classes are affected (Carlson 1984). NA has therefore been called the "broad spectrum" plasma lipid-lowering drug. The general effects of NA in hyperlipoproteinaemia are given in Table 1. Thus, NA lowers elevated plasma concentrations of very low density (VLDL), intermediate-density (IDL), and low-density (LDL) lipoproteins, as well as chylomicrons in type V hyperlipoproteinaemia. It increases plasma high-density lipoprotein (HDL) concentrations and also LDL in hyperlipoproteinaemias characterized by low concentrations such as type V hyperlipoproteinaemia (Carlson et al. 1974).

The effect on plasma lipoprotein concentrations represents a dose–response relationship for all lipoproteins up to high dosages (up to 16 g daily).

a) Very Low-Density Lipoproteins and their Subfractions

The size of the changes in the plasma lipoprotein concentrations induced by NA are given in a recent study by Tornvall and Walldius (1991). The effects of 3 g NA in comparison to 0.75 g daily of the NA derivative acipimox on plasma lipoprotein levels in 31 patients with types IV and IIB hyperlipoproteinaemia were studied in an open, randomized, cross-over study. Treatment periods were 6 weeks. NA and acipimox were about equally effective in decreasing atherogenic plasma lipoprotein concentrations. Plasma triglycerides and VLDL lipid levels decreased with both treatments by about 40% (Table 2). Plasma cholesterol decreased significantly by 14% and 11% after NA and acipimox, respectively. In this study no significant effects by any of the two treatments were achieved on LDL cholesterol levels (see below). Plasma HDL cholesterol increased by 18% and 12% after NA and

Table 1. Principal effects of nicotinic acid on plasma lipoprotein concentrations

Types of hyperlipoproteinemia	Lowered lipoprotein class	Raised lipoprotein class
IIA	LDL	
IIB	VLDL+LDL	HDL_2
III	β-VLDL, IDL	HDL_2
IV	VLDL	LDL[a], HDL_2
V	chylomicrons, VLDL	LDL[a], HDL_2

LDL, low-density lipoprotein; VLDL, very low-density lipoprotein; IDL intermediatedensity lipoprotein; HDL, high-density lipoprotein.
[a] If below about 3.5 mmol/l before treatment (Carlson et al. 1974).

Table 2. Typical effects of nicotinic acid treatment on serum lipoproteins in type IV hyperlipoproteinemia (Tornvall and Walldius 1991)

	Serum Cholesterol (mmol/l)	Serum TG (mmol/l)	VLDL TG (mmol/l)	LDL Cholesterol (mmol/l)	HDL Cholesterol (mmol/l)
Before NA	7.98	5.02	3.96	4.49	1.06
During NA	6.90	3.14	2.39	4.36	1.25
Percent change	−14	−37	−40	−3	+18
p	<0.001	<0.001	<0.001	n.s.	<0.001

NA, nicotinic acid; TG, triglyceride; VLDL, very low density lipoprotein; LDL, low-density lipoprotein; HDL, high-density lipoprotein; n.s., not significant.

acipimox, respectively. Thus, the effects of the two NA drugs on plasma lipoprotein concentrations in hypertriglyceridaemia were equal in this study.

Not only do VLDL concentrations decrease during NA treatment. Distinct changes also occur regarding the composition of VLDL subfractions. Eight patients with hypertriglyceridaemia were studied during NA treatment with regard to compositional effects of large ($S_f > 100$) and small (S_f 100–20) VLDL subfractions (Tornvall et al. 1990). VLDL of hypertriglyceridaemia typically have large VLDL particles that are cleared from the circulation as remnant particles and are to a lesser extent metabolized intravascularly to LDL (Shepherd and Packard 1986). If the small VLDL particle fraction was lipolyzed in vitro, a clear difference was seen from normolipidaemic small VLDL, hypertriglyceridaemic small VLDL giving rise to particles that were less dense and richer in cholesterol than normolipidaemic small VLDL (Fig. 6). It thus seems that small VLDL from hypertriglyceridaemic patients cannot be converted adequately to LDL. Also, small hyperlipidaemic VLDL were enriched in cholesteryl esters. NA treatment was able to normalize this pattern of conversion block of small VLDL to LDL in vitro (Fig. 6), and the cholesteryl ester content of small VLDL was decreased and normalized. Another interesting finding was that NA treatment gave rise to particles that were larger ($S_f > 100$) and richer in triglyceride.

b) Low-Density Lipoprotein Concentrations and Subfractions

Elevated plasma LDL cholesterol concentrations are decreased by NA treatment (Olsson et al. 1986). However, in hyperlipidaemic patients with normal or low LDL cholesterol levels (types III, IV, and V hyperlipoproteinaemia), LDL cholesterol often increases during treatment (Carlson et al. 1977). The LDL rise is usually not excessive, but could occasionally give rise to the lipoprotein profile of type II hyperlipoproteinaemia. Thus, the effect of NA treatment on LDL cholesterol concentration cannot easily be

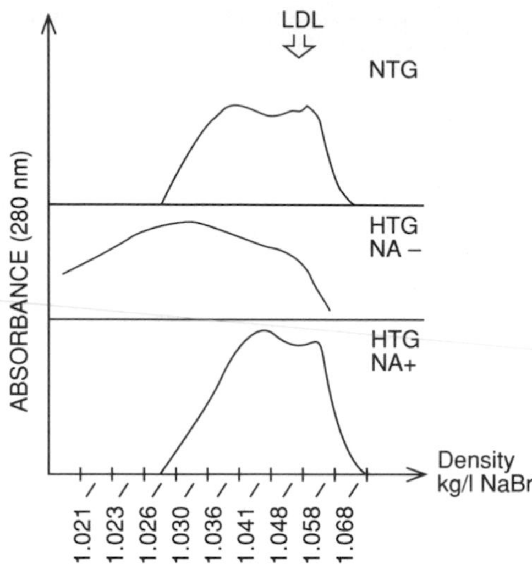

Fig. 6. Effect of nicotinic acid (*NA*) on density distribution of lipolyzed very low density lipoproteins (VLDL) (Tornvall et al. 1990). *Upper panel*, normolipidaemic small VLDL (*arrow* represents peak of native, *NTG*, low-density lipoprotein, LDL, distribution); *middle panel*, hypertriglyceridaemic (*HTG*) small VLDL before NA; *lower panel*, hypertriglyceridaemic small VLDL after NA treatment. The "normalization" of the lipolyzed particles after NA in hyperlipidaemic plasma is evident

expressed in percentages. The baseline concentrations must be taken into account. However, in patients with moderate LDL cholesterol elevations, decreases of about 20% are often noted. In this type of hyperlipoproteinemia, a dose–response relationship is also seen.

c) Apolipoprotein B, C, and E

Carlson and Wahlberg (1978) noted that as NA treatment lowered the triglyceride content of all VLDL subfractions in a patient with severe type V hyperlipoproteinaemia, the relative proportion of apolipoprotein C-II to C-III-1 increased. The ratio of C-II to triglycerides also increased. As apolipoprotein C-II activates plasma lipoprotein lipase activity, it was suggested that this effect of NA might facilitate the action of lipoprotein lipase and thereby the elimination of triglyceride-rich lipoproteins.

Yovos et al. (1982) used gel isoelectric focusing of VLDL apolipoproteins in hyperlipidaemic patients and studied the effect of NA treatment for 6 months. Effects on lipids in plasma VLDL, LDL, and HDL confirmed previous experience of the effect of NA treatment. The results on apolipoprotein C-II and C-III are given in Table 3. Total VLDL protein decreased, but significantly so only after 6 months. The effect on apolipoprotein C subspecies proportions did not change until after 6 months either, when

Table 3. Effects of nicotinic acid treatment on very low density lipoprotein, total protein, and apolipoproteins C-II and C-III in types IIB and IV hyperlipoproteinaemia (YOVOS et al. 1982)

HLP type	Month of treatment	VLDL protein (mg/dl)	ApoC-II (% of total C area)	ApoC-III (% of total C area)	ApoC-II/ApoC-III
IIB	0	44.7 ± 8.2	32.0 ± 3.5	68.1 ± 3.5	0.50 ± 0.1
	3	28.4 ± 3.0	34.9 ± 2.0	65.1 ± 2.0	0.55 ± 0.05
	6	27.1 ± 3.9[a]	47.7 ± 5.1[a]	52.3 ± 5.1[a]	1.02 ± 0.2[a]
IV	0	46.3 ± 7.1	37.5 ± 2.4	62.5 ± 2.4	0.62 ± 0.07
	3	38.6 ± 4.8	36.0 ± 3.9	64.0 ± 3.9	0.60 ± 0.1
	6	30.6 ± 4.9[a]	45.2 ± 3.6[a]	54.8 ± 3.6[a]	0.88 ± 0.13[a]

Values indicated are the mean ± standard error of mean.
HLP, hyperlipoproteinaemia; VLDL, very low density lipoprotein.
[a] Significant difference compared to baseline values.

apolipoprotein C-II showed a significant increase and apolipoprotein C-III showed a significant decrease in both types IIB and IV hyperlipoproteinaemia. Thus, in both types, the apolipoprotein C-II/C-III ratio rose significantly for both types. As apolipoprotein C-II is mainly responsible for the activation of lipoprotein lipase (LAROSA et al. 1970) and apolipoprotein C-III is responsible for its inhibition (BROWN and BAGINSKY 1972), the redistribution of these apolipoproteins could increase the catabolism of triglyceride-rich lipoproteins and increase HDL concentrations. Thus, NA treatment should favour an increased catabolism of triglyceride-rich lipoproteins.

WAHLBERG et al. (1988) studied the effect of NA treatment on the plasma concentrations of apolipoproteins B, C-I C-II, C-III, and E in 24 patients with various types of hyperlipoproteinaemia. Treatment caused a decrease in all types of plasma apolipoprotein C and in apolipoprotein E. The authors concluded that these effects were probably secondary to decreases in VLDL particle concentration induced by NA. In this study, no effort was made to examine the redistribution of the various types of apolipoprotein C between apolipoprotein B-containing and high-density lipoproteins, nor was the relative amount of apolipoprotein C-II after NA treatment given. The data given in the paper do not indicate an increased proportion of apolipoprotein C-II after NA; NA treatment did not normalize apolipoprotein B concentrations, although they decreased significantly. The authors suggest therefore that adjunct therapy may be required in patients with high apolipoprotein B concentrations.

d) Kinetic Studies of Apolipoprotein B-Containing Lipoproteins

GRUNDY et al. (1981) studied the VLDL triglyceride kinetics in hyperlipoproteinaemic patients, of whom most had hypertriglyceridaemia before and

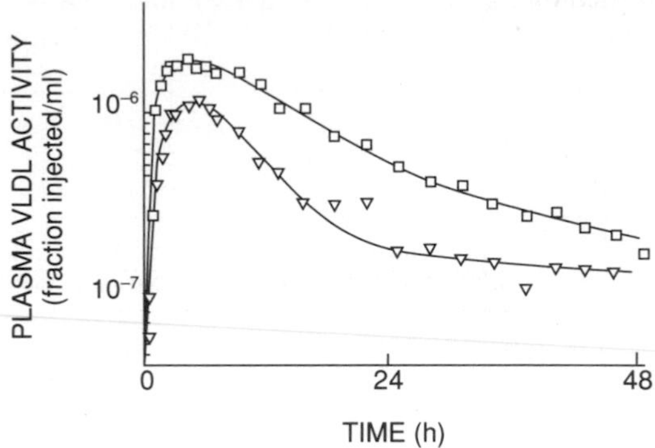

Fig. 7. Typical very low density lipoprotein (VLDL) activity curve for one patient before and during *nicotinic acid* therapy (GRUNDY et al. 1981). △, Nicotinic acid; □, control

during NA treatment. While VLDL triglyceride during 1 month of NA treatment decreased by 36%, the production rate decreased by 21% (Fig. 7). The reduced synthesis of VLDL triglycerides was mostly due to a reduction in particle size, the relative particle size ratio between NA-treated particles and controls being 0.79 ± 0.2, $p < 0.05$, and not to a reduced number of lipoprotein particles, NA/controls being 0.97 ± 0.28 (not significant). In spite of a reduction in particle size, NA caused a decrease in the cholesterol to triglyceride ratio in VLDL. This suggests the formation of small cholesterol-poor VLDL by NA. In addition, the patients also showed an increase of 21% in their fractional catabolic rate. This was probably because the smaller VLDL particles could pass through the delipidation chain more rapidly than larger ones.

GAW et al. (1990) studied the aplipoprotein B metabolism in six patients with moderate hypercholesterolaemia before and after acipimox treatment (1250 mg daily). Fractional catabolic rates and synthetic rates of $VLDL_1$ (S_f 60–400) and $VLDL_2$ (S_f 20–60) as well as mean LDL subfraction distribution were determined before and after treatment.

Acipimox reduced plasma cholesterol and triglycerides by 25% and 33%, respectively. The effect of acipimox treatment on VLDL kinetics is given in Tables 4 and 5. Acipimox decreased the circulating pool of both $VLDL_1$ and $VLDL_2$ apolipoprotein B by one-third. The mechanism of this change differed in fact between the two VLDL subfractions. The synthetic rate of $VLDL_1$ was only 50% of that before treatment with acipimox, while the fractional catabolic rate decreased. In the $VLDL_2$ fraction, the decrease in synthetic rate caused by treatment was less pronounced than in $VLDL_1$. The fractional catabolic rate, however, was increased. This latter finding is in line with that of GRUNDY et al. (1981), who also found an increased fractional catabolic rate and ascribed this to a more rapid metabolism of

Table 4. Effect of acipimox treatment on apolipoprotein B metabolism (GAW et al. 1990)

	$VLDL_1$ (S_f 60–400)			$VLDL_2$ (S_f 20–60)		
	Pool (mg)	FCR (pool/h)	SR (mg/h)	Pool (mg)	FCR (pool/h)	SR (mg/h)
Pretreatment	151	0.512	57.5	413	0.135	55.2
Posttreatment	103	0.264	28.1	278	0.172	44.2

VLDL, very low density lipoprotein; FCR, fractional catabolic rate; SR, synthetic rate.

Table 5. Mean low-density lipoprotein subfraction distribution (GAW et al. 1990)

	LDL-I (mg lipoprotein/dl)	LDL-II (mg lipoprotein/dl)	LDL-III (mg lipoprotein/dl)
Pretreatment	48 (40)	212 (122)	168 (158)
Posttreatment	87 (58)	223 (53)	87 (60)

LDL, low-density lipoprotein.
Values in parentheses indicate standard deviation.

smaller particles than larger through the delipidation process. The mean LDL subfraction distribution changes (Table 5) showed a redistribution in particle density from the small dense LDL-III to the larger more buoyant LDL-I with acipimox therapy. There is evidence that the LDL subclass pattern associated with a preponderance of small, dense LDL is connected with an increased risk for myocardial infarction (AUSTIN et al. 1988). The LDL particle distribution shift induced by acipimox may, therefore, lessen the risk of atherosclerosis.

The effect of NA treatment on the turnover of LDL was studied in two patients with type II hyperlipoproteinemia (LANGER and LEVY 1971). The reduction of circulating LDL concentrations was accompanied by a reduction in the synthetic rate of LDL. It was suggested that the fall in LDL synthesis may result from a reduction in hepatic VLDL synthesis secondary to the inhibition of FFA mobilization.

e) High-Density Lipoprotein Concentrations and Subfractions

Plasma HDL are a heterogeneous plasma lipoprotein class that can be further divided into subfractions according to various physicochemical characteristics. One early subclassification used density differences and two major subfractions were defined: HDL_2 ($d = 1.063–1.125\,kg\,l^{-1}$) and HDL_3 ($d = 1.125–1.210\,kg\,l^{-1}$).

Patients with CHD typically have lower levels of both plasma HDL_2 and HDL_3, the largest and most significant decrease in various studies often being that of HDL_2 (e.g., Wallentin and Sundin 1985). Most authors have claimed that the protective effect against atherosclerosis is more related to the HDL_2 than to the HDL_3 subfraction.

Shepherd et al. (1979) described the effect of 3 g NA daily on HDL_2 and HDL_3 and also examined the influence of the drug on the metabolism of the major HDL apolipoproteins, apolipoprotein A-I (apoA-I) and apolipoprotein A-II (apoA-II). Cholesterol in VLDL and LDL decreased as expected, and HDL cholesterol increased by 23%. Figure 8 shows the results before and after NA treatment on HDL rate zonal ultracentrifugation profiles in one male and one female subject. Before treatment, the charac-

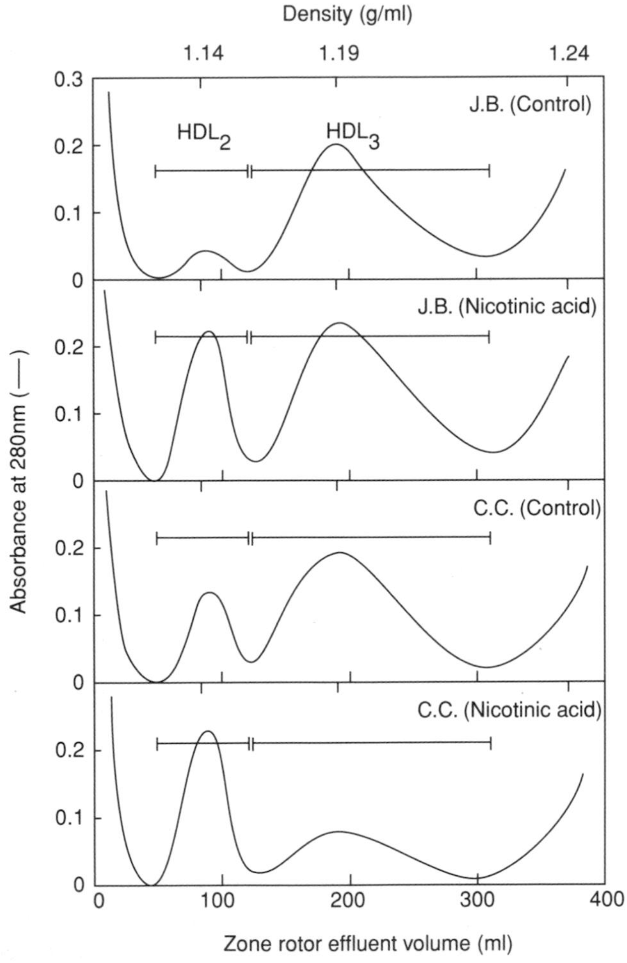

Fig. 8. High-density lipoprotein (*HDL*) rate zonal ultracentrifugation profiles of one male (*J.B.*) and one female (*C.C.*) before and during nicotinic acid treatment (Shepherd et al. 1979)

teristic sex-related difference (PATSCH et al. 1974) in the HDL_2 to HDL_3 ratio is evident, the male ratio being substantially lower than that of the female. In the male subject, NA produced a large increase in HDL_2, whereas in the female the increase in HDL_2 was also accompanied by an evident reduction in HDL_3. Treatment thus gave rise to a consistent and substantial increase in the HDL_2 to HDL_3 ratio. Concomitantly, apoA-I rose by 7% and apoA-II fell by 14%. Consequently, the apoA-I to apoA-II ratio increased markedly. Further analysis showed that the NA-induced increase in plasma HDL_2 and fall in HDL_3 was accompanied by (a) a rise in the plasma apoA-I concentration, (b) an increase in plasma HDL_2-associated apoA-I with a reciprocal change in HDL_3, (c) a decrease in the plasma apoA-II level, (d) a fall in the amount of apoA-II associated with HDL_3 in the plasma, and (e) a significant decrease in the apoA-I to apoA-II ratio in HDL_3. Studies on apoA metabolism demonstrated that NA treatment induced an increase in the whole body apoA-I pool, while that of apoA-II fell. The rise in plasma apoA-I was due to a decrease in its fractional catabolic rate. The synthesis rate of apoA-II fell, while its fractional catabolic rate also decreased, but less so.

We (WAHLBERG et al. 1990) studied the effect of NA on these plasma HDL subfractions in 41 patients with different types of hyperlipoproteinaemia. In these patients treated with NA in the prescribed dose of 4 g daily for 6 weeks, plasma cholesterol and triglyceride concentrations decreased on average by 44% and 21%, respectively. Plasma HDL cholesterol increased from a mean of 1.13 mmol/l to 1.54 mmol/l or by 36%, $p < 0.001$. Increases also occurred in hyperlipoproteinaemia with normal plasma HDL cholesterol concentrations, i.e., type IIa. This increase was completely due to a mean increase in the HDL_2 fraction, which increased from 0.43 to 0.85 mmol/l or by 97%, $p < 0.001$, while the HDL_3 fraction remained constant (0.78 vs. 0.75 mmol/l, respectively).

Therefore, NA seems to influence plasma HDL_2 and HDL_3 concentrations positively in terms of their ability to hinder the development of atherosclerosis, as HDL_2 has been more associated with such a protective faculty than HDL_3.

HDL can also be subdivided into different sizes, using gradient gel electrophoresis (BLANCHE et al. 1981) and protein estimation of different subfractions. Five subfractions have thus been identified; in decreasing order of size, these are: HDL-2b, -2a, -3a, -3b and -3c (Fig. 9). Angiographic investigations of young male survivors of myocardial infarction demonstrated that the largest subclass, i.e., 2b, was strongly inversely related to the severity of coronary atherosclerosis and also to the progression of the disease (JOHANSSON et al. 1991). Furthermore, the subfraction HDL-3b correlated positively with progression of coronary atherosclerosis. JOHANSSON and CARLSON (1990) treated 23 patients with different types of hyperlipoproteinemia with 4 g NA daily for 6 weeks and studied the effect on HDL subclasses as defined by gradient gel electrophoresis. Effects on VLDL and

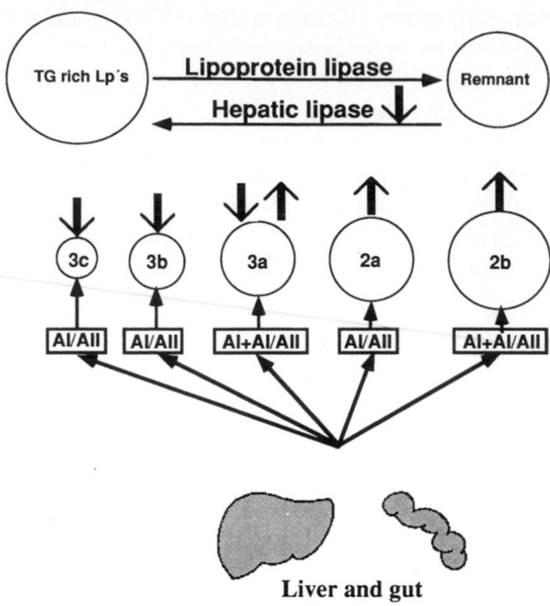

Fig. 9. Schematic representation of the effect of nicotinic acid (NA) on high-density lipoprotein (HDL) metabolism. *Broad arrows* indicate effects induced by NA. *TG*, triglyceride; *Lp*, lipoprotein; *AI*, apolipoprotein A-I; *AII*, apolipoprotein A-II; *2a*, *2b*, *3a*, *3b*, *3c*, subfractions of HDL

LDL lipids confirmed previous experiences with NA. In accordance with the experience of WAHLBERG et al. (1990), HDL_2 cholesterol increased by a mean of 126%, while the HDL_3 cholesterol concentration remained unchanged. The effect of NA on the HDL subclasses is given in Table 6. The largest subclasses, 2b and 2a, both increased, the most pronounced effect being noted in the largest 2b subfraction. Decreases in concentrations were noted for the size subfractions 3b and 3c.

The pronounced increase in HDL-2b and the decrease in HDL-3b induced by NA, therefore, imitates the antiatherogenic HDL particle size profile noted in the study of young myocardial infarction survivors. This suggests that NA may have a protective effect on the progression of coronary atherosclerosis and that NA works in hyperlipoproteinaemic patients towards normalization of plasma lipoprotein concentrations and compositions.

Together with the findings of TASKINEN and NIKKILÄ (1988) (see below) regarding the effect of NA on hepatic lipase, a logical picture of the effect of NA on HDL metabolism appears (Fig. 9).

However, contrasting findings were noted by FRANCESCHINI et al. (1990), who treated 11 males with type IV hyperlipoproteinaemia with acipimox at a dosage of 250 mg t.i.d. for 3 months. HDL subfractionation was performed both by rate zonal ultracentrifugation and by particle size estimation. The

Table 6. Mean protein concentrations of high-density lipoprotein gradient gel electrophoresis subclasses before and after 6 weeks of treatment with nicotinic acid 4 g daily in 23 subjects

HDL subclasses in mg protein/ml					
NA	2a	2b	3a	3b	3c
Before	0.27	0.13	0.44	0.41	0.21
After	0.36	0.37	0.47	0.31	0.18
Change (%)	+36	+183	+8	−24	−18
p	<0.001	<0.001	n.s.	<0.001	<0.05

HDL, high-density lipoprotein; NA, nicotinic acid; n.s., not significant.

effect of treatment on HDL cholesterol was much lower (+11%) than both the JOHANSSON and CARLSON (1990) study (+45%) using plain NA and the TASKINEN and NIKKILÄ (1988) study (+33%) using a higher dose of acipimox. FRANCESCHINI et al. (1990) found minor changes in both HDL_2 and HDL_3 mass, significantly so for HDL_3 cholesterol. Elution profiles showed accumulation of dense HDL_3, in some patients resulting in bimodal distribution. HDL particle size distribution showed only minor variations.

The reason for the different result obtained by FRANCESCHINI et al. (1990) compared with the other two studies is not known. It has to be noted however, that FRANCESCHINI et al. used a lower dose of acipimox than TASKINEN and NIKKILÄ (1988), who also treated patients with more pronounced hyperlipidemia for longer periods of time.

f) Lipoprotein Lp(a)

Lipoprotein Lp(a) is an important, independent risk factor for CHD. Its metabolism is closely linked to LDL and it has strong structural similarities with plasminogen. The effect of plasma lipid-lowering drugs on lipoprotein Lp(a) has in general been disappointing, since most such drugs seem to be inefficient (KOSTNER 1988). The exception is NA, adding further support to idea of the drug as the broad spectrum plasma lipid-lowering drug (CARLSON 1984). Thirty-one hyperlipidaemic patients were treated with 4 g NA daily and lipoprotein Lp(a) levels were determined before and after 6 weeks of treatment (CARLSON et al. 1989). The decrease was 38% after exclusion of massively hypertriglyceridemic subjects. Interesting relations were found to other lipoproteins. Thus, the higher the initial plasma triglyceride level before treatment, the smaller the effect on lipoprotein Lp(a). With high initial plasma triglyceride concentrations, the lipoprotein Lp(a) levels even increased following treatment with NA. This could probably be explained by the negative relation between initial VLDL and LDL concentrations, as lipoprotein Lp(a) is closely related to LDL concentrations. In patients with massive hypertriglyceridaemia, low levels of LDL and HDL are present (OLSSON and CARLSON 1975).

Treatment with NA increases these low LDL cholesterol concentrations (Carlson et al. 1977; Carlson and Olsson 1979). There was also a highly significant relation between the reduction in LDL cholesterol and the reduction in lipoprotein Lp(a) concentration, again illustrating the close relation in metabolism between these two plasma lipoproteins.

NA derivatives have also been demonstrated to lower increased levels of Lp(a). For example, α-tocopheryl nicotinate was studied in 28 patients with hyperlipidaemia (Noma et al. 1990). In those with high Lp(a) levels, a mean decrease of 29% was seen. The daily dose of α-tocopheryl nicotinate corresponded to a daily intake of only 140 mg daily, clearly a low dose, as even 500 mg NA in previous studies had proven to be inefficient (Kostner 1988).

g) Postprandial Lipaemia

Ditschuneit et al. (1992) found that patients with severe CHD react with a pronounced hypertriglyceridaemia after an oral fat load, a response not seen in controls. Acipimox treatment at a dosage of 250 mg twice daily significantly decreased this postprandial triglyceride elevation in CHD patients (Fig. 10). The effect was not seen in controls. The abolished postprandial effects in patients was noted for VLDL triglycerides and cholesterol, for HDL triglycerides, in which an increase was prevented, and for HDL cholesterol, in which a decrease was prevented. The authors concluded that the changes of lipoprotein composition suggest that there is a specific coronary risk

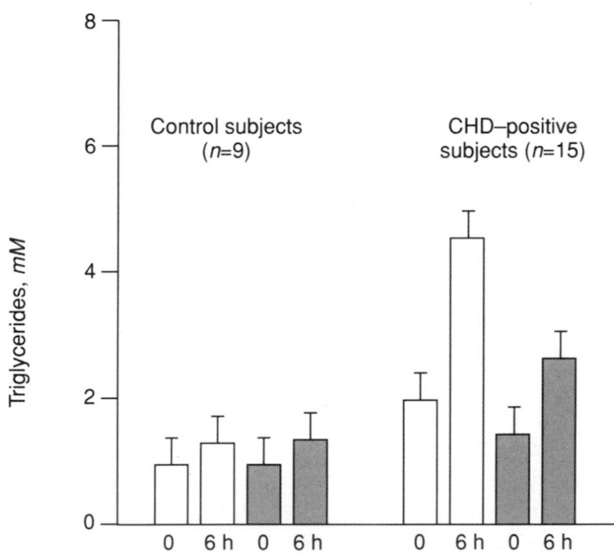

Fig. 10. Plasma triglycerides in healthy controls and patients with coronary heart disease (*CHD*) before and 6 h after ingestion of 100 g fat, and before (*open bars*) and after (*shaded bars*) administration of acipimox. Mean + 1 S.D. (Ditschuneit et al. 1992)

associated with the postprandial state that can be attenuated by treatment with acipimox.

3. Enzyme Activities and Receptor Functions in Lipid Metabolism

a) Hormone-Sensitive Lipase

The effect of NA on HSL is still held to be the most important effect of NA affecting lipid metabolism and is covered above and in Fig. 5.

b) Lipoprotein Lipase and Plasma Triglyceride Removal

The effect of NA treatment on plasma lipoprotein lipase activity was studied in ten hypertriglyceridaemic patients and no effect was noted (BOBERG et al. 1971). On the other hand, NA induced an increase in the fractional removal rate constant for triglycerides following an i.v. injection of a triglyceride emulsion. This is in line with the kinetic findings of GRUNDY et al. (1981), as given above, of an increased fractional catabolic rate of VLDL following NA treatment. The hypothesis that NA really seems to influence the removal side of plasma triglyceride metabolism has gained further support by the observation of an increase in the FIAT induced by NA in patients with hypertriglyceridemia (CARLSON et al. 1973). Further studies are needed to clarify the effect of NA on lipoprotein lipase activity.

c) Hepatic Lipase

The effect of a relatively high dose of acipimox was studied in patients with severe hypertriglyceridaemia and low HDL cholesterol levels as part of a larger study of this compound (TASKINEN and NIKKILÄ 1988). In addition to its effect on plasma lipoprotein fractions and subfractions of HDL, plasma lipoprotein and hepatic lipases were studied as well as muscle and adipose tissue lipoprotein lipase activities. As expected, plasma triglycerides decreased and plasma HDL increased; both HDL_2 and HDL_3 increased. Neither postheparin plasma lipoprotein lipase activity nor skeletal muscle lipoprotein lipase changed. However, adipose tissue lipoprotein lipase activity decreased after 9 months of treatment. Postheparin plasma hepatic lipase activity was reduced from 6 months onwards. Thus, the data do not support the theory that an effect of acipimox on lipoprotein lipase activity is the mechanism of action of the hypolipidemic effect of the drug. However, long-term treatment could affect hepatic lipase by reducing HDL catabolism and thus increasing HDL levels, in particular HDL_2, which is the substrate for hepatic lipase. This finding fits well with the effect on HDL subfractions observed by WAHLBERG et al. (1990) and by JOHANSSON and CARLSON (1990), as given above. An outline of the effect of NA on HDL metabolism is shown in Fig. 9.

Hepatic lipase has also been reported to affect the conversion of IDL to the small, dense LDL subfraction (AUWERX et al. 1989). The concept that

NA should inhibit the effect of hepatic lipase is also supported by the findings of Gaw et al. (1990), who showed that acipimox treatment changes the LDL distribution from small, dense, cholesterol-rich to large, buoyant particles.

d) Other Enzymes and Receptors in Lipid Metabolism

Neither the lecithin cholesteryl acyltransferase (LCAT) activity determined in vitro nor the fractional turnover rate of the plasma cholesteryl esters determined in vivo was significantly affected by NA treatment in a study by Kudchodkar et al. (1978).

The effect of NA on other enzyme systems and receptor functions involved in lipid metabolism such as cholesteryl ester transfer protein (CETP), 3-hydroxy-3-methylglutaryl coenzyme A (HMG-CoA) reductase, and the LDL receptor has been incompletely studied.

In the study by Tornvall et al. (1990), NA treatment induced drastic changes in VLDL particle composition towards particles less rich in cholesteryl esters. In HDL, the cholesterol content increased by 35%. Taken together, these changes point to a decreased lipid transfer activity induced by NA. A decreased cholesteryl ester transfer activity induced by NA could also partly explain the changes described as a decreased hepatic lipase activity by NA.

NA treatment does not seem to influence LDL receptor activity as such. However, it has recently been suggested that the reduction of the LDL triglyceride content following fibrate treatment of hyperlipidemia, for example, may improve receptor-mediated catabolism of LDL (Kleinman et al. 1985). Francheschini et al. (1991) therefore examined whether acipimox administration might alter receptor-binding activity in 11 patients with type IV hyperlipoproteinaemia. Hypertriglyceridaemic LDL had a 30% lower capacity to inhibit receptor-mediated uptake and degradation of labeled LDL in competition experiments. This abnormality was fully corrected after acipimox. This indicates that acipimox treatment in patients with type IV hyperlipoproteinaemia can normalize the defective interaction of hypertriglyceridaemic LDL with the LDL (B,E) receptor. The mechanism probably acts by decreasing the triglyceride content of LDL, which diminished by 46% in this experiment. This conclusion is strengthened by the in vitro observation of lipoprotein lipase-induced enhancement of the interaction between LDL and its receptor (Aviram et al. 1988).

4. Biliary Lipid Metabolism and Cholesterol Balance

Kudchodkar et al. (1978) studied the effect of NA on cholesterol metabolism in five normolipidaemic and ten hyperlipidaemic subjects. NA had no effect on plasma lipids of patients who were on a high-carbohydrate diet, confirming the view that NA exerts its action through inhibition of lipolysis in adipose tissue and not on hepatic synthesis of fatty acids, which is increased on a high-carbohydrate diet. NA did not affect intestinal absorption of cholesterol. Synthesis of cholesterol was markedly inhibited in 11 investigated patients. NA did not reverse the increased synthesis of cholesterol in patients on high-carbohydrate diets, but reversed the increased synthesis caused by cholestyramine. The hypocholesterolaemic effect of NA was associated with mobilization of tissue cholesterol in this study. The authors conclude that the primary mechanism of the hypolipidaemic action of NA appears to be its antilipolytic action on adipose tissue. Its effect on hepatic metabolism of lipids is probably secondary to the reduction of the circulating pool of plasma FFA.

Bile acid kinetics and steroid balance were studied by EINARSSON et al. (1977) in patients with type II and IV hyperlipoproteinaemia before and during NA treatment. In spite of the effect of NA treatment on plasma lipids, treatment was not associated with any significant effects on total bile acid formation in type II patients, but the cholic acid to chenodeoxycholic acid ratio was significantly increased in both the synthesis and the pool of bile acids. In all the type IV patients, who basically have an increased bile acid production when compared to type II patients, bile acid formation decreased by 25%–30%. The net steroid balance did not change in any consistent way.

In the study by GRUNDY et al. (1981), a slight and inconsistent increase in the excretion of neutral steroid was noted on NA. This is in sharp contrast to fibrates, which almost invariably show increases in biliary excretion of neutral steroids. Such an increase in neutral steroid excretion probably reflects a mobilization of cholesterol from tissue pools (GRUNDY et al. 1981). Although a similar marked increment was not observed by GRUNDY et al. (1981) on NA treatment, a decrease in tissue pools by means of protracted, but undetectable, increase in fecal steroids, remains a possibility. MIETTINEN (1968) and MOUTAFIS and MYANT (1971) have noted that NA treatment partially inhibits cholesterol synthesis. Therefore, an unchanged steroid excretion is compatible with a flux of cholesterol out of tissues. In line with this, MAGIDE and MYANT (1975) reported a marked loss of cholesterol from tissues of monkeys treated with NA. Cholesterol balance was unchanged in the study by GRUNDY et al. (1981). However, all patients had an increased cholesterol output during NA treatment. Nevertheless, the lipid composition and saturation indices of stimulated hepatic bile were not increased significantly. This observation concurs with the lack of increase in the rate of gallstone disease in those patients in the Coronary Drug Project (CDP) who were treated with NA. An increase was noted in those treated with clofibrate. Cholesterol absorption was not influenced by NA.

In a study by ERICSSON et al. (1990), the conclusion that NA does not increase the risk for gallstone formation was reinforced. Patients with familial combined hyperlipidaemia were studied. These patients initially had a higher relative concentration of cholesterol in their bile than normal controls. This resulted in a higher cholesterol saturation in bile from the hyperlipidaemic patients and therefore probably a higher risk of gallstone formation. Treatment with acipimox at a dosage of 750 mg daily decreased plasma triglyceride and cholesterol concentrations. Concomitantly, acipimox significantly decreased the cholesterol content in stimulated fasting duodenal bile from 6.5 ± 0.3 to 4.3 ± 0.5 mol% ($P < 0.01$), resulting in a "normalization" of biliary cholesterol saturation from 85 ± 6 to $58 \pm 6\%$ ($P < 0.005$). The results indicate that acipimox treatment of patients with familial combined hyperlipidaemia, which is often associated with supersaturated bile, decreases biliary cholesterol and, presumably, the future risk of gallstone formation.

It should be noted that work from the same group showed a different effect of NA on bile acid metabolism (Angelin et al. 1979).

5. Summary of the Mechanism of the Hypolipidaemic Action of Nicotinic Acid

An outline of the sites of action of NA on lipid metabolism is shown in Fig. 11. The exact mechanism of action of NA on the lipoprotein metabolism is still only incompletely known. After 30 years, the most appealing explanation

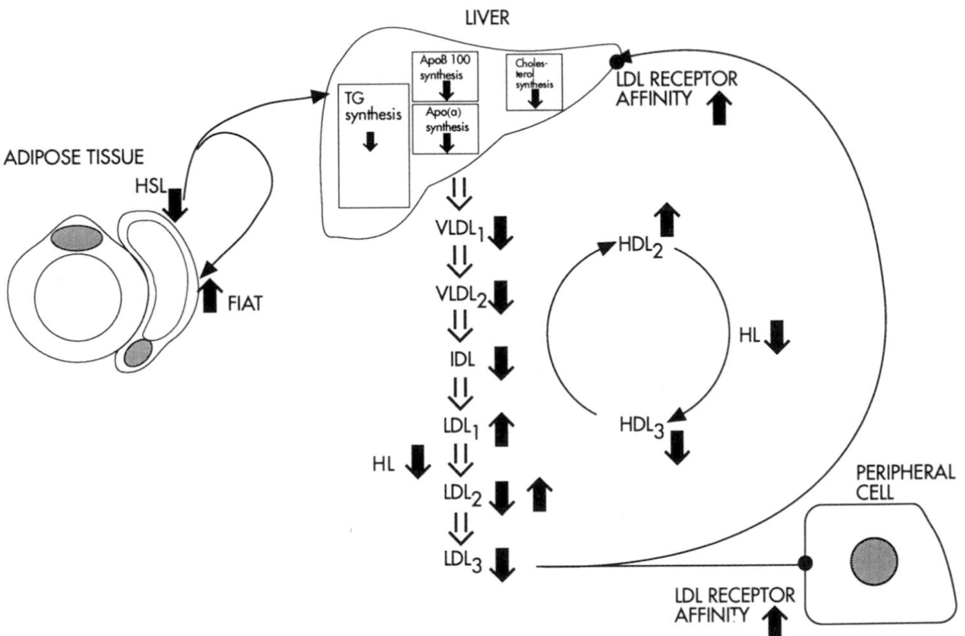

Fig. 11. Summary of the effect of nicotinic acid (NA) on lipoprotein metabolism. *Black arrows* indicate effects induced by NA. *TG*, triglyceride; *HSL*, hormone-sensitive lipase; *FIAT* fatty acid incorporation into adipose tissue; *VLDL*, very low density lipoprotein; *LDL*, low-density lipoprotein; *HDL*, high-density lipoprotein; *IDL*, intermediate-density lipoprotein; *HL*, hepatic lipase

of the mechanism of action of NA on plasma lipid levels in man is still the inhibitory effect on fat-mobilizing lipolysis in adipose tissue (Carlson and Orö 1962). Because of the decreased FFA levels in blood, the liver receives less substrate for VLDL synthesis, which thereby decreases, resulting in lower circulating levels of both plasma triglycerides and cholesterol-including LDL cholesterol (Fig. 11). The kinetic work by Grundy et al. (1981) was the first to demonstrate that a retardation of hepatic triglyceride synthesis is induced by NA treatment.

However appealing this hypothesis is, several lines of evidence question this simplistic explanation. First, as already stated, the 24-hour FFA profile during NA treatment does not point to an overall depressed FFA curve. Second, Tornvall et al. (1990) recently demonstrated that NA increased the triglyceride content and size of large VLDL particles during treatment of hypertriglyceridaemia with NA. Third, the recent finding of a lowering effect by NA on lipoprotein Lp(a) levels does not fit with the concept of a decrease in VLDL triglyceride synthesis being the hypolipidaemic mechanism of NA (Carlson et al. 1989). Instead, it seems more plausible that the effect is on the apolipoprotein B synthesis level, since the effect on LDL cholesterol related very closely to that apolipoprotein. More research is needed into the hepatocellular biosynthesis of triglycerides, cholesterol, and apolipoprotein B and the way different substrates could influence the synthetic rates of these various cellular products.

Fourth, kinetic studies also point to other mechanisms of action. The fact that the fractional catabolic rate increased and the VLDL residence time decreased in the NA-treated subjects in the experiment by Grundy et al. (1981) does not, however, mean that the absolute rate of removal of triglycerides was enhanced by NA. Since VLDL was smaller, absolute clearance rates of triglycerides from plasma were actually lower during treatment. This was only partly counteracted by the fact that smaller particles could pass through the intravascular lipolytic chain more rapidly than larger ones.

The general contention is that NA does not influence plasma lipoprotein lipase activity (Boberg et al. 1971). However, this does not rule out the possibility that an effect of NA also could influence lipid removal from blood. On the contrary, as stated above, other findings indicate that NA affects lipid removal from blood. One example is the redistribution of apolipoprotein C-II and C-III, resulting in a higher apoC-II to C-III ratio during NA treatment and thereby a metabolic situation favouring lipolytic activity and breakdown of triglyceride-rich lipoprotein particles. Another example is the effect of NA on the intravenous fat tolerance, which shows an increased fractional removal rate induced by NA (Boberg et al. 1971). Furthermore, a finding pointing to an increase in removal mechanisms induced by NA is the increased rate of FIAT induced by the drug (Carlson et al. 1973). An inhibitory effect on hepatic lipase has also been pointed out (Fig. 9).

6. Prostanoid Metabolism

We studied the effects of 1 g NA orally on forearm blood flow and FFA concentrations in healthy volunteers with and without pretreatment with indomethacin to see if some effects of NA were mediated through release of prostaglandins (Kaijser et al. 1979). NA decreased plasma FFA concentrations to 24% of the initial level and increased forearm blood flow fourfold. After pretreatment with indomethacin, the blood flow increase was only one-third of that without pretreatment. The decrease of FFA was not altered, but the rebound effect was abolished. Thus the dilatory effect of NA seems to be mediated by release of endogenous prostaglandin while the inhibition of lipolysis by the drug is mainly produced by other mechanisms.

We gave NA to eight female subjects who were not taking any cyclo-oxygenase-inhibiting drugs and followed the urinary excretion of $PGF_{2\alpha}$ and of the prostacyclin metabolite 6-keto-$PGF_{1\alpha}$ before and after 2 and 28 days of treatment (Olsson et al. 1983). The excretion of 6-keto-$PGF_{1\alpha}$ increased significantly after 2 days on NA. After 28 days, excretion was not significantly different from pretreatment values. $PGF_{2\alpha}$ excretion did not increase significantly. The study indicates that NA initially increases prostacyclin production and that this increase is abolished during continued treatment. As it is in the early stages of treatment that flushing occurs, the study supports the view that this action is mediated via release of prostacyclin. Other authors have recently demonstrated that NA causes the release of markedly increased quantities of PGD_2 in vivo in humans (Morrow et al. 1989). Levels of the PGD_2 metabolite 9-a,11-b-PGF_2 increased dramatically following ingestion of 500 mg NA, and the levels correlated with the intensity and duration of flushing. As only minor increases were noted for the urinary excretion of the prostacyclin metabolite 6-keto-$PGF_{1\alpha}$, the authors drew the conclusion that PGD_2 is the most probable mediator of NA-induced vasodilatation in humans. Nicotinamide does not cause flushing and is devoid of plasma lipid-lowering properties. It also did not cause the release of endogenous PGD_2. The authors therefore speculate on the possibility that PGD_2 may play a role in the lipid-lowering properties of NA.

Elevations of PGE_2 have also been shown to correlate with NA-induced flushing, and these levels can be reduced by concomitant administration of aspirin (Nozaki et al. 1987).

III. Effects on Glucose Metabolism

1. Basal Mechanisms

The relation between NA treatment and glucose metabolism depends on the effect of NA on FFA levels in blood. A distinction has to made between the acute plasma FFA-lowering effect and its long-term effect, which, particularly regarding plain NA, results in a long period of high plasma FFA levels, referred to as overshoot or rebound effect.

NA lowers plasma FFA levels acutely (see above), resulting in improved insulin action (GROSS and CARLSON 1968; REAVEN et al. 1988). During long-term administration of NA, however, FFA are increased (GUNDERSEN and DEMISSIANOS 1969). Since the increased availability of FFA markedly impairs glucose utilization in skeletal muscle and other tissues the "Randle effect"; RANDLE et al. 1963; FERRANNINI et al. 1983, this is the likely explanation for the increased glucose levels and insulin resistance associated with long-term NA treatment (MIETTINEN et al. 1969).

It has also been hypothesized that insulin resistance in patients with non-insulin-dependent diabetes mellitus (NIDDM) may be caused by or be partly due to elevated FFA concentrations in plasma and skeletal muscles. Recently, this hypothesis has been further documented by REAVEN et al. (1988). The authors used obese rats that were streptozotocin treated and rendered insulin resistant and insulin deficient. A lowering effect of plasma glucose was demonstrated by two different antilipolytic agents, NA and phenylisopropyladenosine (PIA) when given to the animals. With both agents, plasma glucose concentrations decreased by about 50%, the major change not taking place until the second hour and after the acute FFA decrease. The decrease in plasma FFA concentrations was immediate, dramatic, and maintained throughout the experiment. Insulin levels were not affected and therefore the decreased glucose levels were not likely to be due to increased insulin secretion. Both NA and PIA were shown to increase both basal and insulin-stimulated glucose uptake in isolated adipocytes in vitro. The data suggested that the major effect of NA was to enhance non-insulin-mediated glucose uptake. NA was also shown to increase in vivo, insulin-stimulated glucose uptake in peripheral tissues. The NA-induced decrease in plasma glucose concentration was also associated with a decrease in hepatic glucose output. Data suggest that PIA and NA inhibit lipolysis via different receptors (Fig. 5).

The authors speculate that there are two possible mechanisms by which a reduction in lipolytic rate could lower plasma glucose concentration. First, the authors suggest the "Randle effect," with FFA as an inhibitor of insulin-stimulated glucose uptake. This was evidenced in the present study of both in vitro and in vivo increased glucose uptake after decrease of plasma FFA levels by two independent mechanisms (NA and PIA). Second, there is evidence that an increased flux of FFA into the liver can increase hepatic glucose production in perfused rat liver, and a direct relationship between hepatic glucose production and fasting and nonfasting plasma FFA levels has been documented in patients with NIDDM (GOLAY et al. 1986). Because the NA-induced decrease in plasma glucose concentration was associated with a lowering of both plasma FFA concentration and hepatic glucose production, it is evident that NA and PIA lower plasma glucose in obese rats with streptozotocin-induced diabetes secondary to their antilipopytic effect. The authors also made an effort to quantify which of the two proposed antilipolytic effects was most responsible for the hypoglycemic effect of NA and PIA. They concluded that a decreased hepatic glucose production was probably the major reason for the plasma glucose decrease.

2. Non-Insulin-Dependent Diabetes in Man

a) Short-Term Studies

In an overnight study of humans with NIDDM carried out by the same group of researchers, plasma FFA was increased by giving a neutral fat emulsion and heparin in one group with mild hyperglycaemia and decreased by giving NA in one group with severe hyperglycaemia (Johnston et al. 1990). No effects of these manipulations were noted on plasma glucose concentrations, hepatic glucose production, or glucose disposal. Thus, acute changes in FFA concentrations in patients with NIDDM did not lead to any short-term changes in glucose production or disposal rates.

Against the hypothesis that insulin resistance in patients with NIDDM may be caused by or be partly due to elevated FFA concentrations in plasma, Vaag et al. (1991) studied the acute effect of acipimox on basal and insulin-stimulated glucose metabolism in NIDDM. A marked acute improvement of insulin action during Acipimox was demonstrated. The improved insulin action was due to both a stimulatory effect of acipimox on glucose oxidation and on nonoxidative glucose disposal (Fig. 12). Acipimox

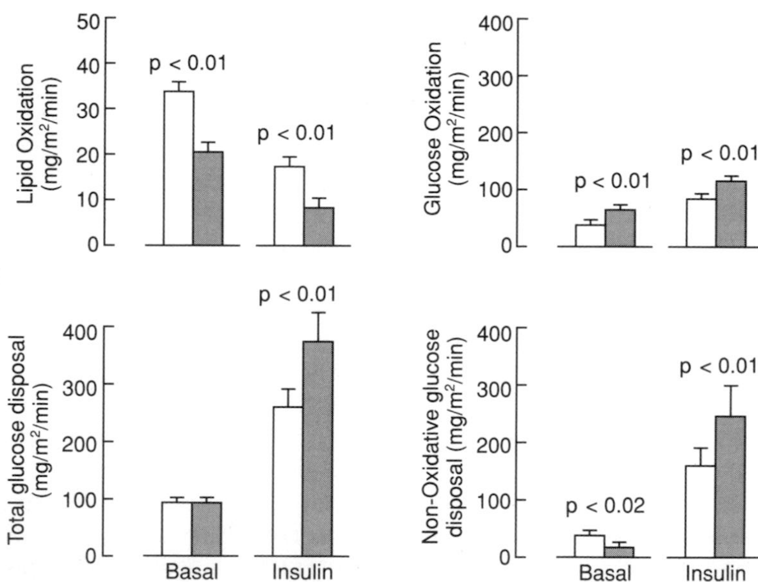

Fig. 12. Effect of acipimox on lipid oxidation, glucose oxidation, total peripheral glucose disposal and non-oxidative glucose disposal in the basal state and during insulin infusion in 12 subjects with non-insulin-dependent diabetes mellitus. *Shaded bars*, acipimox; *white bars*, placebo. Mean ± S.E.M. (Vaag et al. 1991)

acutely inhibited lipid oxidation and stimulated glucose oxidation in the basal state and during insulin stimulation. The effect of acipimox on lipid oxidation was highly negatively correlated with the effect of acipimox on glucose oxidation in both basal state and during insulin infusion. In contrast the effect of the drug on lipid oxidation was not correlated with the effect on insulin-stimulated nonoxidative glucose metabolism. Furthermore, in the basal state acipimox had no effect on peripheral glucose uptake, and thus it was actually found that acipimox decreased nonoxidative glucose disposal in the basal state. The findings support the hypothesis that elevated FFA levels may be of importance for the development of insulin resistance in NIDDM. The increase in nonoxidative glucose disposal was the quantitatively most important effect and was probably caused by an increased activity of glycogen synthase in skeletal muscles. The increase in glucose oxidation could not be attributed to increased activities of pyruvate dehydrogenase and/or phosphofructokinase in skeletal muscles. These enzyme activities are of importance in the glycolytic flux. A decrease in these enzyme activities in heart and skeletal muscle has been suggested by RANDLE et al. (1963) to be responsible for the inhibitory effect of FFA on glycolysis and pyruvate oxidation. The authors conclude that nonmuscle tissues are probably responsible for the increase in glucose oxidation caused by acipimox.

b) Long-Term Studies

NA has been tried in patients with NIDDM and hyperlipidemia (GARG and GRUNDY 1990). Efforts were made to keep antidiabetic treatment constant. NA at a dosage of 4.5 g daily for 8 weeks decreased plasma cholesterol and triglycerides by 24% and 45%, respectively. HDL cholesterol increased by 34%, i.e., the effects in these NIDDM patients were in the same order of magnitude as are seen in hyperlipidaemic normoglycaemic patients. At the same time, mean plasma glucose concentrations increased from 7.8 to 9.1 mmol/l. Glycosylated hemoglobin increased by 21% and marked glucosuria were noted in several patients. Because of the possible untoward effect of decreased glucose tolerance in the development of atherosclerosis, the authors suggested that NA should not be the drug of choice in patients with NIDDM and hyperlipidaemia.

DULBECCO et al. (1989) reported on the effect of acipimox treatment in 12 patients with NIDDM using a double-blind, crossover technique with treatment periods of only 1 week and 2 days washout periods in between. Acipimox significantly decreased fasting plasma levels of triglycerides by 22% and of glucose by 28%.

In a large study of 3009 patients with NIDDM, acipimox was studied with regard to plasma lipid and glucose concentrations (LAVEZZARI et al. 1989). Patients with both types II and type IV hyperlipoproteinaemia were included. Acipimox was given at a dosage of 500–750 mg daily. Plasma total cholesterol and triglyceride levels decreased as expected and plasma HDL cholesterol increased. Mean fasting blood glucose decreased significan from 8.66 mmol/l before treatment to 7.83 and 7.38 mmol/l during and the end of acipimox treatment. Glycosylated hemoglobin also showed reduction of 8.5%. The effect could partly be due to a decrease in bo weight during acipimox treatment, but is in contrast to the common find of increases in plasma glucose levels during plain NA treatment.

3. Differences Between Nicotinic Acid Derivatives

In the study by Tornvall and Walldius (1991; see above), the effects NA and acipimox were studied with regard to glucose tolerance. At the end of each treatment period of the hypertriglyceridaemic patients, 120-min oral glucose tolerance test was performed. No diabetic patients were included in the study, i.e., no subject had a fasting plasma glucose greater than

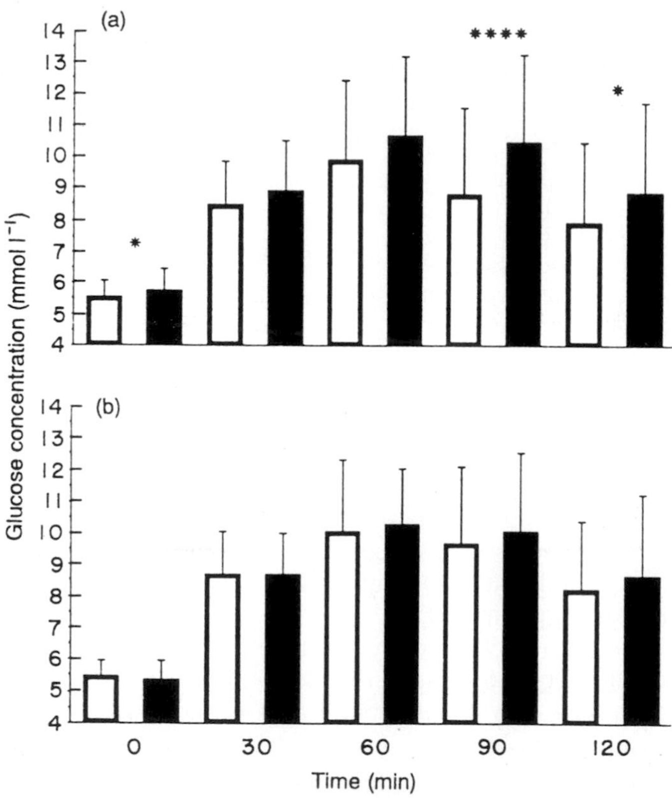

Fig. 13. Glucose concentrations before and 30, 60, 90, and 120 min after an oral glucose load of 75 g before (*white bars*) and after (*black bars*) treatment with **a** 3 g nicotinic acid daily and **b** 0.75 g acipimox daily. $^{*}p < 0.05$, $^{****}p < 0.0001$ (Tornvall and Walldius 1991)

7.7 mmol/l. NA significantly increased the fasting plasma glucose levels (Fig. 13), while acipimox did not have this effect. Oral glucose load before drug treatment demonstrated an impaired glucose tolerance or diabetes mellitus (WHO criteria) in almost half of the hypertriglyceridaemic subjects. NA treatment increased the glucose levels at 90 and 120 min (Fig. 13a). Acipimox did not affect glucose levels after glucose load (Fig. 13b). At 90 min, NA had significantly higher glucose levels than acipimox. A reduced glucose tolerance is regarded as a risk factor for the development of ischemic heart disease and is therefore an unwanted effect of NA. Acipimox has no negative effects on glucose tolerance and, at least with regard to its equal effect on plasma lipoproteins in hypertriglyceridaemia, should represent a good alternative to plain NA. The reason for the noted difference in glucose metabolism may be the glucose fatty acid cycle (RANDLE et al. 1963) and the different response in plasma FFA profile over 24 h induced by NA and acipimox (see above).

4. Conclusions About Effects on Glucose Metabolism

The effect of NA and its derivatives on glucose metabolism is thus varying and complex. It depends on the type of study (short-term or long-term), type of patients (e.g., diabetics or nondiabetics), animal model, NA derivative and doses used, etc. A common pattern seems to be that if the antilipolytic effect is maintained throughout an experiment, avoiding rebound phenomena of FFA levels, a decreased plasma glucose level and insulin resistance is most probable. Therefore, the tendency is that short-term experiments show positive effects on glucose metabolism, while long-term experiments tend to show deteriorating effects.

D. Clinical Approach

I. Side Effects

Side effects of NA treatment were recently reviewed (STEINER et al. 1991). Subjective side effects include mainly facial flushing and gastric irritation (Table 7). Other side effects are increases in plasma activities of hepatic enzymes and occasionally hepatocellular toxicity or hepatitis. The effect of NA on glucose metabolism has been covered above. NA also has side effects on purine metabolism. Adverse ophthalmological effects have also been reported. A good picture of the frequency and severity of side effects caused by NA is given in the Coronary Drug Project, as this is the largest controlled study on patients that have been treated with NA published so far (CORONARY DRUG PROJECT RESEARCH GROUP 1975).

Table 7. Most frequent adverse effects of nicotinic acid in different galenic formulations (from Knopp et al. 1985)

Effect	Unmodified NA (%) ($n = 37$)	Time release (%) ($n = 34$)
Flushing	100	82
Heartburn	9	9
Indigestion	0	12*
Nausea	8	38*
Vomiting	0	18*
Anorexia	3	15
Diarrhea	22	45*
All GI symptoms	30	100
Impotence	3	22*
Dry skin	5	12
Brown skin	0	3
Itching	8	3
Rash	3	3
Fatigue	3	24

GI, gastrointestinal; NA, nicotinic acid. $^*p < 0.05$

1. Flushing

a) Symptoms

The most common side effect of NA is facial flushing and cutaneous vasodilatation, which occurs in virtually all patients starting treatment with pharmacological doses (Table 7). Flushing consists of a skin erythema caused by increased blood flow in the skin. The symptom is similar to a sunburn, but lasts only about 1 h. On continuous intake of NA, flushing rapidly disappears in most individuals (tachyphylaxis). The symptom is harmless, but this side effect remains the most important reason for the underuse of the drug. A comparison between NA and acipimox (Tornvall amd Walldius 1991) showed that acipimox resulted in significantly less flush problems than plain NA both short-term and after 6 weeks of treatment.

b) Preventing Flushing

As NA flushing is subject to tachyphylaxis, the symptom can be avoided by taking the drug regularly and without interruptions. Flushing could be less severe if NA is ingested with meals. As the symptom is prostaglandin mediated, a small dose of a cyclooxygenase inhibitor, e.g., 75–250 mg aspirin, half an hour before NA ingestion will blunt or eliminate the flushing effect.

2. Gastrointestinal Side Effects

Gastrointestinal side effects by NA treatment include heartburn, stomach pain, indigestion, nausea, anorexia, and diarrhea and occur with a frequency of up to 30% of patients (Knopp et al. 1985; Table 8). In addition, gastro-

Table 8. Typical scheme for start of nicotinic acid treatment in inpatients (fast increase of dosage)

Time of day	Day 1 (g)	Day 2 (g)	Day 3 (g)
Morning	(0.25)	0.25	0.5
Noon	0.25	0.5	0.75
Afternoon	0.5	0.5	0.75
Evening	0.5	0.75	1

intestinal symptoms can be blunted by taking the drug with meals. Antiacids could also be tried.

3. Hepatic Side Effects

Plasma transaminase activities are usually not increased during treatment with plain NA or pentaerythritol nicotinate (Olsson et al. 1974). Occasional transitory increases in plasma liver enzyme activities are commonly seen and were first reported by rivin in 1959. Knopp et al. (1985) noted greater increases in serum activity of aspart-aminotransferase (S-ASAT) activities following time-release NA than unmodified, all differences being insignificant. Plasma alkaline phosphatase activities can particularly increase if time-release NA is used (Olsson et al. 1974; Knopp et al. 1985).

Numerous occasional reports on hepatotoxicity exist (Clementz and Holmes 1989; Knopp 1989; Mullin et al. 1989; Henkin et al. 1990; Hodis 1990). This effect seems to particularly occur when NA is given in the sustained-release form and if plain NA is used usually only in dosages above 3 g daily (Etchason et al. 1991). Typically, the patient experiences nausea and vomiting and, in some cases, is jaundiced. In more severe cases, elevations of serum bilirubin and ammonia concentrations and prolonged prothrombin time are observed. The most severe cases (Hodis 1990; Mullin et al. 1989; Clementz and Holmes 1989) have resulted in fulminant hepatic failure and hepatic encephalopathy (Rader et al. 1992).

4. Cutaneous Side Effects

Apart from flushing, a number of cutaneous side effects induced by NA are often reported. These include itching, dry skin, brown skin discoloration, rash, and worsening of cutaneous diseases such as psoriasis (Knopp et al. 1985; Blankenhorn et al. 1987b; Table 8). These symptoms are usually not dose limiting.

5. Lactacidosis

A case of lactacidosis of 9.5 mmol/l (normal range 0.5–2.2 mmol/l) with concomitant toxic delirium was reported by Schwab and Bachhuber (1991) in a man receiving generic NA at a dosage of 3 g daily for hypercholesterolaemia. The symptoms emerged in connection with alcohol consumption. The patient recovered promptly on rehydration and administration of thiamine and magnesium (also see below under slow release NA).

6. Gout

NA causes hyperuricemia and occasionally gout (BERGE et al. 1961; PARSONS 1961). We noted significantly higher mean uric acid concentrations in patients treated with plain NA than with equivalent doses of niceritrol (OLSSON et al. 1974).

There are two possible mechanisms for the higher uric acid levels during NA treatment. The most important reason is probably an impaired excretion of uric acid induced by NA. In 1971 GAUT et al. observed a 75% decrease in clearance during treatment with NA at a dosage of 4.5 g daily. Increased de novo biosynthesis of purine has, however, also been observed (BECKER et al. 1973).

The hyperuricemia induced by NA is readily abolished by the additional treatment with allopurinol.

7. Retinal Edema

Adverse ophthalmological effects have rarely been attributed to NA therapy, the most often reported side effect being retinal edema (GASS 1973; BLANKENHORN et al. 1987b). These effects are reversible after discontinuation of therapy.

8. Time-Release Nicotinic Acid

Time-release NA preparations were tried early in an attempt to counteract the acute subjective side effects of the drug. These preparations, however, turned out to be hepatotoxic and resulted in pronounced elevations in hepatic enzyme activities in blood after just short treatment times (2 days to 7 weeks). More than 30 years ago, CHRISTENSEN et al. (1961) reported a higher incidence of hepatic toxicity with some forms of time-release NA than with equivalent doses of crystalline NA.

This unwanted effect was recently confirmed (ETCHASON et al. 1991) when hepatitis developed in five patients receiving low dosages (3 g/day or less) of time-release NA. In four of the five patients, clinical symptoms of hepatitis developed after the medication had been taken for as short a time as 2 days to 7 weeks. This manifestation of hepatotoxicity seems to differ from the experience with plain NA, which typically occurs at high dosage and after long treatment periods. It was concluded that time-release NA must be used cautiously and discriminately, even in low doses. A case of lactacidosis induced by ingestion of sustained release NA has been reported (EARTHMAN et al. 1991). Prompt recovery followed with glucose-containing fluids and discontinuation of NA therapy.

KNOPP et al. (1985) reported their experience of side effects of conventional-release and slow-release NA (see Table 8). The hepatic toxicity of unmodified and time-release preparations of NA was recently reviewed by RADER et al. (1992). Adverse reactions in six patients resulted from the exclusive use of unmodified NA and in two patients from the exclusive use of time-release preparations. In ten further patients, adverse reactions developed after an abrupt change from unmodified to time-release preparations. Many of these patients were ingesting time-release NA well above the usual therapeutic doses currently recommended. Signs of liver toxicity developed in less than 7 days in four of these ten patients. In doses that achieve equivalent reductions in serum lipids, hepatic toxicity occurred more frequently with time-release preparations than with unmodified preparations. The authors conclude that that an awareness of toxicity associated with ingestion of high doses of time-release NA preparations is important.

A possible exception from the rule of more hepatic toxicity induced by slow-release NA was reported by KEENAN et al. (1991). A wax-matrix sustained-release NA preparation was used in a double-blind, randomized, controlled study in 201 hypercholesterolaemic patients. Daily doses ranged from 1 to 2 g, i.e., comparatively small doses. Nevertheless, the 2.0 g daily dose resulted in total and LDL cholesterol reductions of 18% and 26%, respectively. The corresponding figures for 1.5 g were 13% and 19%. Effects on HDL cholesterol were less impressive. Some drop in patient adherence was noted in the 2-g dose group and the authors suggested that the 1.5 g NA dose of this slow release preparation offered the best combination of efficacy and tolerance.

II. Laboratory Safety Tests

Against the background of the metabolic effects and side effects induced by NA treatment given above, the clinician should monitor possible abberrations in glucose and purine metabolism as well as liver function in each patient. Thus, levels of blood glucose and serum uric acid and plasma activities of ASAT, alaninaminotransferase (ALAT), and alkaline phosphatases should be determined regularly.

III. Start of Treatment

Because of the 100% occurrence of flushing, treatment with NA should be started with small initial dosages and successive increase of the dosage. The patient should be informed by the prescribing doctor about the inevitable cutaneous symptoms before starting treatment. He should also be informed about the tachyphylaxis phenomenon and about the recurrence of flushing if NA is not taken at regular intervals. The flushing symptoms and the high frequency of side effects by time-release NA necessitates that it be administered three or even four times daily.

1. Fast Dosage Increase

Table 8 shows the dosage schedule that we use when treatment is started for patients that are in hospital. In this situation, when the patient is under close supervision and has no own professional or other obligations, it is good (not least for economic reasons) to be able to reach the therapeutic dosage quickly. In our hands, this schedule has worked well for inpatients.

2. Slow Dosage Increase

In outpatients, which is most often the case, dosage increase should be slower. One suggestion has been brought forward by Kane and Malloy (1982), who recommend a starting dosage of 100 mg three times daily. The dosage is increased gradually as tolerated and does not exceed 2.5 g per day by the end of the first month, 5 g at the end of the second month, and 7.5 g at the end of the third month.

We have found the optimal dosage to be 4–6 g daily, and this is the minimum dosage needed to obtain a reasonable effect in familial hypercholesterolaemia.

IV. Treatment Compliance

Figures of compliance during NA treatment are available for most controlled studies of the drug.

In the CD Project, adherence to NA treatment was less than for clofibrate and placebo. For months 56 to 60, 14.2% of patients were taking less than 20% of the maximum dose (3 g NA) and 21.8% were taking less than 80% of the maximum dose. Ninety-percent adherence to maximum dose, i.e., complete adherence to the prescription for 5 years plus 80%–100% adherence to that prescription for every follow-up period, was found in 26% of patients for NA, 36% for clofibrate, and 40% for placebo.

In the 5-year Stockholm Ischaemic Heart Disease Study, 59% took 50% or more of the prescribed 3 g niceritrol and 75% or more of the prescribed 2 g clofibrate.

In the Familial Atherosclerosis Treatment Study (FATS), in which combinations with colestipol were used, the compliance or percentage of patients taking the prescribed dose was 89% in the placebo group and 88% and 86% in lovastatin–colestipol and NA–colestipol groups, respectively. Thus, there were no differences in compliance between lovastatin and NA treatment in the FATS.

V. Combinations with Other Plasma Lipid-Lowering Drugs

NA has been evaluated in combined treatment with many plasma lipid-lowering drugs. In many intervention studies, drug combinations (NA or a NA derivative being one of the drugs) have been used successfully (see below). The following combinations with NA as one form of treatment have been documented.

1. Nicotinic Acid and Bile Acid Sequestrants

a) Nicotinic Acid and Colestipol

The combination of colestipol and NA was studied by KANE et al. (1981) in patients with familial hypercholesterolaemia. Figure 14 clearly shows that the effect of the two drugs are additive with regard to total and LDL cholesterol and that very pronounced decreases and complete normalization can be achieved with this drug combination. The order of magnitude for total and LDL cholesterol is 45% and 55%, i.e., similar or even larger effects compared with those obtained by HMG-CoA reductase inhibitors. From Fig. 14 it is also clear that NA addition to colestipol can counteract the unwanted effect of VLDL increase which is often seen in treatment with bile acid resins (OLSSON and DAIROU 1978). The addition of NA to colestipol treatment also increases HDL cholesterol concentrations.

b) Acipimox and Cholestyramine

SERIES et al. (1990) studied the combination of acipimox and cholestyramine in 32 patients with hypercholesterolaemia. Small dosages (750 mg and 12 g daily, respectively) were administered to avoid the unpleasant gastrointestinal side effects which otherwise often appear with this combination. The treatment produced a dramatic improvement, giving a total 27% fall in

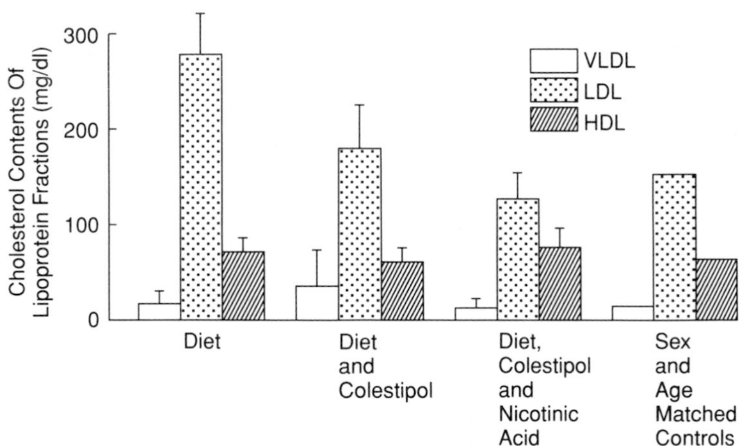

Fig. 14. Effect of colestipol and colestipol plus nicotinic acid treatment on plasma lipoprotein concentrations in familial hypercholesterolaemia. *VLDL*, very low density lipoprotein; *LDL*, low-density lipoprotein; *HDL*, high-density lipoprotein (mean + S.D. (KANE et al. 1981)

plasma cholesterol and 32% decrease in LDL cholesterol. The rise in plasma triglycerides and VLDL seen with cholestyramine alone was not seen with the combination, which resulted in a decrease of plasma triglycerides by 13%. The combination also produced a significant increase in plasma HDL cholesterol and in HDL_2 subfraction mass compared with cholestyramine alone. The authors conclude that this combination of moderate dosages of each drug offers the combined benefits of good plasma lipid regulation and a well-tolerated regimen.

c) Nicotinic Acid and Lovastatin

In a retrospective study Henkin et al. (1991) reported the effect of NA in combination with lovastatin in 22 patients. Mean treatment duration was 12 months. Average NA dosage was 2.5 g daily and average lovastatin dosage was 34 mg/day. In half of the cases, lovastatin was added to a fixed dose of NA, thereby allowing the evaluation of the additive effect of the two drugs (Fig. 15). The addition of lovastatin resulted in a further reduction of total, VLDL, IDL, and LDL cholesterol concentrations, although statistical significance was not reached in the latter value. No excess incidence of adverse effects was noted in this group of patients. In particular, no evidence of muscle damage or elevation of the serum creatine phosphokinase activities was noted.

Eight patients were treated with NA–gemfibrozil combinations, of whom five were also receiving lovastatin. No adverse effects were noted in this group of patients.

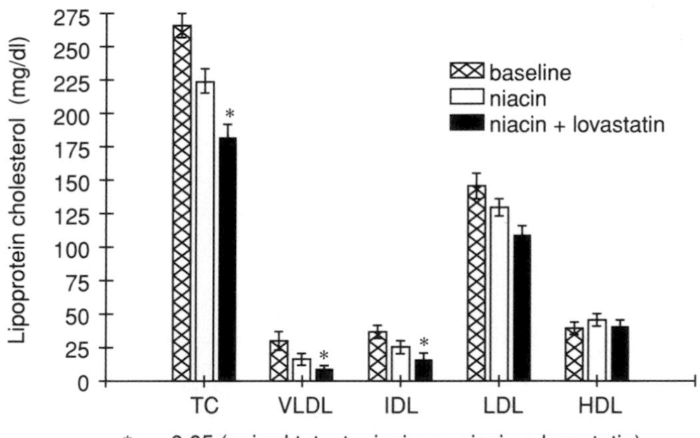

Fig. 15. Lipoprotein levels before niacin therapy, during niacin monotherapy, and after addition of *lovastatin* in 11 patients with hypercholesterolaemia. *TC*, total cholesterol; VLDL, very low density lipoprotein; *IDL* intermediate-density lipoprotein; *LDL*, low-density lipoprotein; *HDL*, high-density lipoprotein (Henkin et al. 1991)

VI. Recommendations for Drug Treatment

1. Policy Statement of the European Atherosclerosis Society

In the Recommendations of the European Atherosclerosis Society (Study Group of the European Atherosclerosis Society 1988), action limits for drug treatment were defined for both plasma cholesterol and triglyceride, whereby groups A, B, C, D, and E were distinguished, thus characterizing borderline hypercholesterolaemia, moderate hypercholesterolaemia, moderate hypertriglyceridaemia, combined moderate hyperlipidaemia, and marked hyperlipidaemia. NA treatment is given as an option in all these groups, emphasizing the broad indications for this drug.

2. The National Cholesterol Education Program

In the American National Cholesterol Education Program (NCEP), in which plasma cholesterol is much emphasized, the drugs of first choice are NA and bile acid sequestrants (The Expert Panel 1988). The reason is that both NA and cholestyramine have been shown to lower CHD risk in clinical trials, and their long-term safety has been established. The Expert Panel, however, points out that these drugs require considerable patient education to achieve effective adherence. NA is the preferred drug in patients with concurrent hypertriglyceridaemia.

E. Clinical Effects

In this section, controlled studies on the effect of NA on atherosclerosis and its manifestations will be covered. A common feature of these controlled studies is the difficulty of maintaining double-blind conditions in studies on NA because of its side effects. Many authors, however, provide circumstantial evidence that observed differences are not due to biases caused by the failure to establish double-blind conditions.

I. Coronary and Femoral Artery Disease

1. Mortality and Morbidity

a) The Coronary Drug Project (CDP)

NA was one of the treatments tested in the CDP. CDP was the first placebo-controlled, multicenter trial of plasma lipid-lowering drugs in the secondary prevention of CHD, i.e., all participants had verified evidence of one or more myocardial infarctions (Coronary Drug Project Research Group 1973). A total of 1119 men aged 30–64 at entry were randomized to NA and 2789 to placebo by the end of recruitment in 1969. The mean length of time from randomization of the patients to scheduled cessation of drug treatment was 74 months (range 5–8.5 years). The dosage prescribed was 3g daily. Of the different plasma cholesterol-lowering treatments used in the CDP (estrogens, clofibrate, and dextrothyroxine), NA was most efficacious, with overall cholesterol lowering of 10% compared with the placebo group. It was also the most efficacious plasma triglyceride-lowering drug in the CDP, with a 27% lower level than the placebo group. The effects on both plasma cholesterol and triglyceride concentrations were maintained throughout the whole study. At the conclusion of the trial in 1975, the NA-treated group exhibited a statistically significantly lower incidence of definite, nonfatal myocardial infarction than the placebo group (CDP Research Group 1975). The 5-year incidence of definite, nonfatal myocardial infarctions was 8.9% for NA and 12.2% for placebo, i.e., a 27% lower incidence in the NA group. There was also a significantly lower incidence of coronary bypass surgery in the NA group than in the placebo group. There was a trend toward improvement in the life table mortality curve, but this was not statistically significant. However, curves on the cumulative mortality rate show a divergence for the NA group, with lower mortality after 68 months.

In 1981 an extended follow-up was carried out concerning mortality rate in the 6008 men in the CDP who were still alive at the end of treatment in 1975 (Canner et al. 1986). Vital status was determined after a mean of 9 years from conclusion of the trial. Death certificates were obtained for 2227 (91%) of the 2451 patients identified as having died after 1975. For others, death reports were received from two or more independent sources. The mortality findings represent a mean patient follow-up period of about 15 years – 6.2 years in the CDP and 8.8 years after termination of the study. In the group previously randomized to NA, the cumulative mortality from all causes was 52.0% compared with 58.2% in the placebo group. There were 69 (11%) fewer deaths than were expected on the basis of mortality in the

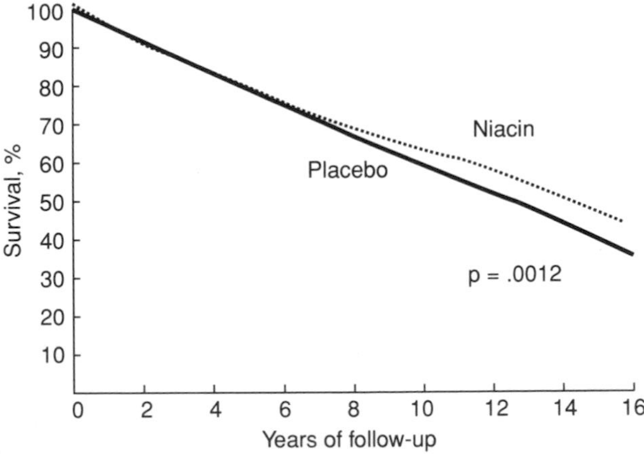

Fig. 16. Survival curves for nicotinic acid and placebo treatment groups in the Coronary Drug Project (Canner et al. 1986)

placebo group (Fig. 16, $p = 0.0004$). The median survival time from entry into the study was 13.03 years for patients in the NA group compared with 11.40 years for those in the placebo group ($p = 0.0012$). The survival benefit in the NA group was primarily evident regarding death caused by CHD. The data also suggested that patients with higher baseline cholesterol experienced greater benefit from NA therapy. This was also the case for those who showed the best cholesterol response to the drug.

The authors suggest that the finding of a significant long-term survival benefit with NA may be explained in part by the earlier decrease in definite nonfatal myocardial infarction. A more likely explanation stems, however, from the cholesterol-lowering effect of NA, which was superior to that of the other drugs studied in the CDP. Thus, it is possible that a 10% reduction in plasma cholesterol, maintained over 5–8 years, may have significantly slowed the progression of coronary atherosclerosis.

b) The Stockholm Ischaemic Heart Disease Study

The Stockholm Ischaemic Heart Disease Study (Carlson and Rosenhamer 1988) was a secondary prevention study comprising 555 patients aged less than 70 years who were all consecutive survivors of a myocardial infarction. Four months after the acute myocardial infarction, patients were randomized to treatment ($n = 279$) or control ($n = 276$) group. Participants in the treatment group received a combined treatment of 3 g niceritrol daily (at month 6) and 1 g clofibrate twice daily. Each patient remained in the study for at least 5 years after the myocardial infarction. Results were evaluated according to the "intention to treat" principle. Randomization was successful and plasma lipid concentrations and distribution of hyperlipoproteinaemias were similar in treatment and control groups. Mean entry plasma cholesterol was 6.35 and 6.51 mmol/l in control and treatment groups, respectively, and the corresponding figures for plasma triglycerides were 2.36 and 2.35 mmol/l. On average, mean plasma cholesterol was lowered by 13% and that of plasma triglyceride by 19% compared to the control group. Total mortality was 61 cases in the treatment group against 82 in the control group ($p < 0.05$), a 26% reduction. For patients over 60 years of age, the corresponding figure was 28%. When major ischaemic heart disease events were considered, the rates per 1000 man-years were 107.8 for the control group and 63.8 for the treatment group. The survival curves are given in Fig. 17.

The reduction in mortality in the treatment group was related to pretreatment plasma lipid levels. Plasma triglyceride concentrations thereby appeared of importance. Practically no effect of treatment was seen if entry plasma triglycerides were below 1.5 mmol/l; almost the entire decrease in mortality rate was confined to patients having entry triglyceride levels above this limit. On the other hand, the pretreatment level of plasma cholesterol did not influence the mortality. In addition, the beneficial effects of treatment on the incidence of CHD death was related to the degree of triglyceride

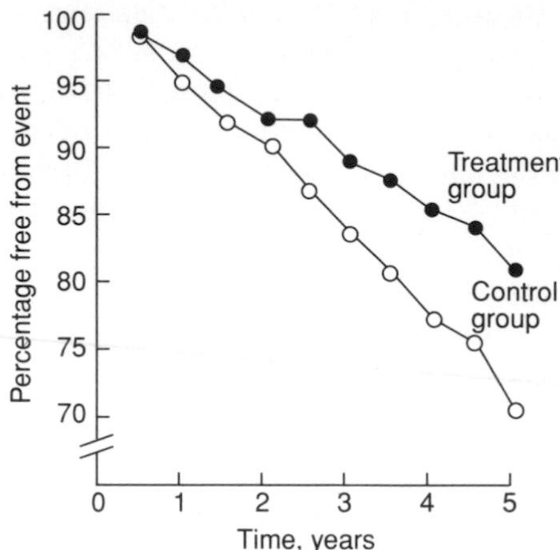

Fig. 17. Curves for survival with time expressed as percentage of patients in the control group and in the treatment group during the Stockholm Ischaemic Heart Disease Secondary Prevention Study (CARLSON and ROSENHAMER 1988).

lowering, but not to plasma cholesterol lowering. In the subgroup in which the plasma triglyceride level was lowered by 30% or more, the ischaemic heart disease mortality was only 9.9% compared to 26.4% in the control group, a reduction of more than 60%. This clear relation between entry plasma lipid levels and lipid effect makes it credible that the effect seen on mortality really is a lipid effect exerted by the combined action of NA and clofibrate.

The effect obtained in the CDP and Stockholm Ischaemic Heart Disease Study diverge in the sense of effect on the mortality rate. To be sure, effects on the mortality rate were seen in both studies, but much later in the CDP. The reasons for this are not known. One reason may be the double treatment of both clofibrate and NA in the Stockholm Study. A more likely explanation is, however, the great differences in patients between the two studies. The Stockholm Study recruited its participants directly after an infarction, while a considerable time had elapsed since the infarction in most cases before randomization into the CDP was made.

2. Regression of Atherosclerosis

a) Coronary Atherosclerosis

α) Cholesterol-Lowering Atherosclerosis Study. In recent years, drug efficacy has been evaluated by monitoring atherosclerosis development over time, a principle named regression studies, and in many of the early studies NA was used. One pioneering study of this stype is the Cholesterol-Lowering Atherosclerosis Study (CLAS; BLANKENHORN et al. 1987a). In this study, several arteries were examined with regard to atherosclerosis development. Results have been published regarding effects on coronary arteries (BLANKENHORN et al. 1987b, 1990; CASHIN-HEMPHILL et al. 1990) and femoral arteries (BLANKENHORN et al. 1991).

The CLAS was a secondary preventive study including 162 non-smoking men of 40–59 years of age who all had undergone coronary bypass graft surgery. Half of them were randomized to a plasma lipid-regulatory treatment by the combination of 3–12 g NA daily and 30 g colestipol daily, and dietary advice was also given. Duration of the study was 2 years. Results of 4 years of follow-up has also been given (CASHIN-HEMPHILL et al. 1990). The primary cardiac end point was the global coronary change score. This was established by a panel of expert angiographers who viewed films without knowledge of a subject's demographic or clinical characteristics, his treatment assignment, or the temporal order of the angiograms. Each reader recorded a global evaluation of the films on a 4-point scale that combined

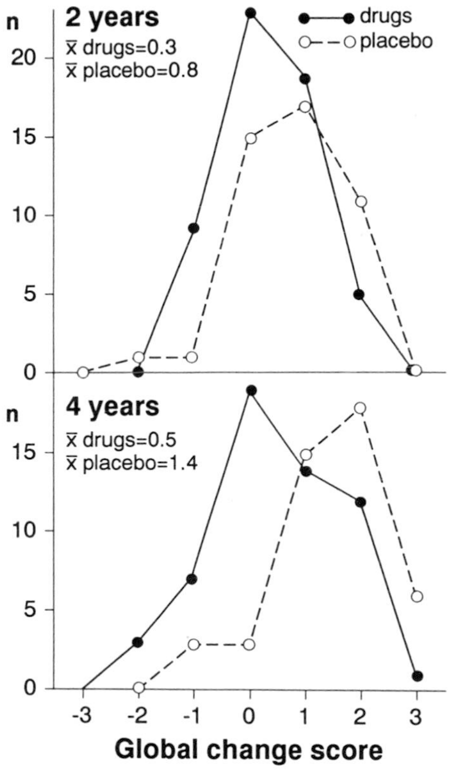

Fig. 18. Global change score at 2 and 4 years in the Cholesterol-lowering Atherosclerosis Study. *0* indicates no change from baseline; *1*, definitely discernible change; *2*, intermediate change; and *3*, extreme change. *Positive numbers* indicate progression, *negative numbers*, regression. Mean global change score in drug and placebo groups are given (BLANKENHORN et al. 1987b; CASHIN-HEMPHILL et al. 1990)

changes in both grafts and native coronary arteries. A score of 0 indicated no demonstrable change, 3, extreme change.

In the treatment group, plasma total cholesterol decreased by 26% and triglycerides by 22%. LDL cholesterol decreased by 43% and the LDL to HDL ratio by 57%. HDL cholesterol increased by 37%.

The global change score is shown in Fig. 18. The distribution for the drug group was shifted towards lower scores, i.e., less disease progression than in the placebo group. Figure 18 also depicts the 4-years results of global change score, which was estimated in a representative majority of the original sample (CASHIN-HEMPHILL et al. 1990). The significant benefit from drug treatment observed at 2 years was more pronounced at 4 years. The mean drug-treated coronary change score was 0.3 at 2 years and 0.5 at 4 years. The mean placebo coronary change score was 0.8 at 2 years and 1.4 at 4 years. The drug-treated group response was more variable at 4 than at 2 years and was slightly shifted towards progression. The placebo-treated group showed a major shift, with 39 subjects (87%) showing definitely discernible to extreme progression at 4 years. At both 2 and 4 years, the beneficial effect of drug treatment was apparent both at high and low levels of entry plasma cholesterol, the division being drawn at 6.22 mmol/l. Among those completing 4 years, 18 experienced clinical events in the drug group versus 23 in the control group. The authors conclude that the results of CLAS has clinical implications insofar as therapy after coronary bypass surgery should routinely include measures to normalize plasma lipid levels.

In another publication, the authors related the global change score to risk factors, i.e., different lipoproteins and apolipoproteins that were monitored during the trial (BLANKENHORN et al. 1990). Univariate analysis in the placebo-treated group confirmed previously known risk factors for CHD to be of importance in the progression of coronary disease: total cholesterol, LDL cholesterol, non-HDL cholesterol, triglycerides, apolipoprotein B, and diastolic blood pressure all had significant effects. Multivariate analysis indicated that the predominant risk factor predicting the probability of global coronary progression in the placebo group was non-HDL cholesterol. In the drug-treated group it was the content of apolipoprotein C-III in HDL which was the only significant risk factor. Apolipoprotein C-III inhibits the enzyme lipoprotein lipase and thereby retards the hepatic uptake of triglyceride-rich lipoproteins and their remnants. Apolipoproteins C also seem to inhibit or modulate the apolipoprotein E-mediated hepatic uptake of triglyceride-rich lipoproteins or their remnants, at least as long as they are integral components of these particles. Apolipoprotein C-III can be used as a marker to estimate the distribution of apolipoprotein C-peptides between VLDL and HDL. It has been suggested (EISENBERG 1983) that larger amounts of apolipoprotein C-III in HDL than in VLDL (a higher C-III ratio) are indicative of recent chylomicron and VLDL clearance. Patients with a consistent increase in the C-III ratio should have decreased concentrations of triglyceride-rich lipoproteins and thus reduced probability of contact between these particles and the arterial wall. It has previously also been demonstrated that patients with severe hypertriglyceridaemia (type V hyperlipoproteinaemia) have a high ratio of apolipoprotein C-III to C-II in S_f100–400 and S_f60–100 (CARLSON and BALLANTYNE 1976). Inhibition of lipoprotein lipase-activated lipolysis by apolipoprotein C-III should prolong the circulation time for these particles. Transfer of apolipoprotein C-III from VLDL to HDL could simultaneously increase the fractional catabolic rate of atherogenic VLDL. In all probability these lipoprotein compositional changes have been induced by NA in the CLAS, as the effect of colestipol is to increase VLDL. It is, therefore, probable that much of the beneficial effect on coronary atherosclerosis development induced by the treatment is an effect of the NA part of the treatment. The findings also points to one mechanism by which NA probably exerts its protective effect against CHD as documented in the CDP and the Stockholm Ischaemic Heart Disease Study.

β) The Familial Atherosclerosis Treatment Study. The FATS was also a secondary preventive study comprising 146 men under 62 years of age, who had documented CHD in their family history and high plasma apolipoprotein B concentrations (BROWN et al. 1990). The study had a double-blind design and a duration of 2.5 years. All patients received dietary counseling and were then randomized to one of three groups: one received NA in increasing dosages up to 4 g daily plus 30 g colestipol daily (NA–colestipol group); one group received colestipol in the same dosage plus 20 or 40 mg lovastatin daily depending on response (lovastatin–colestipol group); and one group (conventional-therapy group), in which placebo and occasionally colestipol was prescribed. Coronary atherosclerosis was estimated by direct visual comparison of pairs of cineangiograms.

Entry mean plasma total cholesterol concentration was 6.99 mmol/l and fell to 6.55 mmol/l in the conventional group and to 5.41 and 4.71 mmol/l in the NA–colestipol and lovastatin–colestipol groups, respectively. Mean plasma LDL cholesterol decreased by 7%, 32%, and 46% in the three groups, respectively, while plasma HDL cholesterol increased by 5%, 43%, and 15%, respectively.

At baseline the average percentage of stenosis caused by the worst lesion was 34%. By conventional therapy, this index increased by 2.1%. By contrast, it decreased by 0.9% in the NA–colestipol group and by 0.7% in the lovastatin–colestipol group. Thus, at the end of the study the lesions were almost 3% less severe in the intensively treated groups in comparison with conventionally treated patients, and this represents almost 10% of the disease present at baseline (34% stenosis). An interesting result of the FATS is also the distribution of clinical events between the groups. Eleven events occurred among the 52 patients originally assigned to the conventional-therapy group. In striking contrast ($p = 0.01$), only two of 48 patients assigned to the NA–colestipol group had events and only three of 46 in the lovastatin–colestipol group had an event. The authors also could correlate worsening symtoms to increased stenosis.

In a multivariate analysis introducing mode of intensive therapy, no difference was noted between the NA–colestipol and lovastatin–colestipol groups, suggesting that the modes of treatment were equally effective in influencing stenosis change. Independent variables significantly predicting stenosis change were percent change in LDL cholesterol, percent change in HDL cholesterol, percent change in systolic blood pressure, and change in ST segment depression at peak effort.

The FATS confirms findings made in the CLAS, namely, that intense plasma lipid-regulatory treatment by drugs, including that by NA, is able to visibly influence the growth rate of coronary atherosclerosis in a relatively short time.

∂) The Specialized Center of Research Study. This regression study of coronary atherosclerosis comprised 72 patients with familial hypercholesterolaemia, of whom 41 were women (KANE et al. 1990). Angiographic evaluation followed the same technique as the FATS. Treatment consisted of diet and up to 30 g colestipol and up to 7.5 g NA. Later, lovastatin was also added to ensure a maximal plasma cholesterol lowering. Of the 40 patients in the treatment group, 36 took NA. Mean plasma LDL cholesterol decreased on treatment from 7.32 to 4.45 mmol/l or by 39%. The mean change in percent area stenosis among controls was +0.80%, indicating progression, while the mean change for the treatment group was −1.53%, indicating regression ($p = 0.039$). Significant regression was also seen in women. The change in percent area stenosis was correlated with LDL cholesterol on trial. The authors conclude that reduction in plasma LDL cholesterol can induce regression of atherosclerotic lesions in coronary arteries in patients with familial hypercholesterolaemia.

b) Femoral Atherosclerosis

The first published observation of angiographically documented cases of regression of human atherosclerosis concerned the femoral arteries in patients with intermittent claudication who were treated with NA (ÖST and STENSON 1967).

We studied the effect of a combined plasma lipid-lowering drug treatment consisting of fenofibrate and NA on femoral atherosclerosis in 45 asymptomatic hyperlipidaemic men (OLSSON et al. 1990). About equal numbers of patients with types IIa, IIb, and IV were included. In addition to diet, NA and fenofibrate were given to the participants in the treatment group ($n = 20$) at dosages of 4 and 0.4 g daily, respectively. The study lasted for 1 year and femoral angiography was performed at the beginning and end of this period. Changes in atheroma were visually judged by two independent, experienced radiologists in four segments of the femoral artery, and an overall atherosclerosis score was then determined. Consensus was reached regarding the change in atherosclerosis.

In the treatment group, plasma VLDL and LDL cholesterol decreased on treatment by a mean of 67% and 36%, respectively, while HDL cholesterol increased by 23%.

Progression of femoral atherosclerosis was found in 24% and 40% of patients in the treatment and control groups, respectively. Regression occurred in 29% in the treatment group, but was not seen in the control group. The difference in distribution of progression/regression in the treatment and control groups was highly statistically significant. Regression correlated significantly and independently with changes in VLDL cholesterol and changes in systolic blood pressure.

The 2-year therapy effect of changes in femoral atherosclerosis was also evaluated in the CLAS (Blankenhorn et al. 1991). For design of study and treatment effect on plasma lipids, see above. In general, effects on femoral atherosclerosis were less consistent than on coronary atherosclerosis in the CLAS. Overall, both treatment and control groups exhibited progression on a per segment basis, although the control group demonstrated a slightly larger progression, which did not differ significantly from treatment group. In subgroups, significant effects of treatment were seen in those with moderate disease and in proximal segments. On a per patient basis, a significant therapy effect was observed with more progressors in the placebo group and more regressors in the treatment group ($p < 0.02$). On-trial HDL cholesterol and ratio of total cholesterol to HDL cholesterol related significantly to regression.

3. Conclusion on the Effect of Nicotinic Acid on Atherosclerosis

Both studies with a clinical end point such as acute myocardial infarction and total mortality and regression studies have documented beneficial effects by NA. It is interesting to note the concordance with these two types of studies, for example, the Stockholm Ischaemic Heart Disease Study and the CLAS, which both included patients with manifest CHD. Dramatic and significant effects were found in both studies. Both studies also show that the triglyceride-rich lipoproteins seem to play the most important role in terms of the benefit of treatment. It is on that particular plasma lipid which NA has it most pronounced effect. Even if combined drug treatments were utilized in most of the performed studies, therefore, the contribution of NA to the overall lipid effect could probably explain a large part of the beneficial effect on disease.

II. Dementia

The NA derivative xanthinol nicotinate has long been prescribed in clinical practice to patients with various types of blood circulation disturbances. The derivative (Fig. 1E) was studied at a dosage of 3g daily in 150 patients with two types of dementia, multi-infarct dementia and senile dementia of Alzheimer type, for 12 weeks in comparison with placebo (Kanowski et al. 1990). The improvement was statistically significant for the clinical global impression, which was the primary criterion of efficacy. Activation of cerebral metabolism and improvement of microcirculation by influencing rheological parameters are claimed to be the underlying pharmacological principles responsible for the efficacy of xantinolnicotinate.

F. General Conclusions

After more than 35 years since the discovery of its plasma cholesterol-lowering effects, NA is still a first-line plasma lipid-regulatory drug. The drug has an effect on a variety of atherogenic and antiatherogenic lipoproteins, and its overall regulatory effects work towards normalizing various types of hyperlipoproteinaemia, including HDL subfractions and Lp(a) concentrations. It is the only plasma lipid-regulatory drug which has shown effects on total mortality in secondary prevention. It can induce atherosclerosis regression and retardation of atherosclerosis development. A recently developed NA analog, acipimox, seems to be devoid of unwanted diabetogenic effects otherwise observed during NA treatment. The exact mechanism of the action of NA on plasma lipoprotein and glucose metabolism is still incompletely understood: in particular, the NA–cellular surface receptor interrelationship (if any) and the hepatocellular events following decreased FFA availability have as yet been poorly investigated and are still not well understood.

In view of the increasing evidence of the importance of triglyceride-rich lipoproteins in atherogenesis, NA and its derivatives will certainly remain first-line plasma lipid-regulatory drugs for many years to come.

References

Aktories K, Jakobs KH (1982) In vivo and in vitro desensitization of nicotinic acid-induced adipocyte adenylate cyclase inhibition. Arch Pharmacol 318:241–245

Aktories K, Schultz G, Jakobs KH (1980) Regulation of adenylate cyclase activity in hamster adipocytes. Naunyn Schmiedebergs Arch Pharmacol 312:167–173

Altschul R, Hoffer A, Stephen JD (1955) Influence of nicotinic acid on serum cholesterol in man. Arch Biochem 54:558

Angelin B, Einarsson K, Leijd B (1979) Biliary lipid composition during treatment with different hypolipidaemic drugs. Eur J Clin Invest 9:185–190

Austin MA, Breslow JL, Hennekens CH, Buring JE, Willett WC, Krauss RM (1988) Low-density lipoprotein subclass patterns and risk of myocardial infarction. JAMA 260:1917–1921

Auwerx JH, Marzetta CA, Hokanson JE, Brunzell JD (1989) Large buoyant LDL-like particles in hepatic lipase deficiency. Arteriosclerosis 9:319–325

Aviram M, Bierman EL, Chait A (1988) Modification of low density lipoprotein by lipoprotein lipase or hepatic lipase induces enhanced uptake and cholesterol accumulation in cells. J Biol Chem 263:116–122

Becker MA, Raivio KO, Meyer LJ, Seegmiller JE (1973) Effects of nicotinic acid on human purine metabolism (Abstr). Clin Res 21:616

Berge KG, Achor WP, Christensen NA et al. (1961) Hypercholesterolemia and nicotinic acid: a long-term study. Am J Med 31:25–36

Blanche PJ, Gong EL, Forte T, Nichols AV (1981) Characterization of human high density lipoproteins by gradient gel electrophoresis. Biochim Biophys Acta 665:408

Blankenhorn DH, Johnson RL, Nessim SA, Azen SP, Sanmarco ME, Selzer RH (1987a) The cholesterol lowering atherosclerosis study (CLAS): design, methods, and baseline results. Contrib Clin Trials 8:354–387

Blankenhorn DH, Nessim SA, Johnson RL, Sanmarco ME, Azen SP, Cashin-Hemphill L (1987b) Beneficial effects of combined colestipol-niacin therapy on coronary atherosclerosis and coronary venous bypass grafts. JAMA 257: 3233–3240

Blankenhorn DH, Alaupovic P, Wickham E, Chin HP, Azen SP (1989) Prediction of angiographic change in native human coronary arteries and aortocoronary bypass grafts. Circulation 81:470–476

Blankenhorn DH, Azen SP, Crawford DW, Nessim SA, Sanmarco ME, Selzer RH, Shircore AM, Wickham EC (1990) Effects of colestipol–niacin therapy on human femoral atherosclerosis. Circulation 83:438–447

Boberg J, Carlson LA, Fröberg S, Olsson A, Orö L, Rössner S (1971) Effects of chronic treatment with nicotinic acid on intravenous fat tolerance and post-heparin lipoprotein lipase in man. In: Gey KF, Carlson LA (eds) Metabolic effects of nicotinic acid and its derivatives. Huber, Bern, p 533

Brown G, Albers JJ, Fisher LD, Schaefer SM, Lin JT, Kaplan C, Zhao XQ, Bisson BD, Fitzpatrick VF, Dodge HT (1990) Regression of coronary artery disease as a result of intensive lipid-lowering therapy in men with high levels of apolipoprotein B. N Engl J Med 323:1289–1298

Brown WV, Baginsky ML (1972) Inhibition of lipoprotein lipase by an apoprotein of human very low density lipoprotein. Biochem Biophys Res Commun 46: 375

Butcher RW, Baird CE, Sutherland EW (1968) Effects of lipolytic and antilipolytic substances on adenosine 3′,5′-monophosphate levels in isolated fat cells. J Biol Chem 243:1705

Canner PL, Berge KG, Wenger NK, Stamler J, Friedman LF, Prineas RJ, Friedewald W, for the Coronary Drug Project Research Group (1986) Fifteen year mortality in coronary drug project patients: long-term benefit with niacin. J Am Coll Cardiol 8:1245–1255

Carlson LA (1963) Studies on the effect of nicotinic acid on catecholamine-stimulated lipolysis in adipose tissue in vitro. Acta Med Scand 173:719

Carlson LA (1984) Effect of nicotinic acid on serum lipids and lipoproteins. In: Carlson LA, Olsson AG (eds) Treatment of hyperlipoproteinemia. Raven, New York, pp 115–119

Carlson LA, Ballantyne D (1976) Changing relative proportions of apolipoproteins CII and CIII of very low density lipoproteins in hypertriglyceridaemia. Atherosclerosis 23:563–568

Carlson LA, Olsson AG (1979) Effect of hypolipidemic drugs on serum lipoproteins. Prog Biochem Pharmacol 15:238–257

Carlson LA, Orö L (1962) The effect of nicotinic acid on the plasma free fatty acids. Demonstration of a metabolic type of sympathicolysis. Acta Med Scand 172: 641

Carlson LA, Orö L (1969) Acute effect of a sustained release pyridyl preparation Ronicol Retard, on plasma free fatty acids. Acta Med Scand 186:337

Carlson LA, Rosenhamer G (1988) Reduction of mortality in the Stockholm Ischaemic Heart Disease Secondary Prevention Study by combined treatment with clofibrate and nicotinic acid. Acta Med Scand 223:405–418

Carlson LA, Wahlberg G (1978) Relative increase in apolipoprotein CII content of VLDL and chylomicrons in a case with massive type V hyperlipoproteinaemia by nicotinic acid treatment. Atherosclerosis 31:77–84

Carlson LA, Walldius (1976) Fatty acid incorporation into human adipose tissue in hypertriglyceridaemia. Eur J Clin Invest 6:195–211

Carlson LA, Orö L, Östman J (1968) Effect of nicotinic acid on plasma lipids in patients with hyperlipoproteinemia during the first week of treatment. J Atheroscler Res 8:667

Carlson LA, Eriksson I, Walldius G (1973) A case of massive hypertriglyceridaemia and impaired fatty acid incorporation into adipose tissue glycerides (FIAT), both corrected by nicotinic acid. Acta Med Scand 194:363–369

Carlson LA, Olsson AG, Orö L, Rössner S, Walldius G (1974) Effects of hypolipidemic regimes on serum lipoproteins. In: Schettler G, Weizel A (eds) Atherosclerosis III. Springer, Berlin Heidelberg New York, p 768

Carlson LA, Olsson AG, Ballantyne D (1977) On the rise in low density and high density lipoproteins in response to the treatment of hypertriglyceridaemia in type IV and type V hyperlipoproteinaemias. Atherosclerosis 26:603–609

Carlson LA, Hamsten A, Asplund A (1989) Pronounced lowering of serum levels of lipoprotein Lp(a) in hyperlipidaemic subjects treated with nicotinic acid. J Intern Med 226:271–276

Cashin-Hemphill L, Mack WJ, Pogoda JM, Sanmarco ME, Azen SP, Blankenhorn DH (1990) Beneficial effects of colestipol–niacin on coronary atherosclerosis. JAMA 264:3013–3017

Christensen NA, Achor RWP, Berge KG, Mason HL (1961) Nicotinic acid treatment of hypercholesteremia: comparison of plain and sustained-action preparations and report of two cases of jaundice. JAMA 177:546–550

Clementz GL, Holmes AW (1989) Nicotinic acid-induced fulminant hepatic failure. J Clin Gastroenterol 9:582–584

Coe FL, Kathpalia S (1991) Hereditary tubular disorders. In: Wilson JD, Braunwald E, Isselbacher KJ, Petersdorf RG, Martin JB, Fauci AS, Root RK (eds) Principles of internal medicine, vol 2. McGraw-Hill, New York, pp 1196–1202

Coronary Drug Project Research Group (1973) The coronary drug project: design, methods, and baseline results. Circulation 1 Suppl 1:1–179

Coronary Drug Project Research Group (1975) Clofibrate and niacin in coronary heart disease. JAMA 231:360–381

Ditschuneit HH, Knispel DC, Flechtner-Mors M, Voisard R, Ditschuneit H (1992) Postprandial lipemia in coronary artery disease: effects of acimipox, Coronary Artery Dis 3:49–56

Duffield RGM, Lewis B, Miller NE, Jamieson CW, Brunt JNH, Colchester ACF (1983) Treatment of hyperlipidaemia retards progression of symptomatic femoral atherosclerosis: a randomised controlled trial. Lancet 2:639–642

Dulbecco A, Albenga C, Borretta G, Vacca G, Milanesi G, Lavezzari M (1989) Effect of acipimox on plasma glucose levels in patients with non-insulin-dependent diabetes mellitus. Curr Ther Res 46:478–483

Earthman TP, Odom L, Mullins CA (1991) Lactic acidosis associated with high-dose niacin therapy. South Med J 84:496–497

Einarsson K, Hellström K, Leijd B (1977) Bile acid kinetics and steroid balance during nicotinic acid therapy in patients with hyperlipoproteinemias types II and IV. J Lab Clin Med 90:613–622

Eisenberg S (1983) Lipoproteins and lipoprotein metabolism: a dynamic evaluation of the plasma fat transport system. Klin Wochenschr 61:119–132

Elvehjem CA (1952) Nutritional interrelations. Int Z Vitaminforsch 23:299

Ericsson S, Eriksson M, Angelin B (1990) Biliary lipids in familial combined hyperlipidaemia: effects of acipimox therapy. Eur J Clin Invest 20:261–265

Etchason JA, Miller TD, Squires RW, Allison TG, Gau GT, Marttila JK, Kottke BA (1991) Niacin-induced hepatitis: a potential side effect with low-dose time-release niacin. Mayo Clin Proc 66:23–28

Ferrannini E, Barret EJ, Bevilacqua S, DeFronzo RA (1983) Effect of fatty acids on glucose production and utilization in man. J Clin Invest 72:1737–1747
Franceschini G, Bernini F, Michelagnoli S, Bellosta S, Vaccarino V, Fumagalli R, Sirtori CR (1991) Lipoprotein changes and increased affinity of LDL for their receptors after acipimox treatment in hypertriglyceridemia. Atherosclerosis 81:41–49
Fredholm BB (1981) Adenosine and lipolysis. Int J Obes 5:643–649
Fuccella LM, Goldaniga G, Lovisolo P, Maggi E, Musatti L, Mandelli V, Sirtori CR (1980) Inhibition of lipolysis by nicotinic acid and by acipimox. Clin Pharmacol Ther 28:790–795
Fumagalli R (1971) Pharmacokinetics of nicotinic acid and some of its derivatives. In: Gey KF, Carlson LA (eds) Metabolic effects of nicotinic acid and its derivatives. Huber, Bern, pp 33–49
Garg A, Grundy SM (1990) Nicotinic acid as therapy for dyslipidemia in non-insulin-dependent diabetes mellitus. JAMA 264:723–726
Gass JDM (1973) Nicotinic acid maculopathy. Am J Ophtalmol 76:500–510
Gaut ZN, Pocelinko R, Solomon HM et al. (1971) Oral glucose tolerance, plasma insulin and uric acid excretion in man during chronic administration of nicotinic acid. Metabolism 20:1031–1035
Gaw A, Griffin BA, Caslake MJ, Collins SM, Lorimer AR, Packard CJ, Shepherd J (1990) Effects of acipomox on apolipoprotein B metabolism and low density lipoprotein subfraction distribution in hypercholesterolaemic subjects. J Drug Dev 3 Suppl 1:107–109
Gey KF (1971) Opening address. In: Gey KF, Carlson LA (eds) Metabolic effects of nicotinic acid and its derivatives. Huber, Bern, pp 27–29
Gey KF, Carlson LA (1971) Metabolic effects of nicotinic acid and its derivatives. Huber, Bern
Golay A, Chen Y-DI, Reaven GM (1986) Effect of differences in glucose tolerance on insulin's ability to regulate carbohydrate and FFA metabolism in obese individuals. J Clin Endocrinol Metab 62:1081–1088
Green A, Milligan G, Belt SE (1991) Effects of prolonged treatment of adipocytes with PGE_1, N^6-phenylisopropyl adenosine and nicotinic acid on G-proteins and antilipolytic sensitivity. Biochem Soc Trans 19:2125
Gross RC, Carlson LA (1968) Metabolic effects of nicotinic acid in acute insulin deficiency in the rat. Diabetes 17:353
Grundy SM, Mok HYI, Zech L, Berman M (1981) Influence of nicotinic acid on metabolism of cholesterol and triglycerides in man. J Lipid Res 22:24–36
Gundersen K, Demissianos HV (1969) The effect of 5-methylpyrazole-3-carboxylic acid and nicotinic acid on free fatty acids (FFA), triglycerides and cholesterol in man. In: Holmes LWL, Carlson LA, Paoletti R (eds) Drugs affecting lipid metabolism. Plenum, New York, p 213
Havel RJ (1990) Role of triglyceride-rich lipoproteins in progression of atherosclerosis. Circulation 81:694–696
Henderson LM (1983) Niacin. In: Darby WJ, Broquist HP, Olson RE (eds) Annual reviews 1983. Palo Alto, pp 289–307
Henkin Y, Johnson KC, Segrest JP (1990) Rechallenge with crystalline niacin after drug-induced hepatitis from sustained-release niacin. JAMA 264:241–243
Henkin Y, Oberman A, Hurst DC, Segrest JP (1991) Niacin revisited: clinical observations on an important but underutilized drug. Am J Med 91:239–246
Hodis HN (1990) Acute hepatic failure associated with the use of low-dose sustained-release niacin. JAMA 264:181
Johansson J, Carlson LA (1990) The effects of nicotinic acid treatment on high density lipoprotein particle size subclass levels in hyperlipidaemic subjects. Atherosclerosis 83:207–216
Johansson J, Carlson LA, Landou C, Hamsten A (1991) High density lipoprotein and coronary atherosclerosis: a strong relation with the largest HDL particles is confined to normotriglyceridaemic subjects. Arteriosclerosis 11:174–182

Johnston P, Hollenbeck C, Sheu W, Chen Y-DI, Reaven GM (1990) Acute changes in plasma non-esterified fatty acid concentration do not change hepatic glucose production in people with type 2 diabetes. Diabetic Med 7:871–875

Kaijser L, Eklund B, Olsson AG, Carlson LA (1979) Dissociation of the effects of nicotinic acid on vasodilatation and lipolysis by a prostaglandin synthesis inhibitor, indomethacin, in man. Med Biol 57:114–117

Kane JP, Malloy MJ (1982) Treatment of hypercholesterolemia. Symposium on lipid disorders. Med Clin North Am 66:537–550

Kane JP, Malloy MJ, Tun P, Phillips NR, Freedman DD, Williams ML, Rowe JS, Havel RJ (1981) Normalization of low-density-lipoprotein levels in heterozygous familial hypercholesterolemia with a combined drug regimen. N Engl J Med 304:251–258

Kane JP, Malloy MJ, Ports TA, Phillips NR, Diehl JC, Havel RJ (1990) Regression of coronary atherosclerosis during treatment of familial hypercholesterolemia with combined drug regimens. JAMA 264:3007–3012

Kanowski S, Fischhof PK, Grobe-Einsler R, Wagner G, Litschauer G (1990) Efficacy of xantinolnicotinate in patients with dementia. Pharmacopsychiatry 23:118–124

Kaplan LM (1991) Endocrine tumors of the gastrointestinal tract and pancreas. In: Wilson JD, Braunwald E, Isselbacher KJ, Petersdorf RG, Martin JB, Fauci AS, Root RK (eds) Principles of internal medicine, vol 2. McGraw-Hill, New York, pp 1386–1393

Keenan JM, Fontaine PL, Wenz JB, Myers S, Huang Z, Ripsin CM (1991) A randomized, controlled trial of wax-matrix sustained-release niacin in hypercholesterolemia. Arch Intern Med 151:1424–1432

Kleinman Y, Eisenberg S, Oschry Y, Gavish Y, Gavish D, Stein O, Stein Y (1985) Defective metabolism of hypertriglyceridemic low density lipoprotein in cultured human skin fibroblasts. Normalization with bezafibrate therapy. J Clin Invest 75:1796–1803

Knopp RH (1989) Niacin and hepatic failure. Ann Intern Med 111:769

Knopp RH, Ginsberg J, Albers JJ, Hoff C, Ogilvie JT, Warnick GR, Burrows E, Retzlaff B, Poole M (1985) Contrasting effects of unmodified and time-release forms of niacin on lipoproteins in hyperlipidemic subjects: clues to mechanism of action of niacin. Metabolism 34:642–650

Kostner GM (1988) The affection of lipoprotein-a by lipid lowering drugs. In: Widhalm K, Naito HK (eds) Recent aspects of diagnosis and treatment of lipoprotein disorders. Liss, New York, p 255

Kudchodkar BJ, Sodhi HS, Horlick L, Mason DT (1978) Mechanisms of hypolipidemic action of nicotinic acid. Clin Pharmacol Ther 24:354–373

Langer T, Levy RI (1971) The effect of nicotinic acid on the turnover of low density lipoproteins in type II hyperlipoproteinemia. In: Gey KF, Carlson LA (eds) Metabolic effects of nicotinic acid and its derivatives. Huber, Bern, p 641

LaRosa JC, Levy RI, Herbert P, Lux SE, Fredrickson DS (1970) A specific apoprotein activator for lipoprotein lipase. Biochem Biophys Res Commun 41:57

Lavezzari M, Milanesi G, Oggioni E, Pamparana F (1989) Results of a phase IV study carried out with acipimox in type II diabetic patients with concomitant hyperlipoproteinaemia. J Int Med Res 17:373–380

Magide AA, Myant NB (1975) Loss of cholesterol from muscle and skin of monkeys treated with nicotinic acid. Atherosclerosis 21:273–281

McCormick DB (1988) Niacin. In: Shils ME, Young VR (eds) Modern nutrition in health and disease. Lea and Febiger, Philadelphia, p 370

McKenzie FR, Milligan G (1990) δ-Opioid-receptor-mediated inhibition of adenylate cyclase in transduced specifically by the guanine-nucleotide-binding protein G_i2. Biochem J 267:391–398

Miettinen TA (1968) Effect of nicotinic acid on catabolism and synthesis of cholesterol in man. Clin Chim Acta 20:43–51

Miettinen TA, Taskinen MR, Pelkonen R, Nikkilä EA (1969) Glucose tolerance and plasma insulin in man during acute and chronic administration of nicotinic acid. Acta Med Scand 186:247

Morrow JD, Parsons WG, Roberts LJ (1989) Release of markedly increased quantities of prostaglandin D2 in vivo in humans following the administration of nicotinic acid. Prostaglandins 38:263–274

Moutafis CD, Myant NB (1971) Effects of nicotinic acid, alone or in combination with cholestyramine, on cholesterol metabolism in patients suffering from familial hyperbetalipoproteinemia in the homozygous form. In: Gey KF, Carlson LA (eds) Metabolic effects of nicotinic acid and its derivatives. Huber, Bern, pp 659–676

Mullin GE, Greenson JK, Mitchell MC (1989) Fulminant hepatic failure after ingestion of sustained-release nicotinic acid. Ann Intern Med 111:253–255

Musatti L, Maggi E, Moro E, Valzelli G, Tamassia V (1981) Bioavailability and pharmacokinetics in man of acipimox, a new antilipolytic and hypolipemic agent. J Intern Med 9:381–386

Noma A, Maeda S, Okuno M, Abe A, Muto Y (1990) Reduction of serum lipoprotein(a) levels in hyperlipidaemic patients with alpha-tocopheryl nicotinate. Atherosclerosis 84:213–217

Nozaki S, Kihara S, Kubo M, Kameda K, Matsuzawa Y, Tarui S (1987) Increased compliance of niceritrol treatment by addition of aspirin: relationship between changes in prostaglandins and skin flushing. Int J Pharmacol Ther Toxicol 25:643–647

Ogata N, Ueda K, Kawaichi, Hayaishi O (1981) Poly(ADP-Ribose) synthetase, a main acceptor of poly(ADP-Ribose) in isolated nuclei. J Biol Chem 256:4135

Olsson AG, Carlson LA (1975) Studies in asymptomatic primary hyperlipidaemia: I. types of hyperlipoproteinaemias and serum lipoprotein concentrations, compositions and interrelations. Acta Med Scand Suppl 580:1–37

Olsson AG, Dairou F (1978) Acute effects of cholestyramine on serum lipoprotein concentrations in type II hyperlipoproteinaemia. Atherosclerosis 29:53–61

Olsson AG, Orö L, Rössner S (1974) Clinical and metabolic effects of pentaerythritol tetra-nicotinate (Perycit) and a comparison with plain nicotinic acid. Atherosclerosis 19:61–73

Olsson AG, Carlson LA, Änggard E, Ciabattoni G (1983) Prostacyclin production augmented in the short term by nicotinic acid. Lancet 2:565–566

Olsson AG, Walldius G, Wahlberg G (1986) Pharmacological control of hyperlipidaemia: nicotinic acid and its analogues–mechanisms of action, effects and clinical usage. In: Fears R (ed) Pharmacological control of hyperlipidaemia. Prous Science, Barcelona, p 217

Olsson AG, Ruhn G, Erikson U (1990) The effect of serum lipid regulation on the development of femoral atherosclerosis in hyperlipidaemia: a non-randomized controlled study. J Intern Med 227:381–390

Öst RC, Stenson S (1967) Regression of peripheral atherosclerosis during therapy with high doses of nicotinic acid. Scand J Clin Lab Invest Suppl 99:241–245

Parsons WB (1961) Studies of nicotinic acid: use in hypercholesterolemia. Arch Intern Med 107:653–667

Patsch JR, Sailer S, Kostner G, Sandhofer F, Holasek A, Braunsteiner H (1974) Separation of the main lipoprotein classes from human plasma by rate zonal ultracentrifugation. J Lipid Res 15:356–366

Rader JI, Calvert RJ, Hathcock JN (1992) Hepatic toxicity of unmodified and time-release preparations of niacin. Am J Med 1:77–81

Randle PLJ, Hales CN, Garland PB, Newsholme EA (1963) The glucose fatty-acid cycle. Its role in insulin sensitivity and the metabolic disturbances of diabetes mellitus. Lancet 1:785–789

Reaven GM, Chang H, Ho H, Jeng C-Y, Hoffman BB (1988) Lowering of plasma glucose in diabetic rats by antilipolytic agents. Am J Physiol 254:E23–E30

Rivin AU (1959) Jaundice occurring during nicotinic acid therapy for hypercholesterolemia. JAMA 170:20881089

Rosenberg IH (1991) Nutrition and nutritional requirements. In: Wilson JD, Braunwald E, Isselbacher KJ, Petersdorf RG, Martin JB, Fauci AS, Root RK (eds) Harrison's principles of internal medicine. McGraw-Hill, New York, p 403

Schwab RA, Bachhuber BH (1991) Delirium and lactic acidosis caused by ethanol and niacin coingestion. Am J Emerg Med 9:363–365
Scott JT (1991) Drug-induced gout. Baillieres Clin Rheumatol 5:39–60
Senogles SE, Spiegel AM, Padrell E, Iyengar R, Caron MG (1990) Specificity of receptor-G protein interactions. J Biol Chem 265:4507–4514
Series JJ, Gaw A, Kilday DK et al. (1990) Acipimox in combination with low dose cholestyramine for the treatment of type II hyperlipidaemia. Br J Clin Pharmacol 30:49–54
Shepherd J, Packard CJ (1986) Apolipoprotein B metabolism in man. Acta Med Scand Suppl 715:61
Shepherd J, Packard CJ, Patsch JR, Gotto AM, Taunton OD (1979) Effects of nicotinic acid therapy on plasma high density lipoprotein subfraction distribution and composition and on apolipoprotein A metabolism. J Clin Invest 63:858–867
Steiner A, Weisser B, Vetter W (1991) A comparative review of the adverse effects of treatments for hyperlipidaemia. Drug Saf 6:118–130
Study Group of the European Atherosclerosis Society (1988) The recognition and management of hyperlipidaemia in adults: a policy statement of the European Atherosclerosis Society. Eur Heart J 9:571–600
Svedmyr N, Hägglund A, Åberg G (1977) Influence of indomethacin on flush induced by nicotinic acid in man. Acta Pharmacol Toxico 41:397–400
Taskinen M-R, Nikkilä EA (1988) Effects of acipimox on serum lipids, lipoproteins and lipolytic enzymes in hypertriglyceridemia. Atherosclerosis 69:249–255
The Expert Panel (1988) Report of the National Cholesterol Education Program Expert Panel on detection, evaluation, and treatment of high blood cholesterol in adults. Arch Intern Med 148:36–69
Tornvall P, Walldius G (1991) A comparison between nicotinic acid and acipimox in hypertriglyceridaemia – effects on serum lipids, lipoproteins, glucose tolerance and tolerability. J Intern Med 230:415–421
Tornvall P, Hamsten A, Johansson J, Carlson LA (1990) Normalisation of the composition of very low density lipoprotein in hypertriglyceridemia by nicotinic acid. Atherosclerosis 84:219–227
Vaag A, Skött P, Damsbo P, Gall M-A, Richter EA, Beck-Nielsen H (1991) Effect of the antilipolytic nicotinic acid analogue acipimox on whole-body and skeletal muscle glucose metabolism in patients with non-insulin-dependent diabetes mellitus. J Clin Invest 88:1282–1290
Wahlberg G, Holmquist L, Walldius G, Annuzzi G (1988) Effects of nicotinic acid on concentrations of serum apolipoproteins B, C-I, C-II, C-III and E in hyperlipidemic patients. Acta Med Scand 224:319–327
Wahlberg G, Walldius G, Olsson AG, Kirstein P (1990) Effects of nicotinic acid on serum cholesterol concentrations of high density lipoprotein subfractions HDL_2 and HDL_3 in hyperlipoproteinaemia. J Intern Med 228:151–157
Walldius G (1976) Fatty acid incorporation into human adipose tissue (FIAT) in hypertriglyceridaemia. Acta Med Scand Suppl 591:1–47
Wallentin L, Sundin B (1985) HDL_2 and HDL_3 lipid levels in coronary artery disease. Atherosclerosis 59:131–136
Wilson JD (1991) Vitamin deficiency and excess. In: Wilson JD, Braunwald E, Isselbacher KJ, Petersdorf RG, Martin JB, Fauci AS, Root RK (eds) Harrison's principles of internal medicine, vol 1. McGraw-Hill, New York, p 434
Yovos JG, Patel ST, Falko JM et al. (1982) Effects of nicotinic acid therapy on plasma lipoproteins and very low density lipoprotein apoprotein C subspecies in hyperlipoproteinemia. J Clin Endocrinol Metab 54:1210

CHAPTER 14

Ion Exchange Resins

P. Schwandt and W.O. Richter

A. Introduction

The bile acid-binding resins cholestyramine (CH) and colestipol (CO) are effective drugs for the treatment of patients with elevated low-density lipoprotein (LDL) cholesterol plasma concentrations without concurrent hypertriglyceridemia. Since the bile acid sequestrants are not absorbed in the gastrointestinal tract, they cannot cause direct systemic side effects. They activate the natural pathway for LDL elimination from the circulation by stimulating the expression of LDL receptors. Thus, the sequestrants are ideal drugs for treatment of LDL hypercholesterolemia from the pharmacologist's point of view. However, their mode of administration as a bulky powder and the gastrointestinal side effects may cause practical problems.

B. Chemistry and Pharmacology

CH (Fig. 1) is the chloride salt of a basic polymer of styrene and divinylbenzene with an average molecular mass of more than 1000 kDa; the trimethylbenzyl ammonium groups represent the ion exchange sites. CO (Fig. 2) is a basic anion exchange polymer of tetraethylene pentamine and epichlorohydrin. Both resins bind bile acids by exchanging them for chloride. The binding is irreversible and independent of temperature or pH changes. The binding capacity of both substances is shown in Table 1.

The bound bile acids are hindered from enterohepatic recirculation and are excreted with the resin in the feces (Hashim et al. 1961; Grundy et al. 1971; Moore et al. 1968; Parkinson et al. 1970). Ion exchange resins are neither metabolized nor absorbed by the gut (Hashim and Van Itallie 1965). It was recommended that they should be administered 1 h before or after meal (Hunninghake 1990), but recently it has been shown that the efficacy of CH does not vary when taken before or during meals (Sirtori et al. 1991).

Both powders are insoluble in water. They are given orally after being suspended in water, juice, skimmed milk, yogurt, etc. to make the drug palatable. However, preexposure of CO to components of tomato juice and orange juice slightly reduced the adsorption of cholate anions (Bilicki et al. 1989).

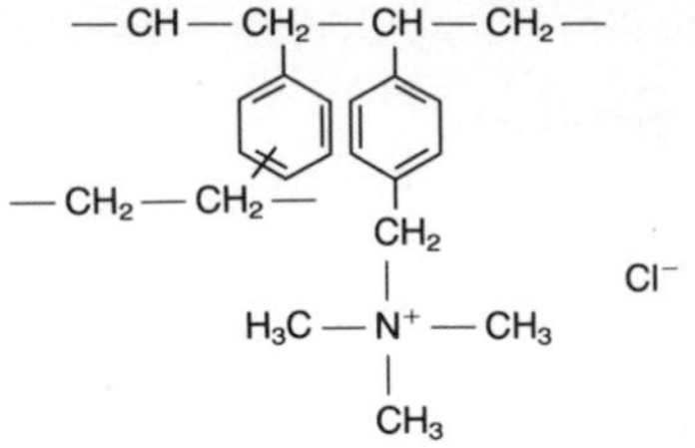

Fig. 1. Structure of cholestyramine

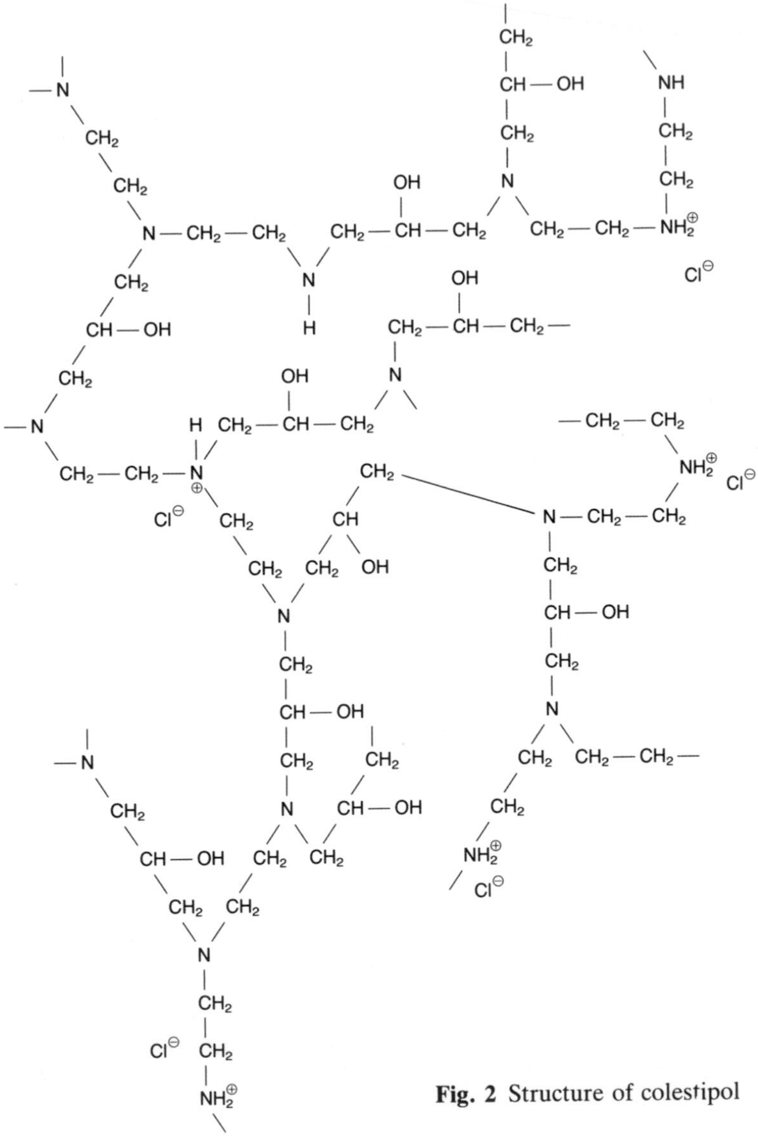

Fig. 2 Structure of colestipol

Table 1. Binding and exchange capacity per gram polymer of bile acid sequestrants (HUNNINGHAKE and PROBSTFIELD 1977)

	Cholestyramine	Colestipol
Bile acid binding capacity	2.2 mEq cholate	1.1 mEq cholate
Anion exchange	4.4 mEq chloride	7.5 mEq chloride

There are ongoing efforts to improve patients' compliance, especially by developing new formulas: a microporous CH analog (Filicol) seems to be somewhat superior regarding tolerability and the potency to bind bile acids (DESIMONE et al. 1978; SIRTORI et al. 1982; ROS et al. 1991). A new formula of CH powder with reduced volume (Questran-Light) looks equally effective and more preferable to the consumer (INSULL et al. 1991; SHAEFER et al. 1990). Efficacy and compliance of a CH bar seem to be comparable to powder (SWEENEY et al. 1991). Low dosages of encapsulated CO (2–10 g/day) together with diet therapy are effective in moderate hypercholesterolemia (LINET et al. 1988; NORRIS et al. 1987). Quaternization of CO-HCl increases the in vitro binding capacity of CO for sodium glycocholate (CLAS 1991).

C. Mode of Action

CH and CO both bind large quantities of bile acids irreversibly, thus preventing their uptake by mucosal cells (HASHIM et al. 1961; GRUNDY et al. 1971; MOORE et al. 1968; PARKINSON et al. 1970). The resulting decrease in hepatic concentrations of bile acids stimulates the activity of the cholesterol 7α-hydroxylase, which is the rate-limiting enzyme in the conversion of cholesterol to bile acids (MYANT and MITROPOULOS 1977; DANIELSSON and SJÖVALL 1975). This results in hepatic cholesterol depletion, which triggers the expression of LDL receptors as well as hepatocellular cholesterol synthesis in the liver cells (SHEPHERD et al. 1980; GOLDFARB and PITOT 1972; REIHENER et al. 1990; BROWN and GOLDSTEIN 1986; GOLDSTEIN and BROWN 1989). In contrast to the 3-hydroxy-3-methylglutaryl coenzyme A (HMG-CoA) reductase inhibitor pravastatin, CH increased serum levels of lathosterol by 125%, but did not change dolichol or ubiquinone levels in a significant manner (ELMBERGER et al. 1991).

In six patients with familial hypercholesterolemia, 24 g/day CH reduced plasma LDL cholesterol by 32%. The LDL synthesis rate remained unchanged, but the fractional catabolic rate was increased (LEVY and LANGER 1972). In addition, LDL composition was shown to be changed by CO treatment: LDL particles became smaller and more dense and were characterized by a decreased cholesterol to protein ratio (YOUNG et al. 1989).

Interruption of the enterohepatic recirculation of bile acids activates the phosphatidic acid phosphatase, which is the key enzyme in the hepatic synthesis of triglycerides. The increased triglyceride synthesis is followed by an enhanced production and secretion of large very low density lipoproteins (VLDL) (SHEPHERD and PACKARD 1983; ANGELIN et al. 1981, 1990; WITZTUM et al. 1976; BEIL et al. 1982).

The slight increase of plasma high-density lipoprotein cholesterol concentrations (THE LIPID RESEARCH CLINICS PROGRAM 1984a,b) is not clearly understood. It was shown that CH increased HDL_2 as well as synthesis of apo-A-I (SHEPHERD 1989), but the levels of apoA-II remained unchanged. Lecithin cholesterol acyltransferase (LCAT) is essentially responsible for all of the cholesteryl esters that are produced in plasma. The LCAT reaction takes place mainly on HDL_3 into HDL_2. CH (WALLENTIN 1978) as well as CO increased LCAT activity expressed as fractional esterification rate (FER) by 30%–50% (CLIFTON-BLIGH et al. 1974). Low-dose CO (10 g/day) further increased FER in patients receiving simvastatin (+33% with 40 mg/day simvastatin; +58% with simvastatin + colestipol; DESAGER et al. 1991).

CH treatment in five type IIa patients decreased lipoprotein lipase activity (WEINTRAUB et al. 1987); there were no changes in the fasting total plasma triglyceride concentrations, but the treatment significantly altered chylomicron metabolism, increasing the peak chylomicron retinol palmitate levels by 89% and the chylomicron retinol palmitate area by 86% (WEINTRAUB et al. 1987).

Lipoprotein (a) serum concentrations are not influenced by CH (WIKLUND et al. 1990; JACOB et al. 1990) or CO treatment (BROWN et al. 1990). Analyzing lipoprotein particles defined by their apolipoprotein composition (lipoprotein A-I, lipoprotein A-II–A-I, lipoprotein E–B, and Lipoprotein C-III–B), lipoprotein A-I was increased by CH by about 20% and lipoprotein A-II–A-I by 8.1%–41.2%. No significant effect on lipoprotein C-III–B was observed, while lipoprotein E–B slightly increased in the first 8

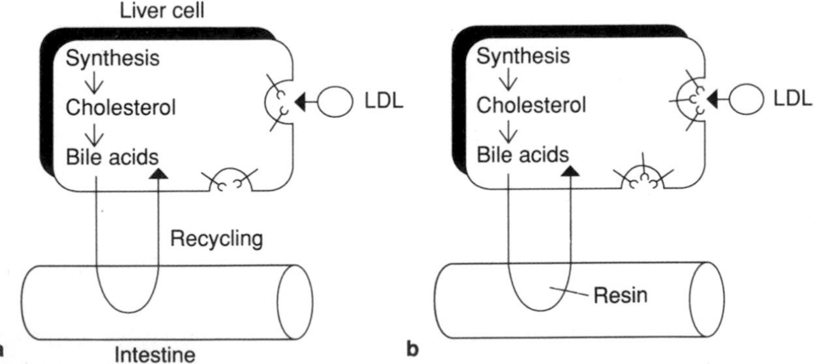

Fig. 3a,b. Mode of action of the resins. **a** baseline situation; **b** low-density lipoprotein (*LDL*)-lowering effect of resins

weeks of treatment and thereafter decreased (−26.2%). Serum apolipoproteins E and C-III were not affected by CH (BARD et al. 1990).

Bile acid sequestrants did not change plasma fibrinogen levels, plasma viscosity, whole blood viscosity, or red cell aggregability and deformability (JAY et al. 1990).

CH did not affect the output of biliary lipids or the cholesterol saturation of bile. Therefore, it should not be associated with any increased risk of gallstone formation (EINARSSON et al. 1991; CARRELLA et al. 1991; Fig. 3).

D. Clinical Experience

Both available bile acid sequestrants CH and CO have been extensively evaluated in well-documented clinical trials, but also in a primary prevention trial, the Lipid Research Clinics Coronary Primary Prevention Trial (LRC-CPPT).

The effect depends, of course, on the doses taken by the patients and the results of the studies are difficult to compare because of differences in the baseline values, duration of treatment, different underlying lipoprotein disorders, age, and sex. Yet it can be stated that the dose–response curves for both CH and CO are not exactly linear.

With doses of 8–12 g CH, reductions of LDL cholesterol between 12% and 18% were reported; for 16 g/day a reduction of up to 24% is typical and with 24 g/day, reductions of up to 34% are possible (SHEPHERD et al. 1980; GLUECK et al. 1972; WEISWEILER et al. 1979, LEVY et al. 1973; SPENGEL et al. 1981; STEIN et al. 1990; STEINHAGEN-THIESSEN et al. 1987; LEVY et al. 1984; THE LOVASTATIN STUDY GROUP III 1988; SCHWARTZKOPFF et al. 1990; BRENSIKE et al. 1984; WITZTUM et al. 1979).

In doses of 15–30 g/day, CO decreased LDL cholesterol by 16%–30% (in one study by 43%) (GROOT et al. 1983, KUO et al. 1979; MORDASINI et al. 1978; MILLER et al. 1973), while in a larger group of studies, total cholesterol was reduced by 7%–34% (GROOT et al. 1983; KUO et al. 1979; MORDASINI et al. 1978; MILLER et al. 1973; HARVENGT and DESAGER 1976; GROSS and FIGUEREDO 1973; RYAN et al. 1975; NYE et al. 1972; MISHKEL and CROWTHER 1977; SEPLOWIZ et al. 1981; SACHS and WOLFMAN 1974; LEES et al. 1976; DORR et al. 1978; NASH et al. 1982).

In most of these studies, a slight increase of HDL cholesterol of between 3%–14% was observed.

When serum triglyceride concentrations were normal at baseline, a mean increase of serum triglycerides of 5%–25% was reported. If the patient has dysbetalipoproteinemia (type III hyperlipoproteinemia) or if baseline triglyceride values exceed 300 mg/dl, there will be a marked increase in triglyceride levels. In addition, in familial combined hyperlipidemia, severe hypertriglyceridemia can occur during treatment with bile acid sequestrants.

In several studies bile acid sequestrants were also evaluated for their effect on atherosclerosis.

In the National Heart, Lung, and Blood Institute (NHLBI) type II Coronary Intervention Study, 59 hypercholesterolemic men were treated with either placebo or up to 24 g/day of CH and followed for 5 years with sequential coronary angiography (BRENSIKE et al. 1984). Coronary artery disease (CAD) progressed in 49% of the placebo-treated patients and in 32% of the CH-treated patients. The mean decrease in LDL cholesterol was 26% in the 5-year period and HDL cholesterol increased by 8%.

In a randomized, placebo-controlled, multiclinic trial, 2278 hypercholesterolemic patients were treated with CO hydrochloride for up to 3 years (DORR et al. 1978). With a dosage of 15 g/day, total cholesterol was decreased from 313 mg/dl by an average of 37 mg/dl over the entire observation period. During the trial, 4.4% of the placebo-treated men (3.3% of women) and 2.4% of the CO-treated men (3.7% of women) had at least one new nonfatal coronary heart disease (CHD) event.

Out of 42 subjects with significant narrowing of a nongrafted coronary artery, 25 CO responders (total cholesterol, −20.5%) were compared to 17 nonresponders (total cholesterol, −1.7%) receiving placebo over 2 years. In the drug-treated group, the angiographic status showed progression in three patients, while in the placebo group eight progressed (NASH et al. 1982).

In a long-term treatment of 25 patients with CO, LDL cholesterol was reduced from 331 mg/dl to 188 mg/dl. HDL cholesterol remained unchanged. Therapy stabilized angiographically visualized atherosclerotic lesions (KUO et al. 1979).

The LRC-CPPT (THE LIPID RESEARCH CLINICS PROGRAM 1984a,b) was a double-blind, placebo-controlled study. All men were first placed on a low-cholesterol, low-saturated fat diet. Those who remained above the 90th percentile for LDL cholesterol after several months were randomized either into placebo or CH groups and followed for 7–10 years. The mean reduction of total cholesterol was 10% and of LDL cholesterol, 19%. This resulted in a 19% difference in the incidence of the first heart attack. A linear relationship between first heart attack rate and decline in LDL cholesterol levels was demonstrated (Table 2).

The suspected responder rate for 12 g CH per day is 70%, for 26 g/day, 90%, and for 20 g/day, 100%. It is suggested that 30% have side effects with

Table 2. Relation between cholesterol lowering by cholestyramine and decreases of risk for coronary heart disease in the Lipid Research Clinics Coronary Primary Prevention Trail (THE LIPID RESEARCH CLINICS PROGRAM 1984a,b)

Cholestyramine (g/day)	Number of patients	Decrease in cholesterol (%)	Decrease in cardiovascular risk (%)
0–8	439	4.4	10.9
8–20	496	11.5	26.1
20–24	965	19.0	39.3

a dosage of 12 g/day and 50% with a dosage of 16 g/day (TIKKANEN and NIKKILÄ 1987).

CH (16 g/day) was also administered for the treatment of hyperlipidemia (ten patients) due to unremitting nephrotic syndrome. It decreased LDL cholesterol by 19% and increased HDL cholesterol by 12%. No serious side effects were observed (RABELINK et al. 1988).

E. Pharmacodynamic and Pharmacokinetic Parameters

Since the water-insoluble resins CH and CO are not absorbed in the gastrointestinal tract, less than 0.05% of ^{14}C-labeled CO is excreted in the urine (AST and FRISHMAN 1990). Thus, any interaction between the resins and other substances nearly exclusively occurs in the gastrointestinal tract.

Ingestion of resins causes a load with chloride anions, which may result in a decrease in the urine pH together with increased excretion of chloride with the urine. In addition, there may be an increase in the urinary calcium ion excretion, which might cause problems in persons at risk for osteoporosis (RUNEBERG et al. 1972). However, studies in three human subjects on a metabolic ward demonstrated a CH-induced increase in urinary calcium, but also a greater decrease in fecal excretion, resulting in an increase in body calcium (BRISCOE and RAGAN 1963). Only slight decreases of serum calcium levels have occassionally been reported (CASDORPH 1976), but could not be found by others (DATTA and SHERLOCK 1963; WELLS et al. 1968). However, such changes obviously do occur only in patients who have impaired renal function (RUNEBERG et al. 1972). In subjects with normal renal function, sodium, potassium, chloride, and carbon dioxide content remained normal (HOWARD et al. 1966). There is on report of a 10-year-old girl with intrahepatic biliary atresia and a decreased insulin clearance who developed hyperchloremic acidosis when receiving 8 g CH per day; withdrawal of the drug and administration of bicarbonate reversed the metabolic abnormality (KLEINMANN 1974).

Long-term treatment of young patients with familial hypercholesterolemia for 5 years did not reveal any adverse effects of CO on the plasma levels of calcium, iron, sodium, parathyroid hormone or water- and fat-soluble vitamins (SCHLIERF et al. 1985). In 18 children with familial hypercholesterolemia treated for 1–2½ years with CH at a mean dosage of 0.6 g/day per kg body weight, normal serum iron, vitamin B_{12}, calcium, protein and prothrombin levels were found. The concentrations of vitamin A and E and of inorganic phosphorus remained within the normal range. However, over the first 2 years mean serum folate concentrations fell from 7.7 mg/dl to 4.4 mg/dl, necessating oral folic acid substitution at a dosage of 5 mg/day (WEST and LLOYD 1975). Similar data have also been reported by others (CASDORPH 1976).

Since bile acids are necessary for the absorption of fat-soluble vitamins, resins might have undesirable effects on the vitamin status. Several investi-

Table 3. Interference of bile acid sequestrants with the absorption of fat-soluble vitamins in man

Design of the study	Vitamin	Result	Reference
4 g CH + test meal 8 g CH + test meal	Vitamin A	Absorption unaffected Absorption decreased	BASU (1965)
Plasma levels before and after 5–9 months CH	Vitamin A Carotene	No effect	BRESSLER et al. (1966)
4 g CH + test meal 8 g CH + test meal (9 h postprandial plasma levels)	Vitamin A	No effect Decrease	LONGENECKER and BASU (1965)
CH, lignin, healthy volunteers	Vitamin A	Absorption decreased by 60%	BARNARD and HEATON (1973)
CO, 5 years plasma levels	Vitamin A, E	Unaffected	SCHLIERF et al. (1985)
0.6 g/day CH, 1–1½ years, children, plasma levels	Prothrombin Vitamin A Vitamin E	Unaffected Decrease Decrease	WEST and LLOYD (1975)
CH long-term, 1-year intervals, plasma levels			
67 determination in 45 HLP patients	Vitamin E	Unaffected	CASDORPH (1976)
172 determination in 72 HLP patients	Vitamin A	Unaffected	
106 determination in 63 HLP patients	Carotene	Unaffected	
107 determination in 71 HLP patients	Prothrombin	In 29 patients 43%–66% decreased	

CH, cholestyramine; CO, colestipol; HLP, hyperlipoproteinemia.

gations in man showed conflicting data (Table 3). There are also a few reports on clinically evident, vitamin-related side effects. A patient with small-bowel disease developed osteomalacia while on CH, which was reversed by the administration of vitamin D (HEATON et al. 1972). There are reports about two patients with biliary cirrhosis who developed a bleeding diathesis after treatment with CH, which could be corrected by vitamin K administration (VISINTINE et al. 1961; GROSS and BROTMAN 1970).

However, in some patients without liver disease, sight prolongations of prothrombin time have been reported (CASDORPH 1976).

In conclusion, there is no great need for concern regarding possible interference of bile acid sequestrants with the absorption of fat-soluble vitamins. Nevertheless, the vitamin status should be evaluated regularly in children receiving long-term treatment with resins.

The reports on the effect of bile acid sequestrants on the alkaline phosphatase activity in serum are also controversial: in children taking CH, an increase of alkaline phosphatase was observed (LEVY 1970), and in five of eight patients 15 g CO per day increased this enzyme activity (MILLER et al. 1973). However, in adults a double-blind study with CH did not show alterations (LEVY et al. 1973). This corresponds to other reports on no changes of the alkaline phosphatase (PARKINSON et al. 1970; GLUECK et al.

1972; NYE et al. 1972; SACHS and WOLFMAN 1973; DUJOVNE et al. 1974) and might reflect enzyme activity accompanying bone growth.

Adrenal and gonadal steroids are derived from cholesterol either delivered by LDL or biosynthesis. Despite effective LDL cholesterol lowering, CH did not effect plasma cortisol responses to tetracosactrin, testosterone, sex hormone-binding globulin, androstenedione, dehydroepiandrosterone sulfate, estradiol, or 17-α-hydroxyprogesterone (JAY et al. 1991).

F. Drug Interactions

Bile acid sequestrants may also bind and decrease the absorption of a variety of anionic drugs when administered simultaneously. However, only a limited number of drugs have been studied systematically. The current recommendation is that other drugs should be administered either 4 h after or 1 h before the resin.

In vitro studies demonstrated that acidic drugs are bound to a variable degree to CH, neutral drugs showed little or no binding, and basic drugs were not bound at all (GALLO et al. 1965). The investigations with several drugs are listed in Table 4. For many drugs the degree of binding to the resins is dose and pH dependent. In vitro CO binds metoclopramide to a significantly greater extent than CH (AL-SHAREEF et al. 1990). In vivo the quarternary ammonium resin CH is more likely to affect the intestinal absorption of anionic drugs than is the tertiary amino polystyrene resin CO (HUNNINGHAKE 1980).

Table 4. Drug interactions with bile acid sequestrants in man

Resin	Drug	Effect	Reference
CH	Amiodaron	−50% 7-h plasma concentration	NITSCH and LÜDERITZ (1986)
CH	Aspirin	Delayed p.L.	HUNNINGHAKE and POLLACK (1977)
CO	Aspirin	Accelerated p.L.	HUNNINGHAKE and POLLACK (1977)
CH	Carbamazepine	None	NEUVONEN et al. (1988)
CO	Carbamazepine	Absorption slightly inhibited	NEUVONEN et al. (1988)
CO	Clofibrate	None	DESANTE et al. (1979)
CO	Chlorothiazide	−58% excretion	KAUFFMAN and AZARNOFF (1973)
CH	Digoxin	Absorption by 30%–40% decreased	BROWN et al. (1978) NEUVONEN et al. (1988)
CO	Digoxin	None	NEUVONEN et al. (1988)
CH	Digitoxin	Decreased	CALDWELL et al. (1971)
CH	Estrogenes	Increased adsorption (half-life reduced from 11.5 to 6.6 days)	ARTS et al. (1991)
CH	Frusemide	−95% absorption	NEUVONEN et al. (1988)
CO	Frusemide	−80% absorption	NEUVONEN et al. (1988)
CO	Gemfibrozil	−30% AUC	FORLAND et al. (1990)
CH	Glipizide	Absorption −29%	KIVISTO and NEUVONEN (1990)

Table 4. *Continued*

Resin	Drug	Effect	Reference
CH	Hydrochlorothiazide	−85% AUC	HUNNINGHAKE and KING (1978)
CO	Hydrochlorothiazide	−35% AUC	HUNNINGHAKE and KING (1978)
CH	Lorazepam	Adsorption 23.7%	HERMAN and CHAUDHARY (1991)
CH	Lorazepam glucuronide	Adsorption 74.3%	HERMAN and CHAUDHARY (1991)
CO	Lorazepam	Adsorption 11.3%	HERMAN and CHAUDHARY (1991)
CO	Lorazepam glucuronide	Adsorption 20.8%	HERMAN and CHAUDHARY (1991)
CO/CH	Metoclopramide	Decreased adsorption	AL-SHAREEF et al. (1990)
CH	Methyldopa	+15% AUC	HUNNINGHAKE and KING (1978), AL-SHAREEF et al. (1990)
CO	Methyldopa	Accelerated p.L.	HUNNINGHAKE and KING (1978)
CH	Paracetamol	Decreased AUC, increased adsorption	DORDONI et al. (1973) AL-SHAREEF et al. (1990)
CO	Paracetamol	Increased adsorption	AL-SHAREEF et al. (1990)
CH	Phenprocoumon	Decreased	HAHN et al. (1972)
CH	Phenytoin	None	CALLAGHAN et al. (1983)
CO	Phenytoin	None	CALLAGHAN et al. (1983)
CH	Piroxicam	−40% average half-life	GUENTERT et al. (1988)
CH	Pravastatin	−49% AUC (at 40 mg/day)	PAN et al. (1990)
CH	Simvastatin	Decreased AUC	GEISEL et al. (1990)
CH	Tenoxicam	−53% average half-life	GUENTERT et al. (1988)
CO	Tetracycline	−56% 48-h excretion	FRIEDMAN et al. (1989)
CH	L-thyroxine	−50% AUC	NORTHCUTT et al. (1969)
CH	Tolbutamide	Delayed p.L.	HUNNINGHAKE and POLLACK (1977)
CO	Tolbutamide	Accelerated p.L.	HUNNINGHAKE and POLLACK (1977)
CH	Warfarin	Delayed p.L., −15% AUC	HUNNINGHAKE and POLLACK (1977)
CO	Warfarin	None	HUNNINGHAKE and POLLACK (1977)
CH	Warfarin	Decreased	ROBINSON et al. (1971) AL-SHAREEF et al. (1990)

CH, cholestyramine; CO, colestipol; AUC, area under the curve; p.L, peak level.

Though there are reports that CH binds vancomycin and teicoplanin in vitro (KING and BARRIERE 1981, PANTOSTI et al. 1985), this is not of clinical relevance. In vitro CH did not inhibit the growth of *Escherichia coli*, bind gentamycin, or affect the antibacterial activity of the antibiotic (COLTMAN et al. 1990). In nine volunteers, 30 g CO significantly impaired tetracycline absorption by more than 50% (FRIEDMAN et al. 1989).

Phenytoin is a weakly acidic drug with limited water solubility and a slow rate of absorption. In six adult male volunteers, CH did not decrease the absorption of phenytoin, while CO produced a slight increase without clinical importance (CALLAGHAN et al. 1983). Since the absorption of paracetamol is grossly reduced by CH, "... either the resin or activated charcoal should be available in every casualty department" (DORDONI et al. 1973), which is also the case for intoxication with digitoxin.

By interruption of the enterohepatic ciruclation of digitoxin by CH, the metabolic disposition of digitoxin is accelerated, abbreviating the physiological response to the glycoside in digitalized human subjects given CH (CALDWELL et al. 1971). The absorption of digoxin is unaffected by CO and reduced by 30%–40% by CH (NEUVONEN et al. 1988). The antiarrhythmic drug amiodaron is also affected by the concomitant application of CH by about 50% because of an enhanced biliary elimination due to the significant reduction of the enterohepatic recirculation (NITSCH and LÜDERITZ 1986). The very slow elimination of the two nonsteroidal anti-inflammatory drugs tenoxicam and piroxicam is accelerated by nearly 100%. No interaction was demonstrated after the concomitant single-dose administration of clofibrate and CO (DESANTE et al. 1979).

A significant reduction of levothyroxine absorption (greater than 50%) occurred when CH was administered within 4–5 h of L-thyroxine. In vitro experiments have shown that approximately 50 mg CH irreversibly bind about 3000 μg of L-thyroxine (NORTHCUTT et al. 1969). Recently, a case report of a patient who received concomitant CH and L-thyroxine described an elevation of the plasma level of thyroid-stimulating hormone (TSH; HARMON and SEIFERT 1991).

G. Adverse Effects of Resins

The range of adverse effects is limited because the resins are not absorbed. Subjective complaints concern taste, texture, and bulkiness of CH and CO with a wide interindividual variation. The most common adverse effect is constipation, especially in the elderly and when administered in large doses. In many patients, constipation can be avoided or made transitory by a slow and stepwise increase in the dose (e.g., 1 sachet per week), the concomitant consumption of higher amounts of fluid, and general measures to relieve constipation, e.g., consumption of more fiber and in severe cases stool softeners. In a double-blind study with CH, constipation was reported by 18% (LEVY et al. 1973). In addition, flatulence and "heartburn", drowsiness, abdominal distension, epigastric fullness, belching, and nausea were the most frequently reported side effects after obstipation (in decreasing frequency of occurrence; DANHOFF 1966).

A rare adverse effect is hyperchloremic acidosis, reported in a child with liver agenesis and renal failure (Kleinmann 1974), in a patient with diarrhea due to ileal resection (Hartline 1976), and in a child with ischemic hepatitis and renal insufficiency (Pattison and Lee 1987). Thus, in patients at risk for hyperchloremia, serum chloride levels should be checked. Furthermore, activation of duodenal ulcer (Chait and Lewis 1972) and a case of intestinal obstruction (Cohen et al. 1969) have been reported. In connection with constipation, exacerbation of preexisting hemorrhoids was observed (Illingworth 1987). The four case reports on calcification of intra-abdominal organs in patients on long-term CH therapy cannot be clearly related to resin treatment (Casdorph 1976). Parathyroid hormone levels and concentrations of the vitamin D metabolites 1,25(OH)2D3 and 25(OH)D3 were not changed during 8 weeks of therapy with 24 g/day CH (Ismail et al. 1990). After 1 year of combined CO and niacin therapy, 75 euthyroid participants in the Cholesterol-Lowering Atherosclerosis Study (CLAS) had lower levels of serum thyroxine (average reduction of 1.5 μg/dl) and increased triiodothyronine uptake ratios in comparison to 79 men in the placebo group remaining completely euthyroid (Cashin-Hemphill et al. 1987).

Extensive loss of dental enamel occurred in a young boy. This was attributed to mixing the CH in Kool-Aid and swirling the mixture in the mouth before swallowing (Curtis et al. 1991).

H. Other Indications

In a patient with erythropoietic protoporphyria and hepatic damage, long-term treatment (1 year) with CH caused a threefold increase in fecal protoporphyrine excretion with concurrent improvement in liver function and photosensitivity (McCullough et al. 1988). Sitosterolemia with xanthomatosis is efficiently treated with CH or CO (Belamarich et al. 1990; Nguyen et al. 1991). This inherited storage disease might be suspected if very high plasma cholesterol concentrations fail to respond to HMG-CoA reductase inhibitors, but are very sensitive to dietary sterol restriction and interruption of the enterohepatic recirculation of bile acids. In primary biliary cirrhosis, the effect of CH treatment on pruritus can be improved by combined phototherapy (Gerio et al. 1987). There are several reports on the use of anion exchange resins for antibiotic-associated pseudomembraneous colitis (Burbige and Milligan 1975; Sinatra et al. 1976; Kreutzer and Milligan 1978; Bartlett et al. 1980; Keighley 1980; Kunimoto and Thomson 1986; Pruksananonda and Powell 1989). However, a critical analysis of all existing data concludes that prospective randomized trials are needed and that until then "resins have no role in the initial management of antibiotic-associated pseudomembraneous colitis, especially if the disease is mild" (Ariano et al. 1990). In the treatment of severe persistent diarrhea in infants, the combination of oral gentamycin and CH is recommended as save and effective (Hill et al. 1986).

I. Children and Adolescents

The most widespread experience with lipid-lowering drugs in childhood is with the bile acid sequestrants which are very effective in familial hypercholesterolemia (HOEG 1991; KWITEROVICH 1990; GLUECK et al. 1976a,b). Numerous studies have demonstrated an 18%–47% response to treatment with CH since 1973 (GLUECK et al. 1973); CO has been shown to have a similar effect (HOEG 1991; MORDASINI et al. 1978; SCHLIERF et al. 1982). The rapidly initiated cholesterol-lowering response remained after more than 8 years (WEST and LLOYD 1980). In 33 children heterozygous for familial hypercholesterolemia, efficacy and safety of CH or CO (8–20 g/day) were studied for 4.3 years, comparing these children to 40 children with familial hypercholesterolemia receiving diet therapy alone for 5.8 years with a positive outcome: "... appears to be safe with regard to physical growth development, does not appear to affect sexual maturation, and reduces plasma total cholesterol levels within ranges previously shown to have efficacy in reducing coronary heart disease events in adult hypercholesterolemic men." The reductions of total cholesterol by 9.6% on diet and 12.5% on diet plus resins should equate with 19%–24% reductions in CHD morbidity and mortality by speculative extrapolation from the LRC-CPPT (GLUECK et al. 1986). Fat-soluble vitamin supplementation was not carried out during resin therapy (TSANG et al. 1978; SCHWARZ et al. 1980; GLUECK et al. 1976a,b).

J. Combined Drug Treatment with Resins

In severe forms of hypercholesterolemia, resin monotherapy often fails to normalize LDL cholesterol plasma concentrations. Even in milder forms, normalization may be not achieved because the full dose of the bile acid sequestrants is not tolerated. Therefore, a combination with drugs acting at different sites (Fig. 4) often results in a complete correction of the disorder in lipoprotein metabolism.

I. Bile Acid Sequestrants and Cholesteral Synthesis Enzyme Inhibitors

Though individual patients are difficult to compare (which is also the case for the drug dosages and times of treatment), the percentage decrease seems to be the best parameter for comparing lipid-lowering effects (Table 5). CH consistently decreased the area under the curve of simvastatin (GEISEL et al. 1990) and pravastatin (PAN et al. 1990), resulting in 2–3 h being the recommended interval between the administration of both drugs. The 38 patients in the lovastatin–colestipol group of the Familial Atherosclerosis Treatment Study (FATS) not only showed a 45% decrease in LDL cholesterol after 2½ years, but also beneficial effects on CAD: only 21% regression (11% in the controls) and clinical events occurred in three out of the 46 patients who

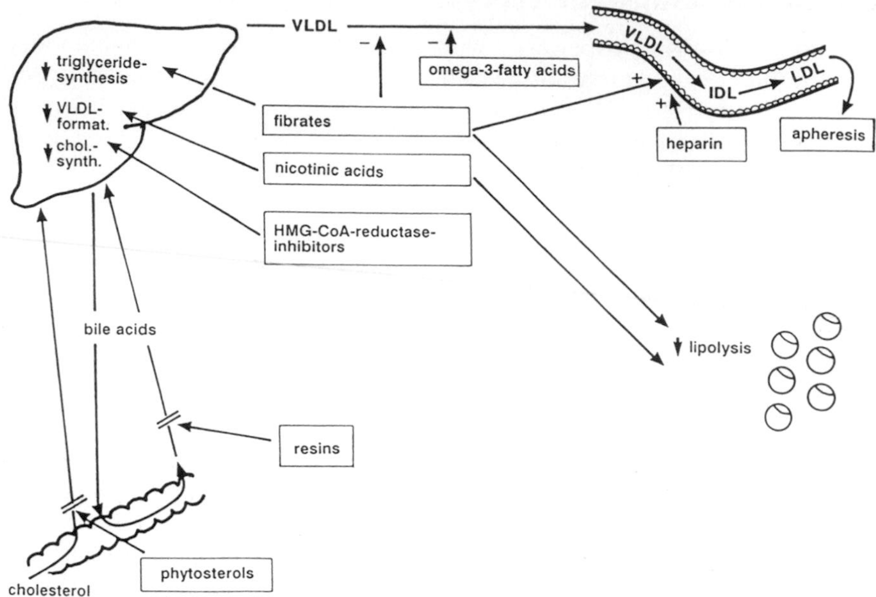

Fig. 4. Sites of action in combined drug treatment. *VLDL*, very low density lipoprotein; *IDL*, intermediate-density lipoprotein; *LDL*, low-density lipoprotein; *HMG-CoA*, 3-hydroxy-3-methylglutaryl coenzyme A. From SCHWANDT (1991)

started in the lovastation–colestipol group and completed the follow-up for 32 months (BROWN et al. 1990).

II. Bile Acid Sequestrants and Nicotinic Acid or Analog

The effect of combined resin plus nicotinic acid treatment is shown in Table 6. The LDL-lowering effect of this combination was between 32%–47% in the different studies. Due to nicotinic acid, a pronounced increase in HDL cholesterol was observed as well as decrease in serum triglyceride concentrations.

In three studies, the influence of the combined resin plus nicotinic acid treatment on coronary artery lesions was evaluated. In the CLAS (BLANKENHORN et al. 1987; CASHIN-HEMPHILL et al. 1990) cholesterol was decreased after 2 years by 43% and after 4 years by 40%, while HDL cholesterol was increased by 37% both after 2 and 4 years. Compared to placebo treatment, there was a significant improvement in atherosclerotic lesions (Fig. 5). The effect on femoral atherosclerosis was less pronounced (BLANKENHORN et al. 1991). CAD was also improved after 26 months in 72 patients with heterozygous familial hypercholesterolemia by a decrease of LDL cholesterol of 39% and an increase of HDL cholesterol by 25%. One arm of the FATS (BROWN et al. 1990) used the combination of CO and

Table 5. Effects of 3-hydroxy-3-methylglutaryl coenzyme A reductase inhibitors plus bile acid sequestrants on serum lipoproteins

Drug regimen	n	Percentage change				Initial LDL (mg/dl)	Reference
		C	LDL	HDL	TG		
Cp (90 mg) + CH (12 g)	10	−39	−52	+44	−34	263	MABUCHI et al. (1983)
Lo (80 mg) + Co (20 g)	10	−45	−54	−2	+7	409	ILLINGWORTH (1984)
Lo (40 mg) + Co (20 g)	7	−37	−46	+10	−17	361	ILLINGWORTH (1984)
Lo (40 mg) + Co (20 g)	8	−43	−52	+29	+1	321	GRUNDY et al. (1985)
Lo (40 mg) + Co (20 g)	10	−36	−48	+17	+8	196	VEGA and GRUNDY (1987)
Lo (80 mg) + CH (8 g)	19	−47	−56	+22	−28	391	LEREN et al. (1988)
Lo (80 mg) + CH (16 g)	19	−51	−61	+22	−25	391	LEREN et al. (1988)
Lo (40 mg) + Co (20 g)	16	−42	−52	±0	−12	325	WITZTUM et al. (1989)
Lo (80 mg) + Co (20 g)	13	−43	−54	+6	−13	325	WITZTUM et al. (1989)
Lo (80 mg) + CH (8 g)	12	−33	−40	+18	−23	303	JACOB et al. (1992)
Pra (40 mg) + CH (8 g)	13	−30	−39	+6	+12	306	JACOB et al. (1992)
Pra (40 mg) + CH (24 g)	9	−33	−53	+18	+8	209	PAN et al. (1990)
Pra (10 mg) + CH (24 g)	11	−38	−56	+11	+5	194	PAN et al. (1990)
Pra (20 mg) + CH (24 g)	7	−32	−47	+15	+13	206	PAN et al. (1990)
Si (40 mg) + CH (24 g)	5	−43	−55	+12	−8	319	YTRE-ARNE and NORDOY (1989)
Si (40 mg) + CH (16 g)	19	−42	−43	+9	−8	355	EMMERICH et al. (1990)
Si (40 mg) + CH (8 g)	43	−42	−49	+13	−19	352	GEISEL et al. (1990)
Si (20–40 mg) + CH (8–24 g)	20	−47	−56	+7	+4	320	MÖLGAARD et al. (1991)

C, cholesterol; CH, cholestyramine; CO, colestipol; Lo, lovastatin; Pra, pravastatin; Si, simvastatin; Cp, compactin; n, number of patients; LDL, low-density lipoprotein; HDL, high-density lipoprotein; TG, triglycerides.

Table 6. Effects of bile acid sequestrants plus nicotinic acid and analog on serum lipoproteins

Drug regimen	n	Percentage change				Initial LDL (mg/dl)	Reference
		C	LDL	HDL	TG		
CH (16 g) + NA (3 g)	6	−41	−48	+25	−37	360	PACKARD et al. (1980)
CH (16 g) + NA (2–12 g)	10	−26	−32	+23	−20	373	ANGELIN et al. (1986)
CH (12 g) + Ac (750 mg)	28	−27	−32	+6	−13	285	SERIES et al. (1990)
CH (12 g) + Ac (750 mg)	9	−29	−39	+31	−8	320	GYLLING et al. (1989)
CO (20 g) + NA (6–7.5 g)	11	–	−45	+17	–	278	KANE et al. (1981)
CO (30 g) + NA (3–8 g)	13	−40	−47	+32	−35	337	ILLINGWORTH et al. (1981)
CO (30 g) + NA (3–7 g)	32	−34	−45	+22	−34	263	KUO et al. (1981)
CO (30 ± 4 g) + NA (4.3 ± 1.3 g)	80	−26	−43	+37	−22	171	BLANKENHORN et al. (1987)
CO (30 ± 6 g) + NA (4.2 ± 2 g)	56	−25	−40	+37	−18	171	CASHIN-HEMPHILL et al. (1990)
CO (30 g) + NA (4 g)	36	−23	−32	+43	−29	190	BROWN et al. (1990)
CO (30 g) + NA (1.5–7.5 g)	21	−30	−36	+25	−20	326	MALLOY et al. (1987)
CO (30 g) + NA (3–6 g)	154	−28	−46	+40	−31	175	CASHIN-HEMPHILL et al. (1987)

C, cholesterol; CH, cholestyramine; CO, colestipol; Ac, acipimox; NA, nicotinic acid; LDL, low-density lipoprotein; HDL, high-density lipoprotein; TG, triglycerides; n, number of patients.

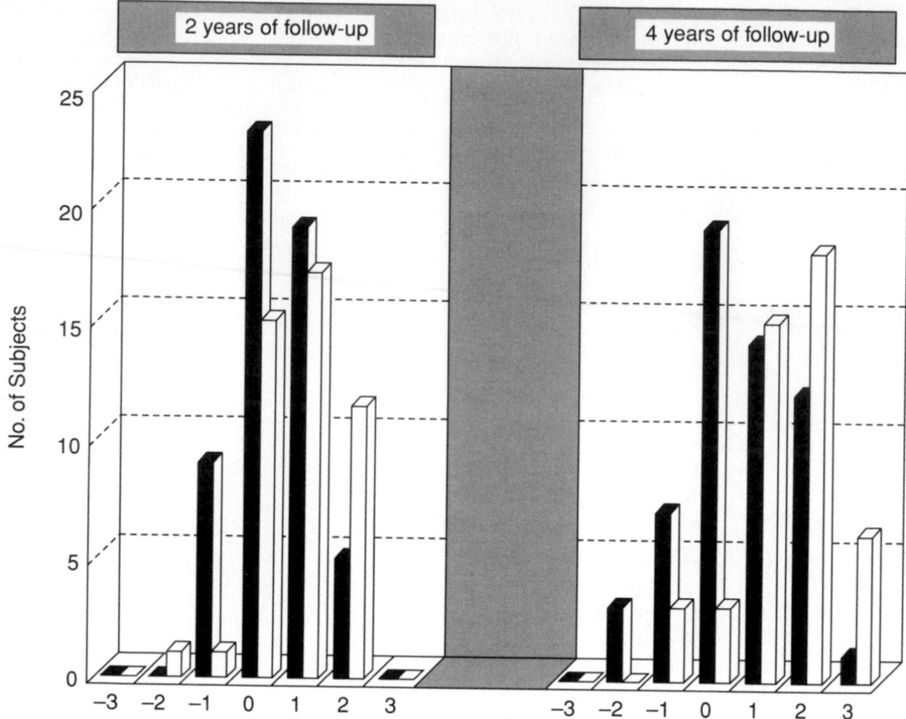

Fig. 5. Coronary global change score after 2 and 4 years of follow-up in the Cholesterol-Lowering Atherosclerosis Study. *Black bars* indicate scores for drug-treated patients, *white bars*, placebo group. Positive values indicate progression, negative values, regression. A score of 0 indicates no change, 1, a definitely discernible change, 2, an intermediate change, and 3, an extreme change (CASHIN-HEMPHILL et al. 1990)

nicotinic acid. LDL cholesterol was decreased by 32% and HDL cholesterol increased by 43%. Only one patient (of 36) had a clinical heart event in the first 2½ years of observation and 39% showed regression of coronary lesions.

III. Bile Acid Sequestrants plus Fibrates

The combination of fibrates to resin therapy leads to decreases in LDL cholesterol by 12%–44% in hypercholesterolemic patients. In patients with type IV hypercholesterolemia, even an increase in LDL cholesterol was observed, clearly showing that bile acid sequestrants are of no value in treating hypertriglyceridemia (Table 7).

Table 7. Effects of bile acid sequestrants plus fibrates on serum lipoproteins

Drug regimen	*n*	Percentage change				Initial LDL (mg/dl)	Reference
		C	LDL	HDL	TG		
CO (12 g) + Cl (2 g)	4[1]	−23	−29	–	−24	192	Rose et al. (1976)
CO (12 g) + Cl (2 g)	11[2]	−5	−14	–	−41	109	Rose et al. (1976)
CO (15 g) + Cl (2 g)	20	−25	–	–	−29	–	Fellin et al. (1978)
CO (20 g) + Cl (2 g)	14	−11	−12	+25	−22	197	Hunninghake et al. (1981)
CO (20 g) + Cl (2 g)	7[3]	−21	−14	+30	−35	146	Hoogwerf et al. (1985)
CH (24 g) + Be (600 mg)	18	−35	−39	+2	−19	278	Curtis et al. (1988)
CH (16 g) + Be (400 mg)	21	−28	−36	+22	−39	228	Series et al. (1989)
CH (24 g) + Be (400 mg)	47	−24	−36	+31	−33	227	Fischer et al. (1990)
CH (20 g) + Ge (1.2 g)	12	−41	−44	−11	–	347	Jones et al. (1988)
CO (15 g) + Fe (250 mg)	6	−36	−41	+14	−30	380	Weisweiler and Schwandt (1986)
CH (20 g) + Fe (300 mg)	10	−28	−36	+8	−20	357	Malmendier et al. (1987)
CO (20 g) + Ge (1.2 g)	9[1]	−22	−20	+11	−44	229	East et al. (1988)
CO (20 g) + Ge (1.2 g)	8[2]	−17	−15	+14	−17	148	East et al. (1988)

C, cholesterol; CH, cholestyramine; CO, colestipol; Be, bezafibrate, Fe, fenofibrate; Ge, gemfibrozil; Cl, clofibrate; LDL, low-density lipoprotein; HDL, high-density lipoprotein; TG, triglycerides; *n*, number of patients.
[1] Type IIb [2] Type IV [3] Type III

Table 8. Effects of bile acid sequestrants plus probucol on serum lipoproteins

Drug regimen	*n*	Percentage change				Initial LDL (mg/dl)	Reference
		C	LDL	HDL	TG		
CO (20 g) + Pr (1 g)	47	−28	−29	−29	+11	242	Dujovne et al. (1984a)
CO (10 g) + Pr (1 g)	22	−29	−29	−40	+14	239	Dujovne et al. (1984b)
CH (16 g) + Pr (1 g)	18	−26	−32	−18	+1	300	Sommariva et al. (1986)
CO (30 g) + Pr (1 g)	44	−49	−53	−30	+15	341	Kuo et al. (1986)

C, cholesterol; CH, cholestyramine; CO, colestipol; Pr, probucol; *n*, number of patients; LDL, low-density lipoprotein; HDL, high-density lipoprotein; TG, triglycerides.

IV. Bile Acid Sequestrants plus Probucol

Probucol also decreases HDL and LDL cholesterol in combination with bile acid sequestrants. Kuo et al. (1981) reported that 30 g CO per day and 1 g probucol per day "stabilized" the progression of established coronary artery lesions after 3.4–4.1 years (Table 8).

Table 9. Effects of the ternary combination of lovastatin, colestipol, and nicotinic acid on serum lipoproteins

Drug regimen	n	Percentage change				Initial LDL (mg/dl)	Reference
		C	LDL	HDL	TG		
Lo (80 mg) + CO (20 g) + NA (3 g)	3	−53	−62	+46	−57	372	ILLINGWORTH and BACON (1989)
Lo (40–60 mg) + CO (30 mg) + NA (1.5–7.5 g)	6	−56	−68	+28	−48	323	MALLOY et al. (1987)

C, cholesterol; Lo, lovastatin; CO, colestipol; NA, nicotinic acid; n, number of patients; LDL, low-density lipoprotein; HDL, high-density lipoprotein; TG, triglycerides.

V. Triple Drug Therapy

CO was used in ternary combination with lovastatin and nicotinic acid. The effect of this combination on serum lipoproteins is shown in Table 9.

References

Al-Shareef AH, Buss DC, Routledge PA (1990) Drug adsorption to charcoals and anionic binding resins. Hum Exp Toxicol 9:95–97

Angelin B, Bjorkhem I, Einarsson K (1981) Influence of bile acids on the soluble phosphatidic acid phosphatase in rat liver. Biochem Biophys Res Commun 100:606–612

Angelin B, Eriksson M, Eriksson K (1986) Combined treatment with cholestyramine and nicotinic acid in heterozygous familial hypercholesterolaemia: effects on biliary lipid composition. Eur J Clin Invest 16:391–396

Angelin B, Leijd B, Hultcrantz R, Einarsson K (1990) Increased turnover of very low density lipoprotein triglyceride during treatment with cholestyramine in familial hypercholesterolaemia. J Intern Med 227:201–206

Ariano RE, Zhanel GG, Harding GKM (1990) The role of anion-exchange resins in the treatment of antibiotic-associated pseudomembranous colitis. Can Med Assoc J 142:1049–1051

Arts CJ, Govers CA, van den Berg H, Wolters MG, van Leeuwen P, Thijssen J (1991) In vitro binding of estrogens by dietary fibre and the in vivo apparent digestibility tested in pigs. J Steroid Biochem Mol Biol 38:621–628

Ast M, Frishman WH (1990) Bile acid sequestrants. J Clin Pharmacol 30:99–106

Bard JM, Párra HJ, Douste-Blazy P, Fruchard JC (1990) Effect of pravastatin, an HMG CoA reductase inhibitor, and cholestyramine, a bild acid sequestrant, on lipoprotein particles defined by their apolipoprotein composition. Drugs 39:917–928

Barnard DL, Heaton KW (1973) Bile acids and vitamin A absorption in man. The effects of two bile acid-binding agents cholestyramine and lignin. Gut 14:316–318
Bartlett JG, Tedesco FJ, Shull S, Lowe B, Chang T (1980) Symptomatic relapse after oral vancomycin therapy of antibiotic-associated pseudomembranous colitis. Gastroenterology 78:431–434
Basu SG (1965) Effect of cholestyramine on absorption of amino acids and vitamin A in man. Master's thesis, University of Texas, Austin
Beil U, Crouse JR, Einarsson K, Grundy SM (1982) Effect of interruption of the enterohepatic circulation of bile acids on the transport of very low density-lipoprotein triglycerides. Metabolism 31:438–444
Belamarich PF, Deckelbaum RJ, Starc TJ, Dobrin BE, Tint S, Salen G (1990) Response to diet and cholestyramine in a patient with sitosterolemia. Pediatrics 86:977–981
Bilicki CV, White JL, Hem SL, Borin MT (1989) Effect of anions on adsorption of bile salts by colestipol hydrochloride. Pharm Res 6:794–797
Blankenhorn DH, Nessim SA, Johnson RL, Sanmarco ME, Azen SP, Cashin-Hemphill L (1987) Beneficial effects of combined colestipol–niacin therapy on coronary atherosclerosis and coronary venous bypass grafts. JAMA 257: 3233–3240
Blankenhorn DH, Azen SP, Crawford DW, Nessim SA, Sanmarco ME, Selzer RH, Shircore AM, Wickham EC (1991) Effects of colestipol–niacin therapy on human femoral atherosclerosis. Circulation 83:438–447
Brensike JF, Levy RI, Kelsey SF, Passamani ER, Richardson JM, Loh IK, Stone NJ, Aldrich RF, Battaglini JW, Moriarty DJ, Fisher MR, Friedman L, Friedewald W, Detre KM, Epstein SE (1984) Effects of therapy with cholestyramine on progression of coronary arteriosclerosis: result of the NHLBI Type II Coronary Intervention Study. Circulation 69:213–324
Bressler R, Nowlin J, Bogdonoff MD (1966) The treatment of hypercholesterolemia and hypertriglyceridemia by anion exchange resin. South Med J 59:1097–1103
Briscoe AM, Ragan D (1963) Enhancement of calcium absorption in man by a bile acid sequestrant. Am J Clin Nutr 13:277–283
Brown DD, Juhl RP, Warner SL (1978) Decreased bioavailability of digoxin due to hypocholesterolemic interventions. Circulation 58:152–172
Brown G, Albers JJ, Fisher LD, Schaefer SM, Lin JT, Kaplan CA, Zhao XQ, Bisson BD, Fitzpatrick VF, Dodge HT (1990) Regression of coronary artery disease as a result of intensive lipid-lowering therapy in men with high level of apolipoprotein B. N Engl J Med 323:1289–1298
Brown MS, Goldstein JL (1986) A receptor-mediated pathway for cholesterol homeostasis. Science 232:34–47
Burbige EJ, Milligan FD (1975) Pseudomembraneous colitis. Association with antibiotics and therapy with cholestyramine. JAMA 231:1157–1158
Caldwell JH, Bush CA, Greenberger NJ (1971) Interruption of the enterohepatic circulation of digitoxin by cholestyramine. J Clin Invest 50:2638–2644
Callaghan JT, Tsuru M, Holtzmann JL, Hunninghake DB (1983) Effect of cholestyramine and colestipol on the absorption of phenytoin. Eur J Clin Pharmacol 24:675–678
Carrella M, Ericsson S, Del Piiano C, Angelin B, Einarsson K (1991) Effect of cholestyramine treatment on biliary lipid secretion rates in normolipidaemic men. J Intern Med 229:241–246
Casdorph HR (1976) Cholestyramine and ion-exchange resins. In: Paoletti R, Glueck CJ (eds) Lipid pharmacology, vol 12, Academic, New York, pp 221–256
Cashin-Hemphill L, Spencer CA, Nicoloff JT, Blankenhorn DH, Nessim SA, Chin HP, Lee NA (1987) Alterations in serum thyroid hormonal indices with colestipol–niacin therapy. Ann Intern Med 107:324–329
Cashin-Hemphill L, Mack WJ, Pogoda JW, Pogoda JM, Sanmarco ME, Azen SP, Blankenborn DH (1990) Beneficial effects of colestipol–niacin on coronary atherosclerosis. JAMA 264:3013–3017

Chait A, Lewis B (1972) The hyperlipidemias – a rational approach to classification and management. S Afr Med J 46:2097–2101
Chiang JYL, Miller WF, Lin G-M (1990) Regulation of cholesterol 7alpha-hydroxylase in the liver. J Biol Chem 265:3889–3897
Clas SD (1991) Quaternized colestipol, an improved bile salt adsorbent: in vitro studies. J Pharm Sci 80:128–131
Clifton-Bligh P, Miller NE, Nestel PJ (1974) Increased plasma esterifying activity during colestipol resin therapy in man. Metabolism 23:437–444
Cohen MI, Winslow PR, Boley SJ (1969) Intestinal obstruction associated with cholestyramine therapy. N Engl J Med 280:1285–1286
Coltman D, Mann MD, Bowie MD (1990) Effect of cholestyramine on activity of gentamycin in vitro. Pediatrics 85:390–391
Curtis DM, Driscoll DJ, Goldman DH, Weidman WH (1991) Loss of dental enamel in a patient taking cholestyramine. Mayo Clin Proc 66:1131
Curtis LD, Dickson AC, Ling KLE, Betteridge J (1988) Combination treatment with cholestyramine and bezafibrate for heterozygous familial hypercholesterolaemia. Br Med J 297:173–175
Danhoff IE (1966) The effect of cholestyramine on fecal excretion of ingested radioiodinated lipids. Am J Clin Nutr 18:343–349
Danielsson H, Sjövall T (1975) Bile acid metabolism. Annu Rev Biochem 44:233–253
Datta DV, Sherlock S (1963) Treatment of pruritus of obstructive jaundice with cholestyramine. Br Med J 1:216–219
Desager JP, Horsmans Y, Harvengt C (1991) Lecithin: cholesterol acyltransferase activity in familial hypercholesterolemia treated with simvastatin and simvastatin plus low-dose colestipol. J Clin Pharmacol 31:537–542
DeSante KA, DiSanto AR, Albert KS, Weber DJ, Welch RD, Vecchio TJ (1979) The effect of colestipol hydrochloride on the bioavailability and pharmacokinetics of clofibrate. J Clin Pharmacol 19:721–725
DeSimone R, Conti F, Lovati MR, Sirtori M, Cocuzza E, Sirtori CR (1978) New microporous cholestyramine analog for the treatment of hypercholesterolemia. J Pharm Sci 67:1695–1698
Dordoni B, Willson RA, Thompson RPH, Williams R (1973) Reduction of absorption of paracetamol by activated charcoal and cholestyramine: a possible therapeutic measure. Br Med J 1:86–87
Dorr AE, Gundersen K, Schneider JC, Spencer TW, Martin WB (1978) Colestipol hydrochloride in hypercholesterolemic patients – effect on serum cholesterol and mortality. J Chronic Dis 31:5–14
Dujovne CA, Hurwitz A, Kauffman MD, Azarnoff DL (1974) Colestipol and clofibrate in hypercholesterolemia. Clin Pharmacol Ther 16:291–296
Dujovne CA, Krehbiel P, DeCoursey S, Jackson B, Chernoff SB, Pittermann A, Garty M (1984a) Probucol with colestipol in the treatment of hypercholesterolemia. Ann Intern Med 100:477–482
Dujovne CA, Chernoff SB, Krehbiel P, Jackson B, DeCoursey S, Taylor H (1984b) Low-dose colestipol plus probucol for hypercholesterolemia. Am J Cardiol 53:1514–1518
East C, Bilheimer DW, Grundy SM (1988) Combination drug therapy for familial combined hyperlipidemia. Ann Intern Med 109:25–32
Einarsson K, Ericsson S, Ewerth S, Reihner E, Rudling M, Stahlberger D, Angelin B (1991) Bile acid sequestrants: mechanisms of action on bile acid and cholesterol metabolism. Eur J Clin Pharmacol 40 Suppl 1:S53–S58
Elmberger PG, Kalen A, Lund E, Reihener E, Eriksson M, Berglund L, Angelin B, Dallner G (1991) Effects of pravastatin and cholestyramine on products of the mevalonate pathway in familial hypercholesterolemia. J Lipid Res 32:934–940
Emmerich J, Aubert I, Bauduceau B, Dachet D, Chanu B, Erlich D, Gautier D, Jacotot B, Rouffy J (1990) Efficacy and safety of simvastatin (alone or in association with cholestyramine). A 1-year study in 66 patients with type II hyperlipoproteinaemia. Eur Heart J 11:149–155

Fellin R, Baggio G, Briani G, Baiocchi MR, Manzato E, Baldo G, Crepaldi G (1978) Long-term trial with colestipol plus clofibrate in familial hypercholesterolemia. Atherosclerosis 29:241–249

Fischer S, Hanefeld M, Lang PD, Fücker K, Bergmann S, Gehrisch S, Leonhardt W, Jaroß W (1990) Efficacy of a combined bezafibrate retard-cholestyramine treatment in patients with hypercholesterolemia. Arzneimittelforschung 40: 469–472

Forland SC, Feng Y, Cutler RE (1990) Apparent reduced absorption of gemfibrozil when given with colestipol. J Clin Pharmacol 30:29–32

Friedman H, Greenblatt DJ, LeLuc BW (1989) Impaired absorption of tetracycline by colestipol is not reversed by orange juice. J Clin Pharmacol 29:748–751

Gallo DC, Bailey KR, Sheffner AL (1965) The interaction between cholestyramine and drugs. Proc Exp Biol Med 120:60–65

Geisel J, Oette K, Burrichter H (1990) HMG-CoA-Reduktase-Inhibitoren bei familiärer Hypercholesterinämie. Fortschr Med 108:69–72

Gerio R, Murphy GM, Sladen GE, MacDonald DM (1987) A combination of phototherapy and cholestyramine for the relief of pruritus in primary biliary cirrhosis. Br J Dermatol 116:265–267

Glueck CJ, Ford S, Schel D, Steiner P (1972) Colestipol and cholestyramine resin. Comparative effects in familial type II hyperlipoproteinemia. JAMA 222:676–681

Glueck CJ, Fallat RW, Tsang RC (1973) Pediatric familial type II hyperlipoproteinemia: therapy with diet and cholestyramine resin. Pediatrics 52:669–679

Glueck CJ, Fallat RW, Tsang RC (1976a) Treatment of hyperlipoproteinemia in children In: Paoletti R, Glueck CJ (eds) Medical chemistry, lipid pharmacology, vol 2. Academic, New York, pp 257–275

Glueck CJ, Fallat RW, Mellies M, Tsang RC (1976b) Pediatric familial type II hyperlipoproteinemia: therapy with diet and colestipol resin. Pediactrics 57:68–74

Glueck CJ, Mellies MJ, Dine M, Perry T, Laskarzewski P (1986) Safety and efficacy of long-term diet and diet plus bile acid-binding resin cholesterol-lowering therapy in 73 children heterozygous for familial hypercholesterolemia. Pediatrics 78:338–348

Goldfarb S, Pitot HC (1972) Stimulatory effect of dietary lipid and cholestyramine on hepatic HMG CoA reductase. J Lipid Res 13:797–801

Goldstein JL, Brown MS (1989) Familial hypercholesterolemia. In: Scriver CR, Beandat AL, Sly WS, Valle D (eds) The metabolic basis of inherited disease, 6th edn. McGraw-Hill, New York, pp 1215–1250

Groot PH, Dijkhuis-Stoffelsma R, Grose WFA, Ambagtsheer JJ, Fernandes J (1983) The effects of colestipol hydrochloride on serum lipoprotein lipid and apolipoprotein B and A-I concentrations in children heterozygous for familial hypercholesterolemia. Acta Paedriatr Scand 72:81–85

Gross L, Brotman M (1970) Hypoprothrombinemia and hemorrhage associated with cholestyramine therapy. Ann Intern Med 72:95–96

Gross L, Figueredo R (1973) Long-term cholesterol-lowering effect of colestipol resin in humans. J Am Geriatr Soc 21:552–556

Grundy SM, Ahrens EH, Salen G (1971) Interruption of the enterohepatic circulation of bile acids in man: comparative effects of cholestyramine and ileal exclusion on cholesterol metabolism. J Lab Clin Med 78:94–121

Grundy SM, Vega GL, Bilheimer DW (1985) Influence of combined therapy with mevinolin and interruption of bile-acid reabsorption on low density lipoproteins in heterozygous familial hypercholesterolemia. Ann Intern Med 103:339–343

Guentert TW, Defoin R, Mosberg H (1988) The influence of cholestyramine on the elimination of tenoxicam and piroxicam. Eur J Clin Pharmacol 34:283–289

Gylling H, Vanhanen H, Miettinen TA (1989) Effects of acipimox and cholestyramine on serum lipoproteins, non-cholesterol sterols and cholesterol absorption and elimination. Eur J Clin Pharmacol 37:111–115

Hahn KJ, Edian W, Schettle M (1972) Effect of cholestyramine on the gastrointestinal absorption of phenprocoumon and acetylosalicylic acid in man. Eur J Clin Pharmacol 4:142–145

Harmon SM, Seifert CF (1991) Levothyroxine–cholestyramine interaction reemphasized. Ann Intern Med 115:658–659

Hartline JV (1976) Hyperchloremia, metabolic acidosis, and cholestyramine, J Pediatr 89:155

Harvengt C, Desager JP (1976) Colestipol in familial type II hyperlipoproteinemia. A three-year trial. Clin Pharmacol Ther 20:310–314

Hashim SA, Van Itallie TB (1965) Cholestyramine resin therapy for hypercholesterolemia. JAMA 192:289–293

Hashim SA, Bergen SS, Van Itallie TB (1961) Experimental steatorrhea induced in man by bile acid sequestrant. Proc Soc Exp Biol Med 106:173–175

Heaton KW, Lever JV, Barnard RE (1972) Osteomalacia associated with cholestyramine therapy for postileectomy diarrhea. Gastroenterology 62:642–646

Herman RJ, Chaudhary A (1991) In vitro binding of lorazepam and lorazepam glucuronide to cholestyramine, colestipol, and activated charcoal. Pharm Res 8:538–540

Hill ID, Mann MD, Househam KC, Bowie MD (1986) Use of oral gentamycin, metronidazole, and cholestyramine in the treatment of severe persistent diarrhea in infants. Pediatrics 77:477–481

Hoeg JM (1991) Pharmacologic and surgical treatment of dyslipidemic children and adolescents. Ann NY Acad Sci 623:275–284

Hoogwerf BJ, Peters JR, Frantz ID, Hunninghake DB (1985) Effect of clofibrate and colestipol singly and in combination on plasma lipids and lipoproteins in type III hyperlipoproteinemia. Metabolism 34:978–981

Howard RF, Brusco OJ, Furman RH (1966) Effect of cholestyramine administration on serum lipids and on nitrogen balance in familial hypercholesterolemia. J Lab Clin Med 68:12–20

Hunninghake DB (1980) Hypolipidemic drugs. In: Cluff LE, Petrie JC (eds) Clinical effects of interaction between drugs. Elsevier North-Holland, Amsterdam

Hunninghake DB (1990) Drug treatment of dyslipoproteinemia. Endocrinol Metab Clin North Am 19:345–360

Hunninghake DB, King S (1978) Effect of cholestyramine and colestipol on the absorption of methyldopa and hydrochlorothiazide. Pharmacologist 20:220

Hunninghake DB, Pollack EW (1977) Effect of bile acid sequestering agents on the absorption of aspirin, tolbutamide and warfarin. Fed Proc 35:996

Hunninghake DB, Probstfield JL (1977) Drug treatment of hyperlipoproteinemia. In: Rifkind BM, Levy RI (eds) Hyperlipidemia: diagnosis and therapy. Grune and Stratton, New York, pp 327–362

Hunninghake DB, Bell C, Olson L (1981) Effects of colestipol and clofibrate, singly and in combination, on plasma lipid and lipoproteins in type IIb hyperlipoproteinemia. Metabolism 30:610–615

Illingworth DR (1984) Mevinolin plus colestipol in therapy for severe heterozygous familial hypercholesterolemia. Ann Intern Med 101:598–604

Illingworth DR (1990) Management of hyperlipidemia: goals for the prevention of atherosclerosis. Clin Invest Med 13:211–218

Illingworth DR (1987) Lipid lowering drugs: an overview of indications and optimum therapeutic use. Drugs 33:259–279

Illingworth Dr, Bacon S (1989) Treatment of heterozygous familial hypercholesterolemia with lipid-lowering drugs. Arteriosclerosis 9 Suppl 1:I–121–134

Illingworth DR, Rapp JH, Phillipson BE, Connor WE (1981) Colestipol plus nicotinic acid in treatment of heterozygous familial hypercholesterolaemia. Lancet 1: 296–298

Insull W, Marquis NR, Tsianco MC (1991) Comparison of the efficacy of questran light, a new formulation of cholestyramine powder, to regular questran in maintaining lowered plasma cholesterol levels. Am J Cardiol 67:501–505

Ismail E, Corder CN, Epstein S, Barbi G, Thomas S (1990) Effects of pravastatin and cholestyramine on circulating levels of parathyroid hormone and vitamine D metabolites. Clin Ther 12:427–430

Jacob BG, Richter WO, Schwandt P (1990) Lovastatin, pravastatin, and serum lipoprotein(a). Ann Intern Med 112:713–714

Jacob BG, Möhrle W, Richter WO, Schwandt P (1992) Short- and long-term effects of lovastatin and pravastatin alone and in combination with cholestyramine on serum lipids, lipoproteins and apolipoproteins in primary hypercholesterolaemia. Eur J Clin Pharmacol 42:353–358

Jay RH, Rampling MW, Betteridge DJ (1990) Abnormalities of blood rheology in familial hypercholesterolaemia: effects of treatment. Atherosclerosis 85:249–256

Jay RH, Sturley RH, Stirling C, McGarrigle HHG, Katz M, Reckless JPD, Betteridge DJ (1991) Effects of pravastatin and cholestyramine on gonodal and adrenal steroid production in familial hypercholesterolaemia. Br J Clin Pharmacol 32:417–422

Jones AF, Hughes EA, Cramb R (1988) Gemfibrozil plus cholestyramine in familial hypercholesterolaemia. Lancet 1:776

Kane JP, Malloy MJ, Tun P, Phillips NR, Freedman DD, Williams ML, Rowe JS, Havel RJ (1981) Normalization of low-density-lipoprotein levels in heterozygous familial hypercholesterolemia with a combined drug regimen. N Engl J Med 304:251–258

Kauffman RE, Azarnoff DL (1973) Effect of colestipol on gastrointestinal absorption of chlorothiazide in man. Clin Pharmacol Ther 14:886–890

Keighley MRB (1980) Antibiotic-associated pseudomembraneous colitis: pathogenesis and management. Drugs 20:49–56

King CY, Barriere SL (1981) Analysis of the in vitro interaction between vancomycin and cholestyramine. Antimicrob Agents Chemother 19:326–327

Kivisto KT, Neuvonen PJ (1990) The effect of cholestyramine and activated charcoal on glipizide absorption. Br J Clin Pharmacol 30:733–736

Kleinmann PK (1974) Cholestyramine and metabolic acidosis. N Engl J Med 290: 861

Kreutzer EW, Milligan FD (1978) Treatment of antibiotic-associated pseudomembraneous colitis with cholestyramine resin. Johns Hopkins Med J 143:67–72

Kunimoto D, Thomson ABR (1986) Recurrent clostridium difficile-associated colitis responding to cholestyramine. Digestion 33:225–228

Kuo PT, Hayase K, Kostis JB, Moreyra AE (1979) Use of combined diet and colestipol in long-term (7–7½ years) treatment of patients with type II hyperlipoproteinemia. Circulation 59:199–211

Kuo PT, Kostis JB, Moreyra AE, Hayes JA (1981) Familial type II hyperlipoproteinemia with coronary heart disease. Chest 79:286–291

Kuo PT, Wilson AC, Kostis JB, Moreyra AE (1986) Effects of combined probucol–colestipol treatment for familial hypercholesterolemia and coronary artery disease. Am J Cardiol 57:43H–48H

Kwiterovich PO (1990) Diagnosis and management of familial dyslipoproteinemia in children and adolescents. Pediatr Clin North Am 37:1489–1521

Lees AM, McCluskey MA, Lees RS (1976) Results of colestipol therapy in type II hyperlipoproteinemia. Atherosclerosis 24:129–140

Leren TP, Hjermann I, Berg K, Leren P, Foss OP, Viksmoen L (1988) Effects of lovastatin alone and in combination with cholestyramine on serum lipids and apolipoproteins in heterozygotes for familial hypercholesterolemia. Atherosclerosis 73:135–141

Levy RI (1970) Dietary and drug treatment of primary hyperlipoproteinemia. Ann Intern Med 77:267–294

Levy RI, Langer T (1972) Hypolipidemic drugs and lipoprotein metabolism. Adv Exp Med Biol 26:155–163

Levy RI, Fredrickson DS, Stone NJ, Bilheimer DW, Brown WF, Glueck CJ, Gotto AM, Herbert PN, Kwiterovich PO, Langer T, LaRosa J, Lux SE, Rider AK,

Shulman RS, Slone HR (1973) Cholestyramine in type II hyperlipoproteinemia – a double-blind trial. Ann Intern Med 79:51–58
Levy RI, Brensike JF, Epstein SE, Kelsey SF, Passamani ER, Richardson JM, Loh IK, Stone NJ, Aldrich RF, Battaglini JW, Moriarty DJ, Fisher ML, Friedman L, Friedewald W, Detre KM (1984) The influence of changes in lipid values induced by cholestyramine and diet on progression of coronary artery disease: results of the NHLBI Type II Coronary Intervention Study. Circulation 69: 325–337
Linet OI, Grzegorczyk CR, Demke DM (1988) The effect of encapsulated, low-dose colestipol in patients with hyperlipidemia. J Clin Pharmacol 28:804–806
Longenecker JB, Basu SG (1965) Effect of cholestyramine on absorption of amino acids and vitamin A in man. Fed Proc 24:375
Mabuchi H, Sakao T, Sakai Y, Yoshimura A, Watanabe A, Wakasugi T, Koizumi J, Takeda R (1983) Reduction of serum cholesterol in heterozygous patients with familial hypercholesterolemia. Additive effects of compactin and cholestyramine. N Engl J Med 308:609–613
Malloy MJ, Kane JP, Kunitake ST, Tun P (1987) Complementarity of colestipol, niacin, and lovastatin in treatment of severe familial hypercholesterolemia. Ann Intern Med 107:616–623
Malmendier CL, Delcroix C, Lontie JF (1987) The effect of combined fenofibrate and cholestyramine therapy on low-density lipoprotein kinetics in familial hypercholesterolemia patients. Clin Chim Acta 162:221–227
McCullough AJ, Barron D, Mullen KD, Petrelli M, Park MC, Mukhtar H, Bickers DR (1988) Fecal protoporphyrin excretion in erythropoietic protoporphyria: effect of cholestyramine and bile acid feeding. Gastroenterology 94:177–181
Miller NE, Clifton-Bligh P, Nestel PJ (1973) Effects of colestipol, a new bile-acid sequestering resin, on cholesterol metabolism in man. J Lab Clin Med 82:876–890
Mishkel MA, Crowther SM (1977) Long-term therapy of diet-resistant hypercholesterolemia with colestipol. Curr Ther Res 22:398–412
Mölgaard J, Lundh BL, von Schenck H, Olsson AG (1991) Long-term efficacy and safety of simvastatin alone and in combination therapy in treatment of hypercholesterolaemia. Atherosclerosis 91:S21–S28
Moore RB, Crane CA, Frantz ID (1968) Effect of cholestyramine on the fecal excretion of intravenously administrated cholesterol-^{14}C and its degradation products in a hypercholesterolemic patient. J Clin Invest 47:1664–1671
Mordasini R, Twelsiek F, Oster P, Schellenberg B, Raetzer H, Heuck CC, Schlierf G (1978) Abnormal low density lipoproteins in children with familial hypercholesterolemia – effect of polyanion exchange resins. Klin Wochenschr 56: 805–808
Myant NB, Mitropoulos KA (1977) Cholesterol 7alpha-hydroxylase. J Lipid Res 18:135–153
Nash DT, Gensini G, Esente P (1982) Effect of lipid-lowering therapy on the progression of coronary atherosclerosis assessed by scheduled repetitive coronary arteriography. Int J Cardiol 2:43–55
Neuvonen PJ, Kivisto K, Hirvisalo EL (1988) Effects of resins and activated charcoal on the absorption of digoxin, carbamazepine and frusemide. Br J Clin Pharmacol 25:229–233
Nguyen LB, Cobb M, Shefer S, Salen G, Ness GC, Tint GS (1991) Regulation of cholesterol biosynthesis in sitosterolemia: effects of lovastatin, cholestyramine, and dietary sterol restriction. J Lipid Res 32:1941–1948
Nitsch J, Lüderitz B (1986) Beschleunigte Elimination von Amiodaron durch Colestyramin. Dtsch Med Wochenschr 111:1241–1244
Norris RM, Dunn GH, Hearron AE (1987) Very low dose colestid in gelatin capsules effectively reduces low density cholesterol in hypercholesterolemic men. Diabetes 36 Suppl 1:180A
Northcutt C, Stiel NJ, Hollifield JW, Stant EG (1969) The influence of cholestyramine on thyroxine absorption. JAMA 208:1857–1861

Nye ER, Jackson D, Hunter JD (1972) Treatment of hypercholesterolemia with colestipol: a bile sequestrating agent. N Z Med J 76:12–16
Packard CJ, Stewart JM, Morgan HG, Lorimer AR, Shepherd J (1980) Combined drug therapy for familial hypercholesterolemia. Artery 7:281–289
Pan HY, DeVault AR, Swites BJ, Whigan D, Ivashkiv E, Willard DA, Brescia D (1990) Pharmacokinetics and pharmacodynamics of pravastatin alone and with cholestyramine in hypercholesterolemia. Clin Pharmacol Ther 48:201–207
Pantosti A, Luzzi I, Cardines R, Gianfrilli P (1985) Comparison of the in vitro activities of teicoplanin and vancomycin against clostridium difficile and their interactions with cholestyramine. Antimicrob Agents Chemother 28:847–848
Parkinson TM, Gunderson K, Nelson NA (1970) Effects of colestipol (U-26597A), a new bile acid sequestrant, on serum lipids in experimental animals and man. Atherosclerosis 11:531–537
Pattison M, Lee SM (1987) Life-threatening metabolic acidosis from cholestyramine in an infant with renal insufficiency. Am J Dis Child 141:479–480
Peters JR, Hunninghake DB (1985) Effect of time of administration of cholestyramine on plasma lipids and lipoproteins. Artery 13:1–6
Pruksananonda P, Powell KR (1989) Multiple relapses of clostridium difficile-associated diarrhea responding to an extended course of cholestyramine. Pediatr Infect Dis J 8:175–178
Rabelink AJ, Hene RJ, Erkelens DW, Joles JA, Koomans HA (1988) Effects of simvastatin and cholestyramine on lipoprotein profile in hyperlipidaemia of nephrotic syndrome. Lancet 2:1335–1338
Reihener E, Angelin B, Rudling M, Ewerth S, Björkhem I, Einarsson K (1990) Regulation of hepatic cholesterol metabolism in humans: stimulatory effects of cholestyramine to HMG-CoA reductase activity and low density lipoprotein receptor expression in gallstone patients. J Lipid Res 31:2219–2226
Robinson DS, Benjamin DM, McCormack JJ (1971) Interaction of warfarin and nonsystemic gastrointestinal drugs. Clin Pharmacol Ther 12:491–495
Ros E, Zambon D, Bertomeu A, Cuso E, Sanllehy C, Casals E (1991) Comparative study of a microporous cholestyramine analogue (Filicol) and gemfibrozil treatment of severe primary hypercholesterolemia. Arch Intern Med 151:301–305
Rose HG, Haft GK, Juliano J (1976) Clofibrate-induced low density lipoprotein elevation. Therapeutic implications and treatment by colestipol resin. Atherosclerosis 23:413–427
Runeberg L, Miettinen TA, Nikkilä EA (1972) Effect of cholestyramine on mineral excretion in man. Acta Med Scand 192:71–76
Ryan JR, Jain AK, McMahon FG (1975) Long-term treatment of hypercholesterolemia with colestipol hydrochloride. Clin Pharmacol Ther 17:83–87
Sachs BA, Wolfman L (1973) Response of hyperlipoproteinemia to colestipol. NY State J Med 73:1068–1070
Sachs BA, Wolfman L (1974) Colestipol therapy of hyperlipidemia in man. Proc Soc Exp Biol Med 1747:694–697
Schlierf G, Mrozik K, Heuck CC, Middelhoff G, Oster P, Riesen W, Schellenberg B (1982) Low-dose colestipol in children, adolescents and young adults with familial hypercholesterolemia. Atherosclerosis 41:133–138
Schlierf G, Vogel G, Kohlmeier M, Vuilleumier JP, Hüppe R, Schmidt-Gayk H (1985) Langzeittherapie der familiären Hypercholesterinämie bei Jugendlichen mit Colestipol: Versorgungszustand mit Mineralstoffen und Vitaminen. Klin Wochenschr 63:802–806
Schwartzkopff W, Bimmermann A, Schleicher J (1990) Comparison of the effectiveness of the HMG CoA reductase inhibitors pravastatin versus cholestyramine in hypercholesterolemia. Drug Res 40:1322–1327
Schwarz KB, Goldstein PD, Witztum JL, Schonfeld G (1980) Fat-soluble vitamin concentrations in hypercholesterolemic children treated with colestipol. Pediatrics 65:243–250

Seplowitz A, Smith FR, Berns L, Eder HA, Goodman DS (1981) Comparison of the effect of colestipol hydrochloride and clofibrate on plasma lipids and lipoproteins in the treatment of hypercholesterolemia. Atherosclerosis 39:35–43

Series JJ, Caslake MJ, Kilday C, Cruickshank A, Demant T, Lorimer AR, Packard CJ, Shepherd J (1989) Effect of combined therapy with bezafibrate and cholestyramine on low-density lipoprotein metabolism in type IIa hypercholesterolemia. Metabolism 38:153–158

Series JJ, Gaw A, Kilday C, Bedford DK, Lorimer AR, Packard CJ, Shepherd J (1990) Acipimox in combination with low dose cholestyramine for the treatment of type II hyperlipidaemia. Br J Clin Pharmacol 30:49–54

Shaefer MS, Jungnickel PW, Miwa LJ, Marquis NR, Hutton GD (1990) Sensory/mixability preference evaluation of cholestyramine powder formulations. DICP Ann Pharmacother 24:472–474

Shepherd J (1989) Mechanism of action of bile acid sequestrants and other lipid-lowering drugs. Cardiology 76 Suppl 1:65–74

Shepherd J, Packard CJ (1983) Mode of action of lipid-lowering drugs. In: Miller NE (ed) Atherosclerosis: mechanisms and approaches to therapy. Raven, New York, pp 169–201

Shepherd J, Packard CJ, Bicker S, Lawrie TDV, Morgan HG (1980) Chole styramine promotes receptor-mediated low-density-lipoprotein catabolism. N Engl J Med 302:1219–1222

Sinatra F, Buntain WL, Mitchell CH, Sunshine P (1976) Cholestyramine treatment of pseudomembraneous colitis. J Pediatr 88:304–306

Sirtori M, Franceschini G, Gianfranceschi G, Montanari G, Cocuzza E, Sirtori CR (1982) Microporous cholestyramine in suspension form. Lancet 2:383

Sirtori M, Pazzucconi F, Gianfranceschi G, Sirtori CR (1991) Efficacy of cholestyramine does not vary when taken before or during meal. Atherosclerosis 88:249–252

Sommariva D, Bonfiglioli D, Tirrito M, Pogliaghi I, Branchi A, Cabrini E (1986) Probucol and cholestyramine combination in the treatment of severe hypercholesterolemia. Int J Clin Pharmacol Ther Toxicol 24:505–510

Spengel FA, Jadhav A, Duffield RGM, Wood CB, Thompson GR (1981) Superiority of partial ileal bypass over cholestyramine in reducing cholesterol in familial hypercholesterolemia. Lancet 2:768–771

Stein E, Kreisberg R, Miller V, Mantell G, Washington L, Shapiro DR (1990) Effects of simvastatin and cholestyramine in familial and nonfamilial hypercholesterolemia. Arch Intern Med 150:341–345

Steinhagen-Thiessen E, Müller S, Holler HD, Lang PD (1987) Effect of bezafibrate and cholestyramine in patients with primary hypercholesterolemia. Arzneimittelforschung 37:726–728

Sweeney ME, Fletcher BJ, Rice CR, Berra KA, Rudd CM, Fletcher GF, Superko RS (1991) Efficacy and compliance with cholestyramine bar versus powder in the treatment of hyperlipidemia. Am J Med 90:469–473

The Lipid Research Clinic Program (1984a) The Lipid Research Clinics Coronary Primary Prevention Trial results: I. Reduction in incidence of coronary heart disease. JAMA 251:351–364

The Lipid Reserach Clinics Program (1984b) The Lipid Research Clinics Coronary Primary Prevention Trial results: II. The relationship of reduction in incidence of coronary heart disease to cholesterol lowering. JAMA 251:365–374

The Lovastatin Study Group III (1988) A multicenter comparison of Lovastatin and cholestyramine therapy for severe primary hypercholesterolemia. JAMA 260:359–366

Tikkanen MJ, Nikkilä EA (1987) Current pharmacologic treatment of elevated serum cholesterol. Circulation 76:529–533

Tsang RC, Roginsky MS, Mellies MJ, Glueck CJ (1978) Plasma 25-hydroxy-vitamin D in familial hypercholesterolemic children receiving colestipol resin. Pedriatr Res 12:980–982

Vega GL, Grundy SM (1987) Treatment of primary moderate hypercholesterolemia with lovastatin (mevinolin and colestipol). JAMA 257:33–38
Visintine RE, Michaels GD, Fukayama G et al. (1961) Xanthomatous biliary cirrhosis treated with cholestyramine. Lancet 2:341–343
Wallentin L (1978) Lecithin: cholesterol acyl transfer rate and high density lipoproteins in plasma during dietary and cholestyramine treatment of type IIa hyperlipoproteinemia. Eur J Clin Invest 8:833–839
Weintraub MS, Eisenberg S, Breslow J (1987) Different patterns of postprandial lipoprotein metabolism in normal, type IIa, type III, and type IV hyperlipoproteinemic individuals. J Clin Invest 79:1110–1119
Weisweiler P, Schwandt P (1986) Colestipol plus fenofibrate versus synvinolin in familial hypercholesterolaemia. Lancet 2:1212–1213
Weisweiler P, Neureuther G, Schwandt P (1979) The effect of cholestyramine on lipoprotein lipids in patients with primary type IIa hyperlipoproteinemia. Atherosclerosis 33:295–300
Wells RF, Knepshield JH, Davis C (1968) Right upper quadrant calcification in a patient receiving long term cholestyramine therapy for primary biliary cirrhosis. Am J Dig Dis 13:86–94
West RJ, Lloyd JK (1975) The effect of cholestyramine on intestinal absorption. Gut 16:93–98
West RJ, Lloyd JK (1980) Long-term follow-up of children with familial hypercholesterolemia treated with cholestyramine. Lancet 2:873–875
Wiklund O, Angelin B, Olofsson SO, Eriksson M, Fager G, Berglund L, Bondjers G (1990) Apolipoprotein(a) and ischaemic heart disease in familial hypercholesterolaemia. Lancet 335:1360–1363
Witztum JL, Schonfeld G, Weidman SW (1976) The effect of colestipol on the metabolism of very-low-density lipoproteins in man. J Lab Clin Med 88:1008–1018
Witztum JL, Schonfeld G, Weidman SW, Giese WE, Dillingham MA (1979) Bile sequestrants therapy alters the composition of low density and high density lipoproteins. Metabolism 28:221–229
Witztum JL, Simmons D, Steinberg D, Beltz WF, Weinreb R, Young SG, Lester P, Kelly N, Juliano J (1989) Intensive combination drug therapy of familial hyperchoelsterolemia with lovastatin, probucol, and colestipol hydrochloride. Circulation 79:16–28
Young SG, Witztum JL, Carew TE, Krauss RW, Lindgren FT (1989) Colestipol-induced changes in LDL composition and metabolism. II. Studies in humans. J Lipid Res 30:225–238
Ytre-Arne K, Nordoy A (1989) Simvastatin and cholestyramine in the long-term treatment of hypercholesterolaemia. J Intern Med 226:285–290

CHAPTER 15
Probucol

J. DAVIGNON

A. Introduction

Probucol is a potent synthetic lipophilic antioxidant which was developed in the 1960s for the plastics and rubber industry by Dow Chemicals Co. In a series of screening assays it was found to lower plasma cholesterol in animals. This effect was reproduced in man and probucol was first licensed for treatment of hypercholesterolemia in Europe in 1976 (Portugal and Spain) and in the United States and Canada in 1978. Although the drug was well tolerated, effective in heterozygous familial hypercholesterolemia (FH), and able to induce xanthoma regression, probucol was initially received with little enthusiasm because it lowered plasma high-density lipoproteins (HDL), an effect considered invariably unfavorable at the time, and had a long plasma half-life. In recent years, however, the interest in probucol was rekindled for several reasons (DAVIGNON 1991a): the HDL-lowering effect was not shown to be deleterious, combination therapy with resins proved to be highly effective at lowering low-density lipoproteins (LDL) and it could induce xanthoma regression in *homozygous* FH and retard progression of atherosclerosis in the Watanabe heritable hyperlipidemic (WHHL) rabbit, an effect ascribed to its antioxidant properties, and further studies uncovered new and rather unexpected potential benefits. The aim of this review is to outline what is known about probucol that is of practical importance for the treatment of hyperlipidemia and atherosclerosis, with an emphasis on antioxidant properties and new observations. For further details, the reader is referred to previous reviews (MURPHY 1977; HEEL et al. 1978; STRANDBERG et al. 1988; BUCKLEY et al. 1989).

B. Chemistry, Physical Properties, and Dosage

Probucol (formerly DH-581, biphenabid) has a structural formula (Fig. 1) which bears little resemblance to any of the current lipid-lowering agents. Its chemical name is 4,4′-(isopropylidene-dithio) bis (2,6-di-tert-butylphenol). Two tertiary butyl phenols are linked together by a dithiopropylidene bridge, resulting in a highly lipophilic, cholesterol-lowering molecule with strong antioxidant properties. It has a molecular mass of 516.8 Da and is 256 000 times more soluble in chloroform than in water. It precipitates from ethanol

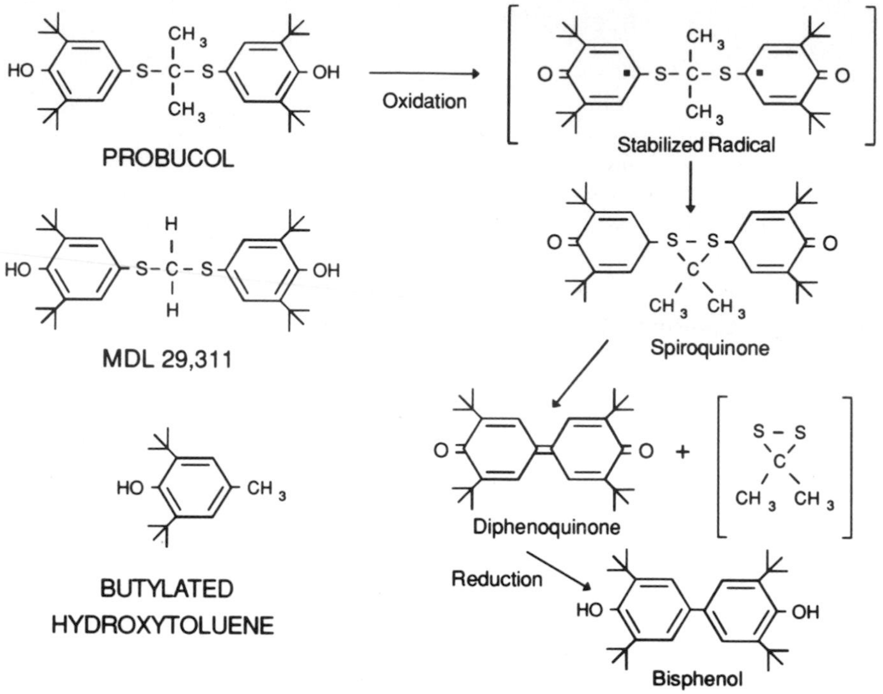

Fig. 1. Chemical structure and biotransformation of probucol. On the *left*, the structure of probucol is contrasted with that of the dinor analog and of butylated hydroxytoluene. Loss of the central two methyl groups in MDL 29311 abolishes the cholesterol-lowering effect, but preserves the antioxidant property (MAO et al. 1991). Probucol is oxidized to a diphenoquinone via a spiroquinone intermediate and reduced to the bisphenol in monkeys and rabbits (MAO et al. 1991; DAGE et al. 1991). The tertiary butyl groups are represented by a *trident*

as a white crystalline solid and has a melting point of 124.5°–126°C. For maximum cholesterol-lowering activity the phenolic hydroxyl groups, the tertiary butyl groups, and the central isopropyl chain are necessary (NEUWORTH et al. 1970; MAO et al. 1991).

Replacement of the latter by a methyl group (Fig. 1) and other substitutions at the disulfide-linked carbon preserve the antioxidant property, but abolish the cholesterol-lowering activity (NEUWORTH et al. 1970; JACKSON et al. 1991). Probucol is structurally related to butylated hydroxytoluene (BHT), an antioxidant widely used as a food additive (Fig. 1). Probucol is measured in body fluids by high performance liquid chromatography (HPLC; COUTANT et al. 1981; SATONIN and COUTANT 1986). It is dispensed as 250 mg tablets and the recommended cholesterol-lowering dose is 2 tablets twice a day with meals, although there is a trend in Japan to use lower doses (500 mg/day; MAEDA et al. 1989; HOMMA et al. 1991; SASAKI et al. 1987).

C. Absorption, Metabolism, and Excretion

Probucol is poorly and variably absorbed from the gastrointestinal tract, less than 10% of an administered dose reaching the circulation (HEEG and TACHIZAWA 1980). Because of its lipophilic nature, factors which interfere with fat absorption may affect probucol absorption. Administration immediately after a meal improves absorption. Following an oral dose of ^{14}C ring-labeled probucol given after 21 days of oral administration of the drug in adult men, ^{14}C appears in plasma after 2 h, peaks between 8 and 12 h, less than 2% appears in urine over the next 4 days, and the radiolabel is still detectable in plasma after 50 days (ARNOLD et al. 1970). Probucol circulates in plasma in association with lipoproteins apparently dissolved in the lipid core (EDER 1982). In vitro, probucol equilibrates between plasma (90%) and erythrocytes (10%); about 95% of the serum fraction distributes among the various lipoproteins as follows: 44.4% in LDL, 38.2% in very low density lipoproteins (VLDL), and 13.2% in HDL (URIEN et al. 1984). After oral administration to hypercholesterolemic patients for 12 weeks, the proportion in lipoproteins is 48.9% in LDL, 13.8% in VLDL, and 37.3% in HDL (FELLIN et al. 1986); after 6 months, it is 13.7%, 74.1%, and 12.2%, respectively (DACHET et al. 1985). After 12 weeks of probucol therapy at the recommended dose, the mean serum and LDL concentrations are 12.8 μg/ml and 6.7 μg/ml, respectively (FELLIN et al. 1986); after 1 year the plasma concentration of probucol averages 19 μg/ml (HEEG and TACHIZAWA 1980) to 40 μg/ml (DACHET et al. 1985) with a wide interindividual variation (13–100 μg/ml). Probucol is readily distributed in the liver and the adrenals and is gradually sequestered in adipose tissue (in monkeys, the concentration in body fat is 100 times that in plasma after 2 years of oral administration as compared to 25 times in adrenal glands and four times in the liver) (BUCKLEY et al. 1989). It can also be found in bile and milk of lactating animals, but does not appear to cross the blood–brain barrier. It has a long plasma half-life and decays with several exponentials during clearance from plasma and adipose tissue. The terminal phase elimination half-life ($t_{1/2\beta}$) was calculated to be 23 days in healthy volunteers (BARNHART et al. 1971) and about 47 days in hyperlipidemic patients (FELLIN et al. 1986). Little is known of the biotransformation of probucol in humans. Work in monkeys (COUTANT et al. 1981) and rabbits (JACKSON et al. 1991) indicates that probucol is oxidized to its diphenoquinone (greenish–yellow compound with a maximal absorbance at 420 nm) via a spiroquinone intermediate and subsequently reduced to the bisphenol (Fig. 1). The physiological and pharmacological roles of these metabolites are as yet unknown. The major route of elimination is via the bile and feces, and renal clearance is negligible (less than 2%). Some enterohepatic recirculation is postulated.

D. Effects on Plasma Lipids, Lipoproteins, and Apolipoproteins

I. Cholesterol and Low-Density Lipoproteins

Probucol lowers LDL cholesterol (LDL-C) and has, in general, little effect on plasma triglycerides (TG) and TG-rich lipoprotein levels; hence, it is primarily indicated in the treatment of hypercholesterolemia (Buckley et al. 1989). It is effective in heterozygous FH and lowers plasma cholesterol by an average of 14% in these patients (Buckley et al. 1989; LeLorier et al. 1977). This effect is accounted for by a fall in both LDL-C (−10%) and HDL cholesterol (HDL-C; −30%), but the response is highly variable among subjects and tends to be more pronounced in milder cases of hypercholesterolemia (Davignon et al. 1988). The effect on plasma cholesterol is sustained over several years (McCaughan 1981) and is additive to the effect of the diet (Fig. 2; Brown and DeWolfe 1974; LeLorier et al. 1977). Occasionally, it may persist for several weeks following discontinuation of the drug (Davignon et al. 1982). A spontaneous escape to the cholesterol-lowering effect is rare (Davignon and LeLorier 1981); it has occurred in FH patients treated with 2 g/day probucol (Miettinen 1972). Probucol lowered plasma cholesterol by 3%–34% (mean 14.3%, 11 trials) in patients with FH, by 31% (mean of two trials) in homozygous FH, and by 8%–35% (mean 17.7%, 21 trials) in patients with "polygenic" hypercholesterolemia (Buckley et al. 1989). Sources of variation may include design of study, duration, dosage, frequency and mode of administration of probucol, dietary factors (fat content), diet responsiveness, concomitant medication, etc. There is recent evidence that variability in response may be affected by apolipoprotein (apo) E genotype. FH patients with the ε4 allele show a greater reduction in cholesterol levels than subjects who do not have this allele, but this effect is not observed in non-FH patients. This study (Nestruck et al. 1987) was confirmed and extended by ETO et al. (1990), who showed that FH patients with apoE-4 had a greater reduction of LDL-C and also of TG than non-E4 subjects. It was speculated that this greater response may represent enhanced catabolism of LDL, perhaps via an enhancement of LDL receptor activity. It may be hypothesized that a gene–gene interaction between the LDL receptor and the apoE locus accounts for this pharmacogenetic observation since (a) the main effect is on LDL, (b) it is observed only in patients with an LDL receptor defect, (c) both genes are on the same chromosome, and (d) an interaction between these two loci has previously been reported (Pedersen and Berg 1989).

II. High-Density Lipoproteins and Reverse Cholesterol Transport

Probucol lowers plasma HDL-C consistently and predictably; 24.6% on average in FH (12 trials, range −13 to −33) and 21.6% in non-FH patients

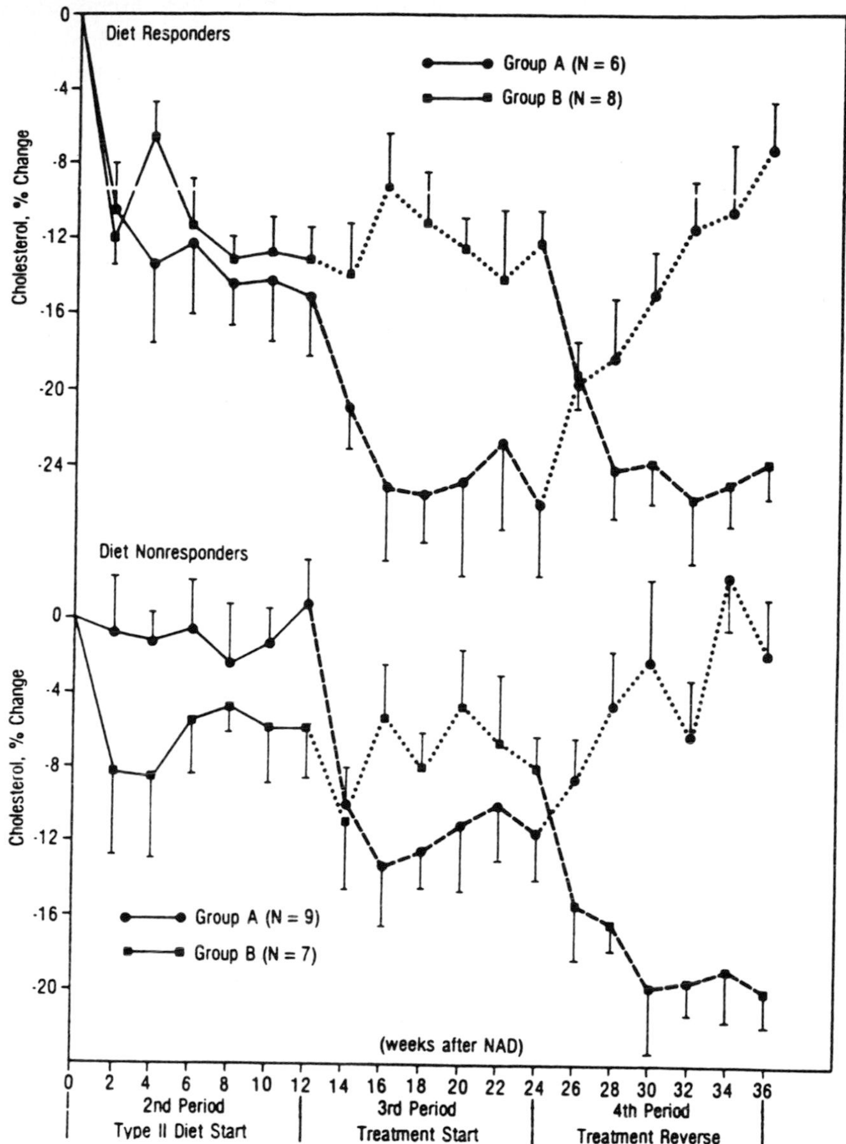

Fig. 2. Additive effect of diet and probucol on plasma cholesterol in familial hypercholesterolemia (FH). Result of a double-blind, crossover trial in 30 FH patients (15 men, 15 women) showing that probucol (1 g/day; *dashed lines*) lowers plasma cholesterol by approximately 14% whether given before or after a placebo period (*dotted lines*) in patients responding (*top panel*) or not responding (*lower panel*) to a low-cholesterol, low-saturated fat diet (*unbroken lines*). Patients were on a standard North American diet (*NAD*) for 8 weeks before onset of dietary treatment. A 25% fall in plasma cholesterol was obtained by the combined effect in diet responders. (From LeLorier et al. 1977)

(six trials, range −16 to −26) (BUCKLEY et al. 1989). There are seemingly conflicting reports regarding which fraction is reduced, but most of these discrepancies can be accounted for by differences in methodology, underlying lipoprotein defect, initial levels of HDL_2 and HDL_3, concomitant medication, duration of treatment, and gender (BUCKLEY et al. 1989; FRANCESCHINI et al. 1989). In general, the largest proportion of the decrease in HDL-C and apoA-I is in the HDL_2 subfraction, except when HDL are initially low (as in endogenous hypertriglyceridemia, type IV), in which case the HDL_3 subfraction may be lowered (and both apoA-I and A-II are reduced). The HDL_3 subfraction (and apo-A-II) is usually unchanged and may even be increased, especially in women who have higher HDL_2 concentrations. Probucol tends to reduce the HDL particle size, especially that of the HDL_2 subfraction (YAMAMOTO et al. 1986a; BAGDADE et al. 1990) and increases the prevalence of small HDL particles, as shown by nondenaturing gradient gel electrophoresis (JOHANSSON et al. 1990; BAGDADE et al. 1990; FRANCESCHINI et al. 1989). Both the largest (HDL_{2b}) and the smallest (HDL_{3c}) subclasses are markedly reduced by probucol treatment of hypercholesterolemic subjects, while intermediate subclasses (HDL_{2a}, HDL_{3a}, or HDL_{3b}) are little affected (JOHANSSON et al. 1990).

The HDL-lowering effect of probucol has been a source of concern in the past, due to the role of HDL in the antiatherogenic process of "reverse cholesterol transport" (MILLER and MILLER 1975; REICHL and MILLER 1989). However, there is a large body of indirect evidence indicating that this effect is not harmful (DAVIGNON and BOUTHILLIER 1982; DAVIGNON 1991a). If the HDL-lowering effect of probucol, often resulting in an increased LDL to HDL ratio, were deleterious, one would expect a markedly enhanced coronary artery disease (CAD) mortality rate on long-term exposure. TEDESCHI et al. (1980) showed that this was not the case; with the life table analysis method, they calculated that the 5-year mortality rate for 373 male subjects exposed to long-term probucol administration was 1.0%, as compared to 1.8% for the 5-year rate observed in the 5296 hypercholesterolemic males in the group II placebo of the WHO cooperative trial. Although this type of comparison is of questionable reliability, the result is favorable to probucol, even though the contrast was biased against the drug since 98% of the patients taking probucol had a cholesterol level above 250 mg/dl (6.46 mmol/l) versus 47% in the WHO placebo group. Similarly, in the Helsinki multifactorial primary prevention trial, where CAD mortality was greater in the treated group than in the two control groups, MIETTINEN et al. (1986) observed that subjects treated with probucol have, in spite of a significant reduction in plasma HDL-C, a low CAD incidence rate and a very low CAD risk ratio (RR, 0.26) compared to that of subjects treated with β-blockers (RR, 5.27) or to that imparted by hypercholesterolemia alone (RR, 4.14). The fact that probucol induces xanthoma regression not only in heterozygous (HARRIS et al. 1974; BEAUMONT et al. 1982; BUXTORF et al. 1985), but also in homozygous FH patients (BAKER et al. 1982; YAMAMOTO et al. 1986a) indicates that probucol has the ability to mobilize cholesterol from tissues. Indeed in one study conducted in homozygous FH patients, reduction in tendon xanthoma thickness correlated paradoxically with the reduction in HDL-C, but not with changes in LDL-C (YAMAMOTO et al. 1986a). The concept emerged that probucol could enhance the mobilization of tissue cholesterol and its return to the liver. This view is supported by the work of GOLDBERG and MENDEZ (1988) showing that probucol, in the presence of apoE-depleted HDL_3, enhances cholesterol efflux from cultured human skin fibroblasts in a dose-dependent fashion up to a 2-μM concentration. BAGDADE et al. (1990) observed that probucol treatment decreases the free cholesterol to phosphatidylcholine (FC to PC) ratio in HDL_2 of hypercholesterolemic subjects. This could be a beneficial effect, since HDL from hypercholesterolemic patients have a high FC to PC ratio, which KUKSIS et al. (1982) found to be a potent predictor of CAD in hyperlipidemic men. A relatively high HDL FC to PC ratio diminishes the ability of HDL to promote tissue cholesterol efflux (JOHNSON et al. 1986). Thus, the capacity of probucol to reverse this alteration in HDL FC to PC ratio may improve cholesterol transport towards the liver. Probucol also affects cholesteryl ester transfer activity (CETA) in plasma, another important factor in reverse cholesterol transport. Both an increase (FRANCESCHINI et al. 1989; MCPHERSON et al. 1991a) and a decrease (BAGDADE et al. 1990) have been reported, and the reason for this discrepancy is unclear (see Sect. F.II. below). Genetic CETA deficiency has been reported in Japan; in one patient it was associated with premature arcus corneae (MATSUZAWA et al. 1988) and in a large family with longevity (INAZU et al. 1990; BROWN et al. 1989). These entities, however, may not necessarily represent the same genetic defect. High CETA and surface lipid changes in HDL that accompany atherogenic dyslipoproteinemias are corrected by probucol (BAGDADE et al. 1990). Although the metabolic implications of CETA changes are not completely established, there is indirect evidence that probucol-modified HDL are functionally efficient in reverse cholesterol transport. It is doubtful that xanthoma and/or atheroma regression, whether related to antioxidant properties or not, could take place to any significant extent in the presence of a markedly impaired reverse cholesterol transport system.

III. Apolipoproteins, Lipoprotein Composition, and Enzymes

The effects of probucol on plasma apolipoproteins and on lipoprotein composition have been studied extensively, especially in recent years (BUCKLEY et al. 1989; HELVE and TIKKANEN 1988; MAEDA et al. 1989; HOMMA et al. 1991; BERG et al. 1991; MCLEAN and HAGAMAN 1988; MCPHERSON et al. 1991b). The most consistent findings are reductions in plasma apoA-I, apoC-II, and apoC-III (Fig. 3). The changes in apoE, A-II, and B have been variable depending on the dose, the duration of exposure to the drug, the number of subjects, and the nature of the hyperlipoproteinemia or associated

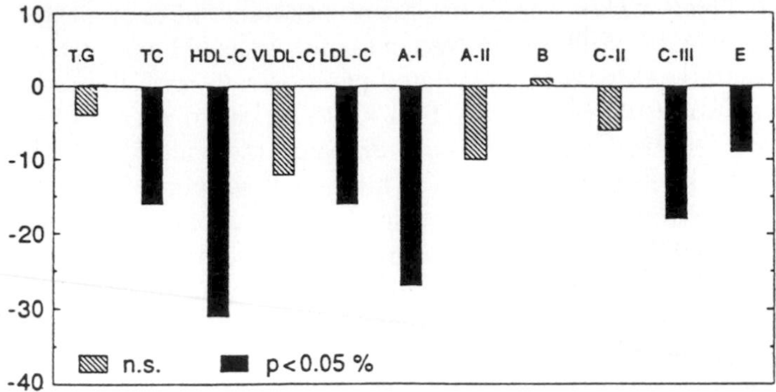

Fig. 3. Effect of probucol on plasma lipids, lipoproteins, and apolipoproteins in hypercholesterolemic men (percent changes). In this study conducted in 14 asymptomatic hypercholesterolemic men (31 ± 6 years), probucol was given for 3 months (1 g/day). Typically, plasma cholesterol (*TC*), high-density lipoprotein cholesterol (*HDL-C*), low-density lipoprotein cholesterol (*LDL-C*), apoA-I (*A-I*), apoC-III (*C-III*), and apoE (*E*) were significantly reduced, but the changes in triglyceride (*TG*), very-low-density lipoprotein cholesterol (*VLDL-C*), apoA-II (*A-II*), and apoC-II (*C-II*) were not significant (*n.s.*). (From BERG et al. 1991)

conditions. In normolipidemic individuals receiving 1 g/day probucol for 12 weeks, apoA-I was lowered by 20% and C-II, 13% (BERG et al. 1988). In hypercholesterolemic subjects, reductions in apoA-I, apoC-III, and apoE were 30%, 18%, and 8%, respectively (BERG et al. 1991). ApoE is also reduced by probucol in patients with cholesteryl ester transfer protein (CETP) deficiency (TAKEGOSHI et al. 1992; MATSUZAWA et al. 1988), but has been reported to be either unchanged (MAEDA et al. 1989; HOMMA et al. 1991) or increased (McPHERSON et al. 1991b) in other studies with hypercholesterolemic patients. The effect on apoA-II is variable with reports of no change (SASAKI et al. 1987; SIRTORI et al. 1988; BERG et al. 1991; McPHERSON et al. 1991b) or a decrease (MORDASINI et al. 1980; ATMEH et al. 1983). In one study, apoA-II was lowered in non-FH, but not in FH patients (HELVE and TIKKANEN 1988). Similarly for plasma apoB, probucol lowers LDL-C by 10%–17% in both FH and non-FH patients, but has no effect on LDL-apoB in FH and reduces LDL-apoB by 13% in non-FH (HELVE and TIKKANEN 1988). In probucol-treated FH patients, the reduction of LDL-C is associated with the appearance of small, dense LDL, enriched in TG, with a relatively low free cholesterol to apoB ratio (BERG et al. 1991; FRANCESCHINI et al. 1989; CORTESE et al. 1982; NARUSZEWICZ et al. 1984). In another study in non-FH patients by HOMMA et al. (1991), probucol significantly reduced plasma concentrations of cholesterol and apoB associated with intermediate-density lipoproteins (IDL) and with buoyant large LDL (LDL_1), but not cholesterol and apoB in small, dense LDL. Probucol lowered total plasma apoB, apoA-I, apoA-II, apoC-II, and apoC-III as well as HDL_2-C, HDL_2-TG, and HDL_2-apoA-I concentrations. HDL_3-C and HDL_3-apoA-I were reduced to a lesser extent. There was no effect on VLDL-C, VLDL-TG, or VLDL-apoB. Lipoprotein lipase (LPL) activity was significantly lowered, as shown previously (STRANDBERG et al. 1981; MIETTINEN 1972; MIETTINEN et al. 1981), but not invariably (HELLER and HARVENGT 1983; KESÄNIEMI and GRUNDY 1984; HELVE and TIKKANEN 1988); however, the activity of hepatic triglyceride lipase (HL) was unchanged, a relatively consistent finding (HELVE and TIKKANEN 1988). The failure of LDL_2 to be reduced by probucol was ascribed to the enhancement of CETA (FRANCESCHINI et al. 1989; MCPHERSON et al. 1991a), which tends to generate smaller and denser LDL particles. The overall reported effects of probucol on plasma TG and TG-rich lipoprotein levels are inconsistent (BUCKLEY et al. 1989), probably due to wide interindividual variations in response as well as heterogeneity of the groups studied, since an occasional patient may be dramatically responsive. We have reported a lack of response to probucol in a 60-year-old woman with well-characterized type III dysbetalipoproteinemia (DAVIGNON 1981), confirmed a posteriori to have the E2/2 phenotype. In contrast, ENGST (1985) reported a dramatic fall in plasma cholesterol and regression of the typical xanthoma striata palmaris in a 47-year-old man with type III. Such differences should be explored further and probucol may be used as a tool to find sources of metabolic heterogeneity in this disease.

Lecithin cholesterol acyltransferase (LCAT) activity has been reported to be increased (HELLER and HARVENGT 1983) as well as decreased (SZNAJDERMAN and KUCZYŃSKA 1981) by probucol. In vitro, probucol causes a significant, dose-related increase in acyl coenzyme A cholesterol acyltransferase (ACAT) activity, measured as the incorporation of [^{3}H]-oleate into cholesteryl esters of rat hepatocytes (BARNHART et al. 1988).

IV. Effect of Combination Therapy with Probucol

Combined drug therapy, superimposed on a cholesterol-lowering diet, is often needed for the treatment of severe and persistent cases of heterozygous FH. This approach allows the potentiation of drug effects, thereby reducing the amount of each drug used, and takes full advantage of complementary modes of action. One of the most effective combinations is probucol with a bile acid-binding resin (DUJOVNE et al. 1984b, 1986; DAVIGNON 1986), which can lower LDL-C by as much as 53% on long-term administration (KUO et al. 1986). In addition to reducing the amount of cholestyramine or colestipol needed for optimal effect (Fig. 4; DUJOVNE et al. 1986), probucol often alleviates the constipating side effect of these resins, resulting in improved compliance. In the rhesus monkey, this combination induces a 35% reduction in aortic atherosclerosis, an effect at least five times greater than that achieved with either drug administered alone (WISSLER and VESSELINOVITCH 1983).

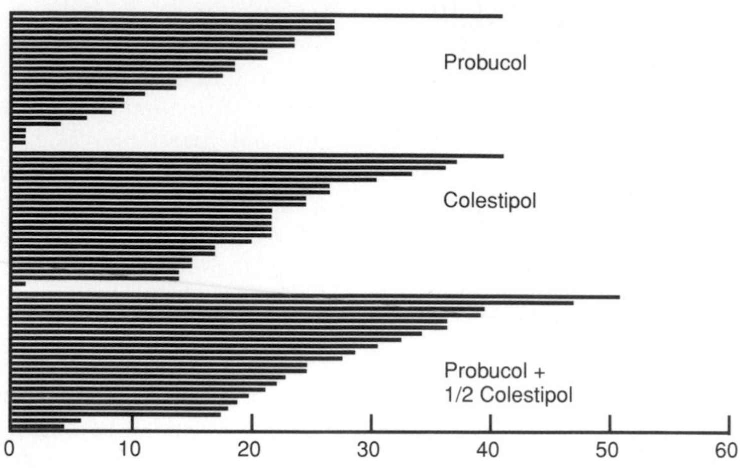

Fig. 4. Effect of *probucol* (1 g/day), *colestipol* (20 g/day), or *probucol plus colestipol* (10 g/day) on individual low-density lipoprotein cholesterol changes in hypercholesterolemic patients. This study was carried out in 22 hypercholesterolemic patients (12 men, ten women) with a median age of 53 years, ten of whom qualified as possibly having familial hypercholesterolemia. Each *bar* represents the percent change from the *diet-placebo baseline* (O) in low-density lipoprotein cholesterol for one patient treated over 19 months; they are arranged in decreasing order of response magnitude. Patients in this study had complained of constipation or dyspepsia on the full dose of colestipol. Most patients achieved the greatest reduction with the combination. Halving the dose of colestipol and adding probucol eliminated the gastrointestinal side effects or greatly reduced their severity. (From Dujovne et al. 1986)

Combination of probucol with lovastatin does not generally result in a further decline in plasma cholesterol, although a marked additive effect has occasionally been observed (Lees et al. 1986; Witztum et al. 1989). In theory however, added benefits are to be expected because of the complementary modes of action and the potential antiatherogenic effect of probucol mediated by its antioxidant properties.

The association of probucol with a fibrate is logical for the treatment of hereditary forms of hypercholesterolemia with associated hypertriglyceridemia (IIb phenotype). Combination of probucol with clofibrate may indeed normalize the lipoprotein profile in some hypercholesterolemic patients. The HDL-reducing effect of probucol is exacerbated in approximately 70% of patients receiving this combination and a severe hypoalphalipoproteinemia may result (Davignon et al. 1986). Paradoxically, this effect does not appear to be harmful; regression of xanthomas has been reported in one FH patient receiving this combination over 3.5 years with an average HDL-C of 11 mg/dl and a LDL to HDL ratio of 24:7. Combination with bezafibrate may cause a less severe reduction in HDL-C than clofibrate (YOKOYAMA et al. 1988).

E. Antioxidant Properties and Antiatherogenic Effects

I. Effect on Low-Density Lipoproteins

When LDL are oxidized they lose their affinity for the LDL receptor and are taken up and degraded preferentially by scavenger receptors ("acetyl LDL receptor" or other more specific receptors for oxidized LDL) on macrophages (see HEINECKE 1987; JÜRGENS et al. 1987; STEINBERG et al. 1989; STEINBRECHER et al. 1990; WITZTUM and STEINBERG 1991 for review). Oxidized LDL, in contrast to native LDL, are cytotoxic and may damage the vascular endothelium. Once they infiltrate the arterial wall, their chemotactic effect induces recruitment of blood monocytes that are eventually transformed into macrophages. Oxidized LDL can also inhibit the motility of resident macrophages preventing their escape from the intima. These processes favor foam cell formation and development of atheromatous plaques; thus, oxidized LDL have potent atherogenic properties. In vitro, LDL may be oxidized by exposure to endothelial cells (EC), or other cell types such as smooth muscle cells, neutrophils, and monocyte/macrophages (STEINBERG et al. 1989), or to Cu^{2+} (or Fe^{2+}) ions. LDL oxidation and its adverse consequences may be prevented by antioxidants such as probucol.

1. In Vitro Studies

Oxidative modification of LDL induces the formation of lipid oxidation products (JÜRGENS et al. 1987; ESTERBAUER et al. 1987) or the loss of reactive amino groups (essentially the ε-amino groups of lysine) on their surface, resulting in an increase in the net negative charge (STEINBRECHER et al. 1987; STEINBRECHER 1987), decreased heparin binding (MCLEAN and HAGAMAN 1989), and a breakdown of LDL phospholipids and apolipoprotein B-100 (PARTHASARATHY et al. 1985; STEINBRECHER et al. 1984; FONG et al. 1987; BELLAMY et al. 1989). The degradation products of apoB bear epitopes that are recognized by the acetyl LDL receptor (PARTHASARATHY et al. 1987). The presence of probucol in LDL prevents such modifications. This protective effect has been studied extensively in vitro.

PARTHASARATHY et al. (1986) showed that probucol, in a dose-related effect, prevents the oxidative modifications of LDL induced by exposure to EC or Cu^{2+} ($5\,\mu M$), i.e., their increase in electrophoretic mobility, peroxide content, and susceptibility to uptake and degradation by macrophages. Moreover, LDL from the plasma of probucol-treated patients resist cell-induced and Cu^{2+}-induced oxidative modification and enhanced macrophage degradation. Similar findings were obtained using LDL from WHHL rabbits treated with probucol (KITA et al. 1987). LDL of untreated WHHL rabbits oxidized with 0.5 μM $CuSO_4$ for 24 h show a 7.4-fold greater quantity of peroxides, measured as thiobarbituric acid reactive substances (TBARS), and a 4.3-fold greater ability to induce cholesteryl ester synthesis in macrophages as compared to LDL from probucol-treated animals. The uptake of β-VLDL, oxidized LDL, and acetylated LDL by macrophages previously exposed to or loaded with probucol is not affected (NAGANO et al. 1989). This absence of a direct effect of probucol on macrophages, also reported in mice (KU et al. 1990), strengthened the notion that the efficacy of probucol in preventing foam cell transformation of macrophage is linked to its inhibitory effect on LDL oxidation. In only one macrophage cell type, the UE-12 cell derived from the human histiocytic line U-937, was it shown that addition of probucol to the culture medium could prevent foam cell transformation by acetyl LDL (YAMAMOTO et al. 1986b). It is important for future studies to look for differences between human and animal cell lines. Preincubation of bovine endothelial cells with probucol, or with serum from a probucol-treated patient, protects these cells against oxidative injury by oxidized LDL (KUZUYA et al. 1991). The extent of protection is proportional to the EC content of probucol. In another study with ultraviolet-treated LDL, probucol (but not vitamin E) was very effective at preventing lipid peroxidation of LDL and their subsequent cytotoxicity to lymphoid cells, whereas lymphoid cells preincubated with vitamin E (but not cells preincubated with probucol) were resistant to the cytotoxic effect of oxidized LDL (NEGRE-SALVAYRE et al. 1991). The way LDL are oxidized and the type of target cells used seem to be an important determinant of the outcome of such experiments.

The effect of probucol on the changes induced in LDL by oxidation has been further documented with a variety of techniques. MCLEAN and HAGAMAN (1989) compared the effect of oxidation with 5 μM Cu^{2+} (1–24 h at 37°) on native LDL and on LDL incubated with 10 μM probucol for 1 h. They found that probucol prevents the loss of β-structure as measured by circular dichroism spectroscopy, reduces the formation of fluorescent degradation products, and, even after 24 h of oxidation, completely prevents the loss of heparin-binding sites and the 40% decrease in reactive surface amino groups (as measured by the trinitrobenzenesulfonic acid (TNBS) method) which occurs when untreated LDL are oxidized. Both hydrolysis (PARTHASARATHY et al. 1987; FONG et al. 1987) and aggregation (BARNHART et al. 1989) of LDL have been reported to occur during oxidative modification. Recently, BELLAMY et al. (1989) used synchrotron X-ray and high-flux neutron solution scattering to examine the effect of probucol on these changes. After a 20-h incubation with Cu^{2+}, LDL associate to form larger aggregate particles. Gel electrophoresis shows that apoB-100 is both degraded and covalently aggregated. Addition of probucol to LDL causes an increase in the polydispersity of LDL with a wide range of particle sizes, without altering its shape. The authors speculated that probucol facilitates changes

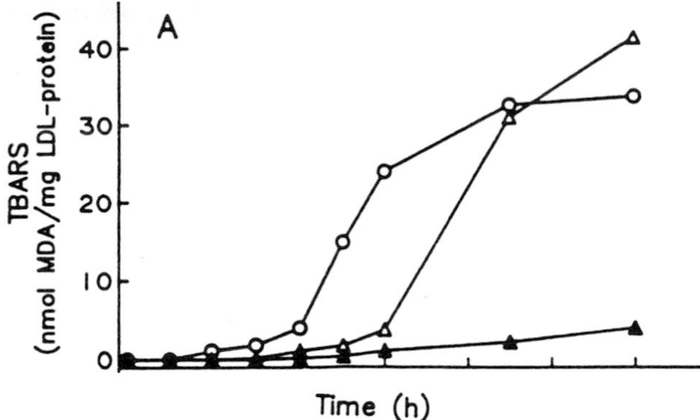

Fig. 5. Protective effect of probucol on Cu^{2+}-induced oxidation of low-density lipoprotein (*LDL*) in vitro. LDL containing O (*open circles*), 0.6 (*open triangles*), or 1.2 mol% (*closed triangles*) probucol were oxidized at 37° with 3 μM $CuSO_4$. At the indicated times, samples were removed and thiobarbituric acid reactive substances (*TBARS*), expressed in malondialdehyde (*MDA*) equivalent, were determined. A low initial oxidation rate is first observed up to 2 h followed by a high rate of TBARS generation in the absence of probucol. With LDL containing 0.6 mol% of probucol, the rapid lipid oxidation phase is not apparent until 3 h. In LDL containing 1.2 mol% probucol, oxidation proceeds at an initial, low rate for up to 6 h. (From BARNHART et al. 1989)

in the size of LDL by the destabilization of the lipid–protein interactions within the particles and that its lipid-lowering effect in vivo may be caused by the preferential catabolism of higher molecular mass forms of LDL thus created.

It has been shown that probucol is carried in the lipid fraction of LDL (EDER 1982; URIEN et al. 1984; DACHET et al. 1985; FELLIN et al. 1986). The kinetics of LDL oxidation and the protective effect of probucol are related to the amount of probucol in the lipoprotein particle. BARNHART et al. (1989) have determined that exposure of LDL to 3 μM Cu^{2+} induces half-maximal lipid oxidation as determined by the formation of TBARS. They showed that in these conditions, the initial low rate of lipid oxidation rises sharply after 2 h as witnessed by an increase in TBARS (Fig. 5) or by the increase in fluorescence due to the formation of aldehyde–protein adducts of 4-hydroxynonenal (peak at 410 nm) or of malondialdehyde (MDA; peak at 470 nm). This lag time is attributed to the presence of natural antioxidants in LDL such as α-tocopherol and β-carotene and is prolonged with probucol. The time required to achieve half-maximal oxidation increases from 130 to 270 min with LDL containing 0.6 mol% of probucol relative to phospholipids (approximately 10 molecules of probucol per LDL particle) and increases with increasing concentrations of probucol in LDL. At 0.6 mol% probucol, rapid oxidation is not apparent until after 3 h, a time at which more than 80% of probucol has been oxidized and converted to a

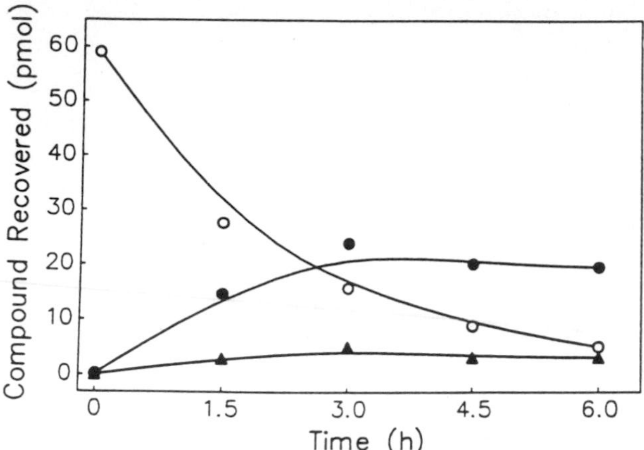

Fig. 6. Probucol transformation during Cu^{2+}-induced oxidation of low-density lipoproteins (LDL) in vitro. LDL containing 0.6 mol% of ^{14}C-probucol were incubated with $3\,\mu M$ $CuSO_4$. At the indicated times, the amount of probucol (*open circles*), spiroquinone (*closed circles*), and diphenoquinone (*closed triangles*) was determined by radioactivity after separation by high-performance liquid chromatography. Rapid oxidation of LDL begins at the 3-h time point (see Fig. 5); at this time 80% of probucol has been oxidized and converted into the spiroquinone and diphenoquinone. (From BARNHART et al. 1989)

spiroquinone and a diphenoquinone (Figs. 1, 6). Greater than 90% depletion of probucol from LDL (leaving about one molecule per LDL particle) is necessary before there is extensive LDL oxidation and loss of apoB-100. At an initial concentration of 4.2 mol% of probucol in LDL, only limited oxidation of these particles occurs even after 24 h. Analogs of probucol devoid of cholesterol-lowering activity also prevent Cu^{2+}-induced oxidation of LDL in vitro in a dose-dependent manner (MAO et al. 1991). Thus, the potent protective antioxidant property of probucol and its analogs appears to be directly related to the amount of drug present in the LDL particles.

The effect of probucol on the content of natural antioxidants dissolved in the lipid core of LDL was addressed by JIALAL and GRUNDY (1991). They measured the content of α-tocopherol, γ-tocopherol, and β-carotene by reversed phase HPLC in LDL after oxidation by $2.5\,\mu M$ Cu^{2+} for 24 h at 37° with or without probucol (5 and $10\,\mu M$) or ascorbate (40 and $60\,\mu M$), a water-soluble, chain-breaking antioxidant. They found that, although both probucol and ascorbate protected LDL against oxidative modification as assessed by peroxide formation and degradation by human monocyte-derived macrophages, only ascorbate prevented the loss of natural antioxidants induced by oxidation. Even at concentrations of $80\,\mu M$, probucol was unable to prevent natural antioxidant consumption. Although these findings are of interest in view of the inverse correlation between plasma ascorbate and the incidence of CAD (GEY et al. 1987), there is so far no evidence that a dietary supplement of ascorbate can inhibit atheroma progression. Furthermore, the in vitro experiment may not necessarily reflect the situation in vivo, since the work of FINCKH et al. (1991) indicates that probucol administration to WHHL rabbits has a sparing effect on plasma α- and γ-tocopherol levels.

2. In Vivo Studies

a) The Watanabe Heritable Hyperlipidemic Rabbit

The results of in vitro studies raised the possibility that probucol might have direct antiatherogenic effects because of its antioxidant properties. Much of the in vivo work has been carried out in the WHHL rabbit (WATANABE 1980). This animal model reproduces most of the features of human homozygous FH. The severe hypercholesterolemia is due to a 12-base pair in-frame deletion in the LDL receptor gene that removes four amino acids from the fourth ligand-binding repeat of the LDL receptor (YAMAMOTO et al. 1986). Lipoprotein changes (HAVEL et al. 1982; BUJA et al. 1983) and development of atherosclerosis (ROSENFELD et al. 1987a,b) in this animal model have been well documented. There is important variability in age of onset of plaque development from litter to litter (FINCKH et al. 1991) and new strains have been bred to alleviate some of the difficulties inherent to using these animals (GALLAGHER et al. 1988).

KITA et al. (1987) demonstrated that atherosclerosis progression can be prevented by feeding 2-month-old WHHL rabbits 1% of probucol mixed with rabbit chow for 6 months (plasma probucol: 43 μg/ml at the end). Compared to the control group, the percent surface area covered with plaques in probucol-treated animals is reduced by 87% in the thoracic aorta and 99.5% in the descending aorta. Plasma cholesterol increased in both groups, but less in the treated animals, so that at sacrifice there was a 22% difference in plasma cholesterol between the two groups with a residual hypercholesterolemia of 584 mg/dl in the probucol group. In a similar experiment, CAREW et al. (1987) established that this protection was not ascribable to the cholesterol-lowering effect of probucol alone, since lovastatin, given to a third group to achieve the same reduction in plasma cholesterol as that induced by probucol, failed to inhibit atheroma progression to the same extent. They also showed that probucol inhibits the rate of both uptake and degradation of LDL by macrophages in areas of atherosclerotic lesions in the treated animals compared to control rabbits. These observations thus link the antiatherogenic effect of probucol to its antioxidant properties. Further work carried out by FINCKH et al. (1991) in the WHHL rabbit showed that atheroma formation is associated with an increase in plasma oxidation products (measured as TBARS) and a reduction in plasma vitamin E concentration. Probucol administration in younger animals reverses these changes while preventing atheroma formation, but does not affect cholesterol nor atheroma formation in 15-month-old animals. Comparing siblings from the same litter, they found smaller areas of arterial involvement after 6 months of probucol than in the "pretreatment controls" (i.e., sacrificed at end of pretreatment period) suggesting that regression of early plaques is possible in young animals after probucol treatment. They attributed the antiatherogenic activity of probucol to its antioxidant properties as well as its ability to lower plasma cholesterol and perhaps enhance reverse cholesterol transport. Further support for a combined cholesterol-lowering and antioxidant effect comes from studies carried out by Mao et al. (1991) in a modified strain of WHHL rabbits produced by crossing British Brown half-lop and Japanese Watanabe rabbits (Gallagher et al. 1988). This model was chosen for its rapid and reproducible development of aortic lesions; 50% surface involvement occurs within 3 months after weaning compared to 1 year with the WHHL strain. Probucol was compared with its analog MDL 29311 (Fig. 1), a potent antioxidant devoid of any cholesterol-lowering activity. Probucol (1%) fed for 70 days resulted in a mean plasma level of 58 μg/ml and a molar content of 10 molecules/LDL, similar to that found in human subjects with FH receiving 1 g probucol per day for 6 months (Dachet et al. 1985), and lowered plasma cholesterol by 23%. MDL 29311 was less effective than probucol in reducing the extent of atherosclerosis. This difference could not be accounted for by a lack of bioavailability; the effect was maximal at a drug concentration of 0.5% in the diet, or 44 μg/ml in the serum. Although LDL isolated from the group fed 1% MDL 29311 contained a greater molar content of antioxidant (12.9/LDL) and had higher antioxidant activity than that from the group receiving 0.5% (8.4/LDL), protection against atherosclerosis was similar in the two groups, despite the observation that the concentration of antioxidant in LDL was positively correlated with the protective effect against oxidative modification in vitro. Neither probucol nor the analog could be detected in the arterial wall in this study. The ability of probucol to induce regression of established plaques was studied by Daugherty et al. (1991) in 9-month-old WHHL rabbits treated with probucol for 6 months. Very high levels of plasma probucol were achieved in this experiment (149 μg/ml), plasma cholesterol was lowered significantly, and LDL of probucol-treated rabbits were resistant to oxidation by Cu^{2+} in vitro, but there was no evidence of atheroma regression. The drug, however, retarded deposition of cholesteryl esters in the lesions without affecting their collagen content.

Histological investigation of the atheromatous lesions during probucol administration to WHHL rabbits primarily revealed a reduction of foam cells in the early period of life (Finckh et al. 1991). O'Brien et al. (1991) probed the pathological changes further with monoclonal antibodies specific for oxidized LDL (OXL 41.1), smooth muscle cells (HHF-35), and macrophages (RAM-11). Although probucol-treated rabbits had much less aortic atherosclerosis than controls, equivalent immunoreactivity for the OXL 41.1 antibody was observed in lesions from both groups of animals. Lesions from probucol-treated animals were smaller and contained fewer cells than those of controls. In the latter group, macrophages predominated whereas smooth muscle cells were the predominant cells in the probucol-treated rabbits. Such findings are in agreement with a reduction of monocyte/macrophage recruitment into and/or retention within lesions in probucol-treated animals as well as with an effect of probucol on the local production of growth factors or cytokines with an impact on the cellular composition of the lesions.

b) Other Animal Models

The antiatherogenic effect of probucol has been tested in other models of experimental atherosclerosis. In the cholesterol-fed (2% wt/wt) New Zealand rabbit, of unspecified sex, DAUGHERTY et al. (1989) showed that probucol (1% in the diet for 60 days) markedly decreased the extent of intimal surface area of the abdominal aorta covered by grossly discernible atherosclerotic lesions (from 55% to 11%) and significantly reduced the deposition of total cholesterol in vascular tissue. In this model, in contrast to the WHHL rabbit, it is the cholesteryl ester-rich β-VLDL that are increased; these constitute a highly atherogenic fraction promoting uncontrolled cholesterol esterification in monocyte-derived macrophages. Probucol administration did not alter the β-VLDL composition or the ability of these lipoproteins to promote incorporation of labeled oleate into cholesteryl esters of cultured macrophages; nor did it lower plasma total cholesterol, unesterified cholesterol, TG, or phospholipid concentrations. Protection from atherosclerosis by probucol in this model was, therefore, ascribed to the antioxidant property of probucol. Plasma probucol levels plateaued at approximately 50 μg/ml after 14 days. This study confirmed previous findings by KRITCHEVSKY et al. (1971) and by TAWARA et al. (1986) in which rabbits were fed 2% and 0.5% cholesterol, respectively. In contrast, STEIN et al. (1989) failed to observe a protective effect of probucol against atheroma formation in male albino New Zealand rabbits fed 1% cholesterol for 14 weeks. Probucol did not affect influx of cholesteryl esters into aortic segments as measured by the uptake of labeled cholesteryl-linoleyl ether. Plasma probucol was not monitored, but the action of probucol was apparent from a significant reduction in plasma apoA-I levels. Dietary cholesterol intake was adjusted during this experiment so that plasma cholesterol levels of the treated rabbits matched those of the controls. The divergent result of this study is difficult to explain, but may be tentatively ascribed to differences in experimental conditions such as rabbit strain, sex of the animals, levels of cholesterol intake, and use of drugs for anesthesia and during sampling. SHANKAR et al. (1989) confirmed the ability of probucol to prevent atherogenesis independently of changes in plasma cholesterol in rats fed an atherogenic diet (4% cholesterol, 1% cholic acid, and 0.5% methylthiouracil). In this study probucol did not prevent monocyte attachment to the arterial endothelium. Finally, the only study carried out in rhesus monkeys fed a high-fat, high-cholesterol diet indicated that probucol could induce a modest regression of aortic atherosclerosis (WISSLER and VESSELINOVITCH 1983). Overall, these animal models provide good evidence for an antiatherogenic effect of probucol which may occur even in the absence of a decrease in plasma cholesterol.

3. Studies in Man

The contribution of lipid peroxidation to atherogenesis in man has been established so far from indirect evidence (reviewed by STEINBERG et al. 1989; YAGI 1987; GEY et al. 1991) and inferences from data in experimental animals. Serum lipid peroxides are elevated in patients suffering from atherosclerosis (YAGI 1987). Oxidized LDL are detected in human atherosclerotic lesions (HOFF and O'NEIL 1991). Plasma levels of physiological antioxidants such as ascorbate (GEY et al. 1987) and α-tocopherol (GEY et al. 1990) are inversely correlated with the incidence of CAD and the latter may be improved by supplementation of the diet with an antioxidant (GAZIANO et al. 1990). These observations are consistent with the demonstration that probucol, a strong antioxidant, is antiatherogenic in animals. To date, the evidence that probucol may be beneficial in human atherosclerosis is anecdotal. An arteriographic study performed by KUO et al. (1986) showed that probucol combined with colestipol arrested the progression of CAD and prevented the development of new lesions in 19 patients treated for 3.4–4.1 years. An open pilot study in 256 hyperlipidemic patients with angina pectoris showed a 59% reduction in mean weekly anginal attack frequency after 8–16 weeks of treatment with probucol (1 g/day) and an associated reduction in weekly nitrate intake in 72% of patients (VON GABLENZ and SCHAAF 1988). Xanthoma formation is another situation where an excess of foam cells is formed; their regression induced by probucol in man (YAMAMOTO et al. 1986a; BAKER et al. 1982) is another indication that probucol might benefit at least a fraction of patients with atherosclerosis. This regression was formerly attributed to its cholesterol-lowering effect alone; it should be reevaluated in terms of free radical scavenging in the light of the experimental evidence recently accrued. Whether probucol can retard progression of atherosclerosis in man is currently being examined in the Probucol Quantitative Regression Swedish Trial (PQRST), a study contrasting the effect of cholestyramine alone with that of cholestyramine in combination with probucol on the progression of femoral atherosclerosis (WALLDIUS et al. 1988). This combination effectively inhibits atheroma formation in the rhesus monkey (WISSLER and VESSELINOVITCH 1983).

II. Studies with Lipoprotein(a)

Lp(a) is an atherogenic circulating lipoprotein consisting essentially of an LDL molecule covalently linked by a disulfide bridge to apolipoprotein (a), a glycoprotein of variable length with homology to plasminogen (SCANU and SCANDIANI 1991). Many of the oxidation-related modifications observed with LDL are reproduced with Lp(a) oxidized with Cu^{2+} or cells: increased negative charge and density, formation of aggregates, changes in immunoreactivity, fragmentation of apoB, and enhanced uptake by scavenger receptors on macrophages (JÜRGENS et al. 1990; SATTLER et al. 1991; NARUSZEWICZ et al. 1992a). Although probucol has no effect on plasma levels of Lp(a) (MAEDA et al. 1989), it has a major impact on its oxidative modification. Probucol associates with Lp(a) and prevents its changes in electrophoretic mobility as well as its uptake and degradation by macrophages (DAVIGNON 1992; NARUSZEWICZ et al. 1992b). The significance of these findings to atherogenesis may relate to several other observations. Lp(a), like LDL, binds to glycosaminoglycans of human aortae (BIHARI-VARGA et al. 1988). Lp(a) accumulates preferentially to LDL in human atheromatous plaques and apo(a) is present in a form that is less easily extracted than most of apoB (PEPIN et al. 1991). Arterial accumulation is proportional to the plasma concentration of Lp(a). Oxidized Lp(a), like oxidized LDL, are detected in atheroma (O'NEIL et al. 1990). Lp(a) is also a major risk factor for CAD in FH (SEED et al. 1990), in which case a presumed prolonged half-life would be conductive to oxidative modification.

F. Mechanisms of Action

I. Mechanism of Low-Density Lipoprotein-Lowering Effect

The mechanism(s) responsible for the LDL-C-lowering effect of probucol remain controversial (BUCKLEY et al. 1989). MIETTINEN (1972) showed that probucol causes a transient increase in fecal bile acids, fat and water content as plasma cholesterol decreases, a reduced cholesterol absorption, and cholesterol synthesis inhibition. NESTEL and BILLINGTON (1981) demonstrated that probucol increases the fractional removal rate of LDL and confirmed the enhanced fecal bile acid excretion. On the other hand, KESÄNIEMI and GRUNDY (1984), though confirming the enhanced fractional catabolic rate of LDL, found no evidence of changes in fecal excretion of neutral, acidic, or total sterols or in bile composition and output after 2–6 months of treatment with probucol. The patients selected for this study were highly heterogeneous with respect to plasma cholesterol (10 of 17 had an LDL-C ranging between 104 and 162 mg/dl at baseline) and LDL-C response to probucol (ranged from −36% to +12%); furthermore, different subsets of subjects were selected among the 17 for specific studies. One-third of the patients had a reduced cholesterol absorption. In marked contrast, the study of MIETTINEN (1972) was conducted in a homogeneous group of patients with severe FH, all responsive to probucol given at twice the recommended dose for 4 weeks (cholesterol lowering ranging from −20% to −39%). These discrepancies plus the knowledge that there is marked interindividual variation in response to probucol should emphasize the importance of studying homogeneous groups of well-characterized subjects to study the mechanism of action of probucol. Probucol may lower plasma cholesterol by one or more of the following mechanisms: inhibition of cholesterol synthesis, reduction of cholesterol intestinal absorption, and enhancement of fecal excretion as bile acids or neutral sterols. Other less likely mechanisms would be a probucol-induced breakdown of the cholesterol phenanthrene structure (for which there is no evidence) or accumulation in adipose tissue (this has not been checked). Most attempts to observe an effect of probucol on cholesterol synthesis in animals have been unsuccessful (STRANDBERG et al. 1981). Evidence of such an effect in man is based on a reduction in serum methyl sterols, an indirect measure of cholesterol synthesis, observed during administration of large doses of probucol in one study (MIETTINEN 1972). Studies with isolated cells indicate that probucol can inhibit sterol synthesis from acetate by 37% in human lymphocytes (ANASTASI et al. 1980) and by 31% in rat hepatocytes (BARNHART et al. 1988). Interestingly, incubation of LDL or HDL with probucol-treated hepatocytes reduced the free cholesterol to protein ratio of these particles by 8%–10% while increasing cellular cholesterol (largely in free and esterified cholesterol fractions for HDL and LDL, respectively). This observed cholesterol shift in vitro is consistent with a proposed increase in the efficacy of reverse cholesterol transport and by probucol and the finding that probucol can induce a twofold increase in HDL-C uptake by the liver in rats fed 0.5% probucol for 3 weeks compared to controls (TAKATA et al. 1986).

There is evidence that the clearance of LDL is enhanced by probucol via the LDL receptor pathway as well as through a mechanism which is non-LDL receptor mediated. The fact that the fractional catabolic rate of apo-LDL is enhanced during probucol administration (KESÄNIEMI and GRUNDY 1984), coupled to an increase in bile acid excretion (MIETTINEN 1972; NESTEL and BILLINGTON 1981) and to a presumed reduction in hepatic cholesterol synthesis (MIETTINEN 1972), is consistent with enhanced LDL receptor activity. On the other hand, the observation that a consistent cholesterol-lowering effect of probucol occurs in both rabbits (KITA et al. 1987; CAREW et al. 1987) and humans (BAKER et al. 1982; YAMAMOTO et al. 1986a) with markedly defective LDL receptors indicates that a non-LDL receptor-mediated pathway may be operative. NARUSZEWICZ et al. (1984) showed that probucol treatment does not affect plasma clearance of native LDL in WHHL rabbits, but that LDL from probucol-treated animals (PB-LDL) are cleared approximately 50% more effectively than native LDL when injected into plasma of probucol-treated or untreated WHHL rabbits or normal rabbits. Similarly, PB-LDL were degraded more effectively than native LDL by fibroblasts from WHHL or normal rabbits. It was postulated that probucol, which gains access to the core of the LDL particles, somehow altered the conformation of its surface apolipoprotein B, favoring enhanced clearance by a pathway independent of the LDL receptor. These findings raised much speculation on the existence of an alternate receptor for cellular uptake and degradation of LDL. Whether this situation prevails in human is currently being investigated. Preliminary results of KESÄNIEMI and MIETTINEN (alluded to in KESÄNIEMI 1991) are in conflict with the animal data. When differentially labeled LDL from probucol-treated and from non-treated subjects were reinjected into patients from both groups, the plasma clearance of both particles was identical. Finally, probucol administration alters LDL composition to form small, dense, TG-enriched particles (BERG et al. 1991; FRANCESCHINI et al. 1989; CORTESE et al. 1982; NARUSZEWICZ et al. 1984), but the functional significance of these changes is not yet established.

II. Mechanism of High-Density Lipoprotein-Lowering Effect

The mechanisms whereby probucol lowers HDL (mainly HDL_2), modifies their composition, and reduces the particle size remain a matter of controversy. Probucol decreases apoA-I synthesis (NESTEL and BILLINGTON 1981; GIADA et al. 1986; ATMEH et al. 1983) and increases CETA (SIRTORI et al. 1988; FRANCESCHINI et al. 1989) and CETP (MCPHERSON et al. 1991a; Fig. 7). In hyperalphalipoproteinemia patients with deficient CETA and high plasma HDL_2 levels, probucol markedly lowers HDL_2 and apoA-I as it increases CETA, whether the defect is due to the presence of a CETP inhibitor in plasma (TAKEGOSHI et al. 1992) or not (MATSUZAWA et al. 1988). These

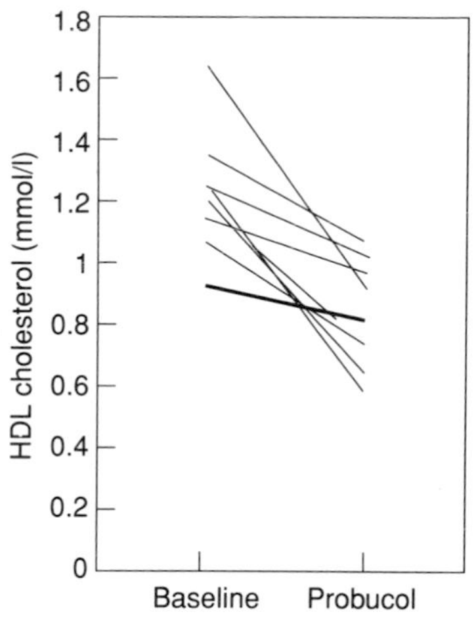

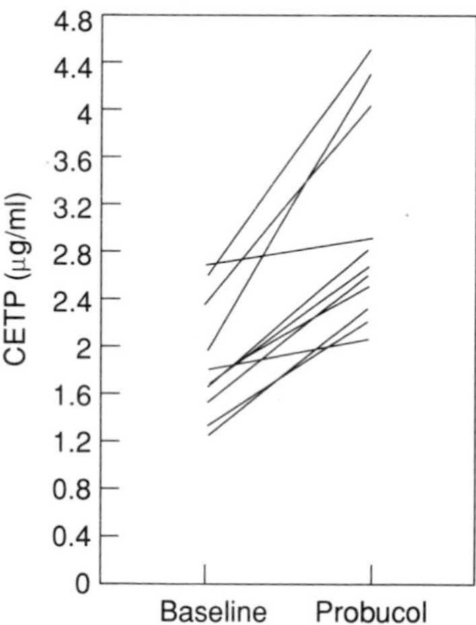

Fig. 7. Effect of probucol on high-density lipoprotein cholesterol (*HDL-C*) and cholesteryl ester transfer protein (*CETP*) mass before and after probucol therapy. The individual responses are given in 11 hypercholesterolemic subjects (four men and seven postmenopausal women) including four familial hypercholesterolemia patients before and after 8 weeks of probucol administration (500 mg twice a day with meals). The fall in fasting plasma HDL-C averaging 29% was paralleled with a 64% increase in CETP. In this study, plasma apoE was increased by 69%. (From MCPHERSON et al. 1991a)

findings suggest an explanation for the observed changes in HDL concentration and composition induced by probucol, but the exact mechanisms are not entirely clear. The fact that FRANCESCHINI et al. (1989) did not observe a significant correlation between the increase in transfer activity and the reduction of HDL-C levels ($r = 0.18$) after probucol indicates that other factors might be involved and warrants further studies. Furthermore, there is still no explanation for the discrepant finding of BAGDADE et al. (1990) that probucol normalizes the high CETA reported in some hypercholesterolemic patients. Possible explanations may relate to the interindividual heterogeneity of HDL particle subspecies (BLANCHE et al. 1981) and to variation in CETP levels as a function of gender, concentration and composition of plasma lipoproteins, and dietary cholesterol intake (e.g., plasma CETP appears to increase as an adaptive response to increased peripheral flux of cholesterol; MCPHERSON and MARCEL 1991). A useful model for the study of the relationship between CETP, HDL, and apoA-I levels and the effect of probucol is the transgenic mouse (a species devoid of CETP). High levels of plasma human CETP in transgenic mice (AGELLON et al. 1991) caused a significant reduction in HDL-C and a decrease in the ratio of free to esterified cholesterol in all plasma lipoprotein fractions, suggesting that high plasma CETA may be a cause of low HDL levels in humans. This effect of CETP on plasma HDL levels is markedly enhanced when both human apoA-I and CETP genes are simultaneously incorporated into the mouse genome, further substantiating the interrelation between CETP activity and HDL metabolism (HAYEK et al. 1991a). In transgenic mice overexpressing the human apoA-I gene alone, probucol was found to reduce the size of HDL particles, to lower plasma apoA-I concentrations by decreasing transport rate (without changing liver apoA-I mRNA), and to lower HDL-C by increasing HDL-C ester fractional catabolic rate (HAYEK et al. 1991b). Hence, the HDL-C-lowering activity of probucol can occur in the absence of CETP. Finally, probucol was reported to lower the activity of LPL in man (MIETTINEN et al. 1980; HOMMA et al. 1991) and in animals (STRANDBERG et al. 1981); this effect could decrease the generation of HDL from TG-rich lipoproteins. An enhanced HL activity could cause a reduction in HDL_2, but probucol has no effect on this enzyme activity (STRANDBERG et al. 1981; HELVE and TIKKANEN 1988; HOMMA et al. 1991). ApoC-III is an inhibitor of LPL (GINSBERG et al. 1986), it may interfere with HDL receptor-mediated uptake of lipoproteins (MITCHEL et al. 1987; VON ECKARDSTEIN et al. 1991), and a mutation in the apoC-III gene has been associated with hyperalphalipoproteinemia (VON ECKARDSTEIN et al. 1991). Probucol lowers plasma apoC-III (BERG et al. 1991; HOMMA et al. 1991) and reduces the rate of association of apoC-III with dimyristoylphosphatidylcholine (DMPC) liposomes (MCLEAN and HAGAMAN 1988). Overall, these findings are compatible with the opinion of HOMMA et al. (1991) that the decrease in HDL_2 could be due to a simultaneous suppression of LPL and stimulation of CETP by probucol. In rabbits, apoE mRNA in the spleen and brain, but not in the liver, is increased by cholesterol feeding, and this effect is enhanced by probucol added to the diet (ABURATANI et al. 1988). Since macrophages secrete both cholesterol and apoE which may be incorporated into plasma HDL and be ultimately removed by the liver through an apoE-mediated receptor process (MAHLEY and INNERARITY 1983), the observations in the rabbit have been interpreted to indicate that probucol may have a beneficial effect on reverse cholesterol transport through this mechanism. It is becoming obvious that the mechanism of HDL reduction under probucol therapy is, to say the least, multifaceted.

III. Antioxidant Effect

Some of the molecular mechanisms accounting for the antioxidant activity of probucol have been studied in vitro. In both probucol and BHT, the hydrogen atom of the phenolic hydroxyl group with its single electron can be readily abstracted (JACKSON et al. 1991), forming a free radical which is poorly reactive with water or other radicals because it is shielded by the bulky tertiary butyl groups (Fig. 1). The unpaired electron delocalizes into the benzene ring, forming a stable quinone structure. Thus, probucol has a free radical chain-terminating activity with the ability to inhibit lipid peroxidation. Tests in vitro with o-dianisidine have demonstrated that, at physiological concentrations, probucol is a potent superoxide (O_2^-) free radical scavenger rather than a general free radical scavenger (BRIDGES et al. 1991). A few studies have looked at the mechanisms whereby probucol protects LDL against oxidative modification and subsequent uptake by macrophages. As mentioned, probucol gains access to the core of the lipoprotein and affords protection against oxidation until it is consumed, thus accounting for the lag time observed before denaturation of the lipoprotein takes place in vitro (BARNHART et al. 1989; MAO et al. 1991; JACKSON et al. 1991). LDL from probucol-treated (but not from cholestyramine-treated) patients in the PQRST were considerably less susceptible to oxidation in vitro as assessed by subsequent binding to fibroblasts LDL receptors, macrophage uptake, and TBARS content. There was wide interindividual variation and patients could be classified as responders (<200% increase in TBARS) and nonresponders (>200%), but there was no difference between these two groups in terms of the effect of probucol on plasma LDL-C, HDL-C, and TG. Surprisingly, in the nonresponsive subset, probucol treatment did not prevent formation of TBARS or a decrease in LDL receptor binding, but almost completely prevented an increases in macrophage uptake (REGNSTRÖM et al. 1990). It will be important to relate these in vitro effects of probucol with the in vivo changes in femoral atherosclerosis when the trial is completed to better understand the protective mechanisms. The individual variations could be partly due to other factors associated with LDL such as the levels of natural antioxidants, but the possibility that the protective effect of probucol is in part mediated by other mechanisms should also be given consideration. Recently, BREUGNOT et al. (1990) compared the protective effect of phenothiazines (chlorpromazine and trifluorerazine), probucol, and BHT against copper- and EC-induced peroxidation of LDL. All had a protective effect, but differed in their antioxidant properties. Vitamin E was more effective against cell-induced than copper-induced oxidation, and only vitamin E reacted with model peroxy radicals produced by irradiation of aerated ethanol. Chlorpromazine is not considered to be a free radical scavenging molecule, but nevertheless had an effect. Phenothiazines are known to interact with phospholipids and to affect membrane microviscosity, and it has been postulated that the physical state of the lipid phase may influence the susceptibility of lipid to peroxidation (NAGATSUKA and NAKAZAWA 1982). Probucol at low concentrations is also known to reduce the phase-transition temperature of phospholipids (GOLDBERG and MENDEZ 1988). The authors proposed that insertion of lipophilic compounds such as phenothiazines or probucol could alter the LDL structure and render the particle less sensitive to the oxidative attack. This hypothesis merits further exploration.

G. Side Effects, Safety, Tolerance, and Drug Interactions

Probucol has been reported to sensitize the myocardium of the dog to epinephrine, an apparently species-specific adverse effect (MARSHALL and LEWIN 1973). Probucol also induced prolongation of the QT interval and sudden death in rhesus monkeys fed an atherogenic diet (EDER 1982). These observations raised concern for a cardiotoxic effect of potential clinical significance. In man, however, the drug has been remarkably well tolerated. In 252 men treated for over 5 years, the rate of discontinuation because of adverse effects was 8% (STRANDBERG et al. 1988; BUCKLEY et al. 1989). The side effects have been infrequent, mild, and mostly gastrointestinal. Diarrhea, loose stools, flatulence, nausea, and abdominal pain have been reported. Rarely, constipation, hyperhydrosis, fetid sweat, headache, dizziness, paresthesia, impotence, eosinophilia, skin rash, and palpitations have been reported. Probucol may lengthen the QT interval in some individuals (BUCKLEY et al. 1989; STRANDBERG et al. 1988; DUJOVNE et al. 1984a) and combination with drugs known to have this effect (KENNY and SUTTON 1985), such as quinidine, β-blockers, tricyclic antidepressants, phenothiazines, and amiodarone should be avoided. Withdrawal of the drug has been advocated if QTc exceeds 470 ms (QT_c is the QT interval corrected for heart rate). *Torsades de pointes*, characterized by short intervals of ventricular tachycardia with progressive varying amplitude and polarity, could lead to ventricular fibrillation and death. This condition is nearly always associated with prolongation of the QT interval and may run in families; it is a contraindication for the use of probucol. There is one case report of torsades de pointes induced by 4 weeks of probucol administration, which prolonged the QT interval to 620 ms in a 36-year-old woman with Romano Ward syndrome (Long QT Syndrome); it resolved upon discontinuation of the drug (MATSUHASHI et al. 1989). Interestingly, β adrenergic blockers are indicated in the treatment of this condition, although their administration results in a slight prolongation of the QT interval. Prolongation of the QT interval does not indicate a cardiotoxic effect of probucol (STRANDBERG et al. 1988; RODEN and WOOSLEY 1985); it merely reflects a mild antiarrhythmic effect shared by many efficient antiarrhythmic drugs of type Ic (such as quinidine) and type III (amiodarone, sotalol), which typically prolong the QT interval as part of their mechanism of action. BROWNE et al. (1984) measured 15 000 QT intervals and monitored the frequency of ventricular premature contractions (VPCs) before and after 6 months of treatment with probucol or placebo in 16 patients with less than 600 VPCs per day. Although they observed a modest prolongation of the QTc interval in the probucol group (22 ± 23 ms in the awake state and 20 ± 18 ms during sleep), there was no increase in the number of VPCs in either group. There is one report of a lowered incidence of VPCs in patients receiving probucol (WEISS et al. 1986).

Little information is available on drug interactions with probucol. In a recent report, probucol was found to lower the area under the plasma concentration time curve (AUC) of cyclosporine in cardiac transplant patients (SUNDARARAJAN et al. 1991), but the exact mechanism of this effect remains to be determined. Probucol causes a 28% decrease in cyclosporin AUC over 11 h, with an increase in apparent clearance rate; whether these changes are associated with changes in absorption or redistribution of cyclosporin is unclear. HELVE and TIKKANEN (1988), in a study of 58 patients (FH and non-FH) who were treated for 14 weeks with probucol, showed that plasma TG were lowered by 13% in the absence of, and increased by 12% in the presence of, concomitant administration of β-blockers ($n = 30$); this effect was not observed during treatment with lovastatin. Interactions during combination with other lipid-lowering agents has been discussed above.

H. Other Properties and Use of Probucol in Specific Diseases

I. Anti-inflammatory Effect

Interleukin-1 (IL-1) is a cytokine that induces the synthesis of acute phase proteins (see KU et al. (1988) for review). Metallothionein is one such protein produced by the liver which binds zinc; its induction is associated with a reduction in plasma zinc levels. A reduction in zinc levels may therefore reflect IL-1 release and has been used as a bioassay for IL-1 release in the zymosan-primed mice challenged by lipopolysaccharides (LPS). IL-1 is a lymphocyte-activating factor which potentiates the response of thymocytes to lectins such as phytohemagglutinin. A thymocyte proliferation assay has also been developed to measure IL-1 release. Since IL-1 is a proinflammatory mediator, it has been surmised that it might play a role in the atherogenic process similar to that of other cytokines, such as platelet derived growth factor. In line with this hypothesis, KU et al. (1988) have presented evidence that modified LDL induce the release of IL-1 and that recombinant IL-1 causes aortic smooth muscle cell proliferation. They also demonstrated that large doses of probucol partially inhibit the decrease in zinc induced by LPS in the zymosan-primed mouse, indirect proof that IL-1 release is partially suppressed in vivo. Direct evidence for inhibition of IL-1 release was obtained in an ex vivo system in which peritoneal macrophages from probucol-treated mice secreted 80%–90% less IL-1 than controls, upon LPS stimulation, using the C3H/HeJ thymocyte proliferation assay. Probucol, however, has no direct influence in vitro on IL-1 release by macrophages. It is believed that probucol may have a beneficial effect in preventing IL-1 release in vivo with its mitogenic effect on arterial wall smooth muscle cells, a release that may be mediated by oxidized LDL in the atherogenic process.

II. Probucol and Diabetes

There is indirect evidence that both anti-inflammatory and antioxidant effects of probucol are involved in affording some protection against experimental diabetes. DRASH et al. (1988) showed that probucol may partially prevent or retard the onset of the spontaneous insulin-dependent diabetes mellitus which develops in the BB/W rat. Using two litter-matched groups of 29 rats each, they showed that diabetes develops in 86.2% of untreated animals at a mean age of 90.4 days, as compared to 62% at a mean age of 99.6 days in rats given probucol over 160 days of life. This is a modest, albeit interesting, effect which was speculatively ascribed to a delay in β-cell destruction through an anti-inflammatory effect of probucol. A similar mechanism has been hypothesized to account for the protective effect of probucol against multiple low-dose streptozotocin-induced diabetes in mice (SHIMIZU et al. 1991), although a contribution of the antioxidant effect cannot be ruled out. In this study, probucol administration also reduced the T-cell subsets of the spleen as measured by specific antibodies. These studies warrant further consideration of probucol for the early prevention of insulin-dependent diabetes, alone or in combination with low-dose cyclosporin.

Several observations indicate that part the protective effect of probucol against experimental diabetes may be due to its antioxidant property. It has been shown that pretreatment with antioxidants such as vitamin E (SLONIM et al. 1983), dimethyl urea (a hydroxyl radical scavenger; SANDLER and

ANDERSSON 1982), and probucol (MATSUSHITA et al. 1989) can prevent streptozotocin- or alloxan-induced diabetes in animals. Exposure of cultured human lymphocytes to oxidized LDL results in a reduced ligand affinity of the insulin receptor (FUKUYAMA et al. 1988), attributed to a direct effect on the membrane. This reduction may be prevented by preincubation of the cells with the natural antioxidant vitamin E. Since insulin binding is decreased in non-insulin dependent diabetes mellitus (NIDDM; KOBAYASHI et al. 1980), where circulating oxidized lipoproteins have been detected (NISHIGAKI et al. 1981), probucol, as an antioxidant, might ameliorate insulin resistance. This possibility has not been tested. Oxidized VLDL and LDL are cytotoxic to cells in culture as are diabetic rat serum and its lipoprotein fractions (MOREL et al. 1983). MOREL and CHISOLM (1989) have shown that streptozotocin-induced diabetes in rats imparts cytotoxicity to the combined VLDL–LDL fraction (but not to HDL) by a peroxidation mechanism taking place in vivo which is independent of the hyperglycemia. Insulin treatment inhibits both oxidation and cytotoxicity of VLDL–LDL on human fibroblasts, whereas probucol or vitamin E treatment inhibits the cytotoxicity without altering the concomitant hyperglycemia and hypertriglyceridemia. For unknown reasons the lipid peroxides in human diabetic plasma reside in the HDL fraction (NISHIGAKI et al. 1981). Chronic streptozotocin diabetes in rats is associated with hypertriglyceridemia and a reduced TG clearance, which are corrected by chronic administration of probucol (4 months; YOSHINO et al. 1991). This effect of probucol is ascribed to normalization of composition and subsequently improved plasma clearance of VLDL. The relevance of these interesting findings to human diabetes remains to be determined.

Probucol has been administered to treat the hyperlipidemia associated with NIDDM. HATTORI et al. (1987) obtained a significant reduction in plasma total cholesterol (−14%), TG (−15%), HDL-C (−17%), apoA-I (−10%), and apoC-II, but not in apoB, with 500 mg/day of probucol over 16 weeks (50 patients). In a smaller group of NIDDM patients with hypercholesterolemia treated with the same dose for 4 weeks, HORIKISHI and SHIMODA (1988) showed a significant reduction in LDL-C, but the trend towards a reduction of HDL-C and TG was not significant (YOSHINO et al. 1986). Similar findings were obtained by YOSHINO et al. (1986) in a small comparative study with pravastatin on seven hyperlipidemic diabetic patients; the 38% fall in TG levels observed was not significant. In 11 mildly hypertriglyceridemic postmenopausal women with NIDDM treated for 2 months with 1 g probucol per day, LANE et al. (1991) observed a significant fall in plasma total cholesterol (−15%), free cholesterol (−28%), HDL-C (−22%), TG (−16%), and in apoAI, B (−12%), and E, without any change in LDL-C and diabetic control as measured by glycosylated hemoglobin A_1. Also reduced were the sphingomyelin to lecithin ratio and the atherogenic free cholesterol to lecithin ratio. In this study, probucol had an impact on HDL subclasses similar to that observed in hypercholesterolemia (TAWARA et al. 1986; FRANCESCHINI et al. 1989; BAGDADE et al. 1990). The plasma TG-lowering effect is apparently more consistent in hyperlipidemic diabetics than in nondiabetic hypercholesterolemic patients.

III. Probucol and Renal Disease

Renal damage induced by acute ischemia and reperfusion is a clinical and experimental syndrome that is characterized by marked reduction in glomerular filtration rate (GFR), extensive tubular damage, tubular cell necrosis, glomerular injury, and signs of tubular obstruction with cell debris. Most of the damage occurs during reperfusion. Studies utilizing superoxide dismutase (SOD) and allopurinol as therapy for experimental acute renal failure (PALLER et al. 1984) suggested that generation of free oxygen radicals contributes to this injury. BIRD et al. (1988) have tested the effect of probucol, selected as an antioxidant for its lipophilicity and presumed efficacy of incorporation into renal lipid membranes, on acute ischemic renal failure in rats subjected to 1 h of renal ischemia followed by 24 h of reperfusion. Probucol induced a significant and persistent improvement in single nephron GFR and a transient increase in proximal tubular reabsorption within a few hours after the ischemic insult. The overall GFR, however, remained low due to extensive tubular backleak as measured by microinjections of insulin, and after 24 h tubular necrosis was more extensive in the probucol-treated rat. It is, therefore, possible that measures to alleviate the ischemia-induced tubular injury could augment the benefit of antioxidant therapy.

Short-term dietary cholesterol supplementation in rats results in hypercholesterolemia and marked constriction of renal vessels. KAPLAN et al. (1990) showed that probucol prevents the renal effect of cholesterol feeding, including the reduction in single nephron GFR and afferent plasma flow and the increase in glomerular capillary pressure, afferent arteriolar resistance, and single nephron filtration fraction. In this experiment, probucol (10 mg/day) partially prevented the rise in VLDL-C, but had no effect on LDL-C. It has been shown that oxidized LDL increase prostaglandin E_2 (PGE_2) and thromboxane A_2 (TXA_2) synthesis by certain cell types in vitro (KITA et al. 1988). Cholesterol feeding is associated with an increase of TXB_2 and PGE_2 in proximal tubular fluid and urine; this increase is prevented by probucol intake. Infusion of a TXA_2 receptor antagonist (SKF 96148) also prevented the renal vasoconstriction. These findings are consistent with the hypothesis that in vivo oxidized LDL mediated the increase in arachidonic acid metabolites and that the vasoconstriction is largely, if not exclusively, due to the increase in TXA_2 (Saralasin, an angiotensin II receptor antagonist, had no effect). HIRANO et al. (1991) have recently shown that probucol treatment of rat puromycin aminonucleoside nephrosis reduces lipoprotein-associated lipid peroxides and proteinuria without affecting blood urea nitrogen and plasma creatinine levels. Plasma TG and HDL-C concentrations were lowered and a significant positive correlation was found between protein excretion and plasma lipid concentration, suggesting a relationship between improvement in proteinuria and the hypolipidemic effect of probucol. Probucol, 500 mg per day for 12 weeks, has been used successfully for the treatment of hypercholesterolemia in 12 patients with nephrotic syndrome by IIDA et al. (1987). There were no drug-related adverse effects. Plasma total cholesterol, TG, HDL-C, and LDL-C were significantly reduced. Parameters of renal function, however, remained unchanged throughout the duration of the study. The potential clinical implications of these observations merit further studies since there is a growing body of evidence that hyperlipidemia may have an influence on impaired renal function in man (JACKSON 1991).

IV. Probucol and the Heart

There is evidence that probucol might improve reperfusion injury in the heart as well as in the kidney. Experimental induction of persistent myocardial contractile dysfunction by a brief period of ischemia followed by reperfusion ("stunned" myocardium) has been attributed to the generation of oxygen-derived free radicals (BOLLI et al. 1989). DAGE et al. (1991) showed in rabbits that probucol feeding (1% in diet, 22 days, plasma probucol of 15 ± 1.2 μg/ml), significantly reduced the regional myocardial dysfunction (measured in vivo), the heart rate, and the number of VPCs induced by a 15-min occlusion of the first marginal branch of the left coronary artery. Interestingly, probucol and its bisphenol derivative (Fig. 1) were present in heart (17.5 ± 2.5 μg/g and 17.5 ± 2.6 μg/g, respectively), but the bisphenol was not detectable in serum. Furthermore, myocardial segment shortening (the measure of cardiac dysfunction) correlated with plasma, but not with myocardial concentrations of probucol. The effect on VPCs suggests a membrane-stabilizing effect of probucol, with reduction of myocardial excitability. Successful thrombolysis for acute myocardial infarction is associated with an increase in venous blood lipid peroxide ascribed to reperfusion injury (DAVIES et al. 1990); another condition that might be benefited by administration of probucol or other antioxidants.

Recently, LEE et al. (1991) presented evidence that probucol treatment beginning 30 days prior to percutaneous transluminal angioplasty (PTCA) for coronary artery stenosis had a preventive effect on restenosis occurring within 5 months of the intervention. In this case, the rate of restenosis was 8.3% (24 subjects), compared to 43.5% (23 subjects) when the treatment was begun only 3 days prior to the intervention. In contrast, pravastatin treatment begun 14 days and 3 days before PTCA resulted in restenosis rates of 35.4% (31 subjects) and 33.3% (33 subjects), respectively. This protective effect occurred in spite of higher cholesterol and lower HDL-C concentrations in the long-term probucol group than in the group of patients treated with pravastatin from 14 days prior to intervention. The results could not be accounted for by differences in Lp(a) levels among the four groups. These findings imply that oxidative and/or inflammatory mechanisms may be involved in post-PTCA restenosis. Probucol, given for 1 month 6 weeks after laser thermal angioplasty of femoral atherosclerosis in hypercholesterolemic rabbits, has no effect on final luminal size (Hsiang et al. 1991), but pretreatment with probucol has not been carried out in this model.

Probucol may have additional beneficial effects on the heart and on the circulation. Berg et al. (1991) reported a significant reduction in heart rate, various indices of cardiac work, and blood lactate accumulation in hypercholesterolemic volunteers undergoing maximum exercise stress tests before and after 12 weeks of probucol administration. They also found that the extent and velocity of adrenaline-induced (but not collagen-induced) platelet aggregation was significantly lowered. This is in agreement with previous work showing that the antioxidant BHT also inhibits platelet aggregation (Alexandre et al. 1986).

I. Conclusions

Probucol has diverse effects which may act synergistically to oppose the atherogenic process: lowering of LDL-C, prevention of LDL and Lp(a) oxidation and associated potential harmful effects, possible enhancement of reverse cholesterol transport, and, perhaps, inhibition of cell proliferation and inflammation in the arterial wall. The ability of probucol to induce xanthoma regression was observed early (Harris et al. 1974) and encouraged continued exploration of the effects and mechanisms of action of this drug (Davignon and Bouthillier 1982; Davignon 1991a). With time, many of the concerns about untoward side effects observed in preclinical studies were dispelled. The HDL-lowering effect has become less threatening and studies with this drug have contributed to improve our knowledge of lipoprotein metabolism. Inhibition of atheroma progression in the WHHL rabbit has given considerable credit to the role of oxidized lipoproteins in atherogenesis and warrants further effort to determine the clinical importance of such findings. Recent studies indicate that probucol may be of potential benefit in ischemia-reperfusion injury, in diabetes, renal disease, and restenosis following transluminal angioplasty. Currently, probucol is a drug of choice among the lipid-lowering agents (Davignon 1991b), being well tolerated, very efficient in combination therapy, and uniquely effective in homozygous familial hypercholesterolemia. What remains to be done is to establish which individuals are most likely to benefit from the antiatherogenic potential of probucol and design tests to refine the indications for antioxidant therapy in atherogenic dyslipoproteinemias. If probucol is especially effective at inhibiting foam cell formation, as seems to be the case in the WHHL rabbit, it is reasonable to consider FH as a prime target for probucol treatment. Probucol-induced xanthoma regression (Buxtorf et al. 1985; Yamamoto et al. 1986a), the richness in foam cells of atheromatous lesions (Buja et al. 1979; Davignon et al. 1991), and the slow clearance of LDL that could enhance the likelihood of oxidation in this disease (Davignon et al. 1991) strongly support this view and suggest the need to start treatment early.

References

Aburatani H, Matsumoto A, Kodama T, Takaku F, Fukazawa C, Itakura H (1988) Increased levels of messenger ribonucleic acid for apolipoprotein E in the spleen of probucol-treated rabbits. Am J Cardiol 62:60B–65B

Agellon LB, Walsh A, Hayek T, Moulin P, Jiang XC, Shelanski SA, Breslow JL, Tall AR (1991) Reduced high density lipoprotein cholesterol in human cholesteryl ester transfer protein transgenic mice. J Biol Chem 266:10796–10801

Alexandre A, Doni MG, Padoin E, Deana R (1986) Inhibition by antioxidants of agonist evoked cytosolic Ca^{++} increase, ATP secretion and aggregation of aspirinated human platelets. Biochem Biophys Res commun 139:509–514

Anastasi A, Betteridge DJ, Galton DJ (1980) Effect of probucol and other drugs on sterol synthesis in human lymphocytes. In: Noseda G, Lewis B, Paoletti R (eds) Diet and drugs in atherosclerosis. Raven, New York, pp 161–163

Arnold JA, Martin D, Taylor HL, Christian DR, Heeg JF (1970) Absorption and excretion studies of the hypocholesterolemic agent 4,4′-(isopropylidenedithio) bis(2,6-di-t-butylphenol) (DH-581) in man. Fed Proc 29:622

Atmeh RF, Stewart JM, Boag DE, Packard CJ, Lorimer AR, Shepherd J (1983) The hypolipidemic action of probucol: a study of its effects on high and low density lipoproteins. J Lipid Res 24:588–595

Bagdade JD, Kaufman D, Ritter MC, Subbaiah PV (1990) Probucol changes lipoprotein surface lipid composition and normalizes cholesterol ester transfer. Atherosclerosis 84:145–154

Baker SG, Joffe BI, Mendelsohn D, Seftel HC (1982) Treatment of homozygous familial hypercholesterolaemia with probucol. S Afr Med J 62:7–11

Barnhart JW, Johnson JD, Rytter DJ, Failey RB (1971) The effect of probucol on cholesterol metabolism. 4th International Symposium on Drugs Affecting Lipid Metabolism, Philadelphia

Barnhart JW, Li DL, Cheng WD (1988) Probucol enhances cholesterol transport in cultured rat hepatocytes. Am J Cardiol 62:52B–56B

Barnhart RL, Busch SJ, Jackson RL (1989) Concentration-dependent antioxidant activity of probucol in low density lipoproteins in vitro: probucol degradation precedes lipoprotein oxidation. J Lipid Res 30:1703–1710

Beaumont JL, Jacotot B, Buxtorf JC, Silvestre M, Beaumont V (1982) Effects of probucol on the cholesterol content of skin in type II hyperlipoproteinemia. Artery 10:71–87

Bellamy MF, Nealis AS, Aitken JW, Bruckdorfer R, Perkins SJ (1989) Structural changes in oxidized low-density lipoproteins and of the effect of the antiatherosclerotic drug probucol observed by synchroton X-ray and neutron solution scattering. Eur J Biochem 183:321–329

Berg A, Frey I, Baumstark M, Keul J (1988) Influence of probucol on lipoprotein cholesterol and apolipoproteins in normolipidemic males. Atherosclerosis 72: 49–54

Berg A, Baumstark MW, Frey I, Halle M, Keul J (1991) Clinical and therapeutic use of probucol. Eur J Clin Pharmacol 40 Suppl 1:S81–S84

Bihari-Varga M, Gruber E, Rotheneder M, Zechner R, Kostner GM (1988) Interaction of lipoprotein Lp(a) and low density lipoprotein with glycoaminoglycans from human aorta. Arteriosclerosis 8:851–857

Bird JE, Milhoan K, Wilson CB, Young SG, Mundy CA, Parthasarathy S, Blantz RC (1988) Ischemic acute renal failure and antioxidant therapy in the rat. The relation between glomerular and tubular dysfunction. J Clin Invest 81:1630–1638

Blanche PJ, Gong EL, Forte TM, Nichols AV (1981) Characterization of human high density lipoproteins by gradient gel electrophoresis. Biochim Biophys Acta 665:408–419

Bolli R, Jeroudi MO, Patel BS, DuBose CM, Lai EK, Roberts R, McCay PB (1989) Direct evidence that oxygen-derived free radicals contribute to postischemic myocardial dysfunction in the intact dog. Proc Natl Acad Sci USA 86:4695–4699

Breugnot C, Mazière C, Salmon S, Auclair M, Santus R, Morlière P, Lenaers A, Mazière JC (1990) Phenothiazines inhibit copper and endothelial cell-induced peroxidation of low density lipoprotein. A comparative study with probucol, butylated hydroxytoluene and vitamin E. Biochem Pharmacol 40:1975–1980

Bridges AB, Scott NA, Belch JJF (1991) Probucol, a superoxide free radical scavenger in vitro. Atherosclerosis 89:263–265

Brown HB, deWolfe VG (1974) The additive effect of probucol on diet in hyperlipidemia. Clin Pharmacol Ther 16:44–50

Brown ML, Inazu A, Hesler CB, Agellon LB, Mann C, Whitlock ME, Marcel YL, Milne RW, Koizumi J, Mabuchi H, Takeda R, Tall AR (1989) Molecular basis of lipid transfer protein deficiency in a family with increased high-density lipoproteins. Nature 342:448–451

Browne KF, Prystowsky EN, Heger JJ, Cerimele B, Fineberg N, Zipes DP (1984) Prolongation of the QT interval induced by probucol: demonstration of a method for determining QT interval change induced by a drug. Am Heart J 107:680–684

Buckley MM-T, Goa KL, Price AH, Brogen RN (1989) Probucol. A reappraisal of its pharmacological properties and therapeutic use in hypercholesterolaemia. Drugs 36:761–800

Buja LM, Kovanen PT, Bilheimer DW (1979) Cellular pathology of homozygous familial hypercholesterolemia. Am J Pathol 97:327

Buja LM, Kita T, Goldstein JL, Watanabe Y, Brown MS (1983) Cellular pathology of progressive atherosclerosis in the WHHL rabbit. Arteriosclerosis 3:87–101

Buxtorf JC, Jacotot B, Beaumont V, Beaumont JL (1985) Action du probucol dans les hypercholestérolémies familiales du type II. Sem Hop Paris 61:837–840

Carew TE, Schwenke DC, Steinberg D (1987) Antiatherogenic effect of probucol unrelated to its hypocholesterolemic effect: evidence that antioxidants in vivo can selectively inhibit low density lipoprotein degration in macrophage-rich fatty streaks and slow the progression of atherosclerosis in the Watanabe heritable hyperlipidemic rabbit. Proc Natl Acad Sci USA 84:7725–7729

Cortese C, Marenah CB, Miller NE, Lewis B (1982) The effects of probucol on plasma lipoproteins in polygenic and familial hypercholesterolaemia. Atherosclerosis 44:319–325

Coutant J, Bargar EM, Barbuch RB (1981) Negative ion chemical ionization mass spectrometry as a structural tool in the determination of the major metabolites of probucol in the monkeys. In: Frigerio A (ed) Recent development in mass spectrometry in biochemistry, medicine and environmental research. Elsevier, Amsterdam, pp 35–38

Dachet C, Jacotot B, Buxtorf JC (1985) The hypolipidemic action of probucol: drug transport and lipoprotein composition in type IIa hyperlipoproteinemia. Atherosclerosis 58:261–268

Dage RC, Anderson BA, Mao SJT, Koerner JE (1991) Probucol reduces myocardial dysfunction during reperfusion after short-term ischemia in rabbit heart. J Cardiovasc Pharmacol 17:158–165

Daugherty A, Zweifel BS, Schonfeld G (1989) Probucol attenuates the development of aortic atherosclerosis in cholesterol-fed rabbits. Br J Pharmacol 987:612–618
Daugherty A, Zweifel BS, Schonfeld G (1991) The effects of probucol on the progression of atherosclerosis in mature Watanabe heritable hyperlipidaemic rabbits. Br J Pharmacol 103:1013–1018
Davies SW, Ranjadayalan K, Wickens DG, Dormandy TL, Timmis AD (1990) Lipid peroxidation associated with successful thrombolysis. Lancet 335:741–743
Davignon J (1981) Les médicaments dans le traitement des hyperlipidémies – nécessité et limites. In: Beaumont JL (ed) Symposium sur le probucol. Excerpta Medica, Amsterdam, pp 38–74
Davignon J (1986) Medical management of hyperlipidemia and the role of probucol. Am J Cardiol 57:22H–28H
Davignon J (1991a) Probucol revisited. In: Descovich GC, Gaddi A, Magri GL, Lenzi S (eds) Atherosclerosis and cardiovascular disease. 7th International Meeting. Kluwer, London, pp 449–459
Davignon J (1991b) Indications for lipid-lowering drugs. Eur J Clin Pharmacol 40 Suppl 1:S3–S10
Davignon J (1992) Modification of Lp(a) and the effect of antioxidants. In: Gotto AM Jr (ed) Atherosclerosis. A decade in perspective. Medical Information Services, New York
Davignon J, Bouthillier D (1982) Probucol and familial hypercholesterolemia. Can Med Assoc J 126:1024–1025
Davignon J, LeLorier J (1981) Utilité du probucol dans le traitement à long-terme des hypercholestérolémies héréditaires. In: Beaumont JL (ed) Symposium sur le probucol. Excerpta Medica, Amsterdam, pp 157–171
Davignon J, Lussier-Cacan S, Dubreuil-Quidoz S, LeLorier J (1982) Experience with probucol in the treatment of hypercholesterolemia. Artery 10:48–55
Davignon J, Nestruck AC, Alaupovic P, Bouthillier D (1986) Severe hypoalphalipoproteinemia induced by a combination of probucol and clofibrate. Adv Exp Med Biol 201:111–125
Davignon J, Xhignesse M, Mailloux H, Nestruck AC, Lussier-Cacan S, Roederer G, Pfister P (1988) Comparative study of lovastatin versus probucol in the treatment of hypercholesterolemia. Atheroscler Rev 18:139–151
Davignon J, Roy M, Dufour R, Roderer G (1991) Familial hypercholesterolemia. In: Steiner G, Shafrir E (eds) Primary hyperlipoproteinemias. McGraw-Hill, New York, pp 201–234
Drash AL, Rudert WA, Borquaye S, Wang R, Lieberman I (1988) Effect of probucol on development of diabetes mellitus in BB rats. Am J Cardiol 62:27B–30B
Dujovne CA, Atkins F, Wong B, Decoursey S, Krehbiel P, Chernoff S (1984a) Electrocardiographic effects of probucol: a controlled prospective clinical trial. Eur J Clin Pharmacol 26:735–739
Dujovne CA, Krehbiel P, Decoursey S, Jackson B, Chernoff SB, Pitterman A, Garty M (1984b) Probucol with colestipol in the treatment of hypercholesterolemia. Ann Intern Med 100:477–482
Dujovne CA, Krehbiel P, Chernoff SB (1986) Controlled studies of the efficacy and safety of combined colestipol-probucol therapy. Am J Cardiol 57:36H–42H
Eder HA (1982) The effect of diet on the transport of probucol in monkeys. Artery 10:105–107
Engst R (1985) Hyperlipoproteinemia type III with apolipoprotein E phenotype 2/2 (in German). Hautarzt 36:629–634
Esterbauer H, Jürgens G, Quehenberger O, Koller E (1987) Autoxidation of human low density lipoprotein: loss of polyunsaturated fatty acids and vitamin E and generation of aldehydes. J Lipid Res 28:495–509
Eto M, Sato T, Watanabe K, Iwashima Y, Makino I (1990) Effects of probucol on plasma lipids and lipoproteins in familial hypercholesterolemic patients with and without apolipoprotein E4. Atherosclerosis 84:49–53

Fellin R, Gasparotto A, Valerio G, Baiocchi MR, Padrini R, Lamon S, Vitale E, Baggio G, Crepaldi G (1986) Effect of probucol treatment on lipoprotein cholesterol and drug levels in blood and lipoproteins in familial hypercholesterolemia. Atherosclerosis 59:47–56

Finckh B, Niendorf A, Rath M, Beisiegel U (1991) Antiatherosclerotic effect of probucol in WHHL rabbits: are there plasma parameters to evaluate this effect. Eur J Clin Pharmacol 40 Suppl 1:S77–S80

Fong LG, Parthasarathy S, Witztum JL, Steinberg D (1987) Nonenzymatic oxidative cleavage of peptide bonds in apoprotein B-100. J Lipid Res 28:1466–1477

Franceschini G, Sirtori M, Vaccarino V, Gianfranceschi G, Rezzonico L, Chiesa G, Sirtori CR (1989) Mechanisms of HDL reduction after probucol. Arteriosclerosis 9:462–469

Fukuyama S, Ogawa T, Shiratori Y, Inaba M, Nagano S (1988) Influence of oxidized LDL on insulin binding by IM-9 lymphocytes. J Clin Biochem Nutr 4:41–47

Gallagher PJ, Nanjee MN, Richards T, Roche WR, Miller NE (1988) Biochemical and pathological features of a modified strain of Watanabe heritable hyperlipidemic rabbits. Atherosclerosis 71:173–183

Gaziano JM, Manson JE, Ridker PM, Buring JE, Hennekens CH (1990) Beta carotene therapy for chronic stable angina. Circulation 82:III-201

Gey KF (1990) The antioxidant hypothesis of cardiovascular disease: epidemiology and mechanisms. Biochem Soc Trans 18:1041–1045

Gey KF, Brubacher G, Stahelin B (1987) Plasma levels of antioxidant vitamins in relation of ischemic heart disease and cancer. Am J Clin Nutr 33:1368–1377

Gey KF, Puska P, Jordan P, Moser UK (1991) Inverse correlation between plasma vitamin E and mortality from ischemic heart disease in cross-cultural epidemiology. Am J Clin Nutr 53 Suppl:326S–334S

Giada V, Valerio G, Bicego L, Padrini R, Moretto R et al. (1986) Probucol with cholestyramine in the treatment of familial hypercholesterolemia: effects on lipoproteins and serum probucol levels. Curr Ther Res 40:975–986

Ginsberg HN, Le NA, Goldberg IJ, Gibson JC, Rubinstein A, Wang-Iverson P, Norum R, Brown WV (1986) Apolipoprotein B metabolism in subjects with deficiency of apolipoprotein CIII and AI. Evidence that apolipoprotein CIII inhibits catabolism of triglyceride-rich lipoproteins by lipoprotein lipase in vivo. J Clin Invest 78:1287–1295

Goldberg RB, Mendez A (1988) Probucol enhances cholesterol efflux from cultured human skin fibroblasts. Am J Cardiol 62:57B–59B

Harris RS Jr, Gilmore HR III, Bricker LA, Kiem IM, Rubin E (1974) Long-term oral administration of probucol [4,4′-(isopropylidenedithio) bis (2,6-di-t-butylphenol)] (DH-581) in the management of hypercholesterolemia. J Am Geriatr Soc 22:167–175

Hattori M, Tsuda K, Taminato T, Nishi S, Fujita J, Tsuji K, Kurose T, Koh G, Seino Y, Imura H (1987) Effect of probucol on serum lipids and apolipoproteins in patients with non-insulin-dependent diabetes mellitus. Curr Ther Res 42: 867–973

Havel RJ, Kita T, Kotite JP, Kane JP, Hamilton RL, Goldstein JL, Brown MS (1982) Concentration and composition of lipoproteins in blood plasma of the WHHL rabbit. Arteriosclerosis 2:467–474

Hayek T, Chajek-Shaul T, Walsh A, Agellon LB, Moulin P, Tall AR, Breslow JL (1991a) Interaction of human apo A-I and CETP genes in transgenic mice results in a profound decrease in HDL cholesterol. Circulation 84:II-17

Hayek T, Chajek-Shaul T, Walsh A, Azrolan N, Breslow JL (1991b) Probucol decreases apolipoprotein A-I transport rate and increases high density lipoprotein cholesteryl ester fractional catabolic rate in control and human apolipoprotein A-I transgenic mice. Arterioscler Thromb 11:1295–1302

Heeg JF, Tachizawa H (1980) Taux plasmatiques du probucol chez l'homme après administration orale unique ou répétée. Nouv Presse Med 9:2990–2994

Heel RC, Brogden RN, Speight TM, Avery GS (1978) Probucol: a review of its pharmacological properties and therapeutic use in patients with familial hypercholesterolaemia. Drugs 15:409–428

Heinecke JW (1987) Free radical modification of low density lipoprotein: mechanisms and biological consequences. Free Radic Biol Med 3:65–73

Heller F, Harvengt C (1983) Effects of clofibrate, bezafibrate, fenofibrate and probucol on plasma lipolytic enzymes in normolipaemic subjects. Eur J Clin Pharmacol 25:57–63

Helve E, Tikkanen MJ (1988) Comparison of lovastatin and probucol in treatment of familial and non-familial hypercholesterolemia: different effects on lipoprotein profiles. Atherosclerosis 72:189–197

Hirano T, Mamo JCL, Nagano S, Sugisaki T (1991) The lowering effect of probucol on plasma lipoprotein and proteinuria in puromycin aminonucleotide-induced nephrotic rats. Nephron 58:95–100

Hoff HF, O'Neil J (1991) Lesion-derived low density lipoprotein and oxidized low density lipoprotein share a lability for aggregation, leading to enhanced macrophage degradation. Arterioscler Thromb 11:1209–1222

Homma Y, Moriguchi EH, Sakane H, Ozawa H, Nakamura H, Goto Y (1991) Effects of probucol on plasma lipoprotein subfractions and activities of lipoprotein lipase and hepatic triglyceride lipase. Atherosclerosis 88:175–181

Horikishi K, Shimoda S (1988) Effect of probucol (Lorelco^R) on the lipid metabolism of diabetes mellitus with hypercholesterolemia. Yak Chiryo 16:2985–2990

Hsiang Y, White RA, Kopchok GE, Rosenbaum D, Guthrie C, Kao J, Zhen E, Peng S-K, Fragoso M (1991) Stenosis following laser thermal angioplasty-A blinded controlled randomized study between aspirin against probucol. J Surg Res 50:252–258

Iida H, Izumino K, Asaka MS, Fujita M, Nishino A, Sasayama S (1987) Effect of probucol on hyperlipidemia in patients with nephrotic syndrome. Nephron 47:280–283

Inazu A, Brown ML, Hesler CB, Agellon LB, Koizumi J, Takata K, Maruhama Y, Mabuchi H, Tall AR (1990) Increased high-density lipoprotein levels caused by a common cholesterly-ester transfer protein gene mutation. N Engl J Med 323:1234–1238

Jackson HR (1991) Chronic renal failure: pathophysiology. Lancet 338:419–423

Jackson RL, Barnhart RL, Mao SJT (1991) Probucol and its mechanisms for reducing atherosclerosis. Adv Exp Med Biol 285:367–372

Jialal I, Grundy SM (1991) Preservation of the endogenous antioxidants in low density lipoprotein by ascrobate but not probucol during oxidative modification. J Clin Invest 87:597–601

Johansson J, Mölgaard J, Olsson AG, Walldius G (1990) Effect of probucol treatment on HDL particle size subclass concentrations as assessed by gradient gel electrophoresis. In: Carlson LA (ed) Disorders of HDL. Smith-Gordon, London, pp 209–214

Johnson WJ, Bamberger MJ, Latta RA, Rapp PE, Rothblatt GH (1986) The bidirectional flux of cholesterol between cells and lipoproteins. Effects of phospholipid depletion of high density lipoproteins. J Biol Chem 261:5766–5776

Jürgens G, Hoff HF, Chisolm GM, Esterbauer H (1987) Modification of human serum low density lipoprotein by oxidation – characterization and pathophysiological implications. Chem Phys Lipids 45:315–336

Jürgens G, Ashy A, Esterbauer H (1990) Detection of new epitopes formed upon oxidation of low-density lipoprotein, lipoprotein (a) and very-low density lipoprotein. Biochem J 265:605–608

Kaplan R, Aynedjian HS, Schlondorff D, Bank N (1990) Renal vasoconstriction caused by short-term cholesterol feeding is corrected by thromboxane antagonist or probucol. J Clin Invest 86:1707–1714

Kenny RA, Sutton R (1985) The prolonged QT interval – a frequently unrecognized abnormality. Postgrad Med J 61:379–386

Kesäniemi YA (1991) Mechanisms of low density lipoprotein lowering by hypolipidemic agents. Ann Med 23:195–198

Kesäniemi YA, Grundy SM (1984) Influence of probucol on cholesterol and lipoprotein metabolism in man. J Lipid Res 25:780–790

Kita T, Nagano Y, Yokode M, Ishii K, Kume N, Ooshima A, Yoshida H, Kawai C (1987) Probucol prevents progression of atherosclerosis in the Watanabe heritable hyperlipidemic rabbit, an animal model for familial hypercholesterolemia. Proc Natl Acad Sci USA 84:5928–5931

Kita T, Nagano Y, Yokode M, Ishii K, Kume N, Narumiya S, Kawai C (1988) Prevention of atherosclerotic progression in Watanabe rabbits by probucol. Am J Cardiol 62:13B–19B

Kobayashi M, Ohgaku S, Iwasaki M, Hirano Y, Maegawa H, Shigeta Y (1980) Evaluation of the method of insulin binding studies in human erythrocytes. Endocrinol Jpn 27:337–342

Kritchevsky D, Kim HK, Tepper SA (1971) Influence of 4,4′-(isopropylidenedithio) bis(2,6-di-t-butylphenol) (DH-581) on experimental atherosclerosis in rabbits. Proc Soc Exp Biol Med 136:1216

Ku G, Doherty NS, Wolos JA, Jackson RL (1988) Inhibition by probucol of interleukin 1 secretion and its implication in atherosclerosis. Am J Cardiol 62: 77B–81B

Ku G, Schroeder K, Schmidt LF, Jackson RL, Doherty NS (1990) Probucol does not alter acetylated low density lipoprotein uptake by murine peritoneal macrophages. Atherosclerosis 80:191–197

Kuksis A, Myher JJ, Geher K, Jones GJL, Breckenridge WC, Feather T, Hewitt D, Little JA (1982) Decreased plasma phosphatidylcholine/free cholesterol ratio as an indicator of risk for ischemic vascular disease. Arteriosclerosis 218:296–302

Kuo PT, Wilson AC, Kostis JB, Moreyra AE (1986) Effects of combined probucol–colestipol treatment for familial hypercholesterolemia and coronary artery disease. Am J Cardiol 57:43H–48H

Kuzuya M, Naito M, Funaki C, Hayashi T, Asai K, Kuzuya F (1991) Probucol prevents oxidative injury to endothelial cells. J Lipid Res 32:197–204

Lane JT, Subbaiah PV, Otto ME, Bagdade JD (1991) Lipoprotein composition and HDL particle size distribution in women with non-insulin-dependent diabetes mellitus and the effects of probucol treatment. J Lab Clin Med 118:120–128

Lee YJ, Yamaguchi H, Daida H, Yokoi H, Miyano H, Takaya J, Sakurai H, Noma A (1991) Pharmacological intervention to modify restenosis. Circulation 84: II-299

Lees AM, Stein SW, Lees RS (1986) Therapy of hypercholesterolemia with mevinolin and other lipid-lowering drugs. Circulation 74:200

LeLorier J, Dubreuil-Quidoz S, Lussier-Cacan S, Huang YS, Davignon J (1977) Diet and probucol in lowering cholesterol concentrations. Arch Intern Med 137: 1429–1434

Maeda S, Okuno M, Abe A, Noma A (1989) Lack of effect of probucol on lipoprotein(a) levels. Atherosclerosis 79:267–289

Mahley RW, Innerarity TL (1983) Lipoprotein receptors and cholesterol homeostasis. Biochim Biophys Acta 737:197–222

Mao SJT, Yates MT, Rechtin AE, Jackson RL, Van Sickle WA (1991) Antioxidant activity of probucol and its analogues in hypercholesterolemic Watanabe rabbits. J Med Chem 34:298–302

Marshall FN, Lewin JE (1973) Sensitization of epinephrine-induced ventricular fibrillation produced by probucol in dogs. Toxicol Appl Pharmacol 24:594–602

Matsuhashi H, Onodera S, Kawamura Y, Hasebe N, Kohmura C, Yamashita H, Tobise K (1989) Probucol-induced QT prolongation and torsades de pointes. Jpn J Med 28:612–615

Matsushita M, Yoshino G, Iwai M, Matsuba K, Morita M, Iwatani I, Yoshida M, Kazumi T, Baba S (1989) Protective effect of probucol on alloxane diabetes in rats. Diabetes Res Clin Pract 7:313–316

Matsuzawa Y, Yamashita S, Funahashi T, Yamamoto A, Tarui S (1988) Selective reduction of cholesterol in HDL_2 fraction by probucol in familial hypercholesterolemia and hyperHDL_2-cholesterolemia with abnormal cholesteryl ester transfer. Am J Cardiol 62:66B–72B

McCaughan D (1981) The long-term effects of probucol on serum lipid levels. Arch Intern Med 141:1428–1432

McLean LR, Hagaman KA (1988) Probucol reduces the rate of association of apolipoprotein C-III with dimyristoylphosphatidylcholine. Biochim Biophys Acta 959:201–205

McLean LR, Hagaman KA (1989) Effect of probucol on the physical properties of low-density lipoproteins oxidized by copper. Biochemistry 28:321–327

McPherson R, Marcel Y (1991) Role of cholesteryl ester transfer protein in reverse cholesterol transport. Clin Cardiol 14:I-31–I-34

McPherson R, Hogue M, Milne RW, Tall AR, Marcel YL (1991a) Increase in plasma cholesteryl ester transfer protein during probucol treatment: relation to changes in high density lipoprotein composition. Arterioscler Thromb 11: 476–481

McPherson R, Mann CJ, Tall AR, Hogue M, Martin L, Milne RW, Marcel YL (1991b) Plasma concentrations of cholesteryl ester transfer protein in hyperlipoproteinemia: relation to cholesteryl ester transfer protein activity and other lipoprotein variables. Arterioscler Thromb 11:797–804

Miettinen TA (1972) Mode of action of a new hypocholesterolaemic drug (DH-581) in familial hypercholesterolaemia. Atherosclerosis 15:163–176

Miettinen TA, Huttunen JK, Ehnholm C, Kumlin T, Mattila S, Naukkarinen V (1980) Effect of long-term antihypertensive and hypolipidemic treatment on high density lipoprotein cholesterol and apolipoproteins A-I and A-II. Atherosclerosis 36:249–259

Miettinen TA, Huttunen JK, Kuusi T, Kumlin T, Mattila S, Naukkarinen V, Strandberg T (1981) Effect of probucol on the activity of postheparin plasma lipoprotein lipase and hepatic lipase. Clin Chim Acta 113:59–64

Miettinen TA, Huttunen JK, Naukkarinen V, Strandberg T, Vanhanen H (1986) Long-term use of probucol in the multifactorial primary prevention of vascular disease. Am J Cardiol 57:49H–54H

Miller GC, Miller NE (1975) Plasma high density lipoprotein concentration and development of ischaemic heart disease. Lancet 1:16–19

Mitchel YB, Rifici VA, Eder HA (1987) Characterization of the specific binding of rat apolipoprotein E deficient HDL to rat hepatic plasma membranes. Biochim Biophys Acta 917:324–332

Mordasini R, Keller M, Riesen WF (1980) Effect of probucol and diet on serum lipids and lipoprotein fractions in primary hypercholesterolemia. In: Noseda G, Lewis B, Paoletti R (eds) Diet and drugs in atherosclerosis. Raven, New York, pp 181–187

Morel DW, Chisolm GM (1989) Antioxidant treatment of diabetic rats inhibits lipoprotein oxidation and cytotoxicity. J Lipid Res 30:1827–1834

Morel DW, Hessler JR, Chisolm GM (1983) Low density lipoprotein cytotoxicity induced by free radical peroxidation of lipids. J Lipid Res 24:1070–1076

Murphy BF (1977) Probucol (Lorelco) in treatment of hyperlipemia. JAMA 238: 2537–2538

Nagano Y, Kita T, Yokode M, Ishii K, Kume N, Otani H, Arai H, Kawai C (1989) Probucol does not affect lipoprotein metabolism in macrophages of Watanabe heritable hyperlipidemic rabbits. Arteriosclerosis 9:453–461

Nagatsuka S, Nakazawa T (1982) Effect of membrane-stabilizing agents, cholesterol and cepharantin, on radiation-induced lipid peroxidation and permeability in liposomes. Biochim Biophys Acta 691:171–177

Naruszewicz M, Carew TE, Pitman RC, Witztum JL, Steinberg D (1984) A novel mechanism by which probucol lowers low density lipoprotein levels demonstrated in the LDL receptor-deficient rabbit. J Lipid Res 25:1206–1213

Naruszewicz M, Selinger E, Davignon J (1992a) Oxidative modification of Lp(a) and the effect of β-carotene. Metabolism 41:1215–1224

Naruszewicz M, Selinger E, Dufour R, Davignon J (1992b) Probucol protects lipoprotein(a) against oxidative modification. Metabolism 41:1225–1228

Negre-Salvayre A, Alomar Y, Troly M, Salvayre R (1991) Ultraviolet-treated lipoproteins as a model system for the study of the biological effects of lipid peroxides on cultured cells: III. The protective effect of antioxidants (probucol, catechin, vitamin E) against the cytotoxicity of oxidized LDL occurs in two different ways. Biochim Biophys Acta 1096:291–300

Nestel PJ, Billington T (1981) Effect of probucol on low density lipoprotein removal and high density lipoprotein synthesis. Atherosclerosis 38:203–209

Nestruck AC, Bouthillier D, Sing CF, Davignon J (1987) Apolipoprotein E polymorphism and plasma cholesterol response to probucol. Metabolism 36:743–747

Neuworth MBR, Laufer J, Barnhart JW, Sefranka JA, McIntosh DD (1970) Synthesis and hypocholesterolemic activity of alkylidenedithio bisphenols. J Med Chem 13:722–725

Nishigaki I, Hagihara M, Tsunekawa HT, Maseki M, Yagi K (1981) Lipid peroxide levels of serum lipoprotein fractions of diabetic patients. Biochem Med 25: 373–378

O'Brien K, Nagano Y, Gown A, Kita T, Chait A (1991) Probucol treatment affects the cellular composition but not anti-oxidized low density lipoprotein immunoreactivity of plaques from Watanabe heritable hyperlipidemic rabbits. Arteriosclerosis 11:751–759

O'Neil JA, Pepin JM, Smejkal G, Gordon EA, Hoff HF (1990) Structural characteristics of Lp(a) extracted from human atherosclerotic plaques. Arteriosclerosis 10:812a

Paller MS, Moidal JR, Ferris TF (1984) Oxygen free radicals in ischemic acute renal failure in the rat. J Clin Invest 74:1156–1164

Parthasarathy S, Steinbrecher UP, Barnett J, Witztum JL, Steinberg D (1985) Essential role of phospholipase A_2 activity in endothelial cell-induced modification of low density lipoprotein. Proc Natl Acad Sci USA 82:3000–3004

Parthasarathy S, Young SG, Witztum JL, Pittman RC, Steinberg D (1986) Probucol inhibits oxidative modification of low density lipoprotein. J Clin Invest 77: 641–644

Parthasarathy S, Fong L, Otero D, Steinberg D (1987) Recognition of solubilized apoproteins from delipidated, oxidized low density lipoprotein (LDL) by the acetyl-LDL receptor. Proc Natl Acad Sci USA 84:537–540

Pedersen JC, Berg K (1989) Interaction between low density lipoprotein receptor (LDLR) and apolipoprotein E (apoE) alleles contributes to normal variation in lipid level. Clin Genet 35:331–337

Pepin JM, O'Neil JA, Hoff HF (1991) Quantification of apo(a) and apoB in human atherosclerotic lesions. J Lipid Res 32:317–327

Regnström J, Walldius G, Carlson LA, Nilsson J (1990) Effect of probucol treatment on the susceptibility of low density lipoprotein isolated from hypercholesterolemic patients to become oxidatively modified in vitro. Atherosclerosis 82:43–51

Reichl D, Miller NE (1989) Pathophysiology of reverse cholesterol transport. Insights from inherited disorders of lipoprotein metabolism. Arteriosclerosis 9: 785–797

Roden DM, Woosley RL (1985) QT prolongation and arrhythmia suppression. Am Heart J 109:411–415

Rosenfeld ME, Tsukada T, Chait A, Bierman EL, Gown AM, Ross R (1987a) Fatty streak expansion and maturation in Watanabe heritable hyperlipidemic and comparably hypercholesterolemic fat-fed rabbit. Arteriosclerosis 7: 24–34

Rosenfeld ME, Tsukada T, Gown AM, Ross R (1987b) Fatty streak initiation in Watanabe heritable hyperlipidemic and comparably hypercholesterolemic fat-fed rabbits. Arteriosclerosis 7:9–23

Sandler S, Andersson A (1982) Partial protective effect of the hydroxyl radical scavenger dimethyl urea in streptozotocin-induced diabetes in the mouse in vivo and in vitro. Diabetologia 23:374–378

Sasaki J, Tanabe Y, Kuwano E, Kawano T, Takii M, Saku K, Arakawa K (1987) The hypocholesterolemic effect of once-daily administration of probucol. Curr Ther Res 41:328–334

Satonin DK, Coutant JE (1986) Comparison of gas chromatography and high-performance liquid chromatography for the analysis of probucol in plasma. J Chromatogr 380:401–406

Sattler W, Kostner GM, Waeg G, Esterbauer H (1991) Oxidation of lipoprotein Lp(a). A comparison with low-density lipoproteins. Biochim Biophys Acta 1081: 65–74

Scanu AM, Scandiani L (1991) Lipoprotein(a): structure, biology, and clinical relevance. Adv Intern Med 36:249–270

Seed M, Hoppichler F, Raeveley D, McCarthy S, Thompson GR, Boerwinkle E, Utermann G (1990) Relation of serum lipoprotein(a) concentration and apolipoprotein(a) phenotype to coronary heart disease in patients with familial hypercholesterolemia. N Engl J Med 322:1494–1499

Shankar R, Sallis JD, Stanton H, Thomson R (1989) Influence of probucol on early experimental atherogenesis in hypercholesterolemic rats. Atherosclerosis 78: 91–97

Shimizu H, Uehara Y, Shimomura Y, Tanaka Y, Kobayashi I (1991) Probucol attenuated hyperglycemia in multiple low-dose streptozotocin-induced diabetic mice. Life Sci 49:1331–1338

Sirtori CR, Sirtori M, Calabresi L, Franceschini G (1988) Changes in high-density lipoprotein subfraction distribution and increased cholesteryl ester transfer after probucol. Am J Cardiol 62:73B–76B

Slonim AE, Surber ML, Page DL, Sharp RA (1983) Modification of chemically induced diabetes in rats by vitamin E. J Clin Invest 71:1282–1288

Stein Y, Stein O, Delplanque B, Lee DM, Alaupovic P (1989) Lack of effect of probucol on atheroma formation in cholesterol-fed rabbits kept at comparable plasma cholesterol levels. Atherosclerosis 75:145–155

Steinberg D, Parthasarathy S, Carew TE, Khoo JC, Witztum JL (1989) Beyond cholesterol. Modification of low-density lipoprotein that increase its atherogenicity. N Engl J Med 320:915–924

Steinbrecher UP (1987) Oxidation of human low density lipoprotein results in derivatization of lysine residues of apolipoprotein B by lipid peroxide decomposition products. J Biol Chem 262:3603–3608

Steinbrecher UP, Parthasarathy S, Leake DS, Witztum JL, Steinberg D (1984) Modification of low density lipoprotein by endothelial cells involves lipid peroxidation and degradation of low density lipoprotein phospholipids. Proc Natl Acad Sci USA 81:3883–3887

Steinbrecher UP, Witztum JL, Parthasarathy S, Steinberg D (1987) Decrease in reactive amino groups during oxidation or endothelial cell modification. Arteriosclerosis 7:135–143

Steinbrecher UP, Zhang H, Lougheed M (1990) Role of oxidatively modified LDL in atherosclerosis. Free Radic Biol Med 9:155–168

Strandberg TE, Kuusi T, Tilvis R, Miettinen TA (1981) Effect of probucol on cholesterol synthesis, plasma lipoproteins and the activities of lipoprotein and hepatic lipase in the rat. Atherosclerosis 40:193–201

Strandberg TE, Vanhanen H, Miettinen TA (1988) Probucol in long term treatment of hypercholesterolemia. Gen Pharmacol 19:317–320

Sundararajan V, Cooper DKC, Muchmore J, Manion CV, Liguori C, Zuhdi N, Novitzky D, Chen P-N, Bourne DWA, Corder CN (1991) Interaction of cyclosporine and probucol in heart transplant patients. Transplant Proc 23:2028–2032

Sznajderman M, Kuczyńska K (1981) Effect of probucol on blood lipid level and lecithin acyltransferase activity in type II hyperlipoproteinemia (in Polish). Pol Arch Med Wewn 65:219–225

Takata K, Okahashi M, Tokumo H, Hirata Y, Kawamura H, Horiuchi I, Kajiyama G (1986) Probucol promotes hepatic uptake of high density lipoprotein. 9th International Symposium on Drugs Affecting Lipid Metabolism, Oct 22–25, Florence

Takegoshi T, Haba T, Kitoh C, Tokuda T, Mabuchi H (1992) Decreased serum cholesteryl-ester transfer activity in a patient with familial hyperalphalipoproteinemia. Jpn J Med 27:295–299

Tawara K, Ishihara M, Ogawa H, Tomikawa M (1986) Effect of probucol, pantethine and their combinations on serum lipoprotein metabolism and on the incidence of atheromatous lesions in the rabbit. Jpn J Pharmacol 41:211–222

Tedeschi RE, Martz BL, Taylor HA, Cerimelle BJ (1980) Etude clinique à long terme (9 ans) de la tolérance et de l'efficacité du probucol. Nouv Presse Med 9:3021–3026

Urien S, Riant P, Albengres E, Brioude R, Tillement J-P (1984) In vitro studies on the distribution of probucol among human plasma lipoproteins. Mol Pharmacol 26:322–327

Von Eckardstein A, Holz H, Sandkamp M, Weng W, Funke H, Assmann G (1991) Apolipoprotein C-III(Lys58→Glu). Identification of an apolipoprotein C-III variant in a family with hyperalphalipoproteinemia. J Clin Invest 87:1724–1731

Von Gablenz E, Schaaf B (1988) Probucol in patients with angina pectoris. 8th International Symposium on Atherosclerosis, Rome, 9–13 October 1988. Abstract Book, p 1005

Walldius G, Carlson LA, Erikson U, Olsson AG, Johansson J, Molgaard J, Nilsson S, Stenport G, Kaijer L, Lassvik C, Holme I (1988) Development of femoral atherosclerosis in hypercholesterolemic patients during treatment with cholestyramine and probucol/placebo: Probucol Quantitative Regression Swedish Trial (PQRST): a status report. Am J Cardiol 62:37B–43B

Watanabe Y (1980) Serial breeding of rabbits with hereditary hyperlipidemia (WHHL-rabbit): incidence and development of atherosclerosis and xanthomas. Atherosclerosis 36:261–268

Weiss R, Leitner EV, Schwartzkopff W (1986) Probucol in the treatment of type IIA and type IIB hyperlipoproteinemia (Abstr). 9th International Symposium on Drugs Affecting Lipid Metabolism, Oct 22–25, Florence

Wissler RW, Vesselinovitch D (1983) Combined effects of cholestyramine and probucol on regression of atherosclerosis in rhesus monkey aortas. Appl Pathol 1:89–96

Witztum JL, Steinberg D (1991) Role of oxidized lipoprotein in atherogenesis. J Clin Invest 88:1785–1792

Witztum JL, Simmons D, Steinberg D, Beltz WF, Weinreb R (1989) Intensive combination drug therapy of familial hypercholesterolemia with lovastatin, probucol and colestipol hydrochloride. Circulation 79:16–28

Yagi K (1987) Lipid peroxides and human diseases. Chem Phys Lipids 45:337–351

Yamamoto A, Matsuzawa Y, Yokoyama S, Funahaashi T, Yamamura T, Kishino B-I (1986a) Effects of probucol on xanthoma regression in familial hypercholesterolemia. Am J Cardiol 57:29H–35H

Yamamoto A, Takaichi S, Hara H, Nishikawa O, Yokoyama S, Yamamure T, Yamaguchi T (1986b) Probucol prevents lipid storage in macrophages. Atherosclerosis 62:209–217

Yamamoto T, Bishop RW, Brown MS, Goldstein JL, Russell DW (1986) Deletion in cysteine-rich region of LDL receptor impedes transport to cell surface in WHHL rabbit. Science 232:1230–1237

Yokoyama S, Yamamoto A, Kurasawa T (1988) A little more information about aggravation of probucol-induced HDL-reduction by clofibrate. Atherosclerosis 70:179–181

Yoshino G, Kazumi T, Uenoyama R, Inui A, Kasama T, Iwatani I, Iwai M, Yokono K, Otsuki M, Baba S (1986) Probucol versus eptastatin in hypercholesterolaemic diabetics. Lancet 2:740–741

Yoshino G, Matsushita M, Maeda E, Nagata K, Morita M, Matsuba K, Tani T, Horinuki R, Kimura Y, Kazumi T (1991) Effect of probucol on triglyceride turnover in streptozotocin-diabetic rats. Atherosclerosis 88:69–75

CHAPTER 16

Miscellaneous Lipid-Lowering Drugs

K.J. LACKNER and E. VON HODENBERG

A. Introduction

A large number of drugs influence lipid or lipoprotein metabolism. The drugs which are most commonly used for lipid lowering are discussed in other sections of this volume. Several agents are used for specific therapeutic purposes other than lipid lowering. Their effects on lipid metabolism are useful or – more often – unwanted side effects. This chapter will focus on the drugs that have been used or may be useful to treat hyper- or dyslipoproteinemia, which have not been covered in other parts of this volume. It would be beyond the scope of this chapter to discuss all the drugs (e.g., β-blockers, diuretics, phenytoin, and others) with side effects on lipoprotein metabolism.

In Table 1 several drugs that have been used for lipid lowering in the past are summarized. Due to the introduction of new and potent lipid-lowering agents over the recent years (e.g., 3-hydroxy-3-methylglutaryl coenzyme A (HMG-CoA) reductase inhibitors, second generation fibrates), some of the drugs used to lower lipids in the 1970s and early 1980s have become obsolete. The major reason is that they have potentially severe side effects which are not found with the newer agents available or that in large trials unfavorable risk benefit ratios for a drug have been observed. Since the rationale to treat hyperlipoproteinemia is the primary or secondary prevention of vascular disease, the profile of side effects is crucial for the recommendation of any drug used.

Estrogen treatment of hyperlipoproteinemia in men after myocardial infarction has been related to excess morbidity. Particularly disturbing was the increased rate of malignancies and the relatively high incidence of thromboembolic events (CORONARY DRUG PROJECT RESEARCH GROUP 1970). In addition, the effect on lipoproteins is unpredictable and at best small. Thus, estrogens cannot be recommended as lipid-lowering agents any more. Other steroid hormones have potentially interesting effects on lipid metabolism, but should not be used for this purpose due to their expected effects. One such example is the lowering of Lp(a) by the anabolic steroid stanozolol (ALBERS et al. 1984).

An increase in total mortality was found by the CORONARY DRUG PROJECT RESEARCH GROUP (1972, 1975) with D-thyroxine, a thyroid hormone

Table 1. Lipid-lowering drugs

Substance	Effects			Comments
	Cholesterol (%)	Triglycerides (%)	HDL-Cholesterol (%)	
Ketokonazole	−20 to −30	Unchanged	Unchanged	Potentially severe hepatotoxicity
Neomycin	−20 to −30	Unchanged	−10 to −20	Potential for nephro- and ototoxicity
β-Sitosterol	−10 to −20	Unchanged	Unchanged	
D-Thyroxine	−10 to −20	−15 to −20	Unchanged	Excess morbidity and mortality in large trial
Estrogen	Slight reduction	Increase (substantial in some patients)	Increase	Excess malignancy and thromboembolic events in men
Lifibrol	−20 to −35	Slight reduction	0 to −13	Not yet approved for clinical use

HDL, high-density lipoprotein.

stereoisomer. This agent should no longer be recommended for the treatment of hyperlipoproteinemia and prevention of cardiovascular disease. The antifungal drug ketoconazole lowers cholesterol, but has the potential for severe hepatotoxicity and several other disturbing side effects. However, its mechanism of action, the inhibition of lanosterol 14α-demethylase, may eventually prove to be an effective way to lower cholesterol without inhibiting other synthetic pathways. Both agents will only be discussed briefly in the following.

Metformin, an oral antidiabetic drug, has been shown to lower triglycerides, but it has not gained any importance in lipid lowering. No larger trials in nondiabetic patients are available.

A new cholesterol synthesis inhibitor, K12.148 (lifibrol), has not yet been approved for lipid-lowering therapy. This agent appears to inhibit cholesterol synthesis at an earlier step than the HMG-CoA reductase inhibitors and lowers cholesterol in a dose-dependent manner by 15%–30% (SCHLIACK et al. 1989; HASIBEDER et al. 1991). Current data indicate that the substance inhibits HMG-CoA synthase. At the same time it induces LDL receptors in skin fibroblasts and HepG2 cells more than one would expect from the inhibition of cholesterol synthesis (MÄRZ et al. 1992 and personal communication). Thus, it appears that the substance may be an interesting cholesterol-lowering agent. However, this agent cannot be discussed in detail before additional data are available.

Similarly, the inhibitors of acyl coenzyme A cholesteryl acyltransferase (ACAT) may become clinically important drugs to lower cholesterol. In theory, they might also inhibit the accumulation of cholesterol ester in the cells of the early atherosclerotic lesion. However, the available agents have not been used clinically, so that no information regarding their efficacy and long-term safety is available. The current status of research in this field has recently been reviewed in detail (BILLHEIMER and GILLIES 1990; SLISKOVIC and WHITE 1991).

The major focus will be on the potentially interesting drugs β-sitosterol and neomycin. Both agents might be reasonable alternatives in patients who cannot tolerate the currently available medications of first choice. However, one should keep in mind that large randomized, blinded trials have not yet been performed for either drug to prove their efficacy in reducing cardiovascular morbidity and mortality.

B. β-Sitosterol

β-Sitosterol is an ubiquitous plant sterol that was shown in the early 1950s to lower cholesterol in animals on a high-fat diet by PETERSON (1951) and POLLAK (1953a). Soon afterwards, POLLAK (1953b) was able to demonstrate that it also lowers serum cholesterol in man. The effects on serum cholesterol are moderate compared to other currently available drugs. The safety of β-sitosterol, however, appears to be excellent. One major problem with the use of β-sitosterol was the disagreeable taste of the commercially available preparations and the ensuing poor patient compliance. Today, some more palatable preparations are available. β-Sitosterol has not gained widespread clinical use for the above-mentioned reasons. Nevertheless, it may be an interesting drug in selected patients.

I. Chemistry, Physical Properties, and Dosage Form

The plant sterol β-sitosterol is closely related to cholesterol (Fig. 1). The only difference is an ethyl group attached to the side chain of the sterol structure at C24. This chemical difference leads to a changed physicochemical behavior. Particularly, the solubility in micellar solution seems altered. Several investigators were able to show that changes in the side chain (e.g., β-sitosterol or campesterol) as well as changes in the steroid core (e.g., β-sitostanol) will increase its affinity to biliary micelles.

Different preparations of β-sitosterol are available. The source of β-sitosterol seems to be important, because phytosterol preparations often contain other sterols such as campesterol that have been implicated in reducing the cholesterol-lowering efficacy. In addition, campesterol may be absorbed in relevant amounts. Preparations from soybean usually contain more than 10% campesterol and other sterols, whereas preparations from pine trees contain only small amounts of campesterol (LEES et al. 1977). This may also explain the different dosage requirements which have been found in the past for β-sitosterol.

β-Sitosterol is available as liquid formula as granulate and recently also as tablets to be chewed. The latter preparation appears to be more agreeable for patients. The recommended daily dose is 1.5–6 g. This may be taken in divided doses, usually before meals.

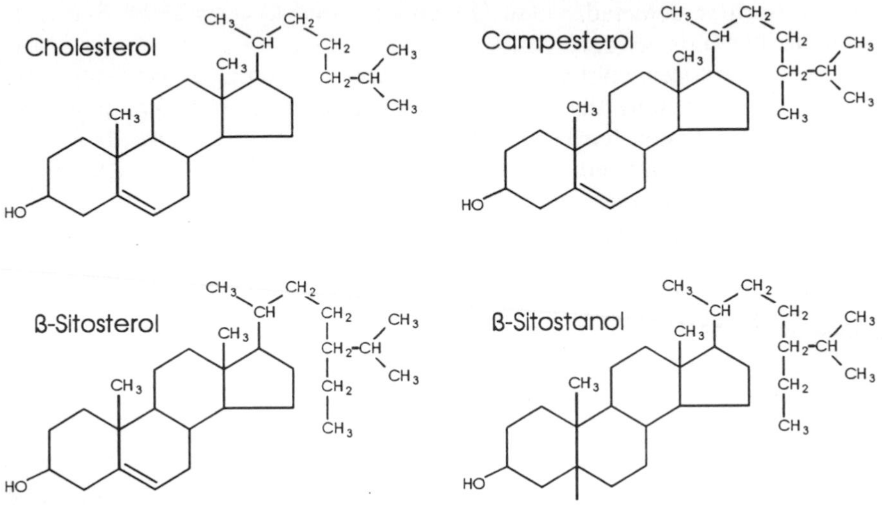

Cholesterol

Campesterol

ß-Sitosterol

ß-Sitostanol

Fig. 1. Structure of cholesterol and plant sterols

II. Absorption, Metabolism, and Excretion

β-Sitosterol is poorly absorbed from the intestine. In man, less than 5% of an oral dose is absorbed (SALEN et al. 1970), whereas about one-third to two-thirds of cholesterol are absorbed (GRUNDY and MOK 1977; SAMUEL et al. 1982; MCNAMARA et al. 1987). In rats, the fraction of β-sitosterol which is absorbed appears to be even smaller (HASSAN and RAMPONE 1979; IKEDA et al. 1988a). The serum level of β-sitosterol usually does not exceed 0.5% of total sterols. Studies in the rat indicate that intestinally absorbed β-sitosterol is transported with lymph chylomicrons. This is similar to cholesterol. β-Sitosterol, however, is only poorly esterified (12% vs. 90% of cholesterol). As expected, free β-sitosterol as well as free cholesterol is found at the surface of chylomicrons, whereas cholesterol esters are part of the chylomicron core (IKEDA et al. 1988b). This is in agreement with data of FIELD and MATHUR (1983), who show that β-sitosterol is a poor substrate for intestinal acyl coenzyme A cholesteryl acyltransferase (ACAT). This has been taken as an explanation for the selective uptake of sterols (GREGG et al. 1986).

In a rare human disease, sitosterolemia, the specificity of sterol uptake is severely reduced so that a much larger percentage of plant sterols is absorbed (BHATTACHARYYA and CONNOR 1974). The percentage of β-sitosterol which is absorbed is increased severalfold to approximately 20%–40%. A sizeable proportion of serum sterol in these patients is in fact β-sitosterol. In addition, other sterols such as campesterol are taken up in large amounts. There are indications that increased esterification of sterols other than cholesterol may occur in sitosterolemia. It should be kept in mind that β-sitosterol is not detected by routine chemistry, because the commercially available methods to determine cholesterol do not differentiate between cholesterol and other sterols. Only chromatographic methods (e.g., gas liquid chromatography) are able to separate the different sterols reliably. Patients with sitosterolemia develop xanthomatosis and early vascular disease. They should not receive β-sitosterol as a drug or dietary supplement.

There are contradictory reports on the fate of absorbed β-sitosterol. Data have been presented which show that β-sitosterol is metabolized to C24 bile acid products. However, more recent data indicated that β-sitosterol is not metabolized efficiently by 7α-hydroxylase, the key enzyme in bile acid synthesis. In vivo, there is no conversion of labeled β-sitosterol into bile acids in man. However, when 7α-hydroxysitosterol was injected, there was conversion to bile acids. These bile acids were found to be different from chenodeoxycholic acid and cholic acid (BOBERG et al. 1990). Thus, β-sitosterol appears to be a poor substrate for 7α-hydroxylase in vivo and is mostly excreted unchanged into the bile. There is, nevertheless, some conversion into water-soluble compounds that have not been further characterized (BOBERG et al. 1986). In sitosterolemia, the plasma half-life of β-sitosterol is increased (SALEN et al. 1992).

III. Effects on Plasma Lipids and Lipoproteins

1. Treatment with β-Sitosterol Alone

The number of patients in clinical trials with β-sitosterol is small. The available data consistently show that β-sitosterol is effective in lowering total serum cholesterol and low-density lipoprotein (LDL) cholesterol. Total cholesterol is usually lowered by 10%–15%, though there have been reports of cholesterol lowering in the range of 30% and more in single patients (POLLAK 1953b; BEST et al. 1954, 1955; FARQUHAR and SOKOLOW 1958; LEHMANN and BENNETT 1958; BERGE et al. 1959; AUDIER et al. 1962; OSTER et al. 1976; LEES et al. 1977; KAFFARNIK et al. 1977; SCHWARTZKOPFF and JANTKE 1978). LDL cholesterol is lowered by 15%–20%. This difference is caused by the fact that β-sitosterol does not appreciably influence high-density lipoprotein (HDL) cholesterol. Again, there have been uncontrolled trials which report an increase in HDL cholesterol. In contrast to ion exchange resins, β-sitosterol does not increase very low density lipoproteins (VLDL) or triglycerides. There appears to even be a small triglyceride-lowering effect, which, however, has not been consistently reported (BEST et al. 1955). The effect of β-sitosterol is already measurable after 1 week and increases over the first month of therapy. In Table 2, a summary of the trials with β-sitosterol is given.

In two studies, the effect of β-sitosterol on serum cholesterol was not accompanied by an effect on serum apolipoprotein B (DREXEL et al. 1981; WEISWEILER et al. 1984). The reason for this discordant behaviour of

Table 2. Trials with β-sitosterol

Author	Number of patients	Dose (g/day)	Duration	Cholesterol reduction (%)
POLLAK (1953b)	25	5–10	1 week–8 months	5–15
BEST et al. (1954)	9	5–6	22 weeks	6–20
BEST et al. (1955)	14	20–25	8–52 weeks	6–29
BERGE et al. (1959)	10	12–18	3 months	13
AUDIER et al. (1962)	60	2.5–6	4 months–2 years	23
LEHMANN and BENNETT (1958)	59	20	3–36 months	n.d.
OSTER et al. (1976)	15	24	8 weeks	12.5
KAFFARNICK et al. (1977)	20	6–18	2–12 months	10–30
LEES et al. (1977)	104	1.5–6 or 9–24[1]	2–36 months	7–12 12
BEGEMANN et al. (1978)	10	12	6–8 weeks	11
WEISWEILER et al. (1984)	10	6	2 months	10
LEDERLE (1983)	20	3	3 months	15

n.d., not determined.
[1] In the study by LEES et al. (1977), different preparations of β-sitosterol were used. The higher dose was with β-sitosterol from soybean that contained several other plant sterols. The lower dose was with tall oil (from pines).

cholesterol and apolipoprotein B is not known. However, it implies that lipoprotein particles of lower density, i.e., particles with a higher ratio of cholesterol to apolipoprotein B are reduced by β-sitosterol.

There are indications that β-sitosterol may be more effective in patients with cholesterol levels above 300 mg/dl than in patients with only moderate cholesterol elevations (AUDIER et al. 1962; BEST et al. 1955).

2. Combined Drug Treatment

In the more recent literature, there are few controlled trials available which systematically investigate the effect of β-sitosterol in combination with other drugs.

BERGE et al. (1959) report the combined treatment of hypercholesterolemia with nicotinic acid and β-sitosterol. The combined effect on serum cholesterol was a 25% reduction compared to a 13% reduction with β-sitosterol alone.

In one very recent study by BECKER et al. (1992), the effect of a combination of β-sitosterol with bezafibrate has been investigated in children with heterozygous familial hypercholesterolemia. β-Sitosterol (6 g/day) was able to lower total and LDL cholesterol by 17% compared to diet alone. It was not as effective as 400 mg/day of bezafibrate. The combination of bezafibrate with β-sitosterol, each at half the dose (200 mg/day and 3 g/day, respectively) of the single drug regimen, lowered total cholesterol by almost 40% and LDL cholesterol by almost 50%. There were no serious side effects detected during the course of 24 months. This indicates that the combination of β-sitosterol with a fibrate may be a highly effective and safe therapy for severe hypercholesterolemia.

IV. Mechanisms of Action

The major action of β-sitosterol is a reduction of cholesterol absorption from the intestine. Both the absorption of dietary cholesterol and reabsorption of biliary cholesterol are inhibited. This may explain the fact that β-sitosterol is still effective when a cholesterol-poor diet is observed. The reduced cholesterol absorption presumably leads to upregulation of LDL receptors in the liver.

The mechanism by which β-sitosterol interferes with cholesterol absorption is not entirely clear. The current hypothesis is that β-sitosterol competes with cholesterol for biliary micelles (Child and Kuksis 1986; Chijiiwa 1987; Ikeda et al. 1989; Armstrong and Carey 1987). It thus reduces cholesterol solubility in the intestine and availability for cellular uptake. The affinity of β-sitosterol to biliary micelles is higher than that of cholesterol.

There have been reports that β-sitosterol interferes with enzymes of cholesterol metabolism. Inhibition of HMG-CoA reductase activity has been described by Brown and Goldstein (1974). This finding could not be reproduced in rats by Boberg et al. (1989). Instead, these investigators found a slight upregulation of HMG-CoA reductase activity and also reported an inhibition of 7α-hydroxylase with intravenous infusion of β-sitosterol. Another important enzyme is the intestinal ACAT, for which β-Sitosterol is a poor substrate, but does not appear to inhibit cholesterol esterfication by the enzyme (Field and Mathur 1983; Ikeda et al. 1988a).

In summary, most data in the literature support the hypothesis that inhibition of cholesterol absorption is related to competition for micellar solubility in the intestinal lumen, whereas specificity of sterol uptake is related to esterification and perhaps a differential uptake at the brush border.

V. Side Effects, Safety, Tolerance, and Drug Interactions

Apart from the unpleasant taste of some preparations, β-sitosterol is well tolerated by most patients. The major complaints encountered regard nonspecific gastrointestinal discomfort (e.g., flatulence, nausea, loose stools, etc.). These symptoms are reported by approximately 5%–10% of patients. They are usually only temporary and limited and are probably not more prevalent than in placebo groups. There are no indications for any relevant long-term toxicity. β-Sitosterol has been used in patients with chronic renal failure without harmful side effects; since the drug is mostly cleared by the liver, no accumulation is to be expected (LEDERLE 1983).

Under therapy with β-sitosterol, the relative cholesterol content of the bile has been reported to be reduced (BEGEMANN et al. 1978). Thus, one would expect a lower frequency of gallstones in patients treated with this drug. Again, there are no large controlled trials available to support this hypothesis.

In vitro BOBERG et al. (1991) showed that β-sitosterol exhibits toxicity towards human umbilical vein endothelial cells in a concentration of 0.7 mmol/l. This is approximately two orders of magnitude higher than the normal plasma level of β-sitosterol. However, it is in the range of β-sitosterol levels in sitosterolemia and may relate to the observed early atherosclerosis in these patients.

One important aspect to consider is that β-sitosterol may be resorbed in some patients. Whether there are patients besides those with classical sitosterolemia who resorb relevant amounts of β-sitosterol and whether this is potentially harmful is not known to date. Parents of patients with sitosterolemia who are considered to be obligate heterozygotes have been reported to absorb approximately 15% of an oral β-sitosterol load compared to 5% in controls. However, there was no accumulation of β-sitosterol, due to a more efficient excretion of the substance, which indicates that decreased clearance of β-sitosterol may also contribute to sitosterolemia (SALEN et al. 1992). β-Sitosterol in plasma was only minimally increased in heterozygotes and remained below 1 mg/dl.

There are indications that the level of phytosterols in serum is correlated to cardiovascular risk. It is not known whether increased phytosterols are only a marker for a subset of patients with hypercholesterolemia and increased sterol absorption or whether they are causally related to atherosclerosis (GLUECK et al. 1991).

There are no known drug interactions of β-sitosterol to date. It should be kept in mind, however, that there are no large controlled trials with β-sitosterol and that no systematic investigation of drug interactions has been performed. There is one report by SCHWARTZKOPFF and JANTKE (1980) which shows that there is no interference with the absorption of digitalis glycosides, as it is known for ion exchange resins.

VI. Conclusions

β-Sitosterol has consistently been shown to reduce serum cholesterol and LDL cholesterol by reducing intestinal cholesterol absorption. The exact mechanism of this action is not known. However, it appears to be dependent on competition with cholesterol for biliary micelles or blockade of resorptive pathways. The composition of β-sitosterol preparations is relevant in this respect: increasing amounts of other plant sterols seem to reduce the effect of β-sitosterol. Triglycerides are either not changed or slightly lowered by treatment with β-sitosterol.

The drug has very few serious side effects and an apparently negligible long-term toxicity, which makes it an interesting alternative or supplement to other cholesterol-lowering agents. The only situation which forbids treatment with β-sitosterol is a rare inherited disorder of sterol metabolism, sitosterolemia. Patient compliance is directly related to the palatability of β-sitosterol preparations. This has been a major problem, but there are preparations available which are more acceptable to patients.

The major application of β-sitosterol alone should be the treatment of mild hypercholesterolemia in patients who, for some reason, cannot tolerate other available drugs or for whom these are considered inappropriate. Thus, β-sitosterol may be used as an alternative to ion exchange resins or in pediatric patients. In addition, β-sitosterol may be useful in combination with HMG-CoA reductase inhibitors or fibrates. Clearly, this drug deserves more attention than it has previously had.

Recently HEINEMANN et al. (1986) reported that β-sitostanol (Fig. 1) has similar effects. However, this drug is not currently available for general therapeutic purposes. It has some potential advantages compared to β-sitosterol. One such advantage is that it is almost unabsorbable. β-Sitostanol appears to inhibit cholesterol absorption even more than β-sitosterol (HEINEMANN et al. 1991). Clinical research with this agent will show whether the pharmacological actions of plant sterols can be improved with this substance.

C. Neomycin

Oral administration of small doses of the nonabsorbable aminoglycoside antibiotic neomycin, which is widely used in the treatment of hepatic coma, can effectively reduce serum cholesterol levels in man, first demonstrated by SAMUEL and STEINER in 1959. However, because of potential serious side effects, neomycin is not a drug of first choice for the treatment of hypercholesterolemia. This section describes the pharmacology, mechanisms of action, and clinical experience of neomycin as a cholesterol-lowering agent.

I. Chemistry, Physical Properties, and Dosage Form

Neomycin was isolated 1949 by WAKSMAN and LECHEVALIER from cultured *Streptomyces fradiae* as an antibiotic complex (neomycin A, B, and C). Among the various froms of neomycin, only neomycin B (also called framycetin) is of clinical importance.

Neomycin – $C_{23}H_{46}O_{13}N_6$ – with a molecular mass of 614.65 Da is a polybasic member of the aminoglycoside series of antibiotics. It is a water-soluble compound which forms amorphous and crystalline salts, as hydrochlorides of sulfates. Neomycın sulfate is highly soluble in water, but is insoluble in organic solvents as aceton, ethanol ether, or chloroform. It is thermostabile and can be stored at room temperature for more than 1 year.

As a lipid-lowering drug, neomycin is administered orally in a dose of 1–2 g/day in divided doses.

II. Absorption, Metabolism, and Excretion

Neomycin is bound to serum proteins up to 45%–55%. Given orally, neomycin is only very little absorbed. After a single dose of 3 g, only 1–4 μg/g of neomycin can be detected in the blood, whereas 97% of the orally administered drug is found unmetabolized in the feces in a concentration of more than 10 mg/g (KUNIN 1960).

After 48 h, renal excretion is between 0.5% and 1% of the oral dose (BREEN et al. 1972). When neomycin is administered intravenously, 30%–50% of the drug is excreted by glomerular filtration. Therefore, one has to be aware that even when given orally, the drug can accumulate in patients with renal insufficiency and cause nephrotoxicity and ototoxicity (HUBMANN 1965; JAWETZ 1956; KUNIN 1960; see side effects).

III. Effects on Plasma Lipids and Lipoproteins

1. Treatment with Neomycin Alone

The first report by SAMUEL and coworkers about the cholesterol-lowering effect of neomycin dates back to 1959 (SAMUEL and STEINER 1959). Other studies performed in the meantime confirmed SAMUEL's findings that in hyperlipidemic patients a daily oral dose of 1.5–6 g neomycin can cause a 20%–30% decrease of total and LDL cholesterol (SAMUEL and WAITHE 1961; SAMUEL et al. 1968, 1973; SEDAGHAT et al. 1975; MIETTINEN 1979; KESÄNIEMI and GRUNDY 1984; HOEG et al. 1984).

The cholesterol-lowering effect is already observed after 1 week of treatment and is still apparent when patients receive long-term treatment of more than 1 year. An increase in the neomycin dose from 1.5 to 2–3 g/day does not lead to a comparable lowering of serum cholesterol. When patients are treated at a dosage of 6 g/day, an additional reduction (9%) of the cholesterol level to an overall fall of 29% can be observed. This is usually accompanied by an increased frequency of gastrointestinal side effects. Even though the cholesterol-lowering efficacy of neomycin is impressive and consistent, it appears surprising that only few studies (often with very few patients) were performed and that ellaborate analyses of the influence on apolipoproteins are missing in the literature.

Whereas neomycin does not seem to effect triglyceride levels, a slight but significant decrease of HDL cholesterol is observed in some studies (HOEG et al. 1984; KESÄMIEMI and GRUNDY 1984).

2. Combined Drug Treatment

The combined treatment of neomycin with other cholesterol-lowering drugs was examined in only very few trials. Although additional cholesterol-lowering effects have been reported for the combination of neomycin with clofibrate, niacin, cholestyramine, and lovastatin, one has to be aware that single studies with few patients were performed for each combination. Therapies of neomycin combined with probucol have to our knowledge not been reported.

a) Neomycin and Niacin

The addition of niacin to the neomycin therapy resulted in an additional 9% and 12% decrease of total and LDL cholesterol (HOEG et al. 1984). This was paralleled by a 17% increase in the HDL cholesterol concentration. However, the decrease of total cholesterol (36%) and LDL cholesterol (45%) in response to the combined treatment was only observed in compliant patients who tolerated this combination. Several patients could not tolerate the additional niacin therapy because of various side effects. Even though the use of combined neomycin and niacin treatment resulted in this impressive cholesterol decrease, only one study of this combination has so far been reported. Other long-term prospective studies on this combination therapy are still missing.

One interesting observation is also that this combination can lower increased levels of Lp(a) (GURAKAR et al. 1985). Two grams neomycin a day alone can lower Lp(a) by approximately 24%. The combination with 3 g niacin per day increases this effect to 45%. So far this is the only reasonable drug combination which will lower Lp(a). This has potentially relevant implications, since Lp(a) is strongly and independently correlated with the risk for cardiovascular disease.

b) Neomycin and Cholestyramine

The combined therapy of neomycin and cholestyramine was investigated in one study with 17 patients (MIETTINEN 1979). This combination treatment resulted in an overall total cholesterol reduction of 38%. The trial was mainly performed to examine the mechanism of action of neomycin and will be described in more detail below. Similar results were obtained by HOEG et al. (1985) with 11 patients. Compared to cholestyramine alone (−32%), total cholesterol was lowered by 39% when cholestyramine and neomycin were combined. The additional LDL-lowering efficacy compared to cholestyramine alone was only minimal (39% vs. 38%). Thus, this combination has no advantage compared to cholestyramine alone.

c) Neomycin and Clofibrate

The individual and combined effect of neomycin and clofibrate on serum cholesterol and triglyceride levels was demonstrated in 16 patients (SAMUEL et al. 1970). Combination of the two drugs caused an effective cholesterol reduction in one-third of the patients of up to 38% from control value. Since triglyceride levels in this study were normal in most patients, the effect of the drugs on triglycerides was moderate. LDL and HDL cholesterol as well as apolipoproteins were not determined in this trial. Since there are no data available regarding the combination of neomycin with one of the second generation fibrates, one has to assume that these combinations are perhaps similarly potent. It is difficult to draw any conclusions or even give recommendations from data of such a limited number of patients; therefore, more clinical trials with a combination therapy of neomycin and fibrates are needed.

d) Neomycin and Lovastatin

The combination of lovastatin and neomycin was tested in a trial by HOEG et al. (1986), because of the complementary mechanisms of these two drugs. It was postulated that the cholesterol-lowering effect of lovastatin, due to its competitive inhibition of the enzyme HMG-CoA reductase catalyzing the rate-limiting step in cholesterol biosynthesis and subsequent upregulation of the hepatic LDL receptors, could be increased by the neomycin-mediated impaired intestinal cholesterol absorption. The addition of neomycin to lovastatin caused a small further reduction of total (5%) and LDL cholesterol (4%), but also resulted in a decrease of HDL cholesterol levels (19%). Both drugs had no influence on VLDL cholesterol or triglyceride levels, whether given alone or in combination. Again no data on the combination with other HMG-CoA reductase inhibitors are available. However, it may be assumed that these combinations will have lipid-lowering effects similar to those of the combination of neomycin with lovastatin.

Because of the small additional effect of neomycin in this combination – as compared to cholestyramine – and the effect on HDL cholesterol levels, this combination therapy cannot be recommended.

IV. Mechanisms of Action

Since neomycin precipitates dihydroxy bile acids in vitro (DESOMER et al. 1964; EYSSEN et al. 1966; VAN DEN BOSCH and CLAES 1967; THOMPSON et al. 1970; CAYEN 1970), it was first postulated that its cholesterol-lowering effect is caused by an increased fecal bile acid excretion. However, the major mechanism leading to a plasma cholesterol reduction by neomycin appears to be a decreased intestinal absorption of cholesterol. When neomycin was administered in low dosages of 2–3 g/day, a marked decrease of exogenous cholesterol as well as an increase in excretion of fecal neutral steroids were demonstrated (MIETTINEN 1979; SEDAGHAT et al. 1975). The decreased cholesterol absorption included not only the cholesterol in the diet, but also the large amount of cholesterol excreted in the bile. Therefore, it was suggested that the increased excretion of neutral steroids was also due to a reduced absorption of endogenous biliary cholesterol. Increased fecal excretion of bile acids could only be observed when high doses of neomycin were used (MIETTINEN 1979). Patients who received high dosages of neomycin (8–12 g/day) were reported to have severe malabsorption syndromes and steatorrhea with possible morphological alterations of the intestinal mucosa and interference with the action of pancreatic lipase and intestinal disaccharidases (JACOBSON et al. 1960a,b).

Turnover studies by KESÄNIEMI and GRUNDY (1984) demonstrated that the neomycin-mediated reduction in cholesterol absorption in the intestine causes a decrease in synthesis of the apolipoprotein of LDL. The decrease in plasma apo-LDL level was positively correlated with the decrease in apo-LDL synthetic rate; an increase in LDL clearance rates was only observed in a few patients. It was also proposed that some of the increment of fecal steroids are explained by a mobilization of cholesterol from tissue pools (SEDAGHAT et al. 1975). The decrease in HDL cholesterol in response to neomycin therapy may be caused by a reduction of cholesterol content of chylomicrons, which are a source of HDL cholesterol during lipolysis.

V. Side Effects, Safety, Tolerance, and Drug Interactions

Earlier studies in the 1950s, when neomycin was tested as an antibiotic drug parenterally, demonstrated that the drug is highly toxic. Elevated plasma levels of neomycin can be nephrotoxic with renal-tubular lesions. In addition, the drug can cause damage to hair cells of the organ of Corti, which sometimes leads to permanent hearing loss or deafness. Therefore, soon after neomycin became available on the market, the parenteral use of the drug was no longer recommended.

However, in cholesterol-lowering trials with the oral administration of low-dose neomycin, only a few side effects have been reported. Among these, diarrhea and nausea were the most frequent. Even though neomycin is only poorly absorbed, this may be sufficient to cause oto- and nephrotoxicity in some patients (HUBMANN 1965). Since plasma neomycin is excreted through the kidneys, the drug should by no means be given to patients with renal insufficiency.

When neomycin was given at larger dosages (6–12 g/day) in patients with hepatic coma or before abdominal surgery, the following side effects have been reported: diarrhea, renal damage, staphylococcal enterocolitis, moniliasis, and multiresistent coliform overgrowth (SAMUEL 1979).

Steatorrhea caused by large doses of neomycin may be accompanied by an increased fecal loss of sodium and potassium; hypokalemia, therefore, may ensue. In addition, oral treatment with neomycin may be accompanied by a reduced absorption of vitamin K and vitamin B_{12}, disaccharides, monosaccharides, and metal ions such as iron. Reduced plasma levels of vitamin A and E have also been observed.

SAMUEL (1979) suggested that before neomycin therapy is started, urinalysis, blood urea nitrogen and creatinine determinations, liver function tests, and complete blood counts should be performed and that these parameters should be controlled monthly under neomycin therapy. It has also been suggested that regular hearing tests should be performed during therapy with neomycin (HOEG et al. 1984).

Neomycin may increase the anticoagulant effects of vitamin K antagonists because of the reduced intestinal absorption of vitamin K. Reduced absorption of digoxin was also reported under neomycin therapy.

VI. Conclusions

Neomycin is a potent drug for the treatment of hypercholesterolemia. It was demonstrated that an oral, low-dose therapy of 1–3 g neomycin per day can cause a 20%–30% reduction of total and LDL cholesterol, probably mediated by a reduced absorption of cholesterol. However, the reduction of total cholesterol is also accompanied by a slight, but significant reduction of HDL cholesterol. Until now, the combination of neomycin with niacin appears to be the only useful drug regimen for increased Lp(a). This potentially relevant action of neomycin clearly deserves further evaluation.

Nevertheless, neomycin is not a drug of first choice. Even though the few lipid-lowering trials with neomycin revealed no serious side effects, it has to be taken into account that the drug can be highly toxic. If plasma levels of neomycin are increased, oto- and nephrotoxicity may occur, which may lead to deafness or impairment of renal function. As lipid-lowering therapy is a lifelong therapy, long-term toxicity must be minimal. Unlike patients being enrolled in clinical trials, the normal ambulatory patient cannot be closely followed up with routine weekly or monthly laboratory tests, as has been suggested for patients on neomycin. In addition, regular audiometric testing may not be feasible for the majority of patients. Thus, neomycin should be restricted to patients who cannot tolerate the first-line drugs or be treated satisfactorily with these drugs. It is important to closely monitor these patients for signs of nephrotoxicity and ototoxicity. This means that neomycin therapy for cholesterol lowering will be a domain of specialized lipid clinics.

D. Ketoconazole

Ketoconazole was introduced as systemic antifungal medication in the late 1970s (THIENPONT et al. 1979). Compared to amphotericin B, the drug was well tolerated and it had a broad antifungal spectrum (VAN CUTSEM 1983). Its cholesterol-lowering effect was detected soon after its introduction (MIETTINEN and VALTONEN 1984; KRAEMER and PONT 1986). Ketoconazole has not gained wide acceptance as a cholesterol-lowering agent due to some potentially serious side effects.

Ketoconazole belongs to a group of antifungal drugs known as azoles. Due to its relatively improved solubility in acidic aqueous solution, it may be used for systemic, rather than topical administration. Its antifungal activity seems to be related to the inhibition of cytochrome P450 enzymes and the concomitant inhibition of fungal sterol and steroid synthesis (LOOSE et al. 1983).

The normal dosage for treatment of fungal infections in adults is 400 mg/day. This may be increased to 800 mg/day if necessary. For inhibition of steroid hormone production in the treatment of prostatic cancer, higher dosages (1.2 g/day) have been used (TRACHTENBERG and PONT 1984). There is not yet a clearcut dosage recommendation for lipid-lowering purposes.

Ketoconazole is absorbed from the intestine in variable amounts. The uptake is dependent on pH, since dissolution requires an acidic environment. Thus, agents that increase gastric pH will reduce the bioavailability of ketoconazole.

In plasma, ketoconazole is bound to proteins (approximately 80%) and red cells (15%). Only 1% is free (CRAVEN et al. 1983). Ketoconazole is eliminated primarily by the liver with a half-life of approximately 6 h. There have been reports of plasma half-lives of between 1 and 11 h. It appears that elimination is dose dependent, so that with higher plasma levels an increased half-life is encountered (DANESHMEND and WARNOCK 1988). Drugs inducing hepatic microsomal enzymes such as barbiturates or rifampin will increase the elimination of ketoconazole.

Ketoconazole has been shown to reduce serum cholesterol by 18%–30% and LDL cholesterol by 20%–40%. The effect appears to be dose dependent. A dosage of 400 mg/day reduced serum cholesterol by almost 20% and LDL cholesterol by 22% in familial hypercholesterolemia (GYLLING et al. 1991). When a dosage of 1.2 g/day was given, cholesterol lowering was at the high end of the given range (MIETTINEN 1988). In the higher dosage range, profound effects on steroid hormone balance have to be expected. Addition of cholestyramine to the high-dose therapy increases the cholesterol-lowering effect by another 40%, resulting in a total reduction of serum cholesterol of 65%. With the lower dosage of 400 mg/day in conjunction with cholestyramine in patients with familial hypercholesterolemia, total cholesterol was lowered by 32% and LDL cholesterol by 41% (GYLLING et al. 1991).

The major mechanism of action appears to be the inhibition of cholesterol synthesis by inhibiting 14α-demethylation of lanosterol (VAN DEN BOSSCHE et al. 1980). Thus, ketoconazole may be regarded as a cholesterol-synthesis inhibitor. At the same time, ketoconazole interferes with the absorption of cholesterol from the intestine (KESÄNIEMI and MIETTINEN 1991). Since it also inhibits steroid hormone synthesis, it interferes with normal hormone production (PONT et al. 1982a,b).

Several potentially serious side effects make the use of ketoconazole as a drug for the prevention of vascular disease questionable. Approximately one-fifth of patients receiving 400 mg/day complain about gastrointestinal discomfort, including nausea. This problem is dose dependent and may be overcome by administration of the drug with meals (SUGAR et al. 1987). A small percentage (3%–5%) of patients experience an allergic rash. Hepatic disturbances reflected in an elevation of transaminases occur in 5%–10% of patients. In rare cases (approximately one in 15 000), drug-induced hepatitis may occur (LAKE-BAKAAR et al. 1987). The cause of this complication is unknown. Since it is not dose related, it has been hypothesized to be immunologic in origin. However, biochemical disturbances of liver function may also explain this side effect. Fatal outcomes have been reported. The symptoms are similar to hepatitis from other causes. Patients should be alerted to this possibility so that ketoconazole can be discontinued immediately. This very serious complication has prompted recommendations not to use ketoconazole for trivial fungal infections.

Ketoconazole is used for pharmacological hormone ablation in prostatic cancer (TRACHTENBERG and PONT 1984). It is known that large doses cause severe disturbances of steroid hormone balance (PONT et al. 1982a,b; BRITTON et al. 1988). In the therapeutic range for fungal infections, one still encounters around 10% of women reporting menstrual cycle abnormalities. Since ketoconazole reduces plasma testosterone, it may lead to hypogonadism during long-term therapy. Gynecomastia has been reported in two male patients on long-term therapy (DEFELICE et al. 1981).

In rats, ketoconazole is teratogenic, so that therapy for hypercholesterolemia with this drug should not be considered in women of child-bearing age unless reliable contraceptive measures are taken.

In summary, even though ketoconazole is effective in cholesterol lowering, it cannot be recommended until larger studies have shown that the known side effects do not outweigh any benefit in terms of reduced cardiovascular risk. Compared to the available drugs, there are no advantages of ketoconazole. Considering the current level of research, more widespread use in the treatment of hyperlipoproteinemia must be discouraged. One other potential problem with ketoconazole, the increase in serum lanosterol (MIETTINEN 1988), is related to its mechanism of action. It is not know whether this is related to any adverse effects.

It would be interesting to know whether it is possible to design related azole drugs with lower toxicity which retain cholesterol-lowering activity. There is one recent report of the development of a novel lanosterol 14α-demethylase inhibitor (MAYER et al. 1991). This drug, designated SKF 104976, reduces cholesterol synthesis in vitro. At the same time, the accumulation of lanosterol appears to decrease HMG-CoA reductase activity. More information is needed, however, before the potential of drugs with this mode of action can be evaluated in more detail. Their safety profile in particular remains to be determined.

E. D-Thyroxine

D-thyroxine is a stereoisomer of the thyroid hormone L-thyroxine. It has been shown to lower cholesterol by approximately 10%–20% in hypercholesterolemic patients (BECHTOL and WARNER 1969). Studies on the cholesterol-lowering mechanisms of D-thyroxine demonstrated that it has retained the ability of L-thyroxine to increase LDL receptor synthesis and expression. On the other hand, most of the effects of thyroid hormones are reduced, i.e., on a molar basis, D-thyroxine is a much less potent thyroid hormone than L-thyroxine. However, in the dosage range used to lower cholesterol (4–8 mg/day), some thyroid hormone activity is still present with the typical effects: restlessness, insomnia, sensitization to β-adrenergic stimuli, weight loss, etc. The influence on catecholamine effects leads to an increased incidence of cardiovascular side effects such as arrhythmias and anginal attacks in patients with CHD. Impaired glucose tolerance has been reported.

In the CORONARY DRUG PROJECT (1972, 1975), a trial of secondary prevention in men after myocardial infarction, treatment with D-thyroxine had to be terminated early due to an increased number of adverse events and reinfarctions and an increased mortality after a mean follow-up of 3 years in this group. Thus, D-thyroxine is not recommended for patients with CHD. It has been proposed that D-thyroxine be restricted to young individuals with severe hypercholesterolemia. There are no long-term data for these patients available, so that it is not known whether the number of adverse effects in this group might also offset any beneficial effect of lipid lowering. In addition, it remains unclear how CHD is excluded in this high-risk group before therapy with D-thyroxine is started.

From the available data on D-thyroxine and the current therapeutic alternatives, it appears to the authors that D-thyroxine should no longer be recommended for treatment of hypercholesterolemia. If, however, it were possible to separate the effects on lipid metabolism of thyromimetics from the cardiac and central nervous system effects, these agents might become potentially relevant drugs. One such report has been made, and it remains to be seen whether this approach will lead to new medications (UNDERWOOD et al. 1986).

References

Albers JJ, Taggart HM, Applebaum-Bowden D, Haffner S, Chestnut C III, Hazzard WR (1984) Reduction of lecithin-cholesterol acyltransferase, apolipoprotein D and the Lp(a) lipoprotein with the anabolic steroid stanozolol. Biochim Biophys Acta 795:293–301

Armstrong MJ, Carey MC (1987) Thermodynamic and molecular determinants of sterol solubilities in bile salt micelles. J Lipid Res 28:1144–1155

Audier M, Pastor J, Pauli AM, Poggi L (1962) Essai de traitement de l'athérosclérose par le beta-sitosterol. Rev Med Fr 43:7–12

Bechtol LD, Warner WL (1969) Dextrothyroxine for lowering serum cholesterol. Angiology 20:565–579
Becker M, Staab D, von Bergmann K (1992) Long-term treatment of severe familial hypercholesterolemia in children: effect of sitosterol and bezafibrate. Pediatrics 89:138–142
Begemann F, Bandomer G, Herget HJ (1978) The influence of β-sitosterol on biliary cholesterol saturation and bile acid kinetics in man. Scand J Gastroenterol 13:57–63
Berge KG, Achor RWP, Barker NW, Power H (1959) Comparison of the treatment of hypercholesterolemia with nicotinic acid, sitosterol, and safflower oil. Am Heart J 58:849–853
Best MM, Duncan CH, van Loon EJ, Wathen JD (1954) Lowering of serum cholesterol by the administration of a plant sterol. Circulation 10:201–206
Best MM, Duncan CH, van Loon EJ, Wathen JD (1955) The effects of sitosterol on serum lipids. Am J Med 19:61–70
Bhattacharyya AK, Connor WE (1974) β-Sitosterolemia and xanthomatosis: a newly described lipid storage disease in two sisters. J Clin Invest 53:1033–1043
Billheimer JT, Gillies PJ (1990) Intracellular cholesterol esterification. In: Esfahani M, Swaney JB (eds) Advances in cholesterol research. Telford, Caldwell, pp 7–46
Boberg KM, Skrede B, Skrede S (1986) Metabolism of 24-ethyl-4-cholesten-3-one and 24-ethyl-5-cholesten-3 β-ol (sitosterol) after intraperitoneal injection in the rat. Scand J Clin Lab Invest Suppl 184:47–54
Boberg KM, Akerlund JE, Bjorkhem I (1989) Effect of sitosterol on the rate limiting enzymes in cholesterol synthesis and degradation. Lipids 24:9–12
Boberg KM, Einarsson K, Bjorkhem I (1990) Apparent lack of conversion of sitosterol into C24-bile acids in humans. J Lipid Res 31:1083–1088
Boberg KM, Pettersen KS, Prydz H (1991) Toxicity of sitosterol to human umbilical vein endothelial cells in vitro. Scand J Clin Lab Invest 51:509–516
Breen KJ, Bryant RE, Levinson JD, Shenker S (1972) Neomycin absorption in man. Ann Intern Med 76:211–218
Britton H, Shebab Z, Lightner E, New M, Chow D (1988) Adrenal response in children receiving high doses of ketoconazole for systemic coccidiomycosis. J Pediatr 112:488–492
Brown MS, Goldstein JL (1974) Suppression of 3-hydroxy-3-methyl-glutaryl coenzyme A reductase activity and inhibition of growth of human fibroblasts by 7-ketocholesterol. J Biol Chem 249:7306–7314
Canner PL, Berg KG, Wenger NK, Stamler J, Friedman L, Prineas RJ, Friedewald W (1986) Fifteen year mortality in coronary drug project patients: long term benefit with niacin. J Am Coll Cardiol 8:1245–1255
Cayen MN (1970) Agents affecting lipid metabolism. 38. Effect of neomycin on cholesterol biosynthesis and bile acid precipitation. Am J Clin Nutr 23:1234–1240
Chijiiwa K (1987) Distribution and partitioning of cholesterol and beta-sitosterol in micellar bile salt solutions. Am J Physiol 253:G268–G273
Child P, Kuksis A (1986) Investigation of the role of micellar phospholipid in the preferential uptake of cholesterol over sitosterol by dispersed rat jejunal villus cells. Biochem Cell Biol 64:847–853
Coronary Drug Project Research Group (1970) The Coronary Drug Project: initial findings leading to modifications of its research protocol. JAMA 214:1303–1313
Coronary Drug Project Research Group (1972) The Coronary Drug Project: initial findings leading to further modifications of its protocol with respect to dextrothyroxine. JAMA 220:996–1008
Coronary Drug Project Research Group (1975) Clofibrate and niacin in coronary heart disease. JAMA 231:360–381
Craven PC, Graybill JR, Jorgensen JH, Dismukes WE, Levine BE (1983) High-dose kotoconazole for treatment of fungal infections of the central nervous system. Ann Intern Med 98:160–167

Daneshmend TK, Warnock DW (1988) Clinical pharmacokinetics of ketoconazole. Clin Pharmacokinet 14:13–34
DeFelice R, Johnson DG, Galgiani JN (1981) Gynecomastia with ketoconazole. Antimicrob Agents Chemother 19:1073–1074
DeSomer P, Vanderhaeghe H, Eyssen H (1964) Influence of basic antibiotics on serum- and liver-cholesterol concentrations in chicks. Nature 204:1306
Drexel H, Breier C, Lisch HJ, Sailer S (1981) Lowering plasma cholesterol with β-sitosterol and diet. Lancet 1:1157–1158
Eyssen H, Evrard E, Vanderhaeghe H (1966) Cholesterol-lowering effects of N-methylated neomycin and basic antibiotics. J Lab Clin Med 68:753–768
Farquhar JW, Sokolow M (1958) Response of serum lipids and lipoproteins of man to beta-sitosterol and safflower oil. A long term study. Circulation 17: 890–899
Field FJ, Mathur SN (1983) β-sitosterol: esterification by intestinal acylcoenzyme A: cholesterol acyltransferase (ACAT) and its effect on cholesterol esterification. J Lipid Res 24:409–417
Glueck CJ, Speirs J, Tracy T, Streicher P, Iμig E, Vandegrift J (1991) Relationships of serum plant sterols (phytosterols) and cholesterol in 595 hypercholesterolemic subjects, and familial aggregation of phytosterols, cholesterol, and premature coronary heart disease in hyperphytosterolemic probands and their first-degree relatives. Metabolism 40:842–848
Gregg RE, Connor WE, Lin DS, Brewer HB Jr (1986) Abnormal metabolism of shellfish sterols in a patient with sitosterolemia and xanthomatosis. J Clin Invest 77:1864–1872
Grundy SM, Mok HYI (1977) Determination of cholesterol absorption in man by intestinal perfusion. J Lipid Res 18:263–271
Gurakar A, Hoeg JM, Kostner G, Papadopoulos NM, Brewer HB Jr (1985) Levels of lipoprotein Lp(a) decline with neomycin and niacin treatment. Atherosclerosis 57:293–301
Gylling H, Vanhanen H, Miettinen TA (1991) Hypolipidemic effect and mechanism of ketoconazole without and with cholestyramine in familial hypercholesterolemia. Metabolism 40:35–41
Hasibeder H, Staab HJ, Seibel K, Heibel B, Schmidle G, März W (1991) Clinical pharmacology of the hypocholesterolemic agent K 12.148 (lifibrol) in healthy volunteers. Eur J Clin Pharmacol 40 Suppl 1:S91–S94
Hassan AS, Rampone AJ (1979) Intestinal absorption and lymphatic transport of cholesterol and beta-sitosterol in the rat. J Lipid Res 20:646–653
Heinemann T, Leiss O, von Bergmann K (1986) Effect of low-dose sitostanol on serum cholesterol in patients with hypercholesterolemia. Atherosclerosis 61:219–223
Heinemann T, Kullak-Ublick GA, Pietruck B, von Bergmann K (1991) Mechanisms of action of plant sterols on inhibition of cholesterol absorption. Comparison of sitosterol and sitostanol. Eur J Clin Pharmacol 40 Suppl 1:S59–S63
Hoeg JM, Maher MB, Bou E, Zech LA, Bailey KR, Gregg RE, Sprecher DL, Susser JK, Pikus AM, Brewer HB Jr (1984) Normalization of plasma lipoprotein concentrations in patients with type II hyperlipoproteinemia by combined use of neomycin and niacin. Circulation 70:1004–1011
Hoeg JM, Maher MB, Bailey KR, Zech LA, Gregg RE, Sprecher DL, Brewer HB Jr (1985) Effects of combination cholestyramine–neomycin treatment on plasma lipoprotein concentrations. Am J Cardiol 55:1282–1286
Hoeg JM, Maher MB, Bailey KR, Brewer HB Jr (1986) The effects of mevinolin and neomycin alone and in combination on plasma lipid and lipoprotein concentrations in type II hyperlipoproteinemia. Atherosclerosis 60:209–214
Hubmann R (1965) Irreversible Ertaubung nach Neomyzinbehandlung bei normaler Nierenfunktion. Urologe 4:27–28
Ikeda I, Tanaka K, Sugano M, Vahouny GV, Gallo LL (1988a) Inhibition of cholesterol absorption in rats by plant sterols. J Lipid Res 29:1573–1582

Ikeda I, Tanaka K, Sugano M, Vahouny GV, Gallo LL (1988b) Discrimination between cholesterol and sitosterol for absorption in rats. J Lipid Res 29: 1583–1591
Ikeda I, Tanabe Y, Sugano M (1989) Effects of sitosterol and sitostanol on micellar solubility of cholesterol. J Nutr Sci Vitaminol 35:361–369
Jacobson ED, Prior JT, Faloon WW (1960a) Malabsorptive syndrome induced by neomycin: morphologic alterations in the jejunal mucosa. J Lab Clin Med 56:245–250
Jacobson ED, Chodos RB, Faloon WW (1960b) An experimental malabsorption syndrome induced by neomycin. Am J Med 28:524–533
Jawetz E (1956) Polymyxin, neomycin, bacitracin. Medical Encyclopedia, New York (Antibiotics monograph, no 5)
Kaffarnik H, Mühlfellner G, Mühlfellner O, Schneider J, Hausmann L, Zöfel P, Schubotz R, Fuchs F (1977) Beta-Sitosterin in der Behandlung essentieller Hyperlipoproteinämien vom Typ II. Fortschr Med 95;2785–2787
Kesäniemi YA, Grundy SM (1984) Turnover of low density lipoproteins during inhibition of cholesterol absorption by neomycin. Arteriosclerosis 4:41–48
Kesäniemi YA, Miettinen TA (1991) Inhibition of cholesterol absorption by neomycin, nenzodiazepine derivatives and ketoconazole. Eur J Clin Pharmacol 40 Suppl 1:65–67
Kraemer FB, Pont A (1986) Inhibition of cholesterol synthesis by ketoconazole. Am J Med 80:616–622
Kunin CM (1960) Absorption of orally administered neomycin. N Engl J Med 262:380–385
Lake-Bakaar G, Scheuer PJ, Sherlock S (1987) Hepatic reactions associated with ketoconazole in the United Kingdom. Br Med J 294:419–422
Lederle RM (1983) Langzeitbehandlung der Hypercholesterinämie bei terminaler Niereninsuffizienz. Med Klin 78:136–137
Lees AM, Mok HYI, Lees RS, McCluskey A, Grundy SM (1977) Plant sterols as cholesterol-lowering agents. Clinical trials in patients with hypercholesterolemia and studies on sterol balance. Atherosclerosis 28:325–338
Lehmann JH, Bennett BM (1958) Effect of sitosterol on survival and recurrence rates in myocardial infarction. Circulation 18:747–748
Loose DS, Kan PB, Hirst MA, Marcus RA, Feldman D (1983) Ketoconazole blocks adrenal P450-dependent enzymes. J Clin Invest 71:1495–1499
März W, Scharnagl H, Biemer G, Schliack M, Siekmeier R, Löser R, Seibel K, Gross W (1992) The mechanism of action of lifibrol (K12.148). 4th International Symposium on treatment of severe dyslipoproteinemia in the prevention of coronary heart disease, Munich
Mayer RJ, Adams JL, Bossard MJ, Berkhout TA (1991) Effects of a novel lanosterol 14α-demethylase inhibitor on the regulation of 3-hydroxy-3-methylglutaryl-coenzyme A reductase in Hep G2 cells. J Biol Chem 266:20070–20078
McNamara DJ, Kolb R, Parker TS, Batvin H, Samuel P, Brown CD, Ahrens EH Jr (1987) Heterogeneity of cholesterol homeostasis in man. J Clin Invest 79: 729–739
Miettinen TA (1979) Effects of neomycin alone and in combination with cholestyramine on serum cholesterol and fecal steroids in hypercholesterolemic subjects. J Clin Invest 64:1485–1493
Miettinen TA (1988) Cholesterol metabolism during ketoconazole treatment in man. J Lipid Res 29:43–51
Miettinen TA, Valtonen VV (1984) Ketoconazole and cholesterol synthesis. Lancet 2:1271
Oster P, Schlierf G, Heuck CC, Greten H, Gundert-Remy U, Haase W, Klose G, Nothelfer A, Raetzer H, Schellenberg B, Schmidt-Gayk H (1976) Sitosterin bei familiärer Hyperlipoproteinämie Typ II. Eine randomisierte, gekreuzte Doppelblindstudie. Dtsch Med Wochenschr 101:1308–1311

Peterson DW (1951) Effect of soybean sterols in the diet on plasma and liver cholesterol in chicks. Proc Soc Exp Biol Med 78:143–145

Pollak OJ (1953a) Successful prevention of experimental hypercholesterolemia and cholesterol atherosclerosis in the rabbit. Circulation 7:696–701

Pollak OJ (1953b) Reduction of blood cholesterol in man. Circulation 7:702–706

Pont A, Williams PL, Azhar S, Reitz RE, Bochra C, Smith ER, Stevens DA (1982a) Ketoconazole blocks testosterone synthesis. Arch Intern Med 142:2137–2140

Pont A, Williams PL, Loose DS, Feldman D, Reitz RE, Bochra C, Stevens DA (1982b) Ketoconazole blocks adrenal steroid synthesis. Ann Intern Med 97:370–372

Salen G, Ahrens EH Jr, Grundy SM (1970) Metabolism of β-sitosterol in man. J Clin Invest 49:952–967

Salen G, Tint GS, Shefer S, Shore V, Nguyen L (1992) Increased sitosterol absorption is offset by rapid elimination to prevent accumulation in heterozygotes with sitosterolemia. Arterioscler Thromb 12:563–568

Samuel P (1979) Treatment of hypercholesterolemia with neomycin. A time for reappraisal. N Engl J Med 301:595–597

Samuel P, Steiner A (1959) Effect of neomycin on serum cholesterol level of man. Proc Soc Exp Biol Med 100:193–195

Samuel P, Waithe WI (1961) Reduction of serum cholesterol concentrations by neomycin, para-aminosalicylic acid, and other antibacterial drugs in man. Circulation 24:578–591

Samuel P, Holtzman CM, Meilman E, Perl W (1968) Effect of neomycin on exchangeable pools of cholesterol in the steady state. J Clin Invest 47:1806–1818

Samuel P, Holtzman CM, Meilman E, Sekowski I (1970) Reduction of serum cholesterol and triglyceride levels by the combined administration of neomycin and clofibrate. Circulation 41:109–114

Samuel P, Holtzman CM, Meilman E, Sekowski I (1973) Effect of neomycin and other antibiotics on serum cholesterol levels and 7-alpha- dehydroxylation of bile acids by the fecal bacterial flora in man. Circ Res 33:393–402

Samuel P, McNamara DJ, Ahrens EH, Crouse JR, Parker T (1982) Further validation of the plasma isotope ratio method for measurement of cholesterol absorption. J Lipid Res 23:480–489

Schliack M, Löser R, Seibel K (1989) Hypolipidemic activity of K 12.148 in rats, marmosets and pigs. Artery 16:90–104

Schwartzkopff W, Jantke H-J (1978) Dosiswirksamkeit von β-Sitosterin bei Hypercholesterinämien der Typen IIa und IIb. MMW 120:1575–1578

Schwartzkopff W, Jantke H-J (1980) Verhalten der β-Acetyldigoxinresorption nach Gabe von β-Sitosterin oder Cholestyramin. Med Welt 33:1183–1188

Sedaghat A, Samuel P, Crouse JR, Ahrens EH Jr (1975) Effects of neomycin on absorption, synthesis, and/or flux of cholesterol in man. J Clin Invest 55:12–21

Sliskovic DR, White AD (1991) Therapeutic potential of ACAT inhibitors as lipid lowering and anti-atherosclerotic drugs. Trends Pharmacol Sci 12:194–199

Sugar AM, Alsip S, Galgiani JN, Graybill JR, Dismukes WE, Cloud GA, Craven PC, Stevens DA (1987) Pharmacology and toxicity of high dose ketoconazole. Antimicrob Agents Chemother 31:1874–1878

Thienpont D, van Cutsen J, van Gerven F, Heeres J, Janssen PAJ (1979) Ketoconazole: a new broad spectrum orally active antimycotic. Experientia 35:606–607

Thompson GR (1989) Lipid related consequences of intestinal malabsorption. Gut 30:29–34 (Festschrift)

Thompson GR, MacMahon M, Claes PJ (1970) Precipitation by neomycin compounds of fatty acid and cholesterol from mixed micellar solutions. Eur J Clin Invest 1:40–47

Thompson GR, Barrowman J, Gutierrez L, Dowling RH (1971) Action of neomycin on the intraluminal phase of lipid absorption. J Clin Invest 50:319–323

Thompson GR, Henry K, Edington N, Trexler PC (1972) Effect of neomycin on cholesterol metabolism in the germ-free pig. Eur J Clin Invest 2:365–371
Trachtenberg J, Pont A (1984) Ketoconazole therapy for advanced prostate cancer. Lancet 2:433–435
Underwood AH, Emmett JC, Ellis D, Flynn SB, Leeson PD, Benson GM, Novelli R, Pearce NJ, Shah VP (1986) A thyromimetic that decreases plasma cholesterol levels without increasing cardiac activity. Nature 324:425–429
Van Cutsem J (1983) The antifungal activity of ketoconazole. Am J Med 74 Suppl 1B:9–15
Van den Bosch JF, Claes P (1967) Correlations between bile salt precipitating capacity of derivatives of basic antibiotics and their plasma cholesterol lowering effect in vivo. Prog Biochem Pharmacol 2:97–104
Van den Bossche H, Willemsen G, Cools W, Cornelissen F, Lauwers WF, van Cutsem JM (1980) In vitro and in vivo effects of the antimycotic drug ketoconazole on sterol synthesis. Antimicrob Agents Chemother 17:922–928
Waisbren BA, Spink WW (1950) Clinical appraisal of neomycin. Ann Intern Med 33:1099–1119
Waksman SA, Lechevalier HA (1949) Neomycin as new antibiotic active against streptomycin resistant bacteria. Science 109:305–307
Weisweiler P, Heinemann V, Schwandt P (1984) Serum lipoproteins and lecithin: cholesterol acyltranferase (LCAT) activity in hypercholesterolemic subjects given β-sitosterol. Int J Clin Pharmacal Ther Toxicol 22:224–226

Subject Index

Springer-Verlag and the Environment

We at Springer-Verlag firmly believe that an international science publisher has a special obligation to the environment, and our corporate policies consistently reflect this conviction.

We also expect our business partners – paper mills, printers, packaging manufacturers, etc. – to commit themselves to using environmentally friendly materials and production processes.

The paper in this book is made from low- or no-chlorine pulp and is acid free, in conformance with international standards for paper permanency.

Printing: Mercedesdruck, Berlin
Binding: Buchbinderei Lüderitz & Bauer, Berlin